ISOPOD SYSTEMATICS AND EVOLUTION

CRUSTACEAN ISSUES

13

General editors
RONALD VONK & FREDERICK R. SCHRAM
Zoological Museum, University of Amsterdam

A.A. BALKEMA / ROTTERDAM / BROOKFIELD / 2001

ISOPOD SYSTEMATICS AND EVOLUTION

Edited by
BRIAN KENSLEY
Smithsonian Institution, Washington D.C., USA
RICHARD C. BRUSCA
Columbia University, Oracle, Arizona, USA

A.A. BALKEMA/ROTTERDAM/BROOKFIELD/2001

Published by
A.A. Balkema, P.O. Box 1675, 3000 BR Rotterdam, Netherlands
Fax: +31.10.4135947; E-mail: balkema@balkema.nl; Internet site: http://www.balkema.nl

A.A. Balkema Publishers, 2252 Ridge Road, Brookfield, VT 05036-9704, USA
Fax: 802.276.3837; E-mail: info@ashgate.com

ISSN 0168-6356
ISBN 90 5809 327 1

Printed in the Netherlands

Contents

3 *Morphology*

Preface

One of our reasons for organizing the Second International Isopod Conference was to recognize the work and career of Thomas Bowman, long-time curator of isopod crustaceans at the National Museum of Natural History, Smithsonian Institution, Washington D.C. The conference took place on 16-17 July 1998, in conjunction with the activities surrounding the Fourth International Crustacean Congress in Amsterdam that year.

Because of the wide-ranging research interests of Tom Bowman, it was most appropriate that the conference attracted attendees from around the world involved in research in all aspects of isopod biology. Tom's career is nicely outlined by Frank Ferrari in this volume, along with a complete bibliography of his publications, a list of the taxa he described, and a separate listing of species that have been named in his honor.

Of course, a central focus of Tom's research resided in the area of taxonomy and systematics, so it is appropriate that the first section of the book is devoted to these subjects. These extend from taxonomic descriptions of new genera (Andreas Leistikow) and species (Sarah F. Riseman, Ana Maria S. Pires & Richard C. Brusca) to synthetic overviews of two of the most fascinating isopod groups, the Anthuridea (Gary Poore) and the Phreatoicidea (George Wilson & Stephen Keable). These last two are especially significant in that they frame their syntheses in a context of comprehensive cladistic analyses and as such will form benchmarks in the literature on crustacean systematics. Closing off this section are two papers dealing with biogeography, one concerning the bopyrid isopods and the thalassinidean decapods upon which they are found (John Markham), and a sweeping review of the biogeographic distribution of isopods in the Indian Ocean (Brian Kensley). Although the collection here is not intended to be exhaustive of all isopod taxa, certainly many of the papers in this section will set the tone for future isopod research and provide sound examples to be followed by workers in other groups.

The second section takes up issues in ecology and evolution of the Isopoda, beginning with an exposition of parental care in wood-boring *Sphaeroma* (Martin Thiel), and ecological distributions in both the Nordic seas (Jörundur Svavarsson, Bjarni Kristjánsson & Torleiv Brattegard) and the southwestern Atlantic (Ana Maria Setubal Pires-Vanin). This section concludes with two papers focused on evolution, including evolutionary rates in groundwater microparasellids in North Africa (Nicole Coineau, Claude Boutin & A. Fakher el Abiari), and a sophisticated treatment of population biology of the marine isopod, *Paracerceis sculpta* (Stephen Shuster, J. William Ballard, Guenivier Zinser, Clay Sassaman & Paul Keim). Individually, these papers take up various aspects of an eco-evolutionary approach to isopod biology. Weaving of these strands into some future reviews of the group will be necessary as we move to a comprehensive understanding of the biology and evolution of Isopoda.

Finally, we include a series of papers dealing with comparative anatomy and functional morphology. These include an overview of respiratory structures in philosciid oniscideans (Andreas Leistikow & Paula Beatriz de Araujo), and aspects of cymothoid anatomy seen in parasites of Amazonian fish (Vernon Thatcher). Finally, we conclude with a consideration of the dynamics of epithelia, both of the integument as well as the digestive system, during molting in oniscideans (Jasna Strus & Andrej Blejec). Although comparative anatomy and functional morphology are sometimes viewed as oldstyle science, modern approaches to systematics, ecology, and physiology seem to only uncover fundamental gaps in our knowledge base about animal form. Studies like these here are absolutely necessary to keep the fires of scientific research stoked.

All in all, we believe this collection of papers represents a fine array of subjects to reflect the rather catholic interests of Tom Bowman.

No conference of this size could happen without the efforts of a well-organized local committee. In this regard we would like to extend our deep thanks to the crustacean department of the Zoological Museum, University of Amsterdam. Despite the demands of dealing with the final preparations for the ICC-4 the following week, the Amsterdam group put together a splendid set of social activities for the isopod conference, and helped coordinate the production of the scientific program. Special thanks in this regard go to Ronald Jenner, Dirk Platvoet, Frederick Schram, Ronald Vonk, and Wil van Zijl.

Brian Kensley, Washington, D.C.
Richard C. Brusca, Oracle, Arizona

Thomas Elliot Bowman III (photo Smithsonian Institution).

The work of Thomas Elliot Bowman III (1918-1995)

FRANK D. FERRARI
Department of Invertebrate Zoology, National Museum of Natural History, Smithsonian Institution, Washington DC, USA

Thomas Elliot Bowman III was curator of Crustacea for thirty years at the National Museum of Natural History until his death on August 10, 1995. Tom was born October 21, 1918, in Brooklyn, New York and grew up there. Immediately after graduating from Harvard College, he spent four years in the US Army (1941-1944). While he was stationed in northern California, Tom attended a summer course held at Stanford University's Pacific Grove Field Station sponsored by the University of California, Berkeley (Reid & Ferrari 1988). After his service Tom returned to Berkeley and took a master's degree with Ralph Smith. In 1948 Tom moved south to Scripps Institution of Oceanography to work under Martin Johnson on a California Current-wide plankton survey begun in 1949 by the California Cooperative Oceanic Fisheries Investigation program. Tom worked primarily on the pelagic amphipods, but identified the copepods as well. The University of California, Los Angeles, with which Scripps was associated for degree granting purposes, awarded his Ph.D.; his dissertation concerned the systematics and distribution of several families of pelagic amphipods. Tom was employed at the Narragansett Marine Laboratory in Rhode Island for six months before assuming the curatorship at the National Museum of Natural History in 1955.

Tom was a small, thin man whose twinkling eyes gave him an elfin, somewhat mischievous, demeanor (Ferrari 1996). He enjoyed controversy and wove a combative wit into the framework of a scholarly career (Brinton & Ohman 1995). Tom wrote papers that were noted for their economy of words, and expected that same economy both as a reviewer for numerous journals and as the crustacean editor over 14 years for the *Proceedings of the Biological Society of Washington*. Tom had a preference for bold to outrageous colors in dress and in office decor, and he could cut a mean figure on the trumpet while following his love of Dixieland music. Tom and his wife Mary Jo raised three lovely daughters. He worked hard all of his life, even after he retired, which somewhat disappointed Mary Jo, who had hoped to share more of his time in retirement.

Beginning with his undergraduate research under John Welsh on chromatophores of crayfish, Tom published 145 scientific papers alone or with colleagues. In addition, 32 other publications, such as his Ph.D. dissertation, faunal lists, nomenclatural opinions, book reviews and newsletter articles, are included in the following bibliography. Articles about crustaceans from earlier editions of the *Encyclopedia Americana*, *Encyclopedia Britannica*, *McGraw-Hill Encyclopedia of Science and Technology*, and Sybil P. Parker's *Taxonomy and Classification of Living Organisms* are not listed here.

The majority of Tom's scientific papers are concerned with taxonomic problems. In these papers he described 116 species new to science (Table 1). [I have not included the pelagic amphipod *Primno abyssalis* which Tom described in Fulton, 1968, because Tom apparently is not an author of the name although his description of the animal was acknowledged.] With exceptions of the suctorian (= Ciliophora) and the chaetognath, all of the new species are crustaceans and most of them quite small crustaceans.

Isopods were the major focus of Tom's research; 55 [47%] of his new species were isopods, as were eight of the 16 new genera he established. Tom did keep an eye on copepods, describing 28 new species, a new subspecies, and two new genera. He also established, with others, a new order of peracaridans, Mictacea. Tom's remaining new names for higher taxa are a subgenus of mysids, a genus and family for the mictacean, and a family, superfamily and infraorder of pelagic amphipods. Tom also proposed seven names for crustacean genera into which he placed previously described species, and five superfamilies for pelagic amphipods. The year 1985 was a banner one for Tom. In addition to a genus and family name for the new peracaridean, Tom named a new genus and species of leptostracan, a new species of mysid, a new genus and species of isopod, two other new species of isopods, and wrote a tribute to his major professor, Martin Wiggo Johnson.

Tom's biological interests extended beyond taxonomy. He studied the seasonal occurrences, as well as regional and geographical distributions, of many different kinds of crustaceans. He also considered significant questions about homologies of the crustacean telson and the evolution of stalked eyes. However, he will be remembered as one of America's premier carcinological taxonomists.

REFERENCES

Brinton, E. & M. Ohman 1995. In Memoriam – Electronic log of Scripps Institution of Oceanography, 32(34) [subscribe to siolog@ucsd.edu]

Ferrari, F.D. 1996. Thomas Elliot Bowman, III, (1918-1995). *Journal of Crustacean Biology*, 16:633-635.

Fulton, J. 1968. A laboratory manual for the identification of British Columbia marine zooplankton. *Fisheries Research Board of Canada Tech. Rpt.* 55: 1-76.

Reid, J. & F. Ferrari. 1988. Thomas E. Bowman at Seventy. *Monoculus* 17:10-21.

PUBLICATIONS OF THOMAS ELLIOT BOWMAN III

Titles with the same date listed alphabetically by author; lower case letter after date indicates titles with new latin names and refers to dates in Table 1.

1942

Bowman, T.E. 1942. Morphological color change in the crayfish. *American Naturalist* 76:332-336.

1949

Bowman, T.E. 1949. Chromatophorotropins in the central nervous organs of the crab, *Hemigrapsus oregonensis*. *Biological Bulletin* 96:238-245.

1953

Bowman, T.E. 1953. The systematics and distribution of pelagic amphipods of the families Vibiliidae, Paraphronimidae, Hyperiidae, Dairellidae, and Phrosinidae from the northeastern Pacific. Unpublished Ph.D. Thesis, University of California, Los Angeles, 430pp.

1955

Bowman, T.E. 1955a. A new genus and species of whale-louse (Amphipoda: Cyamidae) from the false killer whale. *Bulletin of Marine Science of the Gulf and Caribbean* 5:315-320.

Bowman, T.E. 1955b. The isopod genus *Chiridotea* Harger, with a description of a new species from brackish waters. *Journal of the Washington Academy of Science* 45:224-229.

Bowman, T.E. 1955c. A new copepod of the genus *Calanus* from the northeastern Pacific with notes on *Calanus tenuicornis* Dana. *Pacific Science* 9:413-422.

Menzies, R.J., T.E. Bowman & F.G. Alverson 1955. Studies of the biology of the fish parasite *Livoneca convexa* Richardson (Crustacea, Isopoda, Cymothoidae). *The Wasmann Journal of Biology* 13:277-295.

1956

Bowman, T.E. 1956a. Una especie nueva de *Bopyrella* (Crustacea: Isopoda) de Los Roques, Venzuela. *Novedades Cientificas*, Serie Zoológica 19; 3pp, Figura 1.

Fleminger, A. & T.E. Bowman 1956b. A new species of *Candacia* (Copepoda: Calanoida) from the western North Atlantic Ocean. *Proceedings of the United States National Museum* 106:331-337.

Menzies, R.J. & T.E. Bowman 1956c. Emended description and assignment to the new genus *Ronalea* of the idotheid isopod *Erichsonella pseudoculata* Boone. *Proceedings of the United States National Museum* 106:339-343.

1957

Bowman, T.E. 1957a. A new species of *Calanopia* (Copepoda: Calanoida) from the Caribbean Sea. *Proceedings of the United States National Museum* 107:39-45.

Bowman, T.E. 1957b. A new species of *Mysidopsis* (Crustacea: Mysidacea) from the southeastern coast of the United States. *Proceedings of the United States National Museum* 107:1-7.

Bowman, T.E. & C. Diaz-Ungria 1957c. Isopodos quimotoideos parasitos de peces de las aguas Venezolanas. *Sociedad Ciencias Naturales (La Salle), Memoria* 17:112-124.

1958

Bowman, T.E. 1958a. A new species of *Acrocalanus* (Copepoda: Calanoida) from off the southeastern coast of the United States. *Bulletin of Marine Science of the Gulf and Caribbean* 8:118-124.

Bowman, T.E. 1958. First Pacific record of the whale-louse genus *Syncyamus* (Amphipoda: Cyamidae). *Pacific Science* 12:181.

1959

Bowman, T.E., P.A. Chappuis & H. Herbst 1959. Plankton from the Persian Gulf. *Papers of the Peabody Museum of Archeology and Ethnology, Harvard* 52:266 (Appendix J).

1960

Bowman, T.E. 1960a. Description and notes on the biology of *Lironeca puhi*, n. sp. (Isopoda: Cymothoidae), parasite of the Hawaiian moray eel, *Gymnothorax eurostus* (Abbott). *Crustaceana* 1:84-91, pl. II.

Bowman, T.E. 1960. The pelagic amphipod genus *Parathemisto* (Hyperiidea: Hyperiidae) in the North Pacific and adjacent Arctic Ocean. *Proceedings of the United States National Museum* 112:343-392.

1961

Bowman, T.E. 1961. The copepod genus *Acartia* in Chesapeake Bay. *Chesapeake Science* 2:206-207.

Bowman, T.E. & J.G. González 1961a. Four new species of *Pseudocyclops* (Copepoda: Calanoida) from Puerto Rico. *Proceedings of the United States National Museum* 113:37-59.

Kabata, Z. & T.E. Bowman 1961. Revision of *Tracheliastes grandis* Wilson, 1915 (Copepoda, Lernaeopodidae). *Crustaceana* 3:120-126.

1962

Bowman, T.E. 1962a. *Tisbe monozota*, a new harpacticoid copepod from Florida. *Proceedings of the Biological Society of Washington* 75:125-131.

Bowman, T.E. 1962. review of 'Abyssal Crustacea' *American Scientist* 50:430A-431A.

1963

Bowman, T.E., C.D. Meyers & S.D. Hicks 1963. Notes on associations between hyperiid amphipods and medusae in Chesapeake and Narragansett Bays and the Niantic River. *Chesapeake Science* 4:141-146.

Bowman, T.E. 1963. review of 'The Distribution of Pacific Euphausiids' *Quarterly Review of Biology* 38:401.

Bowman, T.E. 1963. Comment on the petition concerning penaeid names. *Bulletin of Zoological Nomenclature* 20:249.

1964

Bowman, T.E. 1964a. *Antrolana lira*, a new genus and species of troglobitic cirolanid isopod from Madison Cave, Virginia. *International Journal of Speleology* 1:229-236, pls. 50-57.

Bowman, T.E. 1964b. *Mysidopsis almyra*, a new estuarine mysid crustacean from Louisiana and Florida. *Tulane Studies in Zoology* 12:15-18.

Bowman, T.E. 1964. review of 'The Natural History and Geography of the Antarctic Krill (*Euphausia superba* Dana)' *Quarterly Review of Biology* 39:193-194.

1965

Bowman, T.E. 1965a. *Cyathura specus*, a new cave isopod from Cuba (Anthuroida: Anthuridae). *Studies on the Fauna of Curacao and Other Caribbean Islands* 22:88-97.

Bowman, T.E. 1965b. *Xilitloniscus*, a new genus for the Mexican troglobitic isopod, *Cordioniscus laevis* Rioja (Oniscoidea: Trichoniscidae). *Proceedings of the Biological Society of Washington* 78:209-215.

Bowman, T.E. 1965. An arostrate population of the copepod, *Acartia lilljeborgii* Giesbrecht (Calanoida, Acartiidae), from St. Lucia, West Indies. *Crustaceana* 8:149-152.

Bowman, T.E. 1965. review of 'Polychaeta Errantia of Antarctica' *Quarterly Review of Biology* 40:303.

Bowman, T.E. & L.J. Lancester 1965. A bloom of the planctonic blue-green alga, *Trichodesmium erythraeum*, in the Tonga Islands. *Limnology and Oceanography* 10:291-293.

Bowman, T.E. & L.W. Peterson 1965. Bibliography and list of new genera and species of amphipod crustaceans described by Clarence R. Shoemaker. *Crustaceana* 9:309-316, pl.22.

González, J.G. & T.E. Bowman 1965c. Planktonic copepods from Bahía Fosforescente, Puerto Rico, and adjacent waters. *Proceedings of the United States National Museum* 117:241-303.

1966

Bowman, T.E. 1966a. *Haptolana trichostoma*, a new genus and species of troglobitic cirolanid isopod from Cuba. *International Journal of Speleology* 2:105-108, pls. 24-27.

1967

Bowman, T.E. 1967a. *Asellus kenki*, a new isopod crustacean from springs in the eastern United States. *Proceedings of the Biological Society of Washington* 80:131-140.

Bowman, T.E. 1967. Bioluminescence in two species of pelagic amphipods. *Journal of the Fisheries Research Board of Canada* 24:687-688.

Bowman, T.E. 1967b. The planktonic shrimp, *Lucifer chacei*, sp. nov., (Sergestidae: Luciferinae), the Pacific twin of the Atlantic *Lucifer faxoni*. *Pacific Science* 21:266-271.

Bowman, T.E. & L.S. Kornicker 1967c. Two new crustaceans: the parasitic copepod *Sphaeronellopsis monothrix* (Choniostomatidae) and its myodocopid ostracod host *Parasterope pollex* (Cylindroleberidae) from the southern New England coast. *Proceedings of the United States National Museum* 123:1-28, pl. 1.

Bowman, T.E. & J.C. McCain 1967. Distribution of the planktonic shrimp *Lucifer* in the western North Atlantic. *Bulletin of Marine Science* 17:660-671.

Bowman, T.E. & J.C. McCain 1967. Variation and distribution of the pelagic amphipod *Cyphocaris challengeri* in the northeast Pacific (Gammaridea: Lysianassidae). *Proceedings of the United States National Museum* 122:1-14.

1968

Bowman, T.E. & L.B. Holthuis 1968. *Lucifer* or *Leucifer*: which spelling is correct? *Crustaceana* 14:216-217.

Bowman, T.E. & L.S. Kornicker 1968. *Sphaeronellopsis hebe* (Copepoda, Choniostomatidae), a parasite of the ostracod, *Pseudophilomedes ferulanus*. *Crustaceana* 15:113-116.

Bowman, T.E. & A. Long 1968. Relict populations of *Drepanopus bungei* and *Limnocalanus macrurus grimaldii* (Copepoda: Calanoida) from Ellesmere Island, N.W.T. *Arctic* 21:172-180.

Bowman, T.E. & R.N. Mariscal 1968a. *Renocila heterozota*, a new cymothoid isopod, with notes on its host, the anemone fish, *Amphiprion akallopisos*, in the Seychelles. *Crustaceana* 14:97-104, pl.1.

Bowman, T.E., R. Prins & B.F. Morris 1968. Notes on the harpacticoid copepods *Attheyella pilosa* and *A. carolinensis*, associates of crayfishes in the eastern United States. *Proceedings of the Biological Society of Washington* 81:571-585.

1969

Kornicker, L.S. & T.E. Bowman 1969a. *Sphaeronellopsis dikrothrix*, a new choniostomatid copepod from the ostracod *Metavargula ampla*. *Crustaceana* 17:282-284, pl. 1.

1970

Bourdon, R. & T.E. Bowman 1970. Western Atlantic species of the parasitic genus *Leidya* (Epicaridea: Bopyridae). *Proceedings of the Biological Society of Washington* 83:409-424.

1971

Bowman, T.E. 1971a. *Cirolana narica*, n. sp., a New Zealand isopod (Crustacea) found in the nasal tract of the dolphin *Cephalorhynchus hectori*. *Beaufortia* 19:107-112.

Bowman, T.E. 1971b. *Excirolana kumari*, a new tubicolous isopod from Malaysia. *Crustaceana* 20:70-76, pl.1.

Bowman, T.E. 1971c. *Palaega lamnae*, new species (Crustacea: Isopoda) from the Upper Cretaceous of Texas. *Journal Paleontology* 45:540-541.

Bowman, T.E. 1971. The case of the nonubiquitous telson and the fraudulent furca. *Crustaceana* 21:165-175.

Bowman, T.E. 1971d. The distribution of calanoid copepods off the southeastern United States between Cape Hatteras and southern Florida. *Smithsonian Contributions to Zoology* 96, 58pp.

Bowman, T.E. 1971e. *Tortanus scaphus* and *Tortanus lophus*, new Pacific planctonic copepods, with notes on *Tortanus murrayi* (Calanoida, Tortanidae). *Pacific Science* 25:521-528.

Heron, G.A. & T.E. Bowman 1971f. Postnaupliar developmental stages of the copepod crustaceans *Clausocalanus laticeps*, *C. brevipes* and *Ctenocalanus citer* (Calanoida: Pseudocalanidae). *Antarctic Research Series* 17:141-165.

1972

Bowman, T.E. 1972a. *Cithadius cyathurae*, a new genus and species of Tachidiidae (Copepoda: Harpacticoida) associated with the estuarine isopod, *Cyathura polita*. *Proceedings of the Biological Society of Washington* 85:249-254.

Bowman, T.E. & R.B. Manning 1972b. Two Arctic bathyal crustaceans: The shrimp *Bythocaris cryonesus* new species, and the amphipod *Eurythenes gryllus*, with in situ photographs from Ice Island T-3. *Crustaceana* 23:187-201, pl. 1.

1973

Bowman, T.E. 1973a. Pelagic amphipods of the genus *Hyperia* and closely related genera (Hyperiidea: Hyperiidae). *Smithsonian Contributions to Zoology* 136, 76pp.

Bowman, T.E. 1973b. Two new American species of *Spelaeomysis* (Crustacea: Mysidacea) from a Mexican cave and land crab burrows. *Association for Mexican Cave Studies, Bulletin* 5:13-20.

Bowman, T.E. 1973c. Two freshwater copepods from Barrancas, Venezuela: *Notodiaptomus cearensis* (Wright) and *N. venezolanus deeveyorum*, new subspecies (Calanoida: Diaptomidae). *Proceedings of the Biological Society of Washington* 86:193-202.

Bowman, T.E. & H.-E. Gruner 1973d. The families and genera of Hyperiidea (Crustacea: Amphipoda). *Smithsonian Contributions to Zoology* 146, 64pp.

Bowman, T.E. & M.W. Johnson 1973. Distributional atlas of calanoid copepods in the California Current Region, 1949 and 1950. *California Cooperative Oceanic Fisheries Investigation (CALCOFI) Atlas* 19, 239pp.

Holsinger, J.R. & T.E. Bowman 1973. A new troglobitic isopod of the genus *Lirceus* (Asellidae) from southwestern Virginia, with notes on its ecology and additional records for the genus in the Appalachians. *International Journal of Speleology* 5:261-271.

1974

Bowman, T.E. 1974a. The 'sea flea' *Dolobrotus mardeni* n. gen., n.sp., a deep-water American lobster bait scavenger (Amphipoda: Eusiridae). *Proceedings of the Biological Society of Washington* 87:129-138.

Bowman, T.E. 1974. The California freshwater isopod, *Asellus tomalensis*, rediscovered and compared with *Asellus occidentalis*. *Hydrobiologia* 44:431-441.

Bowman, T.E. 1974. Review of 'Shallow-water Gammaridean Amphipoda of New England'. *Undersea Biomedical Research* 1:112-113.

Bowman, T.E. & H. Kuehne 1974b. *Cymodetta gambosa*, a new sphaeromatid isopod (Crustacea) from Australia, with notes on its mating behaviour. *Records of the Australian Museum* 29:235-242, pl. 8.

Bowman, T.E. & G.A. Schultz 1974. The isopod crustacean genus *Munnogoniun* George and Strömberg, 1968 (Munnidae, Asellota). *Proceedings of the Biological Society of Washington* 87:265-271.

Williams, A.B., T.E. Bowman & D.M. Damkaer 1974. Distribution, variation, and supplemental description of the opossum shrimp, *Neomysis americana* (Crustacea: Mysidacea). *United States Fishery Bulletin* 72:835-842.

1975

Bowman, T.E. 1975a. A new genus and species of troglobitic cirolanid isopod from San Luis Potosi, Mexico. *Occasional Papers of the Museum Texas Tech University* 27, 7pp.

Bowman, T.E. 1975b. *Oithona colcarva*, n. sp., an American copepod incorrectly known as *O. brevicornis* (Cyclopoida: Oithonidae). *Chesapeake Science* 16:134-137.

Bowman, T.E. 1975c. Three new troglobitic asellids from western North America (Crustacea: Isopoda: Asellidae). *International Journal of Speleology* 7:339-356.

Bowman, T.E. & C. Holmquist 1975d. *Asellus (Asellus) alaskensis*, n.sp., the first Alaskan *Asellus*, with remarks on its Asian affinities (Crustacea: Isopoda: Asellidae). *Proceedings of the Biological Society of Washington* 88:59-72.

Glynn, P.W., D.M. Dexter & T.E. Bowman 1975. *Excirolana braziliensis*, a Pan-American sand beach isopod: taxonomic status, zonation and distribution. *Journal of Zoology* London 175:509-521.

1976

Bowman, T.E. 1976a. *Miostephos cubrobex*, a new genus and species of copepod from an anchialine pool in Cuba (Calanoida: Stephidae). *Proceedings of the Biological Society of Washington* 89:185-190.

Bowman, T.E. 1976b. *Stygiomysis major*, a new troglobitic mysid from Jamaica, and extension of the range of *S. holthuisi* to Puerto Rico (Crustacea: Mysidacea: Stygiomysidae). *International Journal of Speleology* 8:365-373.

Bowman, T.E. & G. Longley 1976c. Redescription and assignment to the new genus *Lirceolus* of the Texas troglobitic water slater, *Asellus smithii* (Ulrich) (Crustacea: Isopoda: Asellidae). *Proceedings of the Biological Society of Washington* 88:489-496.

1977

Bowman, T.E. 1977a. A review of the genus *Antromysis* (Crustacea: Mysidacea) including new species from Jamaica and Oaxaca, Mexico, and a redescription and new records for *A. cenotensis*. *Association for Mexican Cave Studies, Bulletin* 6:27-38.

Bowman, T.E. 1977. Isopod crustaceans (except Anthuridae) collected on the Presidential Cruise of 1938. *Proceedings of the Biological Society of Washington* 89:653-665.

Bowman, T.E. 1977b. *Dendrosomides lucicutiae*, a new species of suctorian from the pelagic calanoid copepod, *Lucicutia*. *Proceedings of the Biological Society of Washington* 89:695-702.

Bowman, T.E. 1977c. *Orchestia vaggala*, a new land-hopper from the Galápagos Islands (Crustacea: Amphipoda: Talitridae). *Proceedings of the Biological Society of Washington* 90:658-668.

Bowman, T.E. 1977d. *Ceratolana papuae*, a new genus and species of mangrove-boring cirolanid isopod from Papua New Guinea. *Proceedings of the Biological Society of Washington* 90:819-825.

Bowman, T.E. 1977. Mysidacea. Pp. 149-150. In S.H. Hurlbert (ed.), *Biota Acuática de Sudaméricana Austral*. San Diego State University, San Diego, California, 342pp.

Bowman, T.E. 1977. Range extension and new hosts for the cymothoid isopod *Anilocra acuta*. *Chesapeake Science* 18:390-393.

Lewis, J.J. & T.E. Bowman 1977e. *Caecidotea carolinensis*, n. sp., the first subterranean water slater from North Carolina (Crustacea: Isopoda: Asellidae). *Proceedings of the Biological Society of Washington* 90:968-974.

1978

Bowman, T.E. 1978. Nomenclatural problems in the cymothoid isopod genera *Ceratothoa*, *Codonophilus*, *Glossobius* and *Meinertia* – their solution by applying the Law of Priority. *Crustaceana* 34:217-219.

Bowman, T.E. 1978. Restoration of the subgenus *Emphylia* Koelbel for the parasitic isopod *Nerocila sundaica* Bleeker (Flabellifera: Cymothoidae). *Crustaceana* 34:33-44.

Bowman, T.E. 1978a. Revision of the pelagic amphipod genus *Primno* (Hyperiidea: Phrosinidae). *Smithsonian Contributions to Zoology* 275, 23pp.

Bowman, T.E. 1978. From Brazil to Jamaica: a range extension of the neritic calanoid copepod, *Pseudodiaptomus acutus*. *Crustaceana* 35:249-252.

Bowman, T.E. 1978. The modified suture between segments 8 and 9 on the first antenna of some calanoid copepods. *Crustaceana* 35:113-118, pl. 1.

Bowman, T.E. & D.C. Beckett 1978. A redescription of the troglobitic isopod, *Caecidotea stygia*, from the environs of Cincinnati, Ohio (Crustacea: Isopoda: Asellidae). *Proceedings of the Biological Society of Washington* 91:294-302.

1979

Bowman, T.E. 1979a. *Notodiaptomus caperatus*, a new calanoid copepod from phreatic groundwater in Barbuda (Crustacea: Diaptomidae). *Bijdragen tot de Dierkunde* 49:219-226.

Bowman, T.E. & B.F. Morris 1979. *Carpias* Richardson, 1902, a senior synonym of *Bagatus* Nobili, 1906, and the validity of *Carpias minutus* (Richardson 1902) (Isopoda: Asellota: Janiridae). *Proceedings of the Biological Society of Washington* 92:650-657.

1980

Bowman, T.E. 1980a. *Antromysis (Surinamysis) merista*, a new freshwater mysid from Venezuela (Crustacea: Mysidacea). *Proceedings of the Biological Society of Washington* 93:208-215.

Ferrari, F. & T.E. Bowman 1980b. Pelagic copepods of the family Oithonidae (Cyclopoida) from the east coasts of Central and South America. *Smithsonian Contributions to Zoology* 312, 27pp.

1981

Bowman, T.E. 1981a. *Calasellus longus*, a new genus and species of troglobitic asellid from Shaver Lake, California (Crustacea: Isopoda: Asellidae). *Proceedings of the Biological Society of Washington* 94:866-872.

Bowman, T.E. 1981. First description of the male opossum shrimp *Heteromysis bermudensis bermudensis* (Crustacea: Mysidacea). *Proceedings of the Biological Society of Washington* 94:458-461.

Bowman, T.E. 1981b. *Thermosphaeroma milleri* and *T. smithi*, new sphaeromatid isopod crustaceans from hot springs in Chihuahua, Mexico, with a review of the genus. *Journal of Crustacean Biology* 1:105-122.

Bowman, T.E. 1981. Mysidacea. P. 27. In S.H. Hulbert, G. Rodriguez & N.D. Santos (eds), *Aquatic Biota of Tropical South America*, Part 1, Arthropoda. San Diego State University, San Diego, California, 323pp.

Bowman, T.E. 1981. Review of 'The Biology of Mysids and Euphausiids'. *Nature* 293:591-592.

Bowman, T.E., N.L. Bruce & J.D. Standing 1981. Recent introduction of the cirolanid isopod crustacean *Cirolana arcuata* into San Francisco Bay. *Journal of Crustacean Biology* 1:545-557.

Lewis, J.J. & T.E. Bowman 1981c. The subterranean asellids (*Caecidotea*) of Illinois (Crustacea: Isopoda: Asellidae). *Smithsonian Contributions to Zoology* 335, 66pp.

1982

Bowman, T.E. 1982a. *Speocirolana pubens* and *S. endeca*, new troglobitic isopod crustaceans from Mexico (Flabellifera: Cirolanidae). *Association for Mexican Cave Studies, Bulletin* 8:13-23.

Bowman, T.E. 1982b. Three new stenasellid isopods from Mexico (Crustacea: Asellota). *Association for Mexican Cave Studies, Bulletin* 8:25-38.

Bowman, T.E. 1982. Mysidacea. Pp. 201-202. In S.H. Hurlbert & A. Villalobos-Figueroa (eds), *Aquatic Biota of Mexico, Central America and the West Indies*. San Diego State University, San Diego, California, 529pp.

Bowman, T.E. & L.G. Abele 1982. Classification of the recent Crustacea. pp. 1-27. In L.G. Abele (ed.), *The Biology of Crustacea* (editor-in-chief D.E. Bliss), vol. I. Systematics, the Fossil Record, and Biogeography. Academic Press, New York, 319pp.

Bowman, T.E., A.C. Cohen & M.M. McGuiness 1982. Vertical distribution of *Themisto gaudichaudii* (Amphipoda: Hyperiidea) in Deepwater Dumpsite 106 off the mouth of Delaware Bay. *Smithsonian Contributions to Zoology* 351, 24pp.

Bowman, T.E. & M.M. McGuiness 1982. Epipelagic amphipods of the family Hyperiidae from the International Indian Ocean Expedition, 1959-1965. *Smithsonian Contributions to Zoology* 359, 53pp.

Bowman, T.E. & R. Franz 1982c. *Anopsilana crenata*, a new troglobitic cirolanid isopod from Grand Cayman Island, Caribbean Sea. *Proceedings of the Biological Society of Washington* 95:522-529.

Bruce, N.L. & T.E. Bowman 1982. The status of *Cirolana parva* Hasen, 1890 (Crustacea, Isopoda, Cirolanidae) with notes on its distribution. *Proceedings of the Biological Society of Washington* 95:325-333.

1983

Bowman, T.E., III & H.H. Hobbs, III. 1983a. *Caecidotea filicispeluncae*, a new troglobitic asellid isopod from Ohio. *Proceedings of the Biological Society of Washington* 96:693-697.

Bowman, T.E. & T.M. Iliffe 1983b. *Bermudalana aruboides*, a new genus and species of troglobitic Isopoda (Cirolanidae) from marine caves on Bermuda. *Proceedings of the Biological Society of Washington* 96:291-300.

Bowman, T.E. & I.U. Tareen 1983c. Cymothoidae from fishes of Kuwait (Arabian Gulf) (Crustacea: Isopoda). *Smithsonian Contributions to Zoology* 382, 30pp.

Orsi, J.J., T.E. Bowman, D.C. Marelli & A. Hutchinson 1983. Recent introduction of the planktonic calanoid copepod *Sinocalanus doerrii* (Centropagidae) from mainland China to the Sacramento-San Joaquin Estuary of California. *Journal of Plankton Research* 5:357-375.

1984

Bowman, T.E. 1984. Stalking the wild crustacean: the significance of sessile and stalked eyes in phylogeny. *Journal of Crustacean Biology* 4:7-11.

Bowman, T.E. 1984a. New records of the troglobitic mysid genus *Stygiomysis*: *S. clarkei*, new species, from the Caicos Islands, and *S. holthuisi* (Gordon) from Grand Bahama Island (Crustacea: Mysidacea). *Proceedings of the Biological Society of Washington* 97:637-644.

Bowman, T.E. 1984. Review of 'Crustacean Issues 1, Crustacean Phylogeny'. *Science* 224:53.

Bowman, T.E. & J.J. Lewis 1984b. *Caecidotea rotunda*, a new troglobitic asellid from Indiana and Ohio (Crustacea: Isopoda: Asellidae). *Proceedings of the Biological Society of Washington* 97:425-431.

Bowman, T.E. & F. Phillips 1984. Bioluminescence in the freshwater amphipod, *Hyalella azteca*, caused by pathogenic bacteria. *Proceedings of the Biological Society of Washington* 97:526-528.

Bowman, T.E. & R.A. Wasmer 1984. The deep-sea amphipod *Paracyphocaris praedator* (Gammarida: Lysianassidae) associated with the pelagic shrimp *Oplophorus novaezeelandiae* as an egg-mimic. *Proceedings of the Biological Society of Washington* 97:844-848.

1985

Bowman, T.E. 1985. The correct identity of the pelagic amphipod *Primno macropa*, with a diagnosis of *Primno abyssalis* (Hyperiidea: Phrosinidae). *Proceedings of the Biological Society of Washington* 98:121-126.

Bowman, T.E. 1985a. *Heteromysoides dennisi*, a new mysidacean crustacean from Cemetery Cave, Grand Bahama Island. *Proceedings of the Biological Society of Washington* 98:945-948.

Bowman, T.E. 1985b. *Thermosphaeroma cavicauda* and *T. macrura*, new sphaeromatid isopods from Mexican hot springs. *Proceedings of the Biological Society of Washington* 98:1042-1047.

Bowman, T.E. 1985. Name endings in ostracode superfamilies – a response to I.G. Sohn. *Lethaia* 18:304.

Bowman, T.E., F.D. Ferrari & A. Fleminger 1985. Martin Wiggo Johnson (1893-1984) An appreciation and bibliography of his works. *Monoculus* (copepod newsletter) 11:7-15.

Bowman, T.E., S.P. Garner, R.R. Hessler, T. M. Iliffe & H.L. Sanders 1985c. Mictacea, a new order of Crustacea Peracarida. *Journal of Crustacean Biology* 5:74-78.

Bowman, T.E. & T.M. Iliffe 1985d. *Mictocaris halope*, a new unusual peracaridan crustacean from marine caves on Bermuda. *Journal of Crustacean Biology* 5:58-73.

Bowman, T.E. & B. Sket 1985e. *Remasellus*, a new genus for the troglobitic swimming Florida asellid isopod, *Asellus parvus* Steeves. *Proceedings of the Biological Society of Washington* 98:554-560.

Bowman, T.E., J. Yager & T.M. Iliffe 1985f. *Speonebalia cannoni*, n. gen., n. sp., from the Caicos Islands, the first hypogean leptostracan (Nebaliacea: Nebaliidae). *Proceedings of the Biological Society of Washington* 98:439-446.

1986

Bowman, T.E. 1986a. *Tridentella recava*, a new isopod from tilefish burrows in the New York Bight (Flabellifera: Tridentellidae). *Proceedings of the Biological Society of Washington* 99:269-273.

Bowman, T.E. 1986b. *Paracymothoa tholoceps*, a new freshwater parasitic isopod from southern Venezuela (Flabellifera: Cymothoidae). *Proceedings of the Biological Society of Washington* 99:753-756.

Bowman, T.E. 1986c. *Halosbaena fortunata*, a new thermosbaenacean crustacean from the Jameos del Agua Marine Lava Cave, Lanzarote, Canary Islands. *Stygologia* 2:84-89.

Bowman, T.E. 1986. Copepoda: Calanoida. Pp. 295-298. In L. Botosaneanu (ed.), *Stygofauna Mundi, a Faunistic, Distributional, and Ecological Synthesis of the World Fauna inhabiting Subterranean Waters (including the Marine Interstitial)*. E.J. Brill/Dr W. Backhuys, Leiden, 740pp.

Bowman, T.E. 1986. Mysidacea. Pp. 405-409. In L. Botosaneanu (ed.), *Stygofauna Mundi, a Faunistic, Distributional, and Ecological Synthesis of the World Fauna inhabiting Subterranean Waters (including the Marine Interstitial)*. E.J. Brill/Dr W. Backhuys, Leiden, 740pp.

Bowman, T.E. 1986. Mictacea. Pp. 410-411. In L. Botosaneanu (ed.), *Stygofauna Mundi, a Faunistic, Distributional, and Ecological Synthesis of the World Fauna inhabiting Subterranean Waters (including the Marine Interstitial)*. E.J. Brill/Dr W. Backhuys, Leiden, 740pp.

Bowman, T.E. 1986d. Freshwater calanoid copepods of the West Indies. *Syllogeus* 58:237-246.

Bowman, T.E. 1986. *Tortanus recticauda*: extension of range to Arabian Gulf (Copepoda, Calanoida, Tortanidae). *Crustaceana* 50:239-242.

1987

Bowman, T.E. 1987a. *Bahalana mayana*, a new troglobitic cirolanid isopod from Cozumel Island and the Yucatan Peninsula, Mexico. *Proceedings of the Biological Society of Washington* 100:659-663.

Bowman, T.E. & T.M. Iliffe 1987b. *Anopsilana lingua*, a new freshwater troglobitic isopod from the Palau Islands (Flabellifera: Cirolanidae). *Proceedings of the Biological Society of Washington* 100:347-352.

Bowman, T.E. 1987. Comment on the proposed precedence of Pseudocalanidae Sars, 1901 (Crustacea, Copepoda) over Clausocalanidae Giesbrecht, 1892. *Bulletin of Zoological Nomenclature* 44:129.

Bowman, T.E., R. Prins & J. Arenas 1987. The occurrence of the freshwater isopod *Heterias (Fritzianira) exul* in the Lakes Region of Chile, with notes on the genus *Heterias* (Asellota: Janiridae). *Hydrobiologia* 146:275-281.

Harrison, E.B. & T.E. Bowman 1987c. *Mysidium rickettsi*, a new species of mysid from the Gulf of California (Crustacea: Mysidacea: Mysidae). *Proceedings of the Biological Society of Washington* 100:674-679.

1988

Bowman, T.E. 1988a. *Nitokra sphaeromata*, a new harpacticoid copepod crustacean associated with the wood-boring isopod, *Sphaeroma peruvianum* in Costa Rica. *Proceedings of the Biological Society of Washington* 101:171-175.

Bowman, T.E. 1988. The input of terrestrial insects and spiders to the nutrient cycle of a 'woodland pond'. *Entomological News* 99:207-216.

Bowman, T.E. & T.M. Iliffe 1988b. *Tulumella unidens*, a new genus and species of thermosbaenacean crustacean from the Yucatan Peninsula, Mexico. *Proceedings of the Biological Society of Washington* 101:221-226.

Lapota, D., T.E. Bowman & J.R. Losee 1988. Observations on bioluminescence in the nauplius of *Metridia longa* (Copepoda, Calanoida) in the Norwegian Sea. *Crustaceana* 54:314-320.

1989

Bowman, T.E. 1989. *Lovenula (Neolovenula) alluaudi* (Guerne & Richard, 1890) in the Canary Islands (Copepoda: Calanoida: Paradiaptominae). *Bijdragen tot de Dierkunde* 59:239-241.

Bowman, T.E. 1989. The high frequency of errors in format in manuscripts submitted to the Proceedings of the Biological Society Washington. *Proceedings of the Biological Society of Washington* 102:1067-1068.

Bowman, T.E. & R. Bieri 1989a. *Paraspadella anops*, new species, from Sagittarius Cave, Grand Bahama Island, the second troglobitic chaetognath. *Proceedings of the Biological Society of Washington* 102:586-589.

Bowman, T.E. & N.L. Bruce 1989b. Species of the parasitic isopod genera *Ceratothoa* and *Glossobius* (Crustacea: Cymothoidae) from the mouths of flying fishes and halfbeaks (Beloniformes). *Smithsonian Contributions to Zoology* 489, 28pp.

Bowman, T.E. & J.J. Lewis 1989 Occurrence of the calanoid copepod *Eurytemora affinis* (Poppe) in the Ohio River at Louisville, Kentucky. *Journal of Crustacean Biology* 9:83-84.

1990

Bowman, T.E. 1990. Groundwater crustaceans of Spain, 13 (Copepoda, Calanoida). *Bijdragen tot de Dierkunde* 60:293-298.

Grygier, M.J. & T.E. Bowman 1990. The correct family-level name for the 'cryptoniscid' isopods (Epicaridea). *Crustaceana* 58:27-32.

Harrison-Nelson, E. & T.E. Bowman 1990. A range extension to the north for *Macrochiridotea giambiagiae* Torti and Bastida (Crustacea: Isopoda: Valvifera). *Proceedings of the Biological Society of Washington* 103:127-130.

Vega-Pérez, L.A. & T.E. Bowman 1990. *Ctenocalanus heronae* espécie nova de copépodo pelágico da regiâo de Ubatuba, Estado de Sâo Paulo. P. 100. In Programa e Resumos, Encontro Brasileiro de Plâncton 4, patrocinado pela Universidade Federal de Pernambuco, Departamento de Oceanografia.

1991

Bowman, T.E. 1991. *Epischura fluviatilis* Herrick, 1883 (Copepoda: Calanoida: Temoridae), first report from South Carolina, with notes on sexual dimorphism in the caudal rami and its possible role in mating. *Journal of Crustacean Biology* 11:460-462.

Bowman, T.E. & T.M. Iliffe 1991a. *Annina fustis*, a new isopod from Phang Nga, Thailand (Crustacea: Isopoda: Cirolanidae). *Proceedings of the Biological Society of Washington* 104:247-252.

Grygier, M.J. & T.E. Bowman 1991. The authorship of Cryptoniscidae (Isopoda, Epicaridea); a correction. *Crustaceana* 61:106-107.

1992

Bowman, T.E. 1992a. Two subterranean aquatic isopod crustaceans new to Texas: *Mexistenasellus coahuila* (Cole & Minckley, 1972) (Asellota: Stenasellidae) and *Speocirolana hardeni*, new species (Flabellifera: Cirolanidae). *Texas Memorial Museum, Speleological Monographs* 3:23-30.

Bowman, T.E. & L. Botosaneanu 1992. First record of a *Tachaea* species from Sulawesi (Indonesia) with description of its manca stage (Isopoda, Flabellifera, Corallanidae). *Bijdragen tot de Dierkunde* 62:111-115.

Bowman, T.E. & J.J. Orsi 1992b. *Deltamysis holmquistae*, a new genus and species of Mysidacea from the Sacramento-San Joaquin Estuary of California (Mysidae: Mysinae: Heteromysini). *Proceedings of the Biological Society of Washington* 105:733-742.

Garcia-Garza, M.E., G.A. Rodriguez-Almaraz & T.E. Bowman 1992. Distribution of the possum shrimp *Taphromysis louisianae* (Mysidacea), including new far inland records. *Journal of Crustacean Biolology* 12:101-103.

Vega-Pérez, L.A. & T.E. Bowman 1992. Description of the pelagic copepod, *Ctenocalanus heronae* Vega-Pérez and Bowman, from off Sâo Paulo, Brazil (Calanoida: Clausocalanidae). *Proceedings of the Biological Society of Washington* 105:97-101.

1993

Bowman, T.E. 1993. Copepodology at the Smithsonian Institution. *Monoculus* (copepod newsletter) 25:10-12.

Saltzman, J. & T.E. Bowman 1993a. *Boreomysis oparva*, a new possum shrimp (Crustacea: Mysidacea) from an eastern tropical Pacific seamount. *Proceedings of the Biological Society of Washington* 106:325-331.

1994

Bowman, T.E. 1994. Mictacea. Pp. 121-122. In C. Juberthie & V. Decu (eds), Encyclopaedia Biospeologia, volume 1. Society of Biospeology, Moulis, France, 880pp.

Falck, C.L. & T.E. Bowman 1994. Commensal life, sexual dimorphism, and handedness in the canuellid harpacticoid *Parasunaristes chelicerata* (Por & Marcus, 1972). *Hydrobiologia* 292/293:455-459.

Holsinger, J.R., D.A. Hubbard & T.E. Bowman 1994. Biogeographic and ecological implications of newly discovered populations of the stygobiont isopod crustacean *Antrolana lira* Bowman (Cirolanidae). *Journal of Natural History* 28:1047-1058.

Williams, E.H., Jr. & T.E. Bowman 1994. *Lironeca* Leach, 1818 (Crustacea, Isopoda): proposed conservation as the correct original spelling. *Bulletin of Zoological Nomenclature* 51:224-226.

1995

Bowman, T.E. 1995. Comments on proposed conservation of *Lironeca* Leach, 1818 (Crustacea, Isopoda) as the correct original spelling. *Bulletin of Zoological Nomenclature* 52:263.

Rodriguez-Almaraz, G.A. & T.E. Bowman 1995a. *Sphaerolana karenae*, a new species of hypogean isopod crustacean from Nuevo Leon, Mexico. *Proceedings of the Biological Society of Washington* 108:207-211.

1996

Garcia-Garza, M.E., G.A. Rodriguez-Almaraz & T.E. Bowman 1996a. *Spelaeomysis villalobosi*, a new species of mysidacean from northeastern Mexico (Crustacea: Mysidacea). *Proceedings of the Biological Society of Washington* 109:97-102.

Lewis, J.J. & T.E. Bowman 1996b. The subterranean asellids of Texas (Crustacea: Isopoda: Asellidae). *Proceedings of the Biological Society of Washington* 109:482-500.

Bowman, T.E., R.C. Brusca, B. Kensley, M. Schotte, G.A. Schultz & G.D.F. Wilson (in press). Crustaceans. *in Common and Scientific Names of Aquatic Invertebrates from the United States and Canada.* American Fisheries Society, Bethesda, Maryland.

Table 1. Latin names that Thomas E. Bowman introduced to biology by himself or with others for animals new to science [brackets], to replace previously existing names 'quotes', or to clarify a classification /slashes/ with date of publication indicated.

SUCTORIDA	
Rhabdophryidae	*Dendrosomides* [*lucicutiae*] 1977b
CHAETOGNATHA	
Spadellidae	*Paraspadella* [*anops*] 1989a
COPEPODA	
Ameiridae	*Nitokra* [*sphaeromata*] 1988a
Calanidae	*Calanus* [*lighti*] 1955c
Candaciidae	*Candacia* [*paenelongimana*] 1956b
Clausocalanidae	*Clausocalanus* [*citer*] 1971f
	Ctenocalanus [*heronae*] 1990a
Diaptomidae	*Mastigodiaptomus* [*nesus*] 1986d
	Notodiaptomus [*caperatus*] 1979a
	Notodiaptomus venezolanus [*deeveyorum*] 1972c
Nicothoidae	*Sphaeronella* [*hebe*] 1967c
	Sphaeronellopsis [*monothrix*] 1967c
	Sphaeronellopsis [*dikothrix*] 1969a
Oithonidae	*Oithona* [*colcarva*] 1975b
	Oithona [*bjornbergae*] 1980b
	Oithona [*fonsecae*] 1980b
	Paroithona [*flemingeri*] 1980b
Paracalanidae	*Acrocalanus* [*andersoni*] 1958a
	Paracalanus [*quasimodo*] 1971d
Pontellidae	*Calanopia* [*biloba*] 1957a
Pseudocyclopidae	*Pseudocyclops* [*cokeri*] 1961a
	Pseudocyclops [*paulus*] 1961a
	Pseudocyclops [*rostratus*] 1961a
	Pseudocyclops [*rubrocintus*] 1961a
Pseudodiaptomidae	*Pseudodiaptomus* [*cokeri*] 1965c
Stephidae	[*Miostephos*] [*cubrobex*] 1976a
Tisbidae	*Tisbe* [*monozota*] 1962a
Tortanidae	*Tortanus* [*compernis*] 1965c
	Tortanus [*scaphus*] 1971e
	Tortanus [*lophus*] 1971e
Tachidiidae	[*Cithadius*] [*cyathurae*] 1972a
LEPTOSTRACA	
Neballiidae	[*Speonebalia*] [*cannoni*] 1985f
MYSIDACEA	
Mysidae	*Antromysis* [*peckorum*] 1977a
	Antromysis [*reddelli*] 1977a
	Antromysis [*merista*] 1980a
	Antromysis ([*Surinamysis*]) 1977a
	Boreomysis [*oparva*] 1993a
	[*Deltamysis*] [*holmquistae*] 1992b

Table 1. Continued.

Mysidae (continued)	*Heteromysoides* [*dennisi*] 1985a
	Mysidium [*rickettsi*] 1987c
	Mysidopsis [*furca*] 1957b
	Mysidopsis [*almyra*] 1964b
Lepidomysidae	*Spelaeomysis* [*cardisomae*] 1972b
	Spelaeomysis [*olivae*] 1972b
	Spelaeomysis [*villalobosi*] 1996a
Stygiomysidae	*Stygiomysis* [*clarki*] 1984a
	Stygiomysis [*major*] 1976b
DECAPODA	
Bythocarididae	*Bythocaris* [*cryonesus*] 1972b
Sergestidae	*Lucifer* [*chacei*] 1967b
THERMOSBAENACEA	
Monodellidae	[*Tulumella*] [*unidens*] 1988b
Thermosbaenidae	*Halosbaena* [*fortunata*] 1986c
[MICTACEA] 1985c	
[Mictocarididae]	[*Mictocaris*] [*halope*] 1985d
AMPHIPODA	
[Anaponoidae] 1973d	
/Lanceoloidea/ superfamily 1973d	
[Lycaeopsoidea] new superfamily 1973d	
/Phronimoidea/ superfamily 1973d	
[Physocephalata] new infraorder 1973d	
/Platysceloidea/ superfamily 1973d	
/Scinoidea/ superfamily 1973d	
/Vibiloidea/ superfamily 1973d	
Cymidae	[*Syncyamus*] [*pseudorcae*] 1955a
Eusiridae	[*Dolobrotus*] [*mardeni*] 1974a
Hyperiidae	*Hyperia* [*crassa*] 1973a
	Hyperia [*leptura*] 1973a
	'*Hyperietta*' 1973a
	Hyperietta [*parvaceps*] 1973a
	Hyperietta [*stebbingi*] 1973a
	Hyperietta [*stephenseni*] 1973a
	'*Hyperionyx*' 1973a
	Lestrigonus [*shoemakeri*] 1973a
Phrosinidae	*Primno* [*brevidens*] 1978a
	Primno [*johnsoni*] 1978a
Scinidae	'*Spinoscina*' 1972c
Talatridae	*Orchestia* [*vaggala*] 1977c
ISOPODA	
Anthuridae	*Cyathura* [*specus*] 1965a
Asellidae	*Asellus* [*alaskensis*] 1975d
	Asellus [*kinki*] 1967a

Table 1. Continued.

Asellidae (continued)	*Caecidotea* [*beattyi*] 1981c
	Caecidotea [*bilineata*] 1996b
	Caecidotea [*carolinensis*] 1977e
	Caecidotea[*chiapas*] 1975c
	Caecidotea [filicispeluncae] 1983a
	Caecidotea [*lesliei*] 1981c
	Caecidotea [*misterae*] 1981c
	Caecidotea [*rotunda*] 1984b
	Caecidotea [*sequoiae*] 1975c
	Caecidotea [*whitei*] 1981c
	[*Calasellus*] [*longus*] 1981a
	'*Lirceolus*' 1976c
	Lirceolus [*hardeni*] 1996b
	Lirceus [*usdagalun*] 1972e
	'*Remasellus*' 1985e
	[*Salmasellus*] [*steganothrix*] 1975c
Bopyridae	*Bopyrella* [*harmopleon*] 1956a
Cirolanidae	*Anopsilana* [*crenata*] 1982c
	Anopsilana [*lingua*] 1987b
	Annina [*fustis*] 1991a
	[*Antrolana*] [*lira*] 1964a
	Bahalana [*mayana*] 1987a
	[*Bermudalana*] [*aruboides*] 1983b
	[*Ceratolana*] [*papuae*] 1977d
	Cirolana [*narica*] 1971a
	Excirolana [*kumari*] 1971b
	[*Haptolana*] [*trichostoma*] 1966a
	[*Mexilana*] [*saluposi*] 1975a
	Speocirolana [*endeca*] 1982a
	Speocirolana [*hardeni*] 1992a
	Speocirolana [*pubens*] 1982a
	Sphaerolana [*karenae*] 1995a
Cymothoidae	*Anilocra* [*monoma*] 1983c
	Catoessa [*gruneri*] 1983c
	Glossobius [*anctus*] 1989b
	[*Joryma*] [*sawayah*] 1983c
	Lironeca [*orinoco*] 1957c
	Lironeca [*puhi*] 1960a
	Nerocila [*arres*] 1983c
	Nerocila [*kisra*] 1983c
	Nerocila [*sigani*] 1983c
	Paracymothoa [*tholoceps*] 1986b
	Renocila [*heterozota*] 1968a
Idoteidae:	*Chiridotea* [*coeca*] 1955b
	'*Ronalea*' 1956c
Sphaeromatidae	*Cymodetta* [*gambosa*] 1974b
	Thermosphaeroma [*cavicauda*] 1985b
	Thermosphaeroma [*macrura*] 1985b
	Thermosphaeroma [*milleri*] 1981b
	Thermosphaeroma [*smithi*] 1981b

Table 1. Continued.

Stenasellidae	*Mexistenasellus* [*colei*] 1982b
	Mexistenasellus [*nulemex*] 1982b
	Etlastenasellus [*confinis*] 1982b
Trichoniscidae	‘*Xilitloniscus*’ 1965b
Tridentellidae	*Tridentella* [*recava*] 1986a
???	*Pelaega* [*lamnae*] 1971c

1 Taxonomy, systematics, biogeography

New genera of terrestrial isopods (Oniscidea: Philosciidae) from South America, with remarks on some species

ANDREAS LEISTIKOW

Universität Bielefeld, Abteilung für Morphologie und Systematik der Tiere, Ruhr-Universität Bochum, Lehrstuhl für spezielle Zoologie, Deutschland

ABSTRACT

When dealing with the genus *Philoscia* Latreille, 1804, one will encounter species that differ considerably from the generic diagnosis. Particularly tropical members of this genus described at the turn of the century are ascribed to this genus because of superficial similarity of the body shape or the like. For a phylogenetic analysis of the Oniscidea on lower taxonomic level, it is necessary to deal with monophyletic taxa. To shed new light on some South American species of the genus *Philoscia*, a re-examination of the following species was done: *Philoscia* (*Ischioscia*) *mineri* van Name, 1936, *P.* (*I.*) *nitida* (Miers 1877), *P.* (*I.*) *debilis* Budde-Lund, 1893, *P. kartaboana* van Name, 1936, *P. inquilina* van Name, 1936, *P. roraimae* Van Name, 1936, *P. omissa* van Name, 1936, *P. moneaguensis* van Name, 1936, *P. seriepunctata* Budde-Lund, 1893, *P. diminuta* Budde-Lund, 1885 and *P. walkeri* Pearse, 1915. Four new monotypic genera are proposed for above mentioned species based on autapomorphies of these species.

1 INTRODUCTION

The systematics of the Oniscidea is full of unresolved questions, although on higher taxonomic level much effort has been made to shed light on the phylogeny (e.g. Erhard 1995, Schmalfuss 1989). On lower taxonomic level, there is still much confusion. Many species are known only from the type series and were described at a time, when no guidelines for focusing on particular characters existed. Therefore, some descriptions deal with characters which are of no value for recognition of the species. On the other hand, generic boundaries were rather wide. A good example is the genus *Philoscia* Latreille, 1804, which formerly was merely a melting pot for crinochete Oniscidea with slender pleon, three-articulate antennal flagellum and without pleopodal lungs. A revision of the genus especially for the European representatives was done by Verhoeff (1908), and a modern diagnosis for *Philoscia* can be found in Schmalfuss (1990). The philosciids from Africa were dealt with by Ferrara & Taiti in several publications, and they gave an overview of the characters that are important for philosciid systematics (Ferrara & Taiti 1985, Taiti & Ferrara 1980). For a phylogenetic analysis it is necessary to use as many characters as possible. There are several species from South America that were ascribed to the genus *Philoscia* at the turn of this century. A re-examination of these species is urgently needed to let *Philoscia* become a monophyletic unit. However, the material is spread over several museums of the world, and some material cannot be located. In the following, a dozen species is reconsidered and for five of them redescriptions are presented. The remaining species are known only from females or material in a horrible condition, they were declared as *nomina nuda* according to articles 12 and 13a ICZN. There is no possibility to discriminate these species correctly from the descriptions or the type material and compare it with future collections in a satisfactory way.

For the American Museum of Natural History the acronym AMNH is used in the following, for the British Natural History Museum, London is used BMNH, for the US National Museum of Natural History, USNM.

2 SYSTEMATIC ACCOUNT

2.1 *Ischioscia* Verhoeff, 1928

A diagnosis for this genus was given by Schmalfuss (1980) and was enlarged by Leistikow (1997).

2.1.1 *Ischioscia mineri* van Name, 1936

Material: AMNH 6558 (paratypes) 2 males, several females; Dominica, Laudat, leg. F.E. Lutz, 12.06.1911; length of dissected male 9.5 mm.

Color: Color has faded to a yellowish brown due to preservation in alcohol. Almost no markings but insertions of musculature on cephalothorax discernible.

Cephalothorax: No linea frontalis, linea supra-antennalis distinct, medially inflexed. Medially of antennal sockets with lamina frontalis, labrum of considerable length. Compound eyes consisting of about 26 ommatidia in four longitudinal rows. Vertex smooth without long setae (Fig. 1, Ctf).

Pereion: Tegument shiny and smooth, some scattered tricorn-like setae, no noduli laterales visible on coxal plates in light microscope at 400× magnification. Those distally rounded in first two segments, pointed in others. Coxal plates more adpressed, thus pereion appearing elongate (van Name 1936).

Pleon: Pleonites 3 to 5 with pronounced neopleurae, distally pointed, but not continuing outline of pereion. Pleotelson pointed, latero-distal margin concave, some sensilla with cuticular sheath, no tricorn-like setae. Uropod protopod not surpassing apex of pleotelson.

Appendages:

Antennula: Relatively slender, three-articulate. Distal joint bearing seven aesthetascs, three set apically, four medially in a row, medial and proximal articles with some small hairs (Fig. 1, An1).

Antenna: Flagellum three-articulate, as long as fourth peduncular article, proximal article the longest, distal bearing terminal bristle of same length as proximal article. Second and third peduncular article subequal, fourth twice as long, fifth one and a half times longer than fourth, distal two articles with sensory spine distally. Both peduncle and flagellum covered with tricorn-like setae (Fig. 1, An2).

Mandible: Molar penicil composed of several branches, pars intermedia with coniform setae, less plentiful on right mandible. Left mandible bearing three, right two penicils (Fig. 2, Mdl/r).

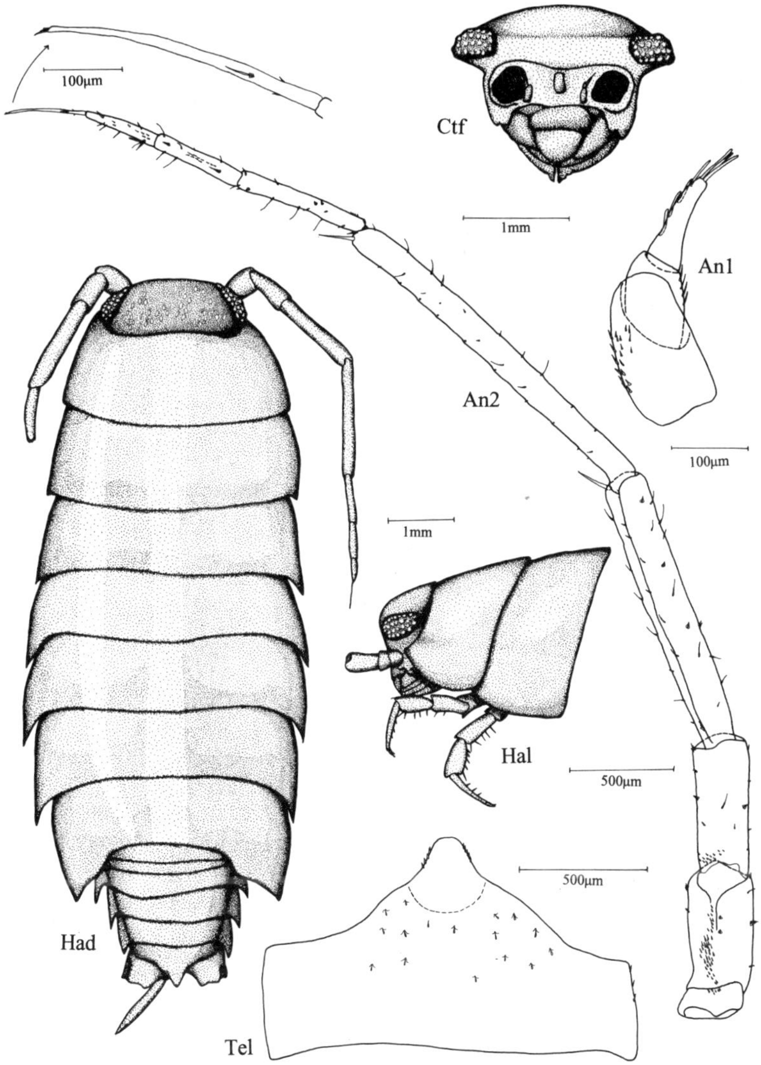

Figure 1. *Ischioscia mineri* van Name 1926, male 9.5 mm. An1 antennula; An2 antenna with detail of apical organ; Ctf cephalothorax, frontal view; Had habitus, dorsal view; Hal habitus, lateral view; Tel pleotelson.

Maxillula: Medial endite latero-distally pointed, with two penicils, one lost in dissected specimen. Lateral endite with 4+6 teeth, 5 apically cleft, hyaline lobe rostrally (Fig. 2, Mx1).

Maxilla: Inner lobe with dense cover of hair-like setae on medial margin, proximally some scattered short setae, outer lobe setose (Fig. 2, Mx2).

Maxilliped: Tricorn-like setae of basis short in comparison to other members of the genus. Endite with knob-like penicil caudally, rostrally setose with two stiff setae. Palp broken in examined specimen, second article (complete but damaged in left maxilliped) with two setal tufts, between them a single seta (Fig. 2, Mxp).

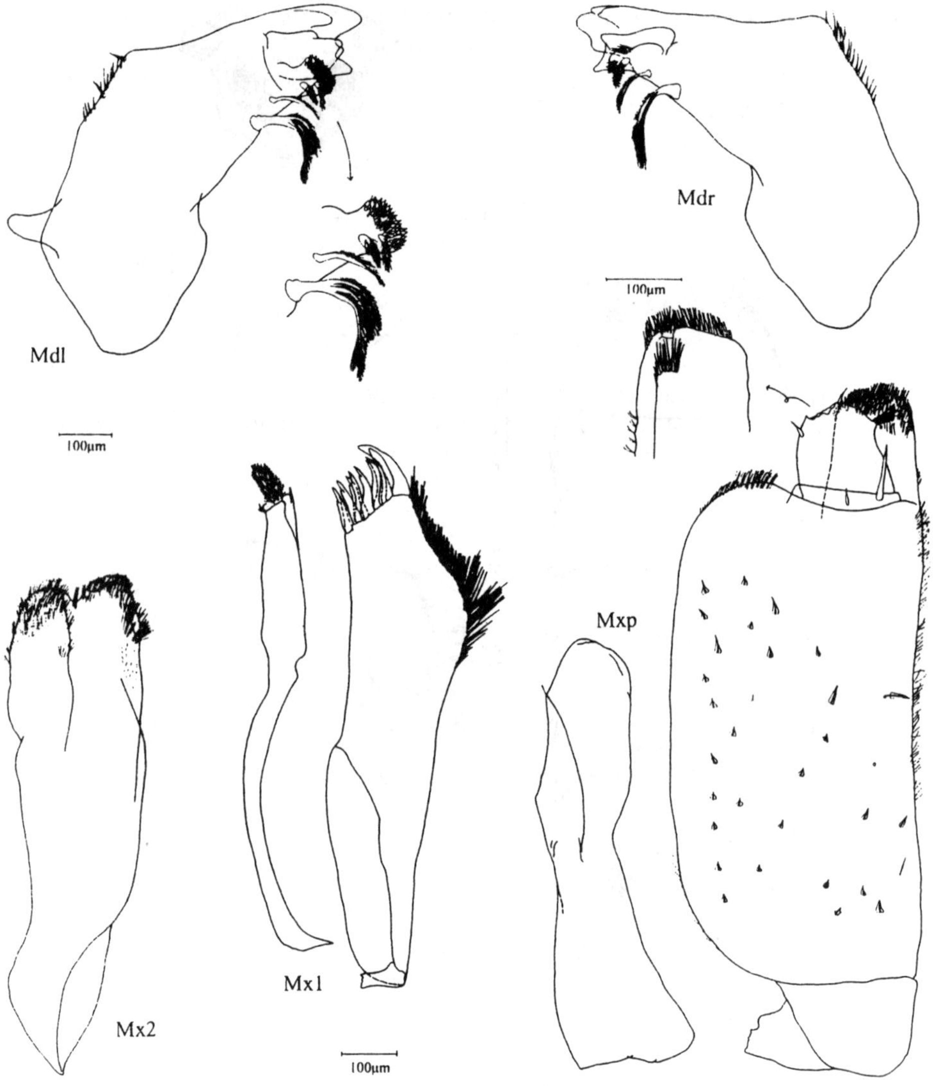

Figure 2. *Ischioscia mineri* van Name 1926, male 9.5 mm. Mdl/r left and right mandible; Mx1 maxillula; Mx2 maxilla; Mxp maxilliped with detail of endite in rostral view.

Pereiopods: In general as in other species of *Ischioscia* Verhoeff, 1928. Dactylus with plumose dactylar seta and two setae between claws (Fig. 3, Dac/Sd1). Carpus of pereiopod 1 with hand-like sensory spine (Fig. 3, Sc1). Pereiopods 5-7 longer than proximal ones, with seventh the longest (Fig. 3, PE1-7).

Sexual differentiation of pereiopods: Male pereiopods 1-3 with enlarged carpus and setal brushes on rostral surface, merus of respective appendages with medial setal brush, too. Pereiopod 1 and 2 subequal, third pereiopod slightly longer with less conspicuous carpus enlargement. Pereiopod 7 ischium more cylindrical, laterally with two sensory

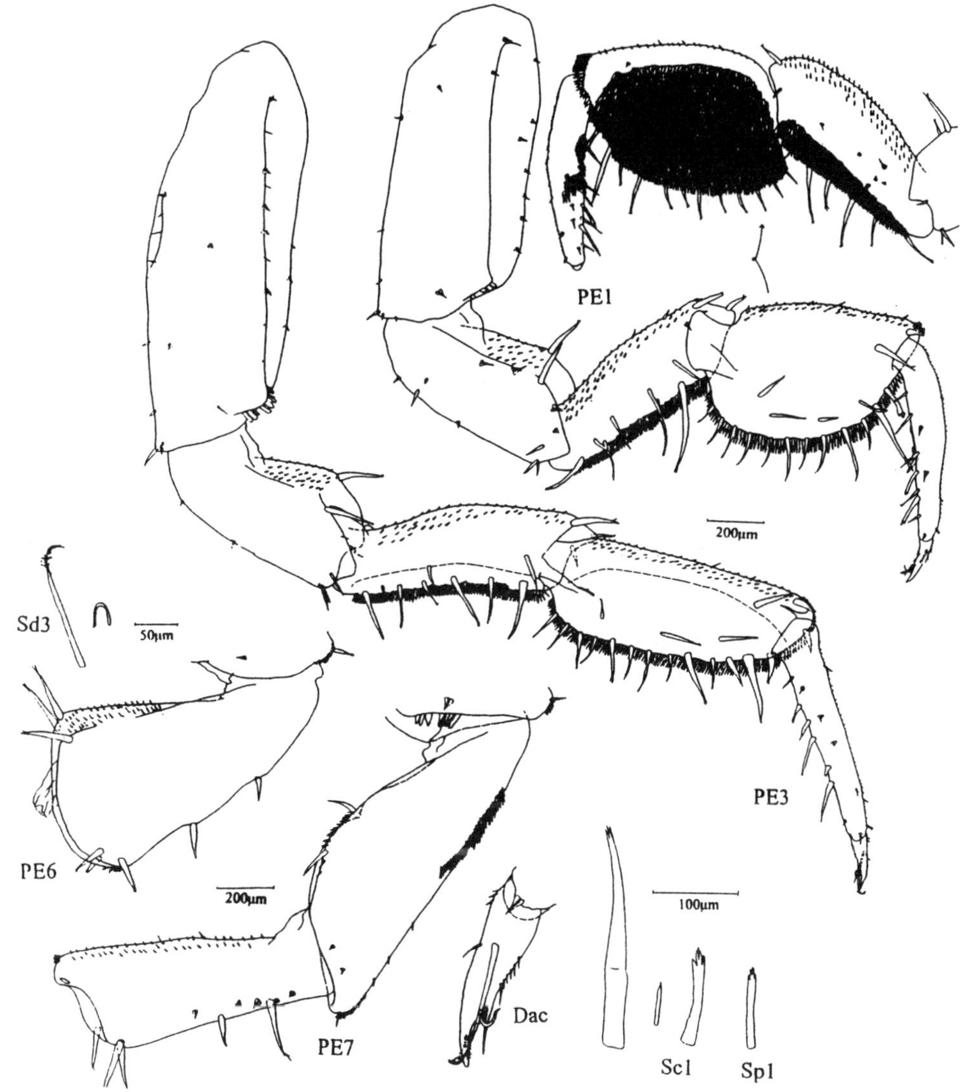

Figure 3. *Ischioscia mineri* van Name 1926, male 9.5 mm. Dac dactylus; PE1/3 pereiopod 1 and 3 in caudal view, with detail of pereiopod 1 in rostral view; PE6 ischium of pereiopd 6; PE7 ischium and merus of pereiopod 7; Sc1 sensory spines and seta of antenna-g rooming brush (middle) of carpus 1; Sd1 interungual (right) and dactylar seta of dactylus 1; Sp1 sensory spine of propus 1.

spines. Medial margin with some short setae and a slight distal protrusion. The latter covered with short setae and a longer sensory spine.

Pleopods: Exopods of pleopods pointed, lateral margin with five sensory spines in third and fourth pair to three in fifth pair. Pleopod 5 exopod with four rows of pectinate scales caudally (Fig. 4, PL1-5).

Sexual differentiation of pleopods: Male first pleopod with somewhat falcate exopod, laterally truncate. Endopod elongate, small accessory process laterally, apex with four

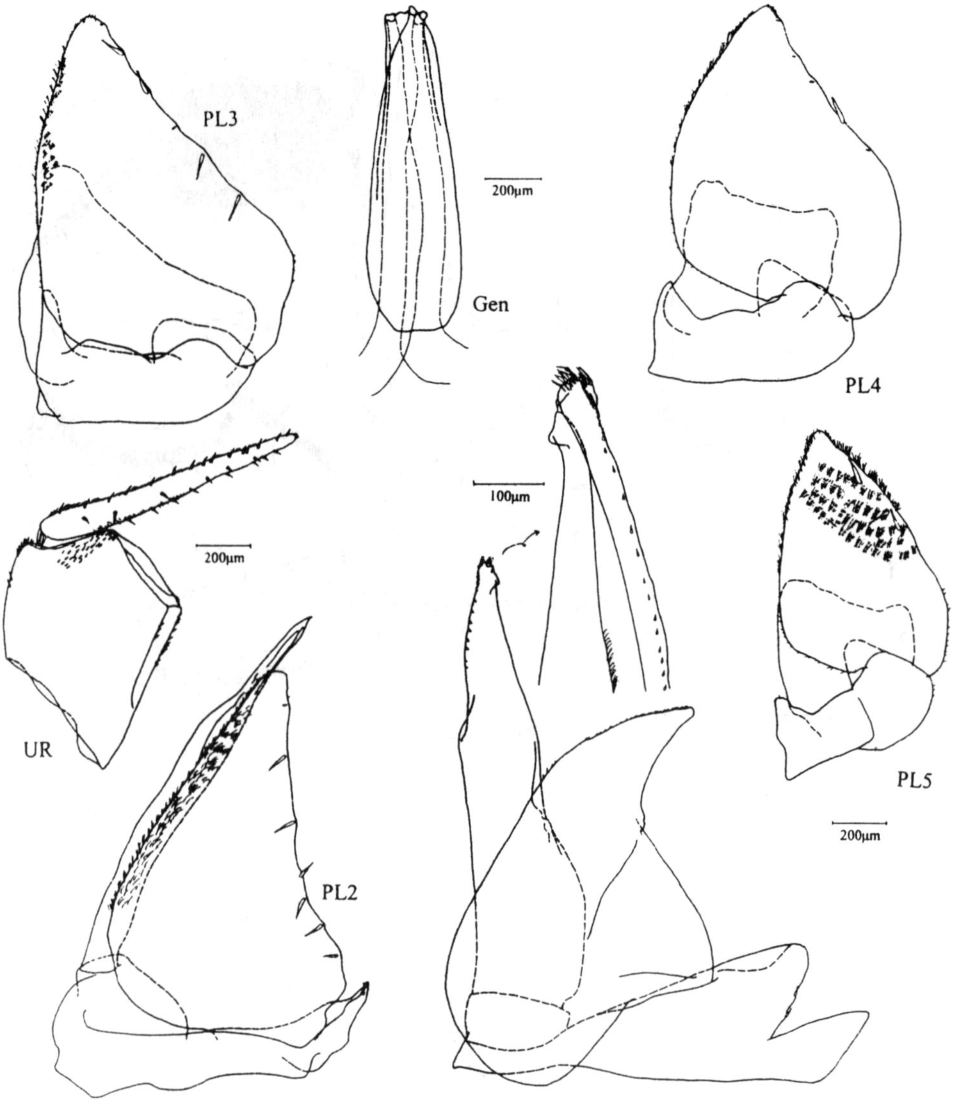

Figure 4. *Ischioscia mineri* van Name 1926, male 9.5 mm. Gen genital papilla; PL1-5 pleopod 1 to 5, with detail of endopod 1 in caudal view; UR uropod.

spiny setae medially and a tuft of hyaline hairs laterally. Male pleopod 2 elongate and slender, surpassing exopod. Lateral border of exopod hardly sinuous, bearing seven sensory spines.

Uropod: Endopod inserting proximally of exopod, protopod medially with some scattered small setae (Fig. 4, UR).

Genital papilla: Without any setation, two ductus ejaculatorii (Fig. 4, Gen).

Discussion: *Ischioscia mineri* van Name, 1936 is similar to the other species lacking a frontal line. The adpressed coxal plates give a comparatively slender appearance. Different from all the other species known in detail, the pleotelson of *I. mineri* shows no tricorn-like setae but small sensilla penetrating the cuticle, accompanied by waxy scales. The setation of the maxillipedal palp is quite distinct, since it bears an individualized seta medially on the second article not present in any species previously investigated, these characters are autapomorphies of the species.

The male pereiopod 1 has considerably fewer sensory spines on caudal surface than *I. variegata* (Dollfus, 1893) from Venezuela. In comparison to other species of the *variegata-martinae-bolivari*-group, pereiopod 7 is closest to *I. martinae* Leistikow, 1997 from Costa Rica but differs in the extent of the setal brush of ischium and the presence of small hairs on the medio-distal edge of basis. Male pleopod 1 is unique among the species of *Ischioscia*, due to the characteristic setation and the shape of the exopod representing apomorphies of this species.

2.1.2 *Species of uncertain taxonomic value*

Ischioscia debilis (Budde-Lund, 1893)

Synonymy: *Philoscia debilis* Budde-Lund, 1893, *P.* (*Ischioscia*) *debilis* van Name, 1936.

Material: BMNH 1921.10.18.2236 and 2239 4 females; Venezuela, Caracas, La Moka.

Remark: *I. debilis* (Budde-Lund, 1893) shows a color pattern similar to *I. variegata*, which occurs in the vicinity of the former species. Thus, it is possible that both species might be synonymous. Since no males of *I. debilis* are available, no further conclusions can be made, and it is declared a *nomen dubium*.

Ischioscia nitida (Miers, 1877)

Synonymy: *Philougria nitida* Miers, 1877, *Philoscia* (*Ischioscia*) *nitida* van Name, 1936.

Material: BMNH 1879.21 (SYNTYPES), 6 females (SYNTYPES); Peru or Guyana (?), South America, leg. Wzesniowsky.

Remark: In contrast to the other herein examined members of *Ischioscia*, in *I. nitida* a linea frontalis is present. Due to the lack of males an identification of this species is almost impossible, so this name is another *nomen dubium*.

2.2 *Yaerikima* gen. n.

Diagnosis: Cephalothorax with linea supra-antennalis and lamina frontalis, no linea frontalis, two sinuous lines on profrons dorsally of antennal sockets. Antennula with numerous aesthetascs distally, antennal flagellum three-articulate, with prominent apical organ.

Mandibles with dichotomized molar penicil, penicils of pars intermedia short. Lateral endite of maxillula with 4+6 teeth, 5 of inner set cleft, medial endite without apical spine. Lateral lobe of maxilla half as broad as medial lobe, lacking setation. Endite of maxillipede with penicil and setal tuft.

Pereiopods slender, antenna grooming-brush on propus and carpus of pereiopod 1, interungual setae subequal in length, dactylar seta prominent, with long plumose area. Coxal plates without sulcus marginalis, neither noduli laterales nor gland pores visible under light microscope.

Pereiopods slender, ornamental sensory spine of carpus 1 apically double-fringed, dactylar seta with plumose apex.

Pleopods without respiratory area. Exopods and endopods pointed, the former bearing sensory spines laterally. Protopod of uropod only slightly grooved, endopod inserting proximally of exopod.

Type species: *Philoscia kartaboana* van Name, 1936, by monotypy.

Etymology: This genus is dedicated to Miss Yae Ri Kim (AMNH), who helped in locating the material and getting access to it.

2.2.1 *Yaerikima kartaboana* (van Name, 1936) comb. n.

Synonymy: *Philoscia kartaboana* van Name, 1936.

Material: AMNH 6557 (paratypes) 2 males, 5 females; Guyana, Kamakusa, leg. H. Lang, 25.10.1922; length of dissected male 6 mm.

Color: More or less faded to yellowish brown, so van Name (1936: 145) is cited: '... light brown or purplish brown usually with a noticably darker median stripe and lateral stripes along the bases of the thoracic epimera. Inside the lateral dark stripe a row of large light spots, one on each thoracic segment on each side, besides the usual light spots and markings on the lateral regions of the back'.

Cephalothorax: Linea frontalis lacking, linea supra-antennalis present, small but distinct frontal lamina. Frons with sinuous carinae above antennal sockets. Eyes composed of 15 ommatidia (Fig. 5, Ctf).

Pereion: Rather slender, tegument smooth and shiny, with very few tricorn-like setae. Coxal plates adpressed, anterior ones with rectangular margin, posterior ones caudally pointed. No sulcus marginalis and gland pores nor noduli laterales discernible in light microscope (Fig. 5, Cx3).

Pleon: Set apart from pereion, with small neopleurae on pleonites 3 to 5. Pleotelson triangular with straight but slightly crenulate margins, some scattered tricorn-like setae along margin.

Appendages:

Antennula: Three-articulate, first article twice as long as second, latter cylindrical in shape; third article coniform, a tuft of 11 aesthetascs apically (Fig. 5, An1).

Antenna: Flagellum composed of three articles of subequal length. Terminal organ two thirds of the length of flagellum. Peduncular article 5 as long as flagellum, peduncular articles 2 and 3 half the length of article 5, article 4 intermediate, bearing a conspicuous tricorn-like seta distally, three times longer than other tricorns (Fig. 5, An2).

Mandible: Pars intermedia with some coniform setae and two short penicils on left and one on right mandible, respectively. Additional plumose seta proximally of pars intermedia rather long. Molar penicil consisting of about 10 plumose branches (Fig. 6, Mdl/r).

Maxillula: Medial endite with two pointed penicils (one in examined specimen broken). Lateral endites of both right and left maxillula missing in examined specimen. Van Name (1936) figured the lateral endite; it may have 4+6 teeth, with five of the inner set cleft (Fig. 6, Mxl/m).

Maxilla: Lateral lobe half as small as medial lobe, without setation save a small tuft. Medial endite latero-distal with styliform setae, distal third bearing hair-like setae (Fig. 6, Mx2).

Maxilliped: Palp three-jointed, articulation between distal joints reduced, bearing two setal tufts, with the medial tuft stalked, two single setae proximally of medial tuft. Proxi-

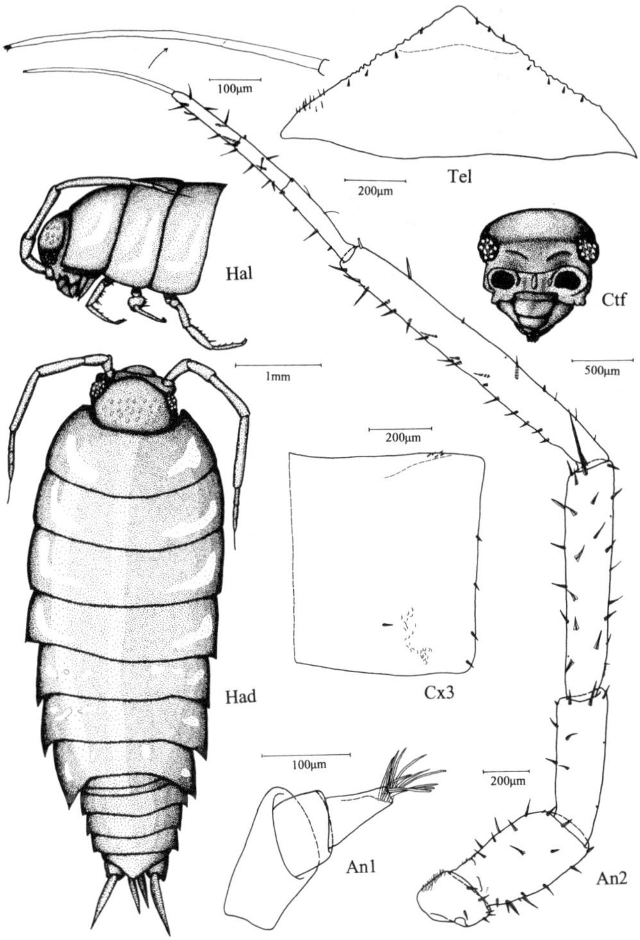

Figure 5. *Yaerikima kartaboana* (van Name, 1936) male 6 mm. An1 antennula; An2 antenna with detail of apical organ; Ctf cephalothorax, fontal view; Cx3 coxal plate 3; Had habitus, dorsal view; Hal habitus, lateral view; Tel pleotelson.

mal article with two setae. Endite bearing a knob-like penicil and a setal tuft on rostral surface and two setae beside some hairs on caudal surface (Fig. 6, Mxp).

Pereiopods: Pereiopods 1 to 3 subequal in length, caudal ones longer with pereiopod 7 the longest. Dactylus with two subequal interungual setae, as long as smaller claw (Fig. 7, Dac), dactylar seta long, distal third plumose (Fig. 7, Sd1). Basis with two rows of tricorn-like setae. Pereiopod 1 propus and carpus with antenna-grooming brush on rostral surface (Fig. 7, PE1). Both articles with ornamental sensory spines, i.e. their apex flattened and characteristically double-fringed. Pereiopod 2 to 6 carpus and merus with medial margin bearing translucent dentation, less extended on pereiopod 6 (Fig. 7, PE1-3; Fig. 8, PE5-7).

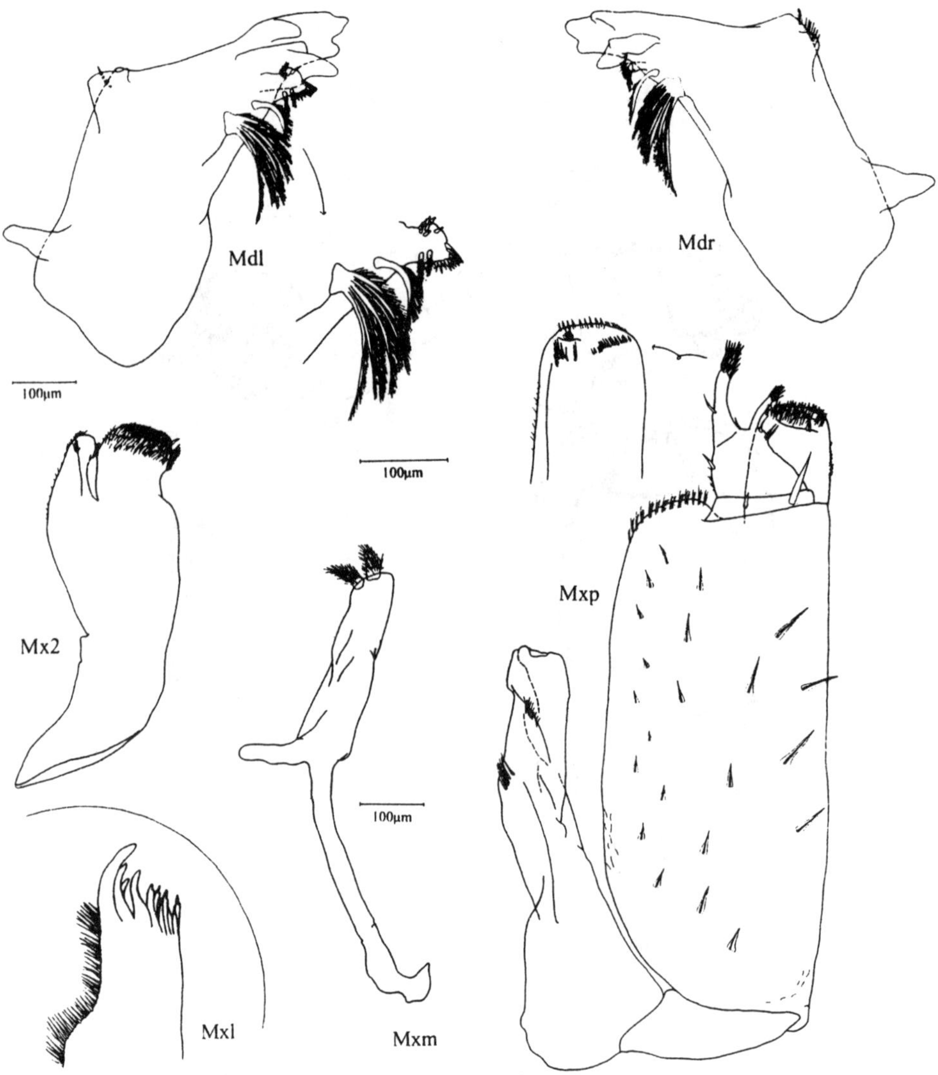

Figure 6. *Yaerikima kartaboana* (van Name 1936) male 6 mm. Mdl/r left and right mandible; M×2 maxilla; Mxl lateral endite of maxillula, redrawn after van Name (1936); Mlm medial endite of maxillula; Mxp maxilliped with detail of endite in rostral view.

Sexual differentiation of pereiopods: Merus of pereiopod 7 bulky, lateral margin proximally swollen, medial margin with area of hair-like setae on proximal edge. Besides this feature no further sexual differentiation observed, at least macroscopically.

Pleopods: Exopods pointed, medial margin with some hair-like setae, lateral margin with three (pleopod 3 and 5) or four sensory spines (pleopod 4). Exopod of pleopod 5 more pointed than other ones, no respiratory areas visible in light microscope. Endopods lacking any setation, medially pointed (Fig. 9, PL1-5).

Sexual differentiation of pleopods: Pleopod 1 exopod without any setation, basally

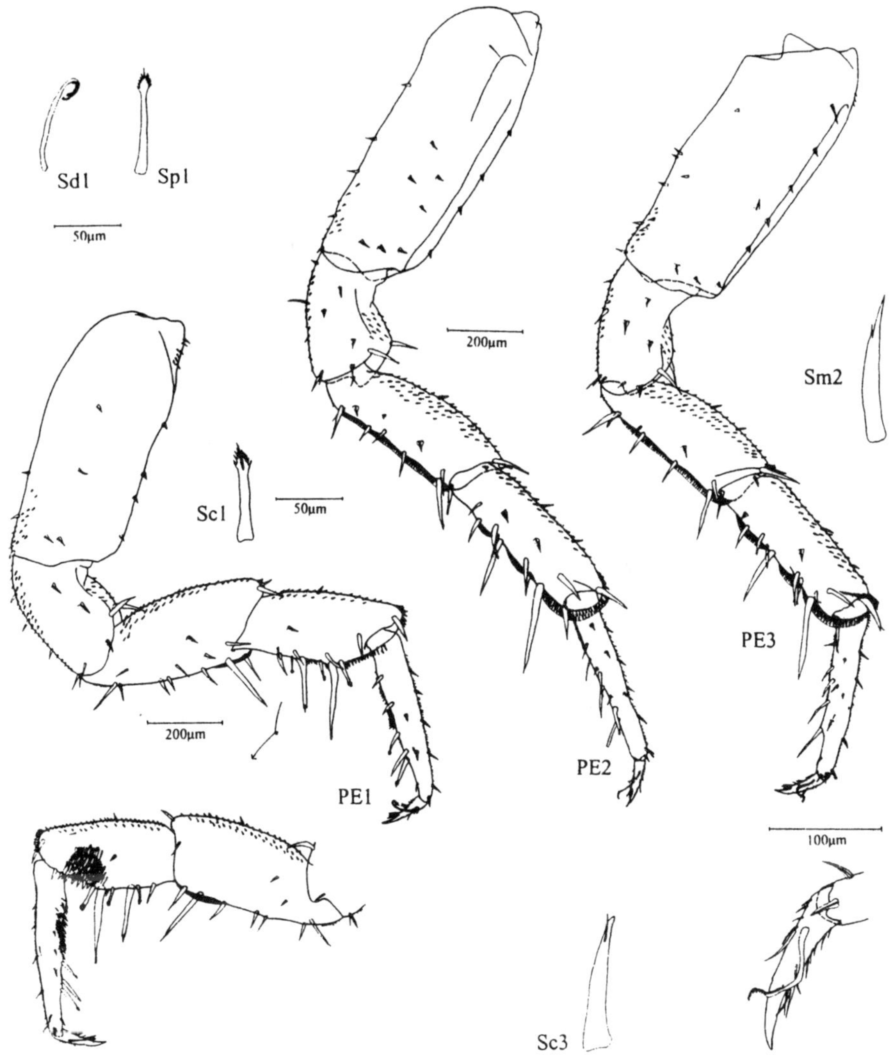

Figure 7. *Yaerikima kartaboana* (van Name 1936) male 6 mm. Dac dactylus; PE1-3 pereiopods 1 to 3, with detail of pereiopod 1 in rostral view; Sc1 ornamental sensory spine of carpus 1; Sc3 sensory spine of carpus 3; Sd1 dactylar seta 1; Sm2 sensory spine of merus 2; Sp1 sensory spine of propus 1.

more or less rounded, laterally hookedly pointed. Endopod elongate, apex pointed, bearing a couple of spiny setae, terminating the caudal setal row. Pleopod 2 exopod pear-shaped, medially with hair-lke setae, laterally with two sensory spines. Endopod slender, surpassing exopod in length about one quarter, distal third tapered, flagelliform.

Uropod: Protopod slightly grooved on lateral margin, endopod inserted somewhat more proximally of exopod, endopod with tricorn-like setae, exopod missing in dissected specimen (Fig. 8, UR).

Genital papilla: Ductus ejaculatorii parallel, orifrices somewhat surpassing ventral shield, joined but unfused, no setation (Fig. 9, Gen).

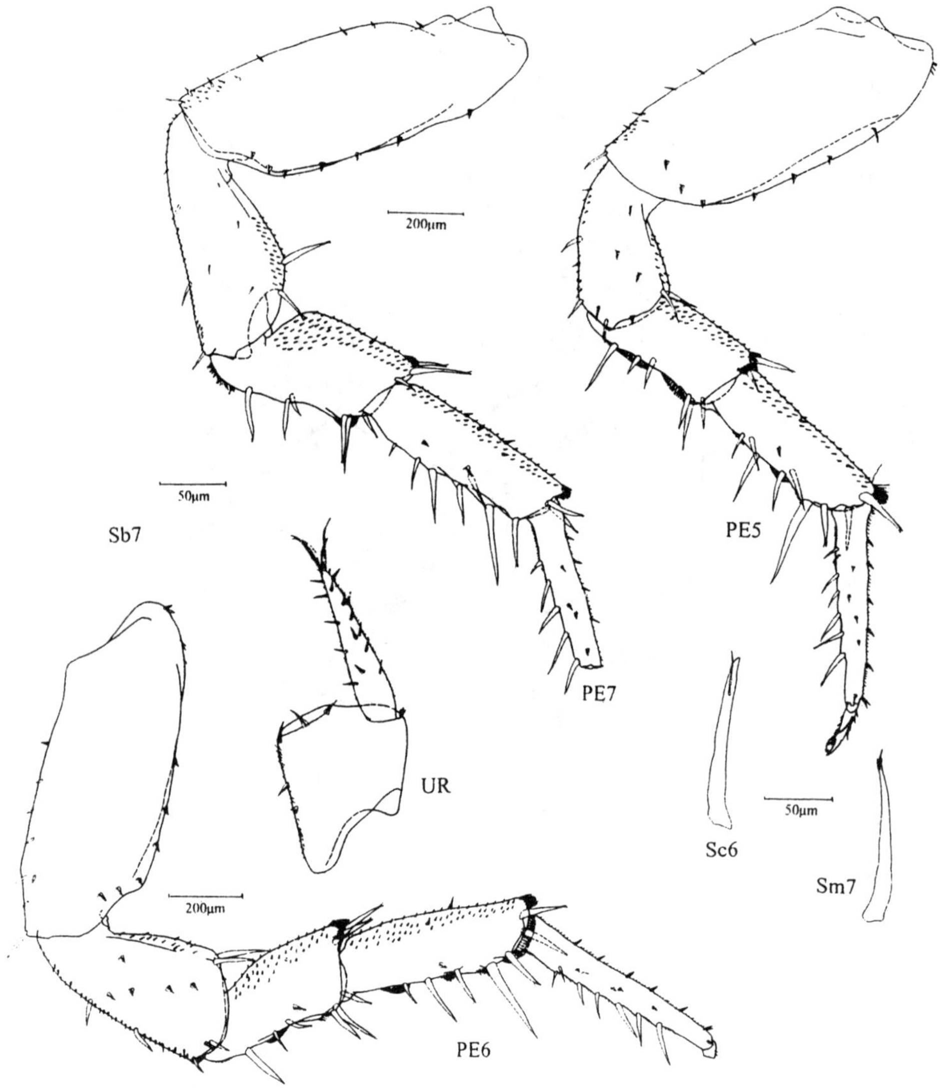

Figure 8. *Yaerikima kartaboana* (van Name 1936) male 6 mm. PE5-7 pereiopods 5 to7; Sc6 sensory spine of carpus 6; Sm7 sensory spine of merus 7; UR uropod.

Discussion: Like many other species of South American Oniscidea, the new genus *Yaerikima* gen. n. lacks conspicuous noduli laterales on its coxal plates. *Yaerikima kartaboana* (van Name 1936) is quite similar to *Formicascia inquilina* (van Name, 1936) but the structure of the cephalothorax differs remarkably in both species: *Y. kartaboana* has a more convex vertex and a lamina frontalis, the clypeus is rather narrow even at its base. In *F. inquilina,* the vertex is more flattened and the basally broad clypeus is medially ridged. The shape of the coxal plates and the pleotelson differs, as does the the shape of the female pleopod 3 exopod.

The relations of *Y. kartaboana* to other philosciid species in South America are diffi-

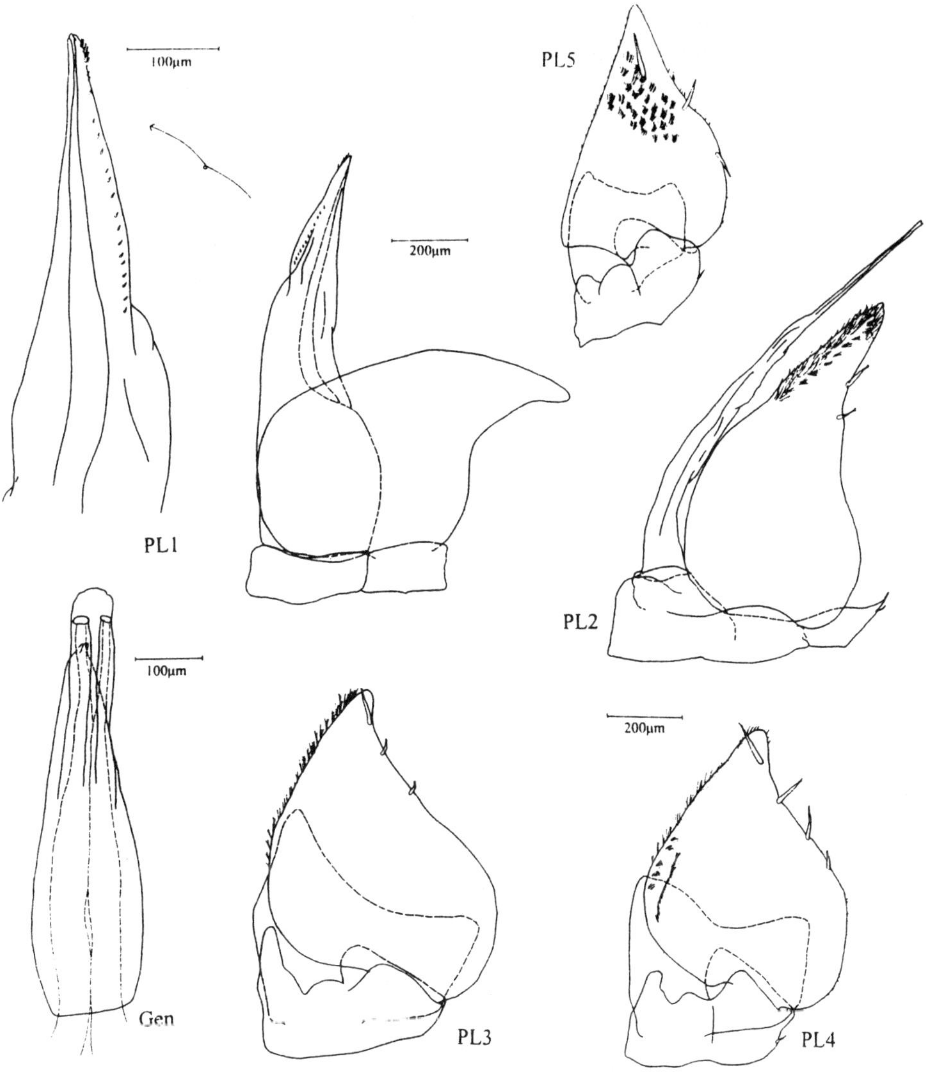

Figure 9. *Yaerikima kartaboana* (van Name 1936) male 6 mm. Gen genital papilla; PL1-5 pleopod 1 to 5, with detail of endopod 1 in caudal view.

cult to establish, as most of the genera known from this region bear only slight differences from the ground pattern. There are some characters which might link *Yaerikima* gen. n. with the genus *Andenoniscus* Verhoeff, 1941, although the similarities are possibly due to convergence. This might be the case in the flagelliform endopod of the male pleopod 2, or the shape of the pleopod 5 exopod with its straight and proximally prolonged medial margin, which is similar to *Pentoniscus* Richardson, 1913, recently redescribed by Leistikow (1998).

Rather interesting, the aesthetascs of the antennula are arranged different to all the other species described herein. There are some inserting apically, but several aesthetascs are inserted more medially, thus diverging from the apical set. This might be a heritage from a common ancestor with *Andenoniscus*, the latter with a clearly bipartite aesthetasc set. It may be the prerequisite for the even more divergent aesthetasc sets separated by a groove of the *Prosekia*-group to which *Andenoniscus* belongs.

2.3 *Formicascia* gen. n.

Diagnosis: Cephalothorax with linea supra-antennalis, lamina frontalis and linea frontalis missing, profrons with two slight depressions, area between profrons and vertex slightly arched, compound eyes with 11 to 13 ommatidia.

Mandibles with dichotomic molar penicil, maxillula with 4+6 teeth, five of inner set cleft, hyaline lobe present, lobes of maxilla subequal in breadth, setose, maxilliped with knob-like penicil rostrally on endite, caudally two teeth, palp with two setal tufts.

Pereiopods with several tricorn-like setae, antenna-grooming brush of carpus 1 very small, ornamental sensory spine with double-fringed apex, dactylar seta slightly plumose, coxal plates distally rounded, without sulcus marginalis, gland pores and noduli laterales invisible in light microscope.

Pleopods with few sensory spines laterally on exopod, female endopod 3 medially prolonged, surpassig exopod, setal creel on pleopod 5 exopod composed of two rows of pectinate scales. No respiratory areas discernible. Protopod of uropod only slightly grooved, endopod inserting proximally of exopod. Ductus ejaculatorii proximally of genital papilla bulbiform.

Type species: *Philoscia inquilina* van Name, 1936, by monotypy.

Etymology: The genus name refers to the habits of the type species, which was found in ant nests.

2.3.1 *Formicascia inquilina* van Name, 1936 n. comb.

Synonymy: *Philoscia inquilina* van Name, 1936.

Material: AMNH 6569, 10 females; Guyana, Bartica, in an ant nest below ground, leg. H. Lang, 10.1922; length of dissected specimen 6.5 mm.

Color: Almost entirely faded, van Name (1936: 149) stated: '...of usual pattern. ... On the epimera the pigment often shades off gradually, leaving their greater part light-colored, while the bars and spots on the lateral regions of the back often become confluent, forming large, irregular light areas that give the body, when seen without much magnification, an appearance of being crossed by imperfect but noticeable transverse bands. There is a dark band across the front of the head between the eyes.'

Cephalothorax: Linea supra-antennalis prominent, slightly bent between antennal sockets, no linea frontalis nor lamina frontalis. Frons somewhat concave, forming two slight depressions above antennal sockets, giving the border frons-vertex a curved pad-like appearance. Clypeus with slight medial longitudinal carina. Compound eyes composed of 11 to 13 ommatidia (Fig. 10, Ctf).

Pereion: Slender as in *Yearikima kartaboana* (van Name 1936), with cephalothorax somewhat set back into pereionite 1. Coxal plates directed ventrally, coxal plates 1 to 4 distally rounded, coxal plates 5 to 7 more and more pointed, but by no means caudally acute but obtuse and rounded. Coxal plates with some long tricorn-like setae, no noduli laterales discernible under light microscope (Fig. 10, Cx4).

Pleon: Set apart from pereion, neopleurae of pleonites 3 to 5 present, distally pointed.

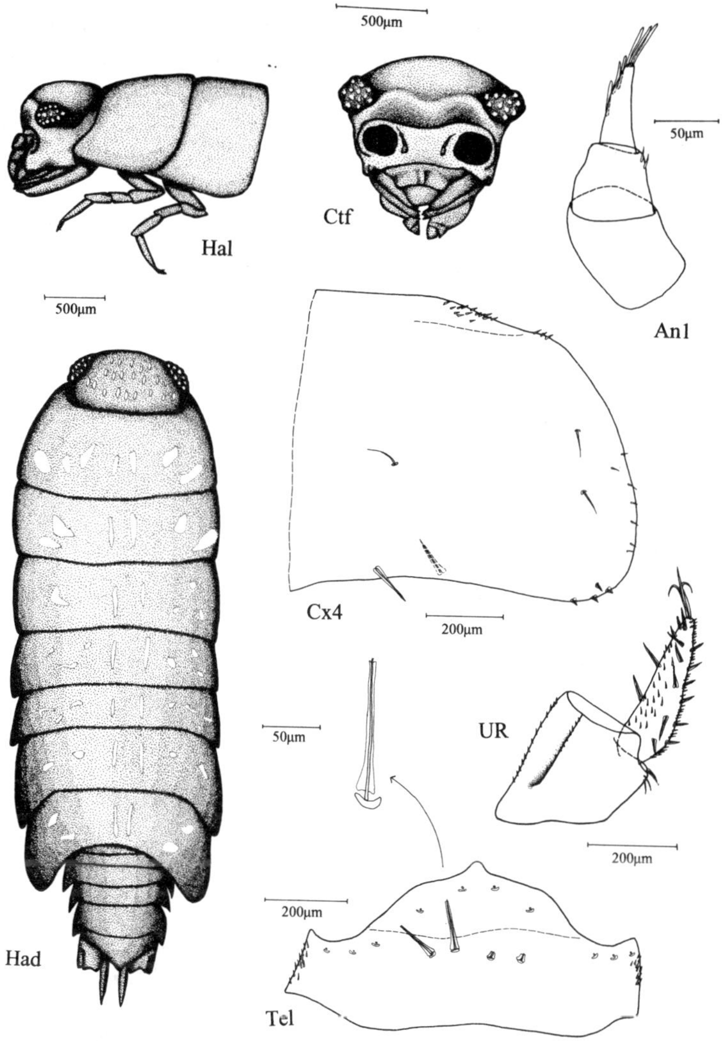

Figure 10. *Formicascia inquilina* (van Name 1936) female 6.5 mm. An1 antennula; Ctf cephalothorax, fontal view; Cx4 coxal plate 4; Had habitus, dorsal view; Hal habitus, lateral view; Tel pleotelson; UR uropod, exopod broken.

Pleotelson two times wider than long, lateral margins somewhat concave, apex small, tegument laterally scaly, dorsally smooth with scattered tricorn-like setae of considerable length.

Appendages:

Antennula: Composed of three articles, all subequal in length, distal article rather slender and coniform, with three aesthetascs apically and three medially (Fig. 10, An1).

Antenna: Slender with three-articulate flagellum and long apical organ, aesthetascs on flagellar articles 2 and 3, length ratio of peduncular articles 1 to 5 is 1:2:2:3:4, all articles bearing long tricorn-like setae (Fig. 13, An2).

Mandible: Molar penicil consisting of about 7 branches, pars intermedia sparsely covered with spiny setae, two penicils on left and one on right mandible, respectively. Additional plumose seta on both mandibles present (Fig. 11, Mdl/r).

Maxillula: Lateral endite with 4+6 teeth apically, 5 of inner set cleft, hyaline lobe present, laterally with fringe of hair-like setae. Medial endite with two slender penicils, lacking laterodistal point (Fig. 11, Mx1).

Maxilla: Both lobes of same size, lateral with some pectinate scales, medial apically with row of stiff setae, enclosed by hair-like setae (Fig. 11, Mx2).

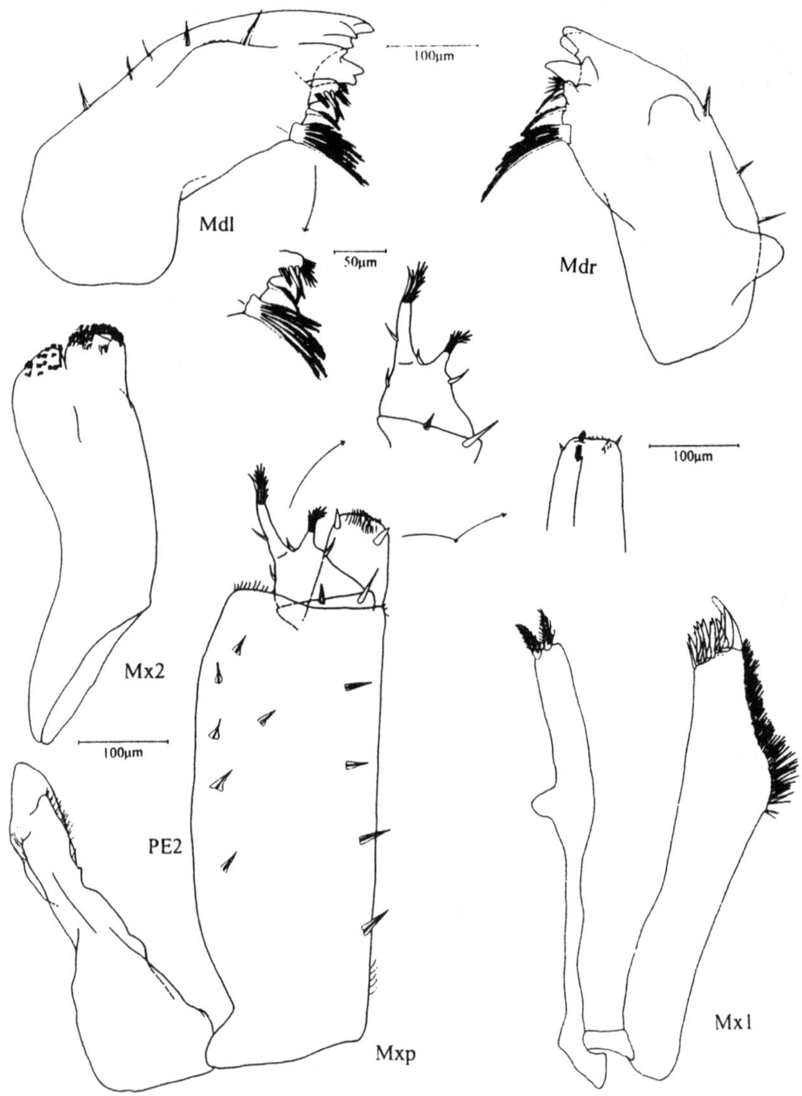

Figure 11. *Formicascia inquilina* (van Name 1936) female 6.5 mm. Mdl/r left and right mandible; Mx1 maxillula; Mx2 maxilla; Mxp maxilliped with detail of endite in rostral view, and palp.

Maxilliped: Endite with knob-like penicil and setal tuft rostrally, caudal surface bearing two teeth and scattered hair-like setae. Palp composed of three articles, distal two rudimentarily articulated, proximal with two setae, medial with setal tuft and two spiniform setae medially, single seta laterally, distal with setal tuft apically and seta laterally, epipodite scarcely covered with hair-like setae (Fig. 11, Mxp).

Pereiopods: Pereiopod 1 carpus and propus with antenna-grooming brush of moderate extension (Fig. 12, PE1), ornamental sensory spine of carpus somewhat resembling this of *Y. kartaboana* (Fig. 12, Sc1). Carpus of following pereiopods – as far as present on dissected specimen – with translucent dentation on distal margin of rostral surface. Dactyli with short inner claw and conspicuous interungual seta (Fig. 12, Dac), dactylar seta with apex only weakly plumose (Fig. 12, Sd1).

Sexual differentiation of pereiopods not observed.

Pleopods: Exopods with about 2 to 3 sensory spines on lateral margin, bearing a prominent sensillum, medial margin only sparsely covered with trichiform setae, endopods of particular shape with the mediodistal apex elongate, in pleopod 3 extraordinarily prolonged, surpassing the tip of exopod, endopod 5 irregularly trapezoidal with incision distally (Fig. 12, PL3-5).

Sexual differentiation of pleopods: Male pleopod 1 exopod with laterally bent apex, without sensory spines, endopod rather slender, acute with row of spines on caudal surface from apex proximally to interlocking area between endopod 1 and genital papilla. Pleopod 2 exopod pointed lateral margin pronounced sinuous but not very elongate, subapically with sensory spine, endopod rather long, tapered at two-thirds of length, distal third slender. Slight differences in length of pleopod 3 endopod, male endopod shorter than female, as long as exopod (Fig. 13, PL1-2).

Genital papilla: Ventral shield somewhat bulky, orifices of ductus ejaculatorii surpassing with tips separated and divergent. No setation discernible (Fig. 13, Gen).

Uropod: Protopod subrectangular with lateral groove, endopod inserting proximally of exopod, this lacking in dissected specimen (Fig. 10, UR).

Discussion: *Formicascia inquilina* (van Name 1936) is another small species of uncertain affinities. The genus is characterized by the shape of the cephalothorax with two slight depressions on the profrons as in *Chaetophiloscia* Verhoeff, 1908, not only the frontal line, but also the lateral lobes are missing. The coxal plates lack a sulcus marginalis and the rounded distal edges are autapomorphies of this genus, whereas the shape and setation of the mouthparts are plesiomorphic. The males have a short genital papilla with the ductus ejaculatorii proximal of the ventral shield much inflated. This character resembles the genus *Halophiloscia* Verhoeff, 1908. Since there are no synapomorphies between those genera, this character is most probably due to convergence. This surely is the case with the pleopod 1 exopod of the male. It is similar to *Yaerikima* gen. n., *Burmoniscus* Collinge, 1914, and *Atlantoscia* Ferrara & Taiti, 1981. From the latter two genera *Formicascia* gen. n. differs in at least the shape of the cephalothorax, the structure of the apical organ of the antenna, the antenna-grooming brush of the pereiopod 1 and several other characters.

For the characters shared with *Roraimoscia* gen. n. see below.

2.4 *Roraimoscia* gen. n.

Diagnosis: Cephalothorax without linea frontalis, linea supra-antennalis and lamina frontalis present, profrons squarish, compound eyes with 25 ommatidia in four rows.

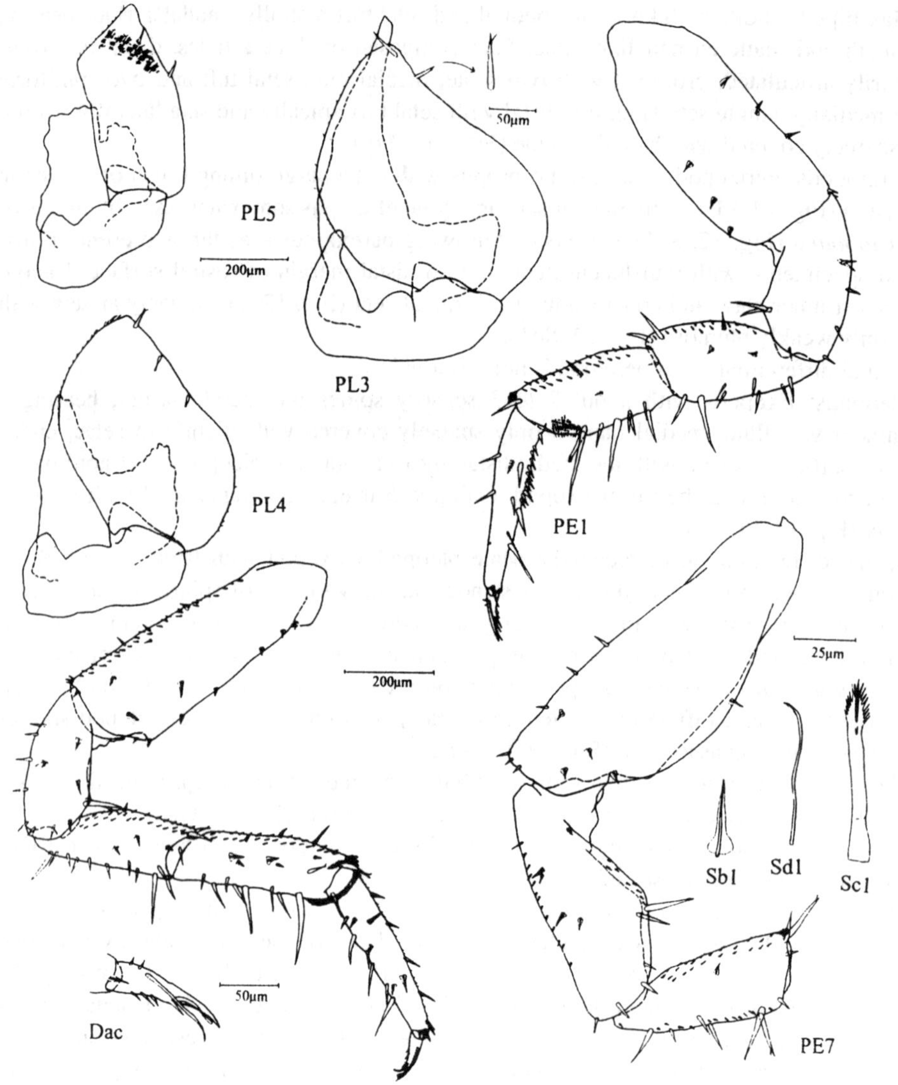

Figure 12. *Formicascia inquilina* (van Name 1936) female 6.5 mm. Dac dactylus; PE1 pereropod 1, rostral view; PE2/7 pereiopods 2 and 7, caudal view; PL3-5 pleopods 3 to 5; Sb1 tricorn-like seta of basis 1; Sc1 ornamental sensory spine of carpus 1; Sd1 dactylar seta 1.

Antennula with pointed distal article, antenna long, apical organ long and slender. Mandible with multibranched molar penicil, maxillula with 4+6 teeth, five of inner set cleft, hyaline lobe present. Medial lobe of maxilla broader than lateral, bearing about 10 cusps, maxilliped with knob-like penicil on endite, medial article of palp with two setal tufts.

Pereiopods slender, basis with waxy plates, ornamental sensory spine double-fringed, dactylar seta apically plumose, coxal plates with several tricorn-like setae and sulcus marginalis. Pleopods with two rows of sensory spines, no respiratory area. Genital papilla

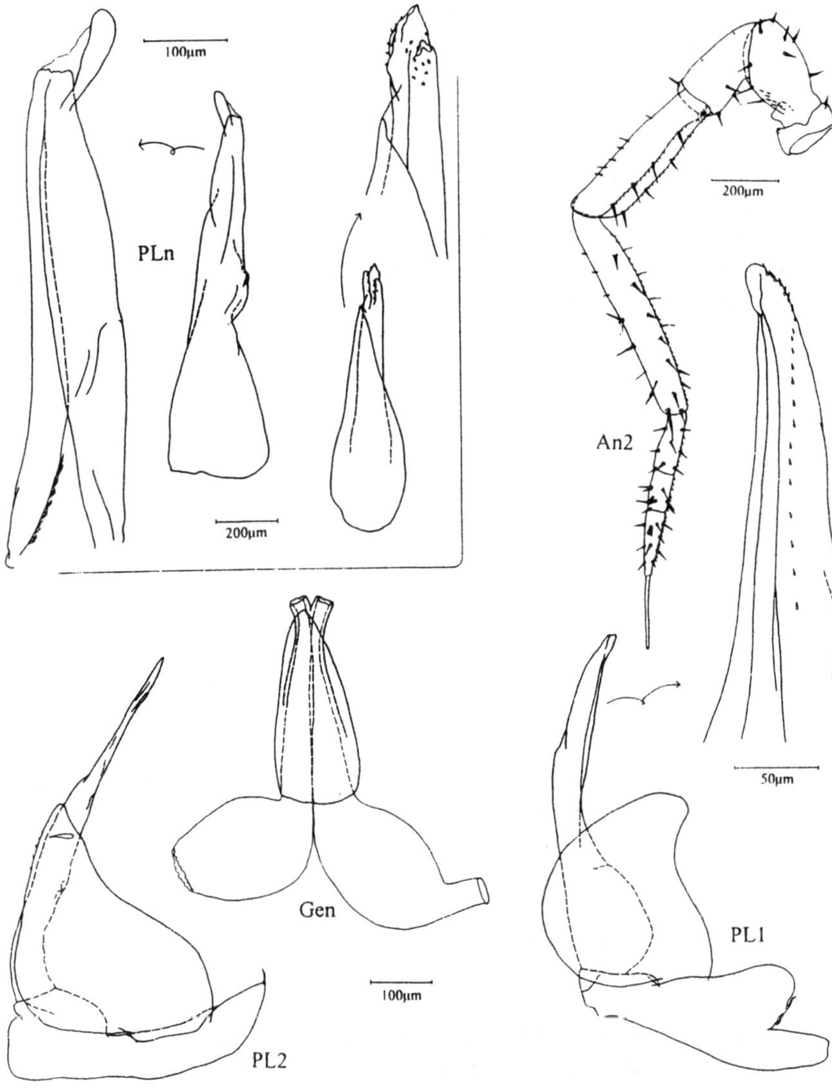

Figure 13. *Formicascia inquilina* (van Name 1936) male 6 mm. An2 antenna; Gen genital papilla; PL1/2 pleopods 1 and 2, rostral view, with detail of endopod 1 in caudal view; *Roraimoscia roraimae* (van Name 1936) male 10 mm: PLn genital papilla (right) and pleopod 1 endopod (middle) with detail in caudal view (left).

with mouth of ductus ejaculatorii semicircular, bearing pectinate scales. Uropod with protopod grooved, endopod inserting proximally of exopod.

Type species: *Philoscia roraimae* van Name, 1936, by monotypy.

Etymology: The genus name is derived form Roraima, a province in southern Venezuela.

2.4.1 *Roraimoscia roraimae* (van Name, 1936) n. comb.

Synonymy: *Philoscia roraimae* van Name, 1936.

Material: AMNH 6535 8 females; Venezuela, Rondon Camp, Mount Roraima; leg, G.H.H. Tate, 12.1927; AMNH 6506 male; same data as above; length of dissected female 11 m.

Color: Now light brown, originally described as follows: 'Coloration is handsome ..., dark brown ground color and light yellowish spots giving a tortoise shell effect. Besides the usual irregular bars on the lateral region of the back, there is a row of small median spots, a row on each side of large squarish spots at the base of the epimera, and usually light spots on the epimera. The abdominal segments have three light spots, median and lateral. There is considerable brown pigment below on the legs, pleopods, etc.' (van Name 1936: 147).

Cephalothorax: Linea frontalis absent, linea supra-antennalis present, curved between antennal sockets. Lamina frontalis prominent, squarish, frons smooth as is vertex. Compound eyes composed of about 25 ommatidia arranged in 4 rows (Fig. 14, Ctf).

Pereion: Cephalothorax set back into pereionite 1, coxal plates of pereionite 1 to 4 straight, more pointed in following pereionites, pereionite 7 acutely pointed, tegument smooth and shiny, some scattered minute tricorn-like setae. Coxal plates with sulcus marginalis prominent, lacking gland pores when examined under light microscope, no trace of noduli laterales (Fig. 14, Cx5).

Pleon: Neopleurae on pleonite 3 to 5 present, pleon retracted from pereion, tegument smooth, pleotelson with some scales on lateral edge, laterodistal margins slightly concave, some scattered tricorn-like setae.

Appendages:

Antennula: Three-articulate, proximal article only slightly larger than medial, distal article as long as others, coniform, bearing a set of aestethascs apically and medially (Fig. 14, An1).

Antenna: Flagellum composed of 3 articles, subequal in length, medial shortest, terminal bristle broken in dissected specimen, as are setae of peduncle. Peduncular article 5 one and a half as long as fourth, two times longer than article 3, article 2 shorter than third (Fig. 14, An2).

Mandible: Molar penicil dichotomized, composed of about 10 branches, pars intermedia with coniform setae, 2 penicils on left, one on right mandible, additional plumose seta of left mandible in dissected specimen most probably broken (Fig. 15, Mdl/r).

Maxillula: Medial endite with 2 stout penicils apically, laterodistal tip lacking, lateral endite with 4+6 teeth apically, 5 of inner set cleft, hyaline lobe rostrally, lateral margin with fringe of hairlike setae, proximal ones the longest (Fig. 15, Mx1).

Maxilla: Both lobes subequal, apically covered with hair-like setae, lateral lobe proximally bearing some pectinate scales, medial lobe laterodistally with row of stiff short setae (Fig. 15, Mx2).

Maxilliped: Palp three-articulate, in dissected specimen apex of distal article broken, medial article with setal tuft and group of 4 setae proximal of it, proximal article bearing 2 setae. Endite with 2 teeth and some rows of pectinate scales on caudal surface, rostral surface bearing knob-like penicil, setal tuft and transversely directed setal field (Fig. 15, Mxp).

Pereiopods: Basis of pereiopods (Fig. 16, PE1-7) present in dissected specimen laterally furrowed, surface covered with small ellipsoid to hexagonal wax plates, tricorn-like

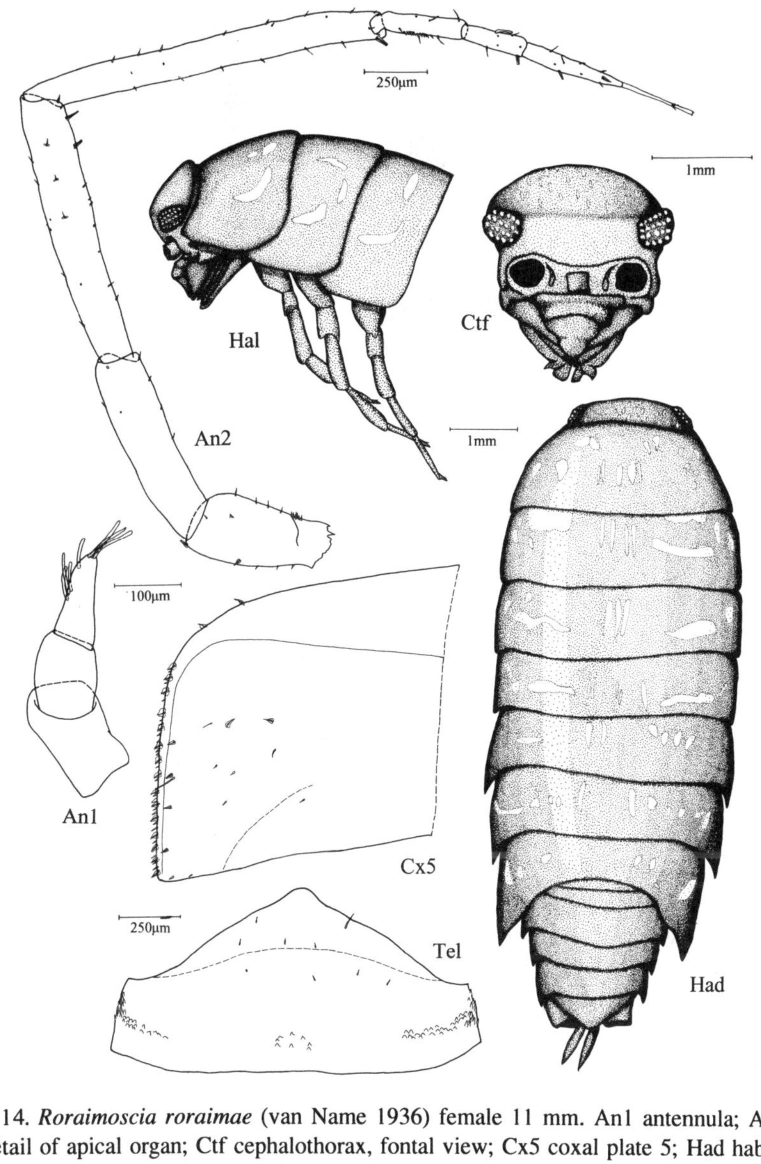

Figure 14. *Roraimoscia roraimae* (van Name 1936) female 11 mm. An1 antennula; An2 antenna with detail of apical organ; Ctf cephalothorax, fontal view; Cx5 coxal plate 5; Had habitus, dorsal view; Hal habitus, lateral view; Tel pleotelson.

setae present, ischium with long sensory spine medially, merus and carpus bearing translucent dentation medially, save in carpus 1, antenna-grooming brush on carpus and propus of pereiopod 1, both articles with ornamental sensory spines apically fringed or double-fringed (Fig. 16, Sc1). Proximally of antenna-grooming brush, tegument of carpus 1 scaly, scales pointed. Dactylus with curved interungual seta (Fig. 16, Dac), dactylar seta apically plumose (Fig. 16, Sd1).

Sexual differentiation of pereiopods not observed.

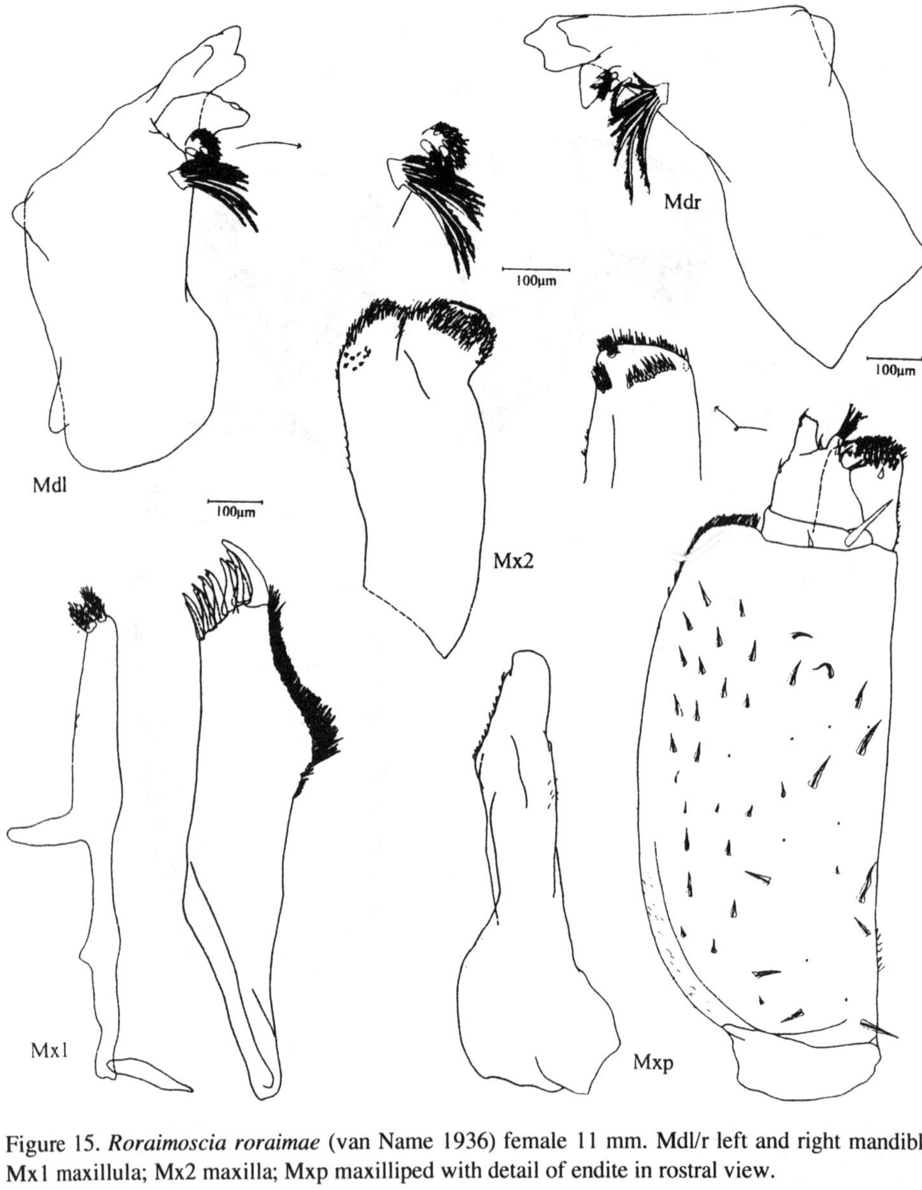

Figure 15. *Roraimoscia roraimae* (van Name 1936) female 11 mm. Mdl/r left and right mandible; Mx1 maxillula; Mx2 maxilla; Mxp maxilliped with detail of endite in rostral view.

Pleopods: Setation of exopods characteristic exopod 1 with single sensory spine apically, exopod 2 with prominent sensory spine apically, lateral margin with some small sensory spines, submarginally second row of sensory spine, one of spines as shaped as apical spine, exopod 3 to 4 equally bearing 2 rows of sensory spines, none in shape of pleopod 2, exopod 5 with 2 prominent and 5 smaller sensory spines on rostral surface. Endopods of pleopod 3 and 5 conspicuously shaped: endopod 3 mediodistally pointed, endopod 5 subrectangular (Fig. 16, PL1-5).

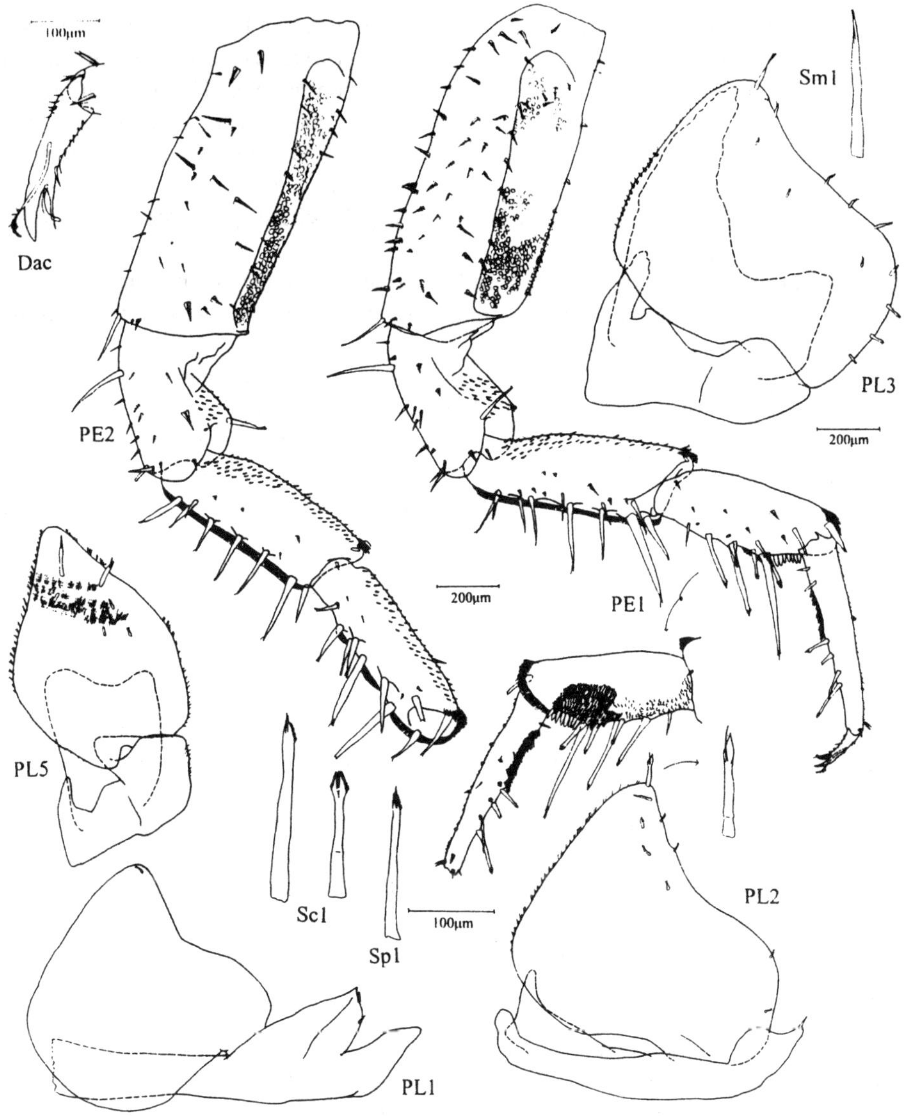

Figure 16. *Roraimoscia roraimae* (van Name 1936) female 11 mm. Dac dactylus; PE1/2 pereiopod 1 and 2, caudal view, with detail of pereiopod 1 in rostral view; PL1-5 pleopods 1 to 5, rostral view; Sc1 sensory spines of carpus 1; Sm1 sensory spine of merus 1; Sp1 sensory spine of propus 1.

Sexual differentiation of pleopods: Only male pleopod 1 endopod known, elongate with basal part slightly triangular, apex truncate, medially with dactyliform process, connected with lateral edge of apex by hyaline membrane, longitudinal row of spines lacking (Fig. 13, PLn).

Genital papilla: Mouths of ductus ejaculatorii surpassing ventral shield, covered with minute pectinate scales (Fig. 13, PLn).

Uropod: Compare generic diagnosis.

Discussion: This large species is, like the preceeding ones, only known from the type series collected in the 20s of this century. *Roraimoscia roraimae* is quite different from all philosciids hitherto known, so it is separated in a genus on its own. The autapomorphies of the genus and of the type species, at the moment the only species known, are: waxy plates covering the groove of the basis of the pereiopods, a very prominent sensory spine on the medial margin of the ischium and two rows of sensory spines on the lateral margin of the pleopod exopods. Probably, the male genital papilla with its pectinate scales are another autapomorphy of *R. roraimae*.

The coxal plates are covered with tricorn-like setae of different shape, one especially slender. This might be a nodulus lateralis, but to get more evidence, new material for ultrastructural investigations has to be found.

Roraimoscia gen. n. has some characters in common with other genera. Possibly the elongate medial part of the female pleopod 3 exopod is a synapomorphy shared with *Formicascia* gen. n. from Guyana and this may be true for the shape of the pleon with its 'undulated' margin. On the other hand, *R. roraimae* bears several character which are plesiomorphic, especially the setation of the mouth parts is shared with many basal philosciid genera.

2.5 *Parischioscia* Lemos de Castro, 1967

Diagnosis: Cephalothorax with linea supra-antennalis, no linea frontalis but a slight bulge between the compound eyes, no lamina frontalis. Mandibles with multibranched molar penicil, lateral endite of maxillula with 4+6 teeth, 5 of inner set cleft, medial endite without apical tip, maxilla with outer lobe narrow, acute. Maxilliped with knob-like seta on endite, distally dense setation.

Pereiopods slender, carpus of pereiopod 2 to 6 with transverse row of scales on midlength, dactylar organ plumose, inner claw long, surpassing interungual seta. Coxal plates with sulcus marginalis, lacking gland pores and distinct noduli laterales.

Pleopods without respiratory area on exopod. Lateral border of exopods with 2 to 3 sensory spines. Uropods missing in examined specimen.

Type species: *Parischioscia omissa* (van Name, 1936), by monotypy.

2.5.1 *Parischioscia omissa* (van Name 1936)

Synonymy: *Philoscia omissa* van Name, 1936, *Philoscia nitida* van Name, 1925, non Miers 1877, *Philoscia olfersi* Pearse, 1917, non Brandt 1833.

Material: PARATYPE male 6.5 mm, Guyana, Kamakusa, at the mouth of the Meama river, leg. F.E. Lutz, AMNH 6560.

Color: Examined specimen faded, ground colour described as purplish with unpigmented spots in the median line and on the coxal plates, lateral margin of coxal plates light. For detailed description see van Name (1936).

Cephalothorax: Without linea frontalis and lamina frontalis, linea supra-antennalis distinctly bent between antennal sockets, vertex smooth, compound eyes composed of up to 30 ommatidia (van Name 1936), in the dissected specimen about 15 arranged in 4 rows (Fig. 17, Ctf).

Pereion: Tegument rather smooth and shiny, coxal plates 1 to 3 caudally rounded, distal end of coxal plates 4 to 7 caudally pointed, no noduli laterales nor gland pores visible, sulcus marginalis present (Fig. 21, Cx3).

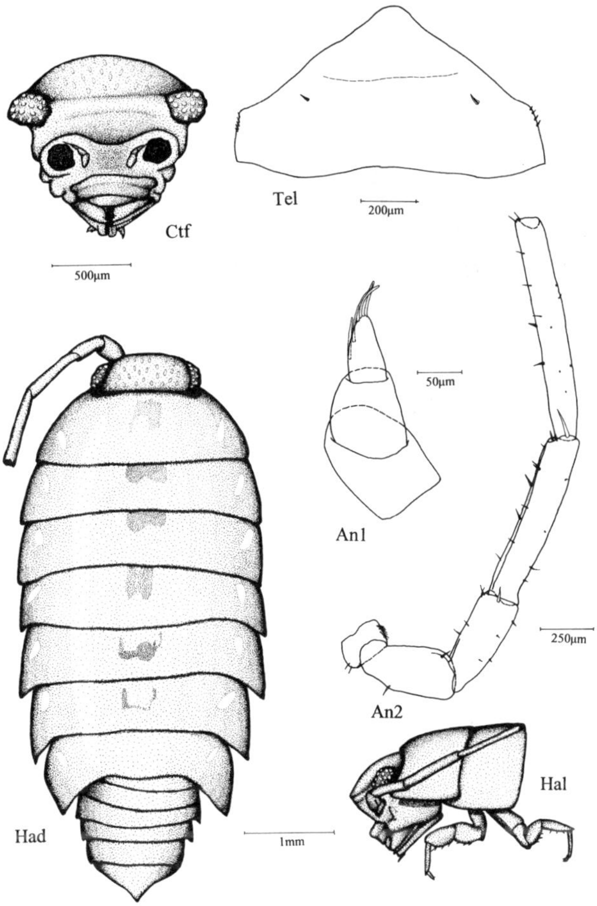

Figure 17. *Parischioscia omissa* (van Name 1936) male 6.5 mm. An1 antennula; An2 peduncle of antenna; Ctf cephalothorax, frontal view; Had habitus, dorsal view; Hal habitus, lateral view; Tel pleotelson.

Pleon: Retracted from pereion, neopleurae of pleonites 3 to 5 narrow but visible in dorsal view, pleotelson triangular with rather straight margins.

Appendages:

Antennula: Composed of 3 articles, distal one bearing a set of aesthetascs, proximal article with somewhat prolonged distal margin (Fig. 17, An1).

Antenna: Antenna of examined specimen lacking flagellum, van Name (1936) depicted a three-articulate flagellum. Peduncle with articles 5 and 4 two times longer than article 3, article 2 subequal in length to article 3 (pl 17, An2).

Mandible: Molar penicil composed of about 7 branches, pares intermediae with acute spiniform setae and 2 penicils on left and 1 penicil on right mandible, additional plumose seta between pars intermedia and molar penicil (Fig. 18, Mdl/r).

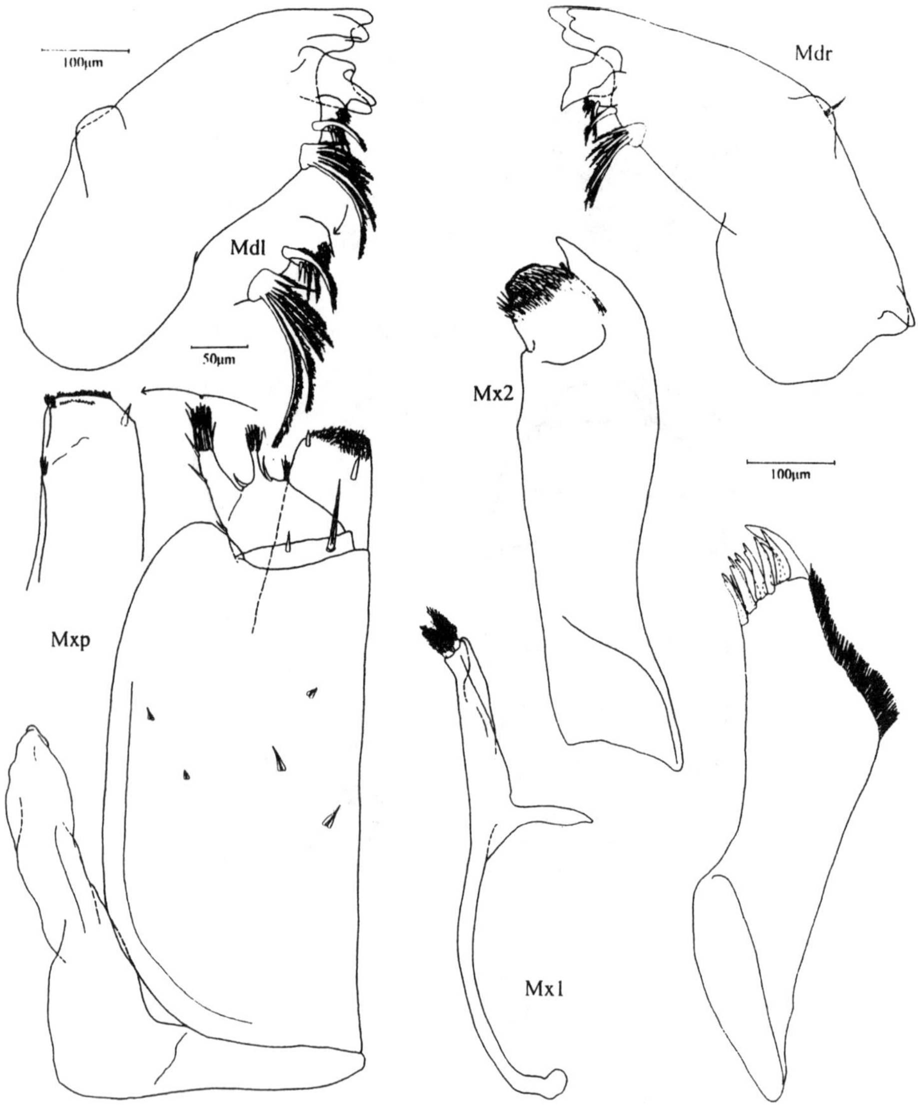

Figure 18. *Parischioscia omissa* (van Name 1936) male 6.5 mm. Mdl/r left and right mandible; Mx1 maxillula; Mx2 maxilla; Mxp maxilliped with detail of endite in rostral view.

Maxillula: Medial endite with two penicils, latero-distal tip lacking, lateral endite with 4+6 apical teeth, 5 of inner set cleft, lateral margin with fringe of trichiform setae (Fig. 18, Mx1).

Maxilla: Medial lobe apically cuspidate, medio-proximally with stiff trichiform setae, densely covered with fine trichiform setae, lateral endite narrow, acute, lacking any setation (Fig. 18, Mx2).

Maxilliped: Endite with knob-like penicil and setal tuft rostrally, a row of scales parallel to distal margin, caudal surface densely setose apically, 2 teeth present. Palp with prominent setal tuft on distal article, medial article with less prominent setal tuft on dactyliform protrusion, 4-5 setae proximally of tuft, proximal article with rather long seta medially, with lateral seta one third of its length (Fig. 18, Mxp).

Pereiopods: Rather slender (Fig. 19, PE1-4; Fig. 20, PE5-7), carpus of pereiopod 1 with antenna-grooming brush rostrally, a row of sensory spines on medial margin of propus and merus 1, pereiopods 2 to 4 with translucent dentate lamina on medial margin of merus, smaller areas of same structure on carpus 2 to 6 and merus 5 to 7 (carpus 7 unknown). Transverse row of scales on midlength of carpus 2 to 6 caudally, rostrally on carpus 5 with sensory spine (Fig. 20, Sc5). Dactylus with long inner claw and interungual seta (Fig. 20, Dac), dactylar seta with distal part plumose (Fig. 19, Sd1).

Sexual differentiation of pereiopods: As only the dissection of a male specimen was allowed, no statement can be made.

Pleopods: Exopods with some sensory spines on distal half of lateral margin, endopods lobate, endopod 3 with medial lobe elongate (Fig. 21, PL1-5).

Sexual differentiation: Dissected specimen lacking exopods 1, endopod 1 acute, row of spiniform setae mediocaudally on distal third. Pleopod 2 exopod with lateral margin very sinuous, bearing 2 sensory spines, endopod straight, slender, apex in examined specimen broken.

Genital papilla: Rectangular terminal spatula surpassing ventral shield, orifices laterally (Fig. 21, Gen).

Uropod: Uropods lacking in dissected specimen.

Discussion: This species was originally described by van Name (1936) on specimens collected at several sites in Guayana, e.g. Kartabo, Kamakusa, Mouth of Meama River. Further records are from Brazil (Lemos de Castro 1967). The largest specimens were about 11 mm, so the male paratype which could be examined is rather small. Nonetheless, it is a mature specimen since there are spermatids within the ductus ejaculatorii of the genital papilla.

In a contribution to the Oniscidea of the Amazon region, Lemos de Castro (1967) separated this species in the monotypic genus *Parischioscia* mainly because of the great laterally protruding compound eyes and postulated a close relationship to *Ischioscia*. Since the pereiopods 1 to 3 have no plate-like carpus with prominent setal fields, *Parischioscia* should be considered more primitive in this character. The close relationship between *Ischioscia* and *Parischioscia* must be doubted, because no synapomorphies can be found: *Parischioscia* has no ventral pit on the pleotelson, the shape of the antennula is different.

Three characters may indicate a close relationship, i.e. the morphology of the cephalothorax with the distinct shape of the eyes, the distally prolonged proximal article of the antennula and the ornamental sensory spine of carpus 1 with its hand-like apex. The latter character can be found in several other genera and, therefore, must be considered a synapomorphy of a wider group. On the other hand, the morphology of the pleotelson represents a plesiomorphic character state with rather triangular shape and without the semicircular ventral depression, an autapomorphic character of the *Ischioscia*-group. In addition, a second character which is apomorphic for the *Ischioscia*-group is missing in *Parischioscia*: the shield-like basal article of the antennula. Thus, other species might be better candidates to be the sister group of *Ischioscia*. For example, in *Tropiscia* Vandel, 1968 shares some apomorphic characters with it. The morphology of its cephalothorax fits the pattern known from *Ischioscia* much better.

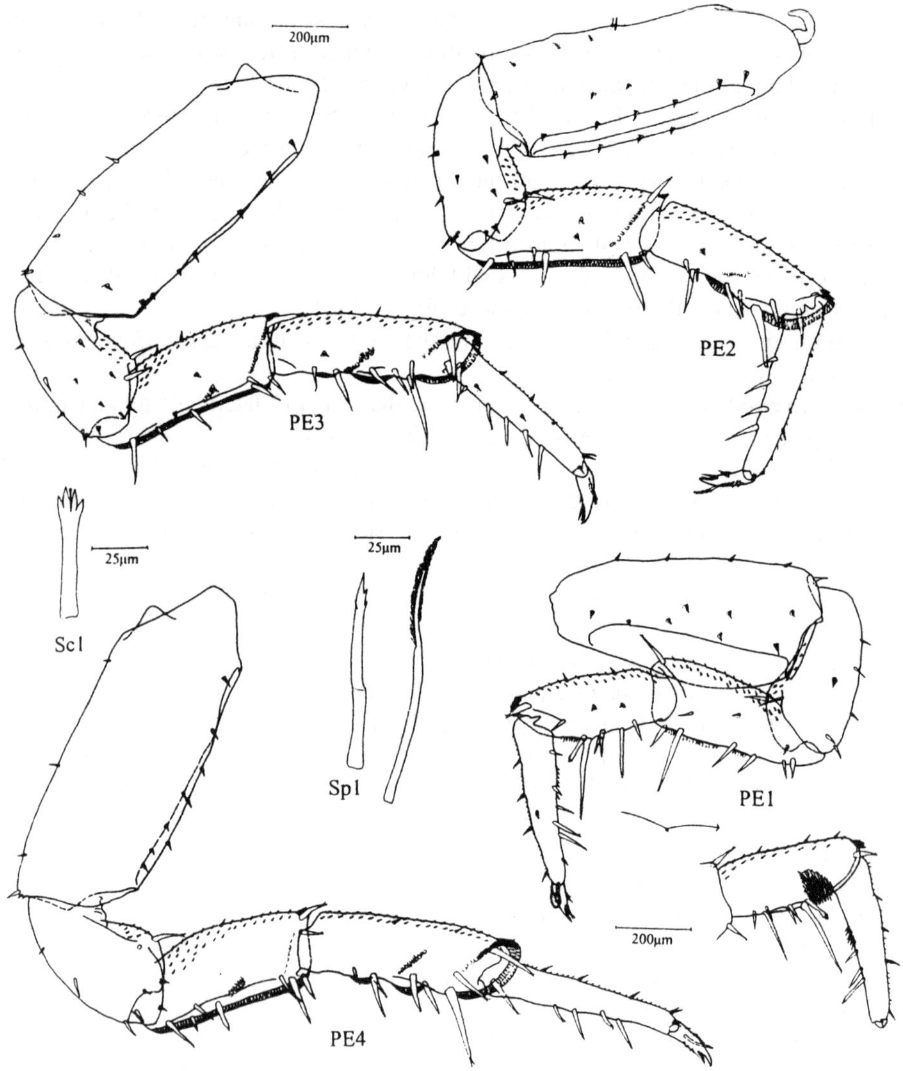

Figure 19. *Parischioscia omissa* (van Name 1936) male 6.5 mm. PE1-4 pereiopods 1 to 4, with detail of pereiopod 1 in rostral view; Sc1 ornamental sensory spine of carpus 1; Sd1 dactylar seta 1; Sp1 sensory spine of propus 1.

2.6 *Doubtful species from South America*

The following species remain enigmatic. The type material is in a too poor condition to draw conclusions on their phylogenetic relationships:

2.6.1 *'Philoscia' diminuta* Budde-Lund, 1885

Material: BMNH 1921.10.18.1785 and 1786 2 females (SYNTYPES); Venezuela, Caracas, La Moka.

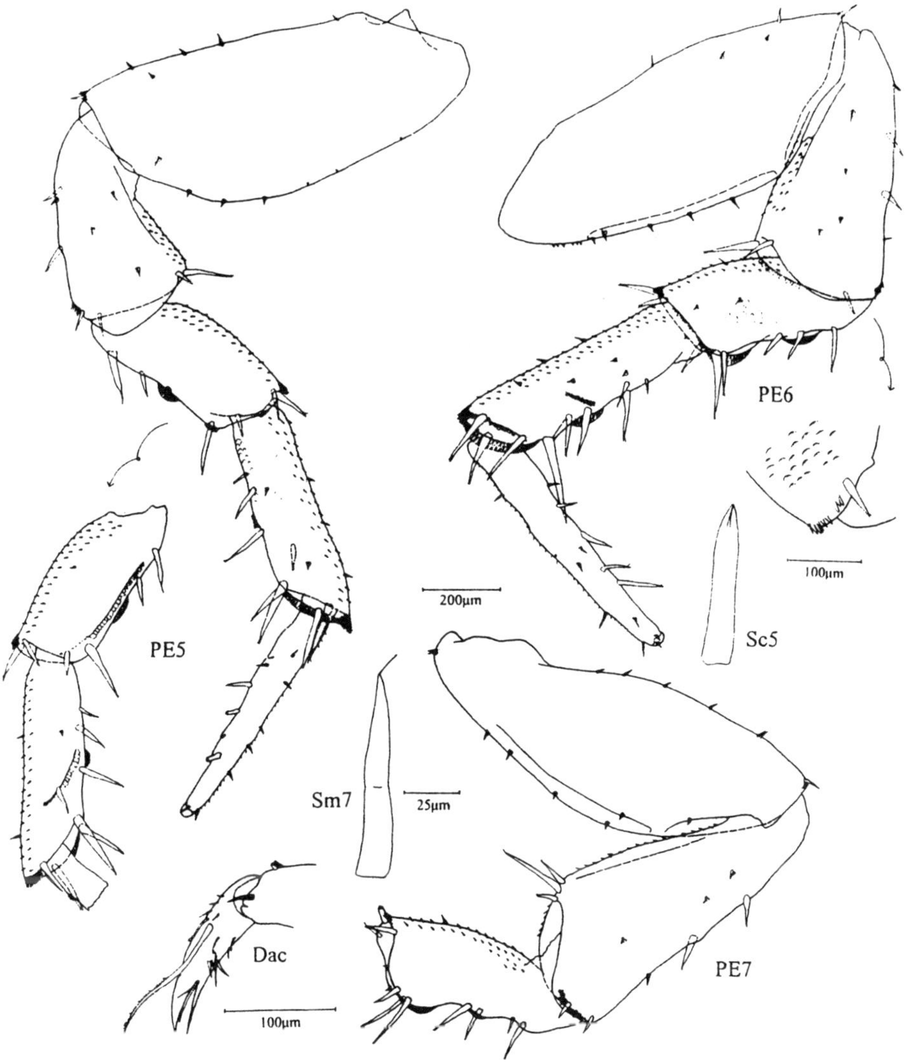

Figure 20. *Parischioscia omissa* (van Name 1936) male 6.5 mm. Dac dactylus; PE5-7 pereiopods 5-7 with details of rostral surface; Sc5 sensory spine of carpus 5; Sm7 sensory spine of merus 7.

Remark: The syntypes lack all appendages and the cephalothorax. From the material and the description of Budde-Lund (1885), this species is not determinable, so it should be treated as a *nomen dubium*.

2.6.2 *'Philoscia' moneaguensis* van Name, 1936

Material: AMNH 6511 female (TYPE) 3.5 mm; Jamaica, Moneague, somewhat off the coast.

Remark: The only specimen is a female, so the recognition of this species in not ensured, at least since the material is not accessible to a thorough re-examination Therefore, it is a *nomen dubium*.

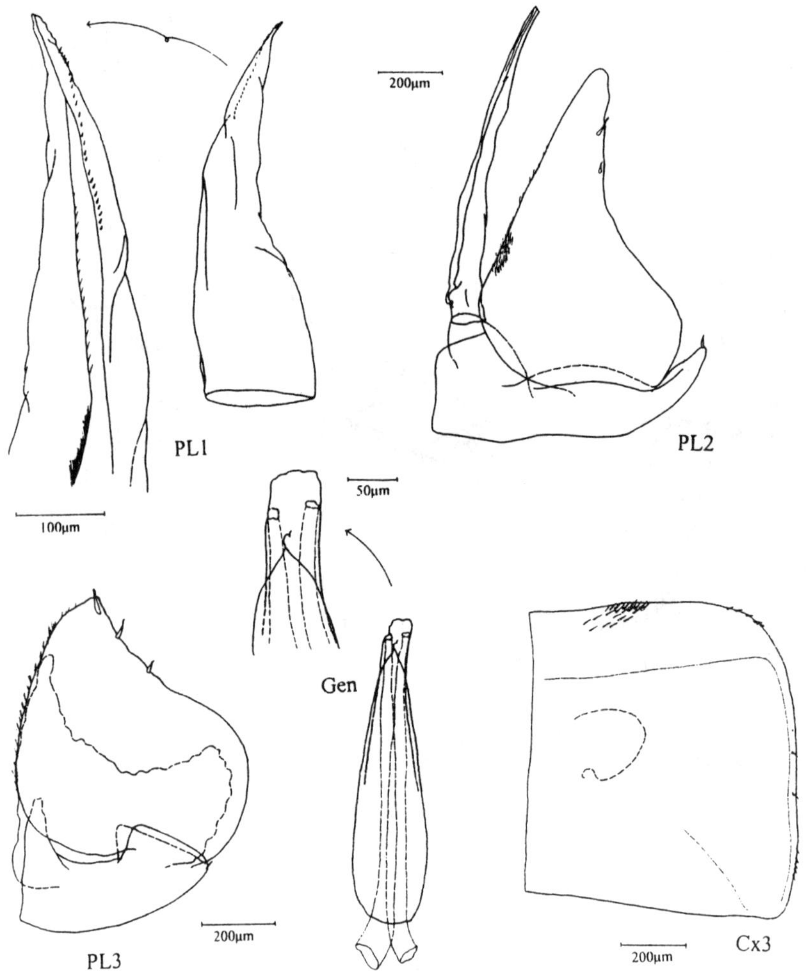

Figure 21. *Parischioscia omissa* (van Name 1936) male 6.5 mm. Cx3 coxal plate 3; Gen genital papilla; PL1-3 pleopods 1 to 3, rostral view, with detail of endopod 1 in caudal view.

2.6.3 *'Philoscia' seriepunctata* Budde-Lund, 1893

Material: BMNH 1921.10.18.1964 female (TYPE); Venezuela, Caracas.

Remark: The type lacks pereiopods and cephalothorax; the description of Budde-Lund (1893) is too poor to recognize this species. Thus, it is another *nomen dubium*.

2.6.4 *'Philoscia' walkeri* Pearse, 1915

Material: USNM 2 females (SYNTYPES).

Remark: The types lack pereiopods and cephalothorax. Lemos de Castro (1967) placed this species within the genus *Chaetophiloscia* but his doing so is by no means reasonable. He did not mention that he re-examined the type material and as recognition of the species based on the poor original description given by Pearse (1915) would be unlikely, the best solution is to treat this species as *nomen dubium*.

ACKNOWLEDGEMENTS

For the loan of the material, the author thanks Mrs. Yae Ri Kim (ANMH), Dr Miranda Lowe (BMNH), and Dr Marilyn Schotte (USNM) and Prof. Dr J. Wolfgang Wägele for fruitful discussions on the material.

REFERENCES

Budde-Lund, G. 1885. Isopoda Terrestria per familias et genera et species descripta. Hauniae. 320pp.

Budde-Lund, G. 1893. Landisopoder fra Venezuela indsamlede af Dr F. Meinert. *Entomol. Meddel.* 1893: 111-129.

Erhard, F. 1995. Das pleonale Skelet-Muskel-System und die phylogenetisch-systematische Stellung der Familie Mesoniscidae. *Stuttg. Beitr. Naturk. Ser. A* 538: 1-40.

Ferrara, F. & S. Taiti 1985. New genera, species and records of Philosciidae from southern Africa and Madagaskar. *Ann. Natal Mus.* 26: 559-600.

Leistikow, A. 1997. Terrestrial Isopoda from Costa Rica and redescription of *Ischioscia variegata* (Dollfus, 1893) from Venezuela. *Can. J. Zool.* 75: 1415-1464.

Leistikow, A. 1998. Considerations about the genus *Pentoniscus* Richardson, 1993 (Crustacea: Isopoda: Oniscidea) with description of a new species. *J. Nat. Hist.* 32: 1339-1355.

Lemos de Castro, A. 1967. Isopodos terrestres da Amazonia Brasileira. *Atas simp. Biota Amazon.* 3: 311-336.

Pearse, A.S. 1915. An account of the Crustacea collected by the Walker Expedition to Santa Marta. *Proc. US Natl. Mus.* 49: 531-556.

Schmalfuss, H. 1980. A revision of the neotropical genus *Ischioscia* Verhoeff, 1928 with description of four new species. *Stud. Neotrop. Fauna. Environm.* 15: 125-139.

Schmalfuss, H. 1989. Phylogenetics in Oniscidea. *Mon. Ital. Zool. (N.S.)* 4: 3-27.

Schmalfuss, H. 1990. Die Landisopoden Griechenlands. 12. Beitrag: Die Gattungen *Lepidoniscus* und *Philoscia*. *Stuttg. Beitr. Naturk. Ser. A* 448: 1-11.

Taiti, S. & F. Ferrara 1980. The family Philosciidae (Crustacea: Onisicoidea) in Africa, south of the Sahara. *Mon. ital. Zool. (N.S.)* SupFig. 13. 4: 53-98.

van Name, W.G. 1936. American terrestrial and fresh water Isopoda. *Bull. Amer. Mus. Nat. Hist.* 71: 1-520.

Verhoeff, K.W. 1908. Zur Kenntnis der Oniscinen-Gattungen und über neue Oniscinen (15. Isopoden-Aufsatz). *Archiv Biont.* 2: 339-360.

A new species of *Politolana* (Flabellifera: Cirolanidae) from the south Brazilian shelf

SARAH F. RISEMAN
University of Charleston, Grice Marine Biological Laboratory, Charleston, USA

ANA MARIA S. PIRES-VANIN
Universidade de São Paulo, Instituto Oceanográfico, São Paulo, Brazil

RICHARD C. BRUSCA
Columbia University Biosphere 2 Center, Oracle, and Department Ecology & Evolutionary Biology, University of Arizona, Tucson, USA

ABSTRACT

A new species of cirolanid isopod has been collected on the inner Brazilian continental shelf, from sandy bottoms at depths of 15 m to 48 m. *P. tricarinata* n.sp. occurs sympatrically with *P. eximia*, the only other *Politolana* species recorded from the south Atlantic. This new species is herein described and figures of all appendages are provided. *P. tricarinata* can be distinguished from other species of *Politolana* by the presence of three transverse cuticular ridges, or carinae, on the dorsal surface of the cephalon. Along with this autapomorphy, a combination of characters unique to *P. tricarinata* is provided in the species diagnosis. A key separating the two known Brazilian species of *Politolana* is included.

1 INTRODUCTION

A broad oceanographic investigation was carried out during 1985-1987 on the south Brazilian continental shelf and slope offshore of Ubatuba, Brazil (Pires 1992). Benthic samples were obtained with grabs and dredges at depths from 10 m to 600 m. The examination of these samples has revealed a new species of cirolanid isopod belonging to the genus *Politolana*. The majority of *Politolana* species are known from the north Atlantic and Gulf of Mexico, and until now only a single species, *P. eximia* (Hansen 1890), had been recorded from South American waters. Both *P. eximia* and the new species were collected in the Ubatuba area, often times in the same sample, indicating that they occur sympatrically in this region. Sympatry is not unusual within *Politolana*; in the northwestern Atlantic several species in this genus occur sympatrically. *P. tricarinata* n.sp. is thus far known only from the Ubatuba region on the inner continental shelf at depths of less than 50 m, while *P. eximia* has been collected at much greater depths of up to 380 m, and is known from the waters of southern Brazil to northern Argentina (about 22°S to 35°S). In light of the sympatric distribution of these congeneric species, and their potential ecological importance as benthic scavengers and as prey of epibenthic fish, a clear understanding of their taxonomy and distribution is necessary.

A detailed taxonomic, phylogenetic, and biogeographic monograph of *Politolana* is in preparation (Riseman & Brusca). In the present paper, we describe the new Brazilian species, *Politolana tricarinata*, and present a key to the two species of *Politolana* found in this region. Because a genus diagnosis is not given here, but will follow in the *Politolana* monograph, the species description is somewhat lengthy, including characters of both specific and generic importance. However, the species diagnosis includes characters solely of specific importance which, in combination, distinguish *P. tricarinata* from all other species of *Politolana*. A redescription of *P. eximia* will follow in the monograph as well.

The terms used to describe appendage orientation in the description follow Brusca et al. (1995). To avoid confusion, setal terminology has been kept as simple as possible, naming the setae by a combination of terms describing length, robustness, complexity, and acuteness, for example: 'long slender plumose seta' or 'short acute robust seta.' In addition, unusual setae are figured in detail separately.

2 KEY TO BRAZILIAN SPECIES OF *POLITOLANA*

Uropod exopod short and broad, about 2.3 times longer than wide, not reaching level of endopodal notch; antenna reaching middle of first pereionite; cephalon with transverse cuticular ridge between frontal ridge and interocular furrow.................. *P. tricarinata* n.sp.

Uropod exopod long and narrow, about 5.3 times longer than wide, reaching beyond distal margin of endopod and pleotelson; antenna reaching second pereionite; cephalon without transverse cuticular ridge between frontal ridge and interocular furrow ....*P. eximia*

3 TAXONOMIC TREATMENT

Order Isopoda Latreille, 1817
Suborder Flabellifera Sars, 1882
Family Cirolanidae Dana, 1853
Politolana Bruce, 1981

Synonymy: Bruce, 1981:958-959, figs. 1j, 1k, 2g, 2h, 3g, 3h, 4d, 5g, 5h; 1986:10; 1991:270; 1996:161-165, figs. 10-12; Kensley & Schotte, 1989:140-143, figs. 63-64; Wetzer et al., 1987:1-11.

Type species: *Aega polita* Stimpson, 1853 (by designation; Bruce 1981).
Species Included:

P. concharum (Stimpson, 1853)
P. crosnieri Bruce, 1996
P. dasyprion Bruce, 1991
P. eximia (Hansen, 1890)
P. impressa (Harger, 1883)
P. micropthalma (Hoek, 1882)
P. obtusispina (Kensley, 1975)
P. polita (Stimpson, 1853)
P. tricarinata n.sp.
P. wickstenae Wetzer et al., 1987

Politolana tricarinata n.sp.: Figures 1-7.

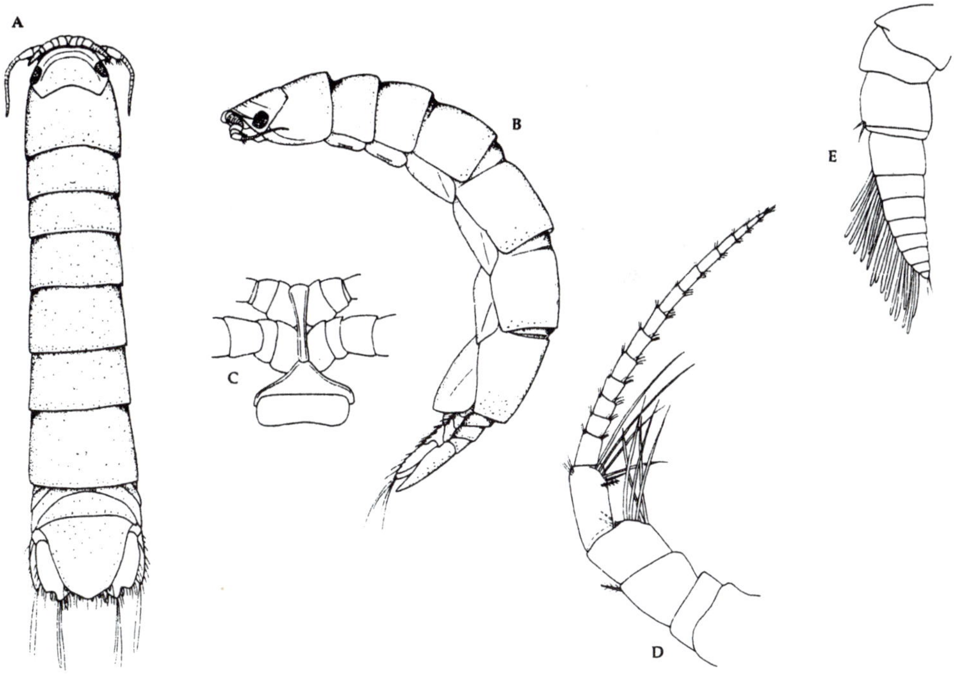

Figure 1. *P. tricarinata* female: (A) dorsal aspect. *P. tricarinata* male holotype: (B) lateral aspect; (C) frontal lamina; (D) antennae 2; (E) antennae 1.

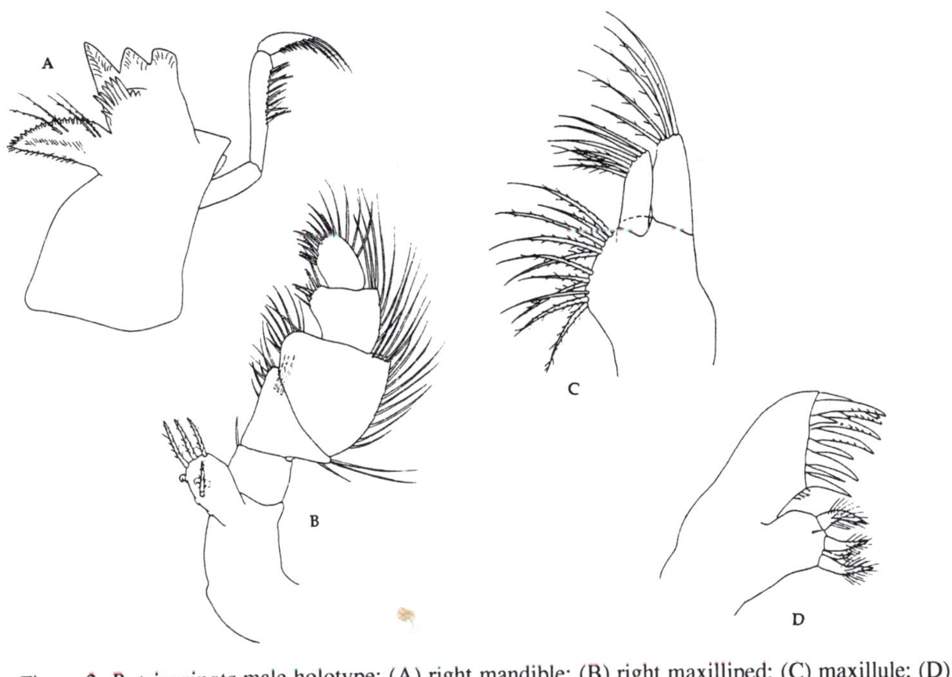

Figure 2. *P. tricarinata* male holotype: (A) right mandible; (B) right maxilliped; (C) maxillule; (D) maxilla.

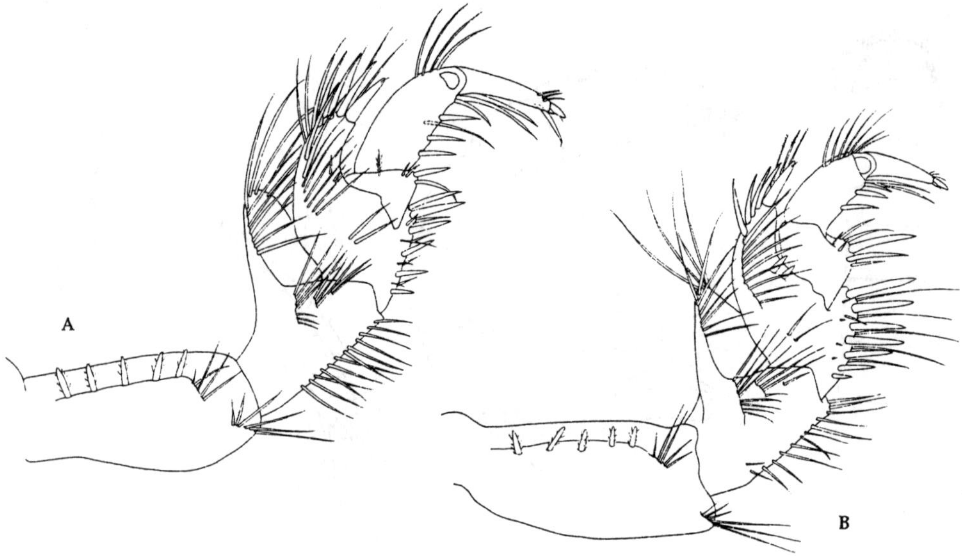

Figure 3. *P. tricarinata* male holotype: (A) pereiopod 1; (B) pereiopod 2.

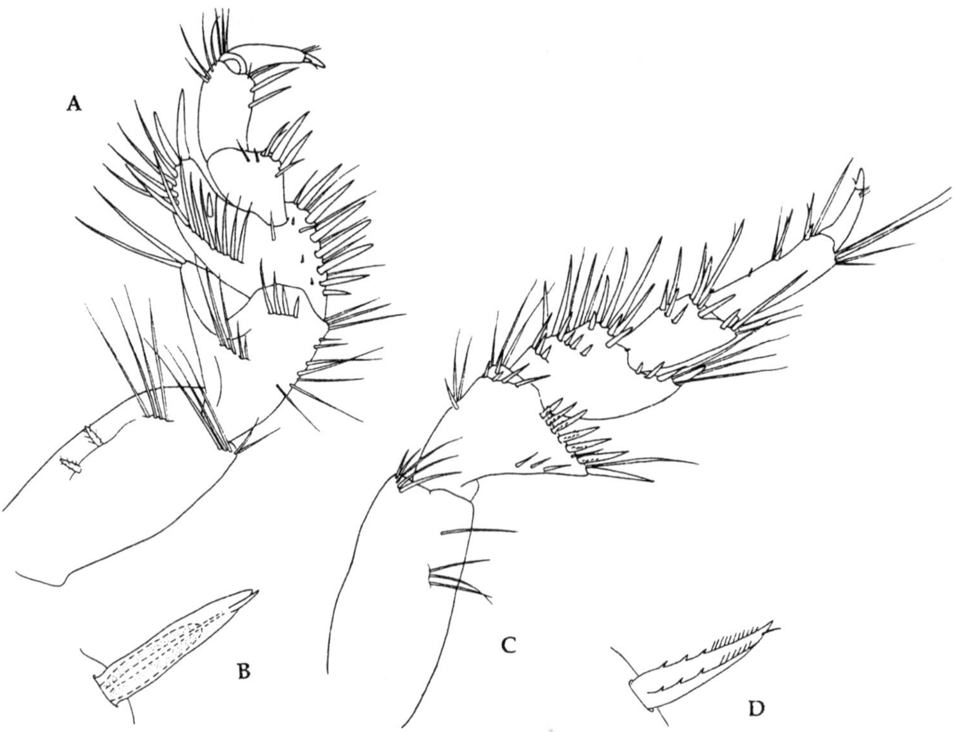

Figure 4. *P. tricarinata* male holotype: (A) pereiopod 3; (B) detail of robust setae on inferior margin of merus, Pereiopods 1-3; (C) pereiopod 4; (D) detail of robust studded-serrate setae on ischium superior margin, pereiopods 4-7.

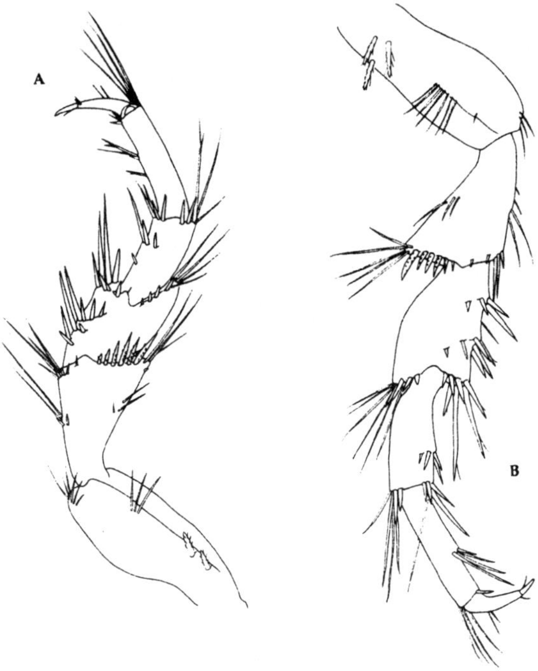

Figure 5. *P. tricarinata* male holotype: (A) pereiopod 5; (B) pereiopod 6.

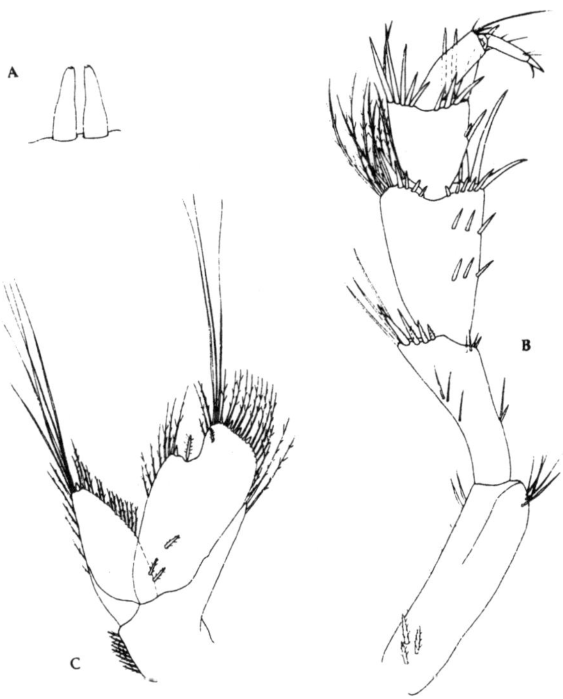

Figure 6. *P. tricarinata* male holotype: (A) penes; (B) pereiopod 7; (C) uropod.

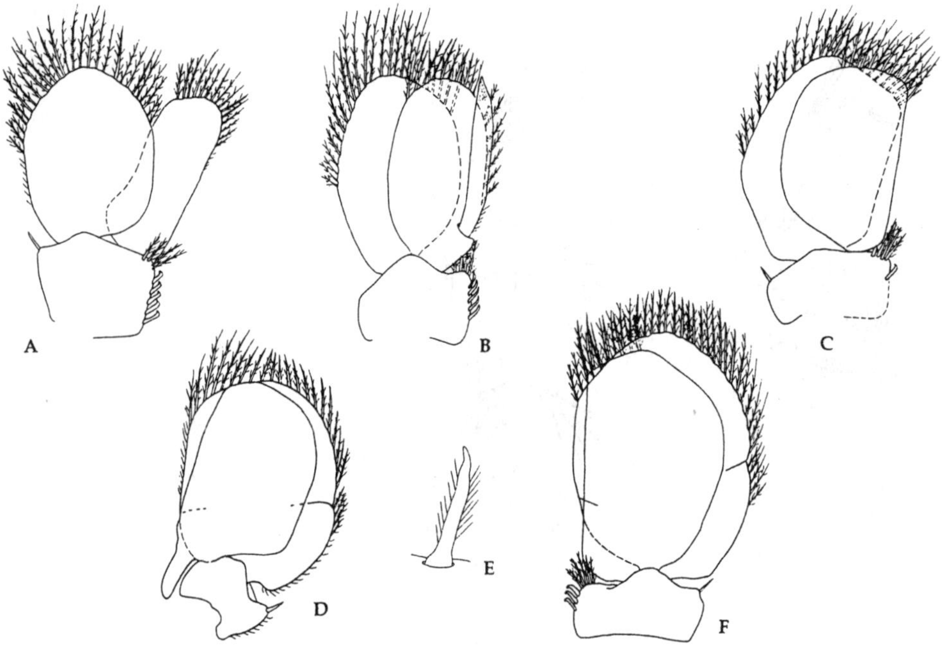

Figure 7. *P. tricarinata* male holotype: (A-C) pleopods 1-3 respectively; (D) pleopod 5; (E) detail of plumose coupling hook. *P. tricarinata* female: (F) pleopod 4.

Type material: Male holotype, deposited at Museu de Zoologia da Universidade de São Paulo (MZSP) cat. no. 12820.

Type locality: Ubatuba, northern São Paulo State, Brazil, 23°38'S, 44°49'W, 22 January 1986, 48 m.

Additional material examined (paratypes): MZSP#12821: Ubatuba, São Paulo State, Brazil, 23°50'S, 45°10'W, 27 October 1985, 38-40 m; 2 males, 9 females, 7 juveniles. MZSP#12822: Ubatuba, São Paulo State, Brazil, 23°34'S, 45°06'W, 9 July 1986, 25 m; 1 male, 1 female, 1 juvenile. USNM#288407: Ubatuba, São Paulo State, Brazil, 23°39'S, 44°53'W, 20 April 1986, 45-48 m; 2 males, 2 females, 3 juveniles. USNM#288408: Ubatumirim, São Paulo State, Brazil, 23°22.50'S, 44°53'W, May 1993, 15 m; 1 female.

Nontype material examined: USNM #243773: single specimen from the gut contents of a dolphin, *Pontoporia blainvillei*, off the coast of Uruguay.

Diagnosis: Cephalon with transverse cuticular ridge between frontal ridge and interocular furrow. Antennae extend to middle of pereionite 1. Frontal lamina narrow, anteriorly expanded, spatulate; visible in dorsal aspect projecting between antennular peduncles. Pereionite 7 longest. Pleotelson posterior margin narrow, slightly subacute; with several plumose marginal setae (pms) and 4-6 very small robust setae. Pereiopods 4-7 posterodistal margin of ischium with robust studded-biserrate setae (Fig. 4D). Pleopod 2 appendix masculina broad and straight, of relatively constant width until apex where it narrows to a point. Uropod with cluster of very long simple setae at apices of both endopod and exopod; exopod shorter than, or just reaching endopod notch.

Description of Holotype: Male, 10.5 mm long; body length about 5.75 times width.

Pereional segments loosely articulated; body long, cylindrical, with lateral margins subparallel in dorsal aspect, slightly constricted in region of pereionites 4 and 5.

Cephalon: About 1.5 times wider than long, laterally enclosed by pereionite 1; anterior margin convex, medially flattened, with minute rostral point; cuticle of frontal margin raised into thickened frontal ridge, the lateral margins of which extend over and divide the eyes; interocular furrow complete and evenly convex; additional transverse cuticular ridge present between frontal ridge and the interocular furrow. Eyes subquadrate, darkly pigmented, with about 6 rows of 6-8 ommatidia, with small notch of missing ommatidia where divided by frontal ridge; cuticular ommatidial facets weakly developed. Frontal lamina narrow, length about 4 times greatest width; anteriorly expanded and spatulate, with raised ridge along margin; visible in dorsal aspect projecting between antennular peduncles; posterior margin abutting clypeus, not projecting. Clypeus not projecting; subtriangular, width greater than length; lateral margins with raised ridge, curved around lateral margins of labrum. Labrum wider than long, posterior margin slightly concave.

Antennules: Short, reaching posterior margins of eyes. Peduncle of 3 articles, plus minute 4th article; articles 1 and 2 short, subequal in length, article 3 longest; peduncle article 2 posterior distal angle with 2-3 short circumplumose setae. Flagellum length subequal to peduncle length, of 8 articles, each with 5-6 aesthetascs; first flagellar article length subequal to width; subsequent flagellar articles shorter, width twice length.

Antennae: Reaching middle of pereionite 1; peduncle of 5 articles, articles 1 and 2 shortest, wider than long; articles 3 and 4 subequal in length and subquadrate, or with length only slightly less than width; article 5 longest, approximately 2 times longer than wide and width about half the basal width of article 4; posterior distal margin of peduncle articles 4 and 5 with row of approximately 5-10 long stiff simple setae. Flagellum of 15 articles, proximally subquadrate, lengthening distally; each article with clusters of short fine setae on distal angles.

Mandible: Incisor tridentate; left mandible incisor teeth less robust than on right mandible. Molar process and setal row well developed; setal row with about 7 long robust setae; molar process with short widely spaced spines along entire anterior margin, submarginal row of slender lightly plumose setae on dorsal surface; molar process covered with short fine setae. Palp 3-articulate; article 2 longest, with dense row of short simple setae; article 3 shortest, narrowing distally, with simple and slender biserrate marginal setae.

Maxillule: Lateral lobe with fine bifid setae along medial margin; gnathal surface with approximately 11 large, heavily scleratized, robust setae, several anterior ones with notched surfaces. Medial lobe with three giant circumplumose setae and a smaller robust plumose seta between the two proximal-most giant setae.

Maxilla: Lateral lobe with simple and lightly plumose slender setae; middle and lateral lobes free and articulating, with long simple and lightly plumose setae.

Maxilliped: Endite with distal cluster of approximately 3 long robust circumplumose setae; right endite with 2 coupling hooks and 3 cuticular bumps on the medial margin (possibly newly forming coupling hooks), left endite with 1 coupling hook. Palp of 5 articles; lateral margins with simple setae, medial margins of distal articles with serrate setae in addition to simple setae; articles 4 and 5 with distomedial angle weakly produced.

Pereion: Cuticle highly polished, without spines, tubercles or setae, but with scattered chromatophores on pereion and pleon, concentrated posteriorly and laterally. Pereionites 1-6 with dorsal medial round depression housing two minute pores; pores present, but without depression on pereionite 7; medially, pereionites 2 and 3 shortest, 4 and 5 subequal and longer than 1, 7 longest. Pereionite 1 only slightly narrowed anteriorly, anterolateral margins straight, forming blunt angles reaching or partially covering eyes; with impression along lateral margin. Coxae narrow, not forming sternal plates; coxae 2 and 3 with impression parallel to lateral margin, posterior margins rounded; coxae 4-6 with oblique impression, posterior margins progressively more oblique; coxae 7 large, triangular, with oblique impression and blunt posterior angle; postero-ventral angle of pereionite 7 extended slightly beyond posterior margin of coxae. Penes long, not fused basally; slightly flattened and tapering.

Pereiopods 1-3: Ambulatory, robust and highly setose as figured. Basis superior submarginal setal row short, with less than 10 long simple setae. Ischium of pereiopods 1 and 2 with three oblique rows of long simple setae on posterior face; ischium of pereiopod 3 with two oblique rows; superior distal angle produced into a large scoop-shaped lobe. Merus with superior distal angle produced into broad recurved lobe extending beyond base of propodus, with 1-2 robust apical setae; inferior margin of merus and carpus with long acute robust setae and long simple setae; posterior face of merus with small cluster of short setae near superior join of carpus. Pereiopods 1-3 shortening posteriorly, with ischial and meral lobes slightly less produced. Pereiopod 1 carpus triangular; carpus of pereiopods 2 and 3 subquadrate. Pereiopod 1 propodus with simple seta inserted on mid-anterior face; propodus with row of simple setae on distal half of superior margin. Dactyl length subequal to, or only slightly less than propodal length; with small, blunt secondary ungui.

Pereiopods 4-6: Longer, more slender, and less setose than pereiopods 1-3; flattened on the anterior-posterior plane. Ischium superior margin with sparse long simple setae; posterior distal margin with studded-biserrate setae. Merus and carpus with short robust setae arranged in rows perpendicular to and extending around inferior margin onto posterior face. Carpus longer than wide. Propodus superior margin bare.

Pereiopod 7: Longer than pereiopod 6. Ischium width less than merus width; merus superior distal angle bearing long slender plumose setae, long slender distally serrate setae, and short robust acute setae. Carpus distal margin expanded, about 3 times as wide as proximal propodal width, slightly longer than wide. Propodus length subequal to carpus length, shorter than in preceding pereiopods.

Pleon: Of 5 free segments (plus pleotelson), first two or three loosely overlapped by pereionite 7; pleon more dorso-ventrally compressed than pereion; cuticle slightly less polished than pereion cuticle. Pleonites 1-4 with epimeres that are produced posteriorly and flared laterally; epimeres 2-4 with dense fringe of plumose setae on lateral margin; pleonite ventral flanges produced and visible, not hidden by dorsal epimeres; ventral flanges with ventral posterior angle rounded.

Pleopods: Rami with pms as figured. Peduncle of pleopods 1-4 with slender plumose coupling hooks and cluster of short plumose setae on distomedial angle. Pleopod 1 peduncle subquadrate, slightly wider than long, with five coupling hooks; endopod narrow, about one-half exopod width; exopod large and circular. Pleopod 2 peduncle with five coupling hooks; endopod slightly more narrow than exopod; appendix masculana arising sub-basally on endopod, broad and straight, of relatively constant width, narrowing abruptly to a point, apex cuticle slightly more scleratized; reaching the distal margin of exopod. Pleopods 3 and 4 with peduncle wider than long, exopod larger than endopod; pleopod 4 exopod with incomplete transverse suture. Pleopod 5 peduncle reduced, without coupling hooks; endopod with large proximomedial lobe, incomplete transverse suture, without pms.

Uropod: Peduncle medial margin produced distally to level of endopodal notch, distally acute, with 3 long plumose setae at apex; peduncle lateral margin with row of plumose setae, curving ventrally onto distal margin. Endopod distally truncate, distal margin with pms and 4 short robust setae of relatively equal size; apex with cluster of long simple setae, much longer than other marginal setae; with deep notch in distolateral margin bearing a short palmate seta inside, and a short robust seta lateral to the notch. Exopod shorter than endopod, not reaching the endopod notch, flat, with lateral margins convex, peltate; with pms on lateral and medial margins; distal medial margin with 2 small robust setae; apex with a small robust seta and cluster of very long simple setae.

Pleotelson: Dorsal surface flat, without setae, tubercles, pits or ridges; posterior margin narrow, slightly subacute; with pms and 6 very small robust setae.

Variation: Males and females are similar, although the two males examined have the pereional segments more loosely articulated. Length/width ratio of body (both sexes) varies from 4.5 to 5.75. Antennule flagellar articles, including the first, appear wider in females; flagellar article counts range from 8 to 12. Antenna flagellum of 9-11 articles. Maxilliped endites with 1-2 coupling hooks; additional cuticular bumps on endite present in 2 males, not present in the females examined. Uropod endopod distal margin with 4-7 robust setae. Pleotelson distal margin with 4-6 small robust setae.

Etymology: The trivial name *tricarinata* is a Latin derived word referring to the three transverse carinae, or ridges on the cephalon.

Remarks: While a generic revision is forthcoming, this species can be confidently placed in *Politolana* by the combination of diagnostic characters provided by Bruce (1981), Wetzer et al. (1987), and Kensley & Schotte (1989). However, the proportions of the antennal peduncle articles of *P. tricarinata* differ from the proportions given in the above diagnoses. Characters of particular utility for placing this species within *Politolana* are: a long, polished, unornamented body form; narrow frontal lamina; uropod endopod with disto-medial emargination (although the degree of emargination may vary with age); pereiopods 1-3 with the superior distal angles of the ischium and merus produced.

P. tricarinata can be distinguished from all other species of *Politolana* by the combination of characters provided in the diagnosis. The third carina, or transverse cuticular ridge on the cephalon is an autapomorphy of this species. Among the species of *Politolana*, *P. tricarinata* is also unique in having studed-biserrate setae on the postero-distal margin of the ischium on the fourth pereiopod. These setae occur on pereiopods 5-7 of several other species, but are absent on the fourth pereiopod. *P. tricarinata* can be easily distinguished from the only other South American *Politolana* species, *P. eximia*, by its short, broad, uropod exopods (about 2.3 times longer than wide) and short antennae reaching the middle of the first pereionite. *P. tricarinata* is quite similar to *P. wickstenae* in body form, being long and narrow with loosely articulated pereionites. However, the absence of eyes, the longer antennae, and the slender appendix masculina in the latter clearly distinguishes these two species.

In addition to the holotype and paratypes, a specimen (USNM 243773) from the gut contents of a dolphin (*Pontoporia blainvillei*) was also examined. The dolphin was collected off the coast of Uruguay, though this locality could be a significant distance from where the isopod was ingested. This specimen agrees with the species diagnosis with the exception of the transverse ridge on the cephalon between the frontal margin and the interocular furrow, which it apparently lacks. However, the absence of this ridge could be due to the partially digested condition of the specimen. Because of the above uncertainties, this specimen should not be considered a part of the type series.

Distribution: *P. tricarinata* is known from the type locality and from the São Sebastião coast, south of Ubatuba. Specimens were collected only on the inner shelf, from 15 m to 48 m depth, on sandy bottoms where medium and fine grains predominate, in water temperatures of 16°-22°C. These isopods were present in in relatively low numbers, generally from 1 to 5 individuals per 0.1 square meter, in approximately 25% (34 of 138) of the vanVeen grab and rectangular dredge samples collected in this region.

ACKNOWLEDGEMENTS

This research was supported by a National Science Foundation P.E.E.T. grant (DEB-9896145) to R.C. Brusca and B. Kensley, and by a grant from the Brazilian Research Council (CNPq – 30.5600/78-OC) to A.M. Pires. Figures drawn by the first author were inked by Rachel Taylor. Thanks go to Brian Kensley and the anonymous reviewers for their helpful comments and suggestions. This is contribution #155 of the Grice Marine Biological Laboratory.

REFERENCES

Bruce, N.L. 1981. Cirolanidae (Crustacea: Isopoda) of Australia: diagnoses of *Cirolana* Leach, *Metacirolana* Nierstrasz, *Neocirolana* Hale, *Anopsilana* Paulian & Debouteville, and three new genera - *Natatolana, Politolana*, and *Cartetolana*. *Aust. J. Mar. Freshwater Res.* 32: 945-966.

Bruce, N.L. 1986. Cirolanidae (Crustacea: Isopoda) of Australia. *Rec. Aust. Mus.* Supplement 6: 1-239.

Bruce, N.L. 1991. New records of marine isopod crustaceans (Sphaeromatidae, Cirolanidae) from south-eastern Australia. *Mem. Mus. Victoria* 52: 263-275.

Bruce, N.L. 1996. Crustacea Isopoda: Some Cirolanidae from the Musorstom cruises off New Caledonia. In A. Crosnier (ed.), Resultats des Campagnes Musorstom, Volume 15. *Mem. Mus. Natn. Hist. Nat.* 168: 147-166.

Brusca, R.C., R. Wetzer & S. France 1995. Cirolanidae (Crustacea: Isopoda: Flabellifera) of the tropical Eastern Pacific. *Proc. San Diego Soc. Nat. Hist.* 30: 1- 96.

Dana, J.D. 1853. Crustacea, Part II. In United States Exploring Expedition during the Years 1838, 1839, 1840, 1841, 1842, under the command of Charles Wilkes, U.S.N., Vol. 14: 689-1618. Philadelphia: C. Sherman.

Hansen, H.J. 1890. Cirolanidæ et familiæ nonnullæ propinquæ Musei Hauniensis. *Det Kongelige Danske Videnskabernes Selskab Skrifter, Naturvidenskabelig og Mathematissk* 6: 331-333.

Harger, O. 1883. Reports on the results of dredging, under the supervision of Alexander Agassiz, on the east coast of the United States, during the summer of 1880, by the US Coast Survey Steamer 'Blake', Commander J.R. Bartlett, U.S.N., commanding. *Bull. Mus. Comp. Zool. Harv. Univ.* 11(4): 91-104.

Hoek, P.P.C. 1882. Die Crustaceen gesammelt während der Fahrten des 'Willem Barents' in den Jahren 1878 und 1879. *Niederländisches Archiv fuer Zoologie* Supplement 1: 1-75.

Kensley, B. 1975. Marine Isopoda from the continental shelf of South Africa. *Ann. S. Afr. Mus.* 67: 35-89.

Kensley, B. & M. Schotte 1989. *Guide to the marine isopod crustaceans of the Caribbean.* Washington, D.C: Smithsonian Institution Press.

Latreille, P.A. 1817. Les Crustacés, les Arachnides, et les Insectes. In: G.L.C.F.D. Cuvier, *Le Regne Animal, distribué d'après son organisation, pour servir de base à l'histoire naturalle des animaux et d'introduction à l'anatomie comparée* vol. 3. Paris.

Pires, A.M.S. 1992. Structure and dynamics of the benthic megafauna from the continental shelf offshore of Ubatuba, southeastern Brazil. *Mar. Ecol. Progr. Ser.* 86: 63-76.

Sars, G.O. 1882. Oversigt af Norges Crustacea. *Forhandlinger i Videnskabsselskabet i Kristiania 1882* 18: 1-124.

Stimpson, W. 1853. Synopsis of the marine Invertebrata of Grand Manan. *Smithson. Contrib. Knowl.* 6: 1-66.

Wetzer, R., P.M. Delaney & R.C. Brusca 1987. *Politolana wickstenae* new species, a new cirolanid isopod from the Gulf of Mexico, and a review of the '*Conilera* genus-group' of Bruce (1986). *Contributions in Science, Natural History Museum of Los Angeles County* 392: 1-10.

Families and genera of Isopoda Anthuridea

GARY C.B. POORE
Museum Victoria, Melbourne, Australia

ABSTRACT

The Anthuridea Monod, 1922 are a suborder of Isopoda characterized by an elongate cylindrical body form, without dorsal coxal plates, and with a uropodal exopod attached to the peduncle proximally and dorsally. Over 500 species are described and these are assigned to 57 genera after synonymy of a further 30 genera. The taxonomic history of the group since 1814 is briefly reviewed. A cladistic analysis supports recognition of six families of Anthuridea, two new: Anthuridae Leach, 1814 (24 genera); Antheluridae Poore & Lew Ton, 1988 (3 genera); Expanathuridae fam. nov. (7 genera); Hyssuridae Wägele, 1981 (6 genera); Leptanthuridae fam. nov. (11 genera); and Paranthuridae Menzies & Glynn, 1968 (6 genera). The significant taxonomic changes proposed here are: erection of the new family Expanathuridae for seven genera previously included in Hyssuridae, Paranthuridae and Anthuridae; erection of the new family Leptanthuridae for 11 genera previously included in Paranthuridae; synonymy of *Indonesanthura* Negoescu, 1997 with *Caenanthura* Kensley, 1978; synonymy of *Filanthura* Wägele, 1981 with *Indanthura* Pillai & Eapen, 1966; synonymy of *Centranthura* Wägele, 1981 and *Monodanthura* Wägele & Platvoet, 1982 with *Notanthura* Monod, 1927; synonymy of *Agulanthura* Kensley, 1975 with *Quantanthura* Menzies & George, 1972; synonymy of *Stellanthura* Wägele, 1979, *Chalixanthura* Kensley, 1984 and *Licranthura* Kensley & Schotte, 1987 with *Heptanthura* Kensley, 1978; and synonymy of *Neoanthura* Menzies, 1956 with *Leptanthura* Sars, 1897. All families and genera are diagnosed using characters from a DELTA database and their relationships discussed. I offer dichotomous keys for their identification. All valid species are listed in an appendix. Biogeography at the family and genus level is briefly discussed.

1 INTRODUCTION

The genus *Anthura* and family Anthuridae were first defined by Leach (1814) for *Oniscus gracilis* Montagu, 1808, the family as 'Last segment of the tail very short, the last narrow, elongate, with two elongate lamellae on each sides; antennae nearly equal, inserted one behind the other in nearly horizontal line.' By the end of the nineteenth century the number of genera had grown to 11 and of species to 25. Now, a century later more than 500 species have been described in 87 nominal genera.

There has never been any doubt about the monophyly of the Anthuridea, the suborder erected by Monod (1922) for these isopods. All are elongate cylindrical isopods in which the uropodal exopod is attached to the peduncle dorsally and more proximally than the terminal endopod. The exopod is held erect or parallel to the telson. There are never extended coxal plates, as in many flabelliferan families, and maxilla 2 is fused with the lower lip as a 'hypopharynx.'

Norman & Stebbing (1886) recognized two sections within the Anthuridae, section A for those species with 'biting' mouth parts and section B for those with 'piercing' mouth parts. This division was followed by Barnard (1925) in the first major synthesis of the group which he considered 'a preliminary study only.' He discussed morphology at length and provided keys to 24 genera. He diagnosed all the species then known, adding 21 new ones and seven new genera. Barnard (1925) was unable to unambiguously define sections A and B on characters other than the general form of the mouth parts but he generalized about the statocyst, unguis of pereiopod 1, the relative size of pereiopods 1 and 2-3, and the telson. Notably he correlated paired statocysts (when present) with normal mouth parts and unpaired (single) statocyst (when present) with piercing mouth parts. This correlation was first questioned by Kussakin (1967) with discovery of *Austranthura* (=*Ananthura*) *elegans* in Antarctica. This species has normal mouth parts and a single statocyst. Menzies & Glynn (1968), probably ignorant of Kussakin's discovery, elevated Norman & Stebbing's sections A and B to family rank, Anthuridae and Paranthuridae respectively.

Wägele's thesis (Wägele 1980b), published in the next year (Wägele 1981d), is the most recent detailed treatment of the Anthuridea. He reported on many observations of living animals and correlated feeding strategies and habitats with internal and external anatomy. His evolutionary analysis derived anthurideans from the flabelliferan isopods, from a hypothetical cirolanid-like ancestor. In this work, Wägele (1981d) he proposed the new family Hyssuridae. The Hyssuridae was elaborated by Wägele (1981b) and redefined by Poore & Lew Ton (1988c). In the same year a fourth family was erected, Antheluridae, which accounted for Kussakin's anomalous species and members of three genera (Poore & Lew Ton 1988b). Wägele (1989) revisited the phylogeny of the Anthuridea as part of a study of the Isopoda as a whole. His view of the Anthuridea as a highly derived group of flabelliferans was confirmed by Brusca & Wilson (1991).

Beginning almost simultaneously in the mid- to late-1970s, B. Kensley in South Africa, N. Nunomura in Japan, J.-W. Wägele in Germany, and I (with H. Lew Ton after 1985) in Australia have described numerous new species and genera. H.-G. Müller added many species during the 1990s to what is now understood to be a diverse and speciose taxon. A list of species in 1984 (Negoescu & Wägele 1984) listed 321 species and its authors pointed out the many taxonomic and evolutionary questions that the group presented.

Present-day nomenclatural complexity and the lack of modern keys to genera stimulated this revision. I observed that, Wägele's (1981d) revision notwithstanding, the absence of a modern synthesis was allowing a proliferation of genera whose species seemed similar to those already in the literature.

The aim of this study is, therefore, to present a new classification of the families and genera based on phylogenetic principles. An evolutionary scenario is hypothesized in cladistic analyses of genera and families. While two new families are presented no new genera are erected. Opportunities for this exist and these are pointed out. All families and genera are diagnosed and illustrated dichotomous keys for their identification are offered. The described species are listed in an appendix.

2 METHODS

2.1 *Sources*

The observations reported in this paper come from museum specimens and a literature of over 200 papers. Many species, examples of most genera, are represented in the collections of Museum Victoria. Others were observed in other museums or borrowed for detailed examination. Illustrations in the literature provided the bulk of the data used. Although I was reluctant to rely on my or other people's published statements and illustrations, many were supported in multiple publications. Type species were the primary source for characterization of genera and conclusions based on these were corroborated with many other species if available. New observations or corrections of earlier statements are reported in the taxonomic section.

A taxonomic list of species of isopods compiled by B. Kensley and M. Schotte, National Museum of Natural History, Washington, DC, is available (http://nmnhgoph.si.edu/gopher-menus/Isopods.html). It lists over 500 species of anthurideans and served as a basis for reviewing the literature. The list has been updated in the light of this revision in Appendix 1. Eighty-seven nominal genera have been assigned to Anthuridea; type species and descriptions were scanned for probable synonyms. Several were documented before this revision commenced but others were discovered, especially of small genera, in the course of it. Taking into account synonymies already published the number of genera was reduced to 57. New synonymies are justified in the taxonomic section. These 57 are diagnosed and are the basis of all phylogenetic analyses.

2.2 *DELTA treatment*

All genera were coded using DELTA (Descriptive Language for Taxonomy; Dallwitz et al. 1997). Separate character suites were selected to diagnose families within the suborder Anthuridea and for diagnosing the genera within the six families. The number of characters necessary to do this was 18 for the families and ranged from 10 to 37 for genera. In all families except Antheluridae, some characters were defined as having *implicit* states. That is, the state was shared by all genera except one. Characters are listed in generic diagnoses only when the implicit state does not apply. Instead, the implicit states are listed separately immediately following keys to genera.

2.3 *Phylogeny*

Hennigian phylogenetic (cladistic) methods were used to generate cladograms as hypotheses of the relationships of anthuridean families and genera that form the basis of the new classification. The program HENNIG86 version 1.5 was used and lists of apomorphies for each clade generated in CLADOS and PAUP 3.1. Analysis of a first data matrix of 57 genera by over 120 characters lead to two conclusions:

1. Many traditionally used characters were highly homoplasous, and
2. In spite of this, four major branches of the cladogram were remarkably robust while the relationships between genera within them were relatively unresolved. A high level of homoplasticity in several characters hindered resolution of otherwise well-defined clades. Resolution of generic relationships seemed impossible while homoplastic char-

acters were treated as potential synapomorphies across what were obviously different family groups. It is intuitively obvious (and an assumption of cladistic treatment of any assumed monophyletic clade) that characters like fusion of maxillipedal palp articles, loss of mandibular palp articles, elongation or shortening of pereiopodal articles etc. have occurred many times in the history of the Anthuridea, as they have in many other crustacean taxa. Continuing to treat characters such as these as equal to more fundamental characters would not be fruitful.

Five subsequent cladistic analyses were therefore performed, one of each of the four strongly supported branches separately and one of the six families so hypothesized. Each analysis used its own character suite. The character suite was derived from that used in the DELTA descriptions except that some highly homoplastic characters were omitted. Characters were almost always treated as unordered; exceptions are noted in the character discussion.

In all cases, a hypothetical ancestor was used. The anthurideans are a highly specialized derived group with relationships to flabelliferan isopod families (Wägele 1989, Brusca & Wilson 1991). The hypothetical ancestor is based therefore on more generalized families such as Cirolanidae and Sphaeromatidae. In practice, absence of a real out-group is not critical in view of most characters being binary and unordered.

2.4 *Characters*

Each cladistic analysis is based on its own character suite. This section discusses the significant differences between character states in turn across all families and refers to Figures 1-11. Characters and their states are listed in Tables 1-5 and states coded in Table 6. The characters are essentially the same as those used in diagnoses produced by DELTA (see above) but are sometimes simplified. The tables include synapomorphies and generic autapomorphies.

Sexual dimorphism (Figs 1, 3): Life histories of few anthurideans have been studied. The best understood genera are *Cyathura* (Cleret 1960, Wägele 1979a) and the atypical *Eisothistos* (Wägele 1979c). Museum collections of most species are undifferentiated juveniles and most descriptions in the literature are of this life stage. Ovigerous females are less common and adult males usually rare. Observations on mature adults are usually consistent with what is known for *Cyathura*, the only exception known to me being species of *Haliophasma*, one of the most commonly encountered genera in Australia, where ovigerous females have never been reported. In general, males are slightly or considerately more elongate than juveniles or females. They may be longer generally or the pleon may be elongate. Males have more elongate pereiopods, pleopods, uropods, and telson and have much larger eyes. Males of *Haliophasma* are sometimes especially elongate (Poore 1975). The flagellum of antenna 1 is usually elongate and swollen with each article bearing numerous aesthetascs, a callynophore (Lowry 1986). In some genera (*Cyathura*, *Expanathura*), the flagellum is shorter rather than longer. Pleopod 2 of adult males bears a medial appendix masculina as in all isopods but medioventral penes on pereionite 7 are minute or obsolete if present at all. Males are often more setose than other stages, especially on the mesial face of pereiopod 1 and sometimes on the uropods and telson. In the large diverse genus *Paranthura*, the males of some species have broadened rather than narrower posterior pereiopods (Poore 1984b). Because males have not been described for many small genera male characters are not relied on heavily in the cladistic analyses. Most male characters are autapomorphies.

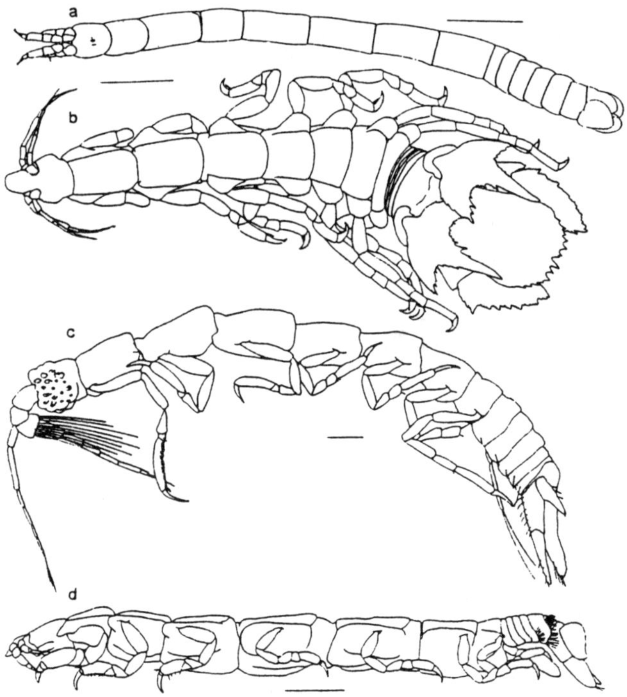

Figure 1. a, Hyssuridae: *Kupellonura*; b, Expanathuridae: *Eisothistos* (juvenile); c, Expanathuridae: *Eisothistos* (male); d, Expanathuridae: *Panathura*. (scale bars 0.5 mm for a, b; 0.1 mm for c; 1 mm for d).

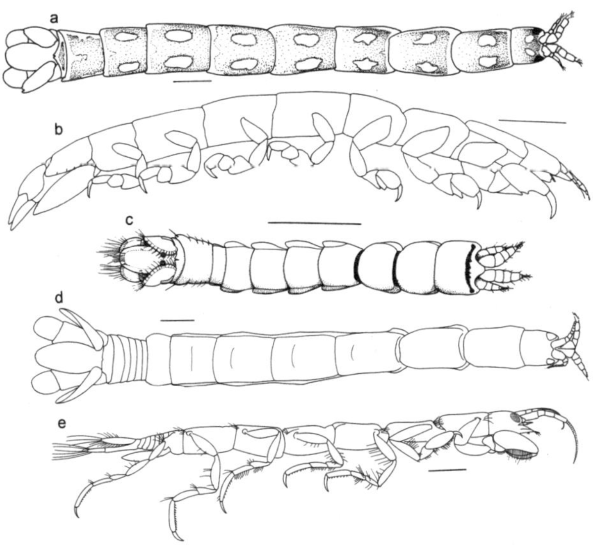

Figure 2. a, Anthuridae: *Mesanthura*; b, Anthuridae: *Apanthura*; c, Anthuridae: *Sauranthura*; d, Paranthuridae: *Paranthura*; e, Leptanthuridae: *Accalathura*. (scale bars 1 mm for a, b, d, e; 0.5 mm for c).

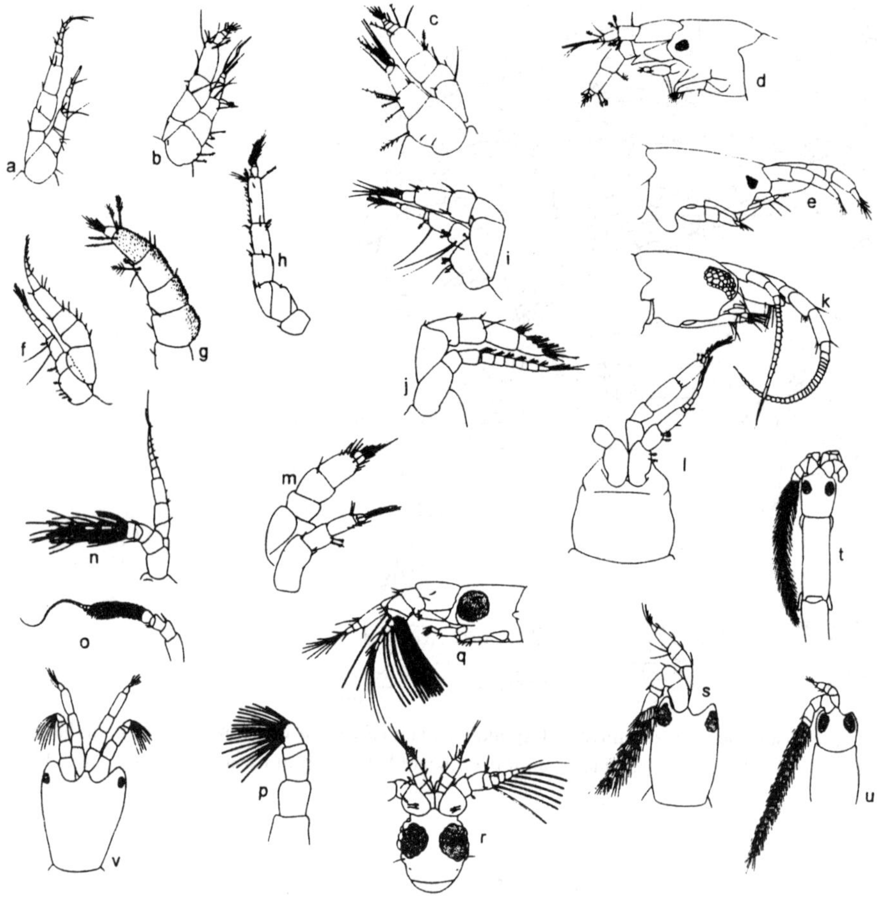

Figure 3. Antenna 1 and 2. a, *Kupellonura*; b, *Xenanthura*; c, *Panathura*; d, *Expanathura* (with biting mouth parts); e, *Apanthura* (with biting mouth parts); f, *Quantanthura*; g, *Pendanthura*; h, *Stygocyathura*; i, *Apanthura*; j, *Paranthura*; k, *Accalathura* (with piercing mouth parts); l, *Pseudanthura*; m, *Colanthura*. Male antennae 1 and 2. n, *Kupellonura*; o, *Accalathura* (antenna 1 only); p, *Skuphonura*; q, *Expanathura*; r, *Panathura*; s, *Apanthura*; t, *Haliophasma*; u, *Apanthuropsis*; v, *Skuphonura*.

Body form (Figs 1, 2): Anthurideans are generally narrow cylindrical isopods. Although it is sometimes possible to characterize some genera as broader than others this is difficult to quantify. *Pendanthura* and *Sauranthura* are particularly small and broad, *Nemanthura* especially thin. Hyssurids are much narrower than members of other families and are notable for the elongation of pereionite 7 and pleonites 1-6. Pereionite 7 is shorter than the others, less so in Hyssuridae but especially so in the neotenous paranthurid genera which lack pereiopod 7 (Poore 1984a). Most species are without pigmentation but chromatophores are common among the paranthurids and a consistent pattern is a feature of *Mesanthura* and *Chelanthura*.

Eyes (Figs 1-3): Paired pigmented eyes are typical but these often fade in preserved specimens and so go unreported. Truly blind species occur in deepwater (*Pseudanthura*) and interstitial (*Curassanthura*) genera.

Antenna 1 (Fig. 3): The first antenna lies lateral to the second. As in all isopods the peduncle is of three articles, the first longer than the others. The flagellum comprises a first very short ring-like article, usually with a lateral brush-seta, plus more distal articles decreasing in size distally. The shortest antennal flagellum (basal article, longer second and minute third) is a frequently seen state. The last two articles bear 1-2 aesthetascs. Longer flagella are seen in most families, the extreme situation seen in *Accalathura*. The number of articles seems to vary between species within genera and possibly with age. I have not relied heavily on antenna 1 in diagnoses or phylogenetic analysis. Antenna 1 is sexually dimorphic, highly modified in males. Usually the number of aesthetascs and aesthetasc-bearing articles is much greater in males than in juveniles or females. In other cases (e.g. *Expanathura*), one article only is swollen and carrying numerous aesthetascs.

Antenna 2 (Fig. 3): The peduncle of five articles usually exceeds antenna 1 in length. It lies medial to antenna 1 and the second article is slotted to fit antenna 1. The number of articles of the flagellum varies considerably but in generic diagnoses and the phylogenetic analysis only two states are usually allowed. Considerable fusion of articles in seen in genera such as *Paranthura*.

Mouth parts (Figs 3d, e, k): The traditional division of the Anthuridae s. l. by Norman & Stebbing (1886) was into two sections, A comprising genera with 'biting' mouth parts and B with 'piercing' mouth parts (Barnard 1925, Menzies & Glynn 1968). The compact mouth parts of section A is the plesiomorphic form while the anteriorly directed mouth parts of section B is apomorphic. In some genera with otherwise biting mouth parts there is a tendency for promotion of the mouth parts (*Apanthuroides*, *Cetanthura*, *Eisothistos*), which lead Wägele (1981d) to include some of them within Paranthuridae. I view these genera as showing some convergence but not the acute incisor or absence of molar characteristic of this family.

Mandible (Fig. 4a-w): The shape of the body of the mandible reflects difference between 'biting' and 'piercing' mouth parts. In the former the incisor is transverse and weakly toothed, the lamina dentata is present and toothed and in some cases shows remnants of the plesiomorphic lacinia mobilis (Poore & Lew Ton 1988c). The molar may be a truncated or sharp tooth or a hinged cirolanid-type blade (Poore & Lew Ton 1988c) or lost. In those families with 'piercing' mouth parts, Paranthuridae and Leptanthuridae, the incisor is a simple acute stiletto and the molar a sharp ridge (except in *Colanthura* and relatives where a curved blade exists). The plesiomorphic mandibular palp is of three articles of which the second is longest. Loss or fusion of palp articles occurs in most families and within some genera (e.g. *Leptanthura*). Setation of the terminal article ranges from a regular row of a dozen or more setae (sub-terminal always longest) to few apical setae. Some genera are consistent but others more variable (Negoescu 1994).

Maxilla 1 (Figs 4x-B): Usually only the outer plate of maxilla 1 has been drawn by most authors, but a small inner plate with a terminal seta exists. Terminal spines are always fused if they exist at all and the degree of flexion mirrors that of the mouth parts as a whole. In Paranthuridae and Leptanthuridae, the outer plate is a finely toothed stylet. Maxilla 1 is not used as a character is these analyses.

Maxilla 2: Maxilla 2 is fused with the lower lip into a hypopharynx in anthurideans and is rarely illustrated (Wägele 1979b). It is not used in these analyses.

Maxilliped (Fig. 5): Barnard (1925) relied heavily on the number of articles in the max-

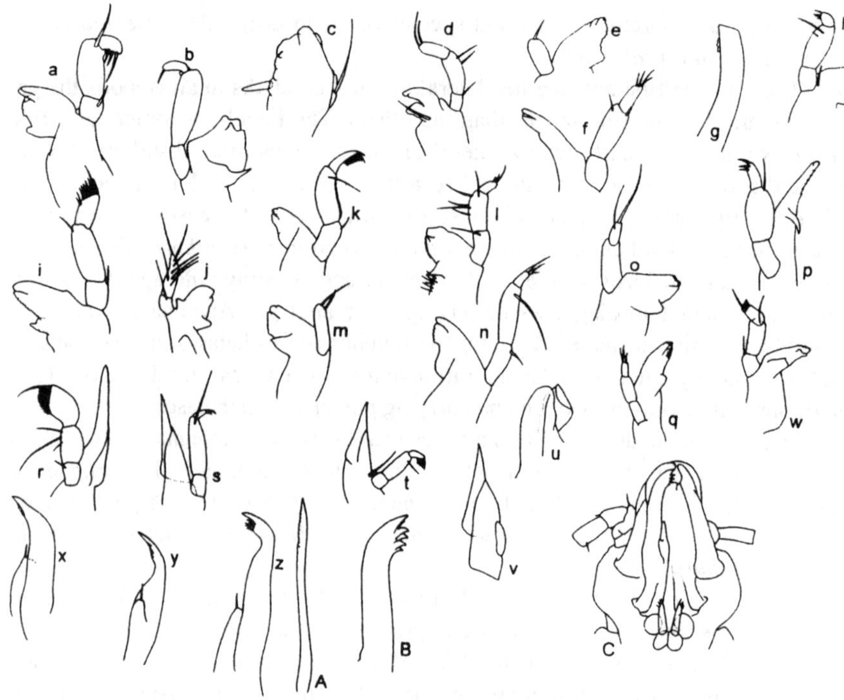

Figure 4. Mandible. a, *Kupellonura*; b, *Neohyssura*; c, *Xenanthura*; d, *Galziniella*; e, *Rhiganthura*; f, *Expanathura*; g, *Eisothistos*; h, *Panathura*; i, *Haliophasma*; j, *Cortezura*; k, *Mesanthura*; l, *Notanthura*; m, *Ptilanthura*; n, *Apanthura*; o, *Caenanthura*; p, *Apanthuroides*; q, *Cetanthura*; r, *Accalathura*; s, *Leptanthura*; t, *Aenigmathura*; u, *Colanthura*; v, *Ulakanthura*; w, *Ananthura*. Maxilla 1. x, *Kupellonura*; y, *Expanathura*; z, *Haliophasma*; A, *Accalathura*; B, *Eisothistos*. Ventral view of head (left maxilla removed). C, *Eisothistos*.

illiped to define genera in his critical work on the anthurideans and many authors have followed him closely. As Barnard (1925) pointed out the first article (coxa) of the maxilliped is fully fused to the head and never seen. Article 2 (basis) is usually the longest; an oval epipod applied to the side of the head lies alongside its proximal end. There is usually a suture between the basis and ischium, i.e. between the maxillipedal protopod and palp. An endite may project beyond the end of the basis as a solid structure or as a weak triangular lobe or be absent altogether. Some authors have reported the number of articles of the whole maxilliped (maximum 7) without differentiating between them. Here and in most recent studies only the numbers of articles in the palp are counted. The maximum number of palp articles, five, is seen in all families recognized here (except Paranthuridae) and each article can be characterized. Article 1 is very short and with few if any setae, 2 longer and with distomesial setae, 3 the largest and with several distomesial and few distolateral setae, 4 is tapering and with distomesial setae, and 5 is minute and with four distal setae. Fusion of palp articles is common and usually a character of generic significance but an appreciation of article homologies is essential. A simple count of visible 'articles' does not convey the way in which the plesiomorphic five articles are fused in anthurideans. The convergence of article fusion is more apparent than real as a comparison of the

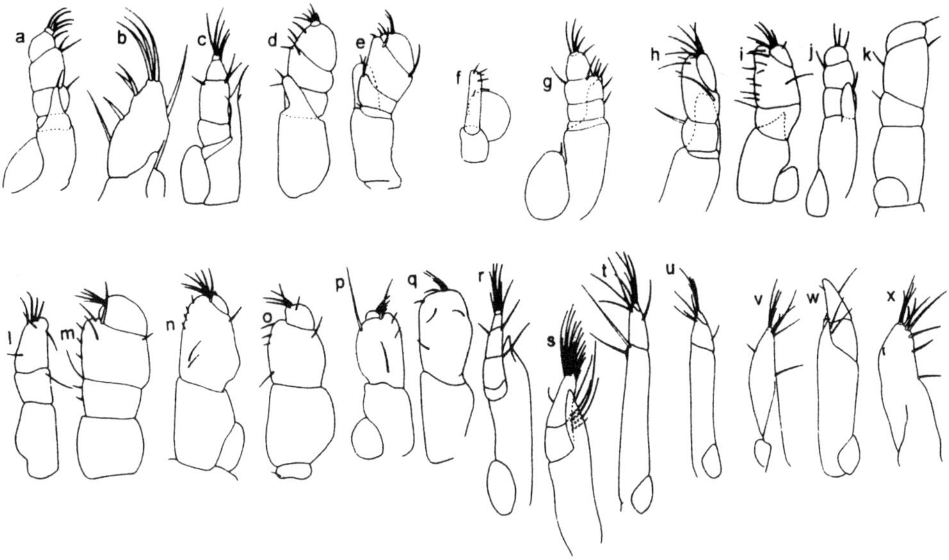

Figure 5. Maxilliped. a, *Kupellonura*; b, *Xenanthura*; c, *Expanathura*; d, *Rhiganthura*; e, *Ananthura*; f, *Eisothistos*; g, *Panathura*; h, *Quantanthura*; i, *Apanthura*; j, *Apanthuroides*; k, *Indanthura*; l, *Cetanthura*; m, *Mesanthura*; n, *Haliophasma*; o, *Cyathura*; p, *Caenanthura*; q, *Anthura*; r, *Bourbonanthura*; s, *Accalathura*; t, *Leptanthura*; u, *Ulakanthura*; v, *Cruregens*; w, *Aenigmathura*; x, *Colanthura*.

various states in different families demonstrates (Tables 1-5). In general, articles 1-2 and 4-5 are fused in many genera of Anthuridae (3-segmented palp) while in Paranthuridae and Leptanthuridae article 5 is often free while others have fused. Fusion of articles is uncommon in Hyssuridae, Expanathuridae and Antheluridae. The size and angle of suture of the terminal articles 4-5 of Anthuridae is of generic importance.

Pereiopod 1 (Figs 6-7): Pereiopod 1 is generally referred to as subchelate, the propodus being more or less swollen and the dactylus closing on a palm. In some genera the propodus is more elongate but few synapomorphies are evident. Specially modified setae of the mesial setal row define some genera (*Heptanthura*, *Caenanthura*), toothed palms others (*Mesanthura*, *Apanthura*). The mesial setal row is composed of simple or complex robust setae in Leptanthuridae and in members of this family the palm is defined by a proximal angle or tooth. Pereiopod 1 is sexually dimorphic often being more elongate in males than in females and juveniles and in genera such as *Apanthura* grossly modified and enlarged. The mesial setal row is replaced by a field of short stiff setae.

Pereiopod 2 (Figs 8a-n): Pereiopod 2 varies more between some related genera than does pereiopod 1, e.g. in Hyssuridae. Palmar robust setae on the carpus of pereiopods 2-3 provide a valuable higher taxon character, remarkably consistent within genus groups. They are numerous in Leptanthuridae and Paranthuridae, few in Hyssuridae and single in Anthuridae. A distally projecting carpus is characteristic of Hyssuridae.

Pereiopod 3 (Figs 8o-q): Pereiopod 3 is most often similar to pereiopod 2 but in isolated genera differs (e.g. *Xenanthura*, *Belura*, *Psittanthura*).

Pereiopods 4-7 (Fig. 9): Barnard (1925) made much of the shape of the carpus, noting

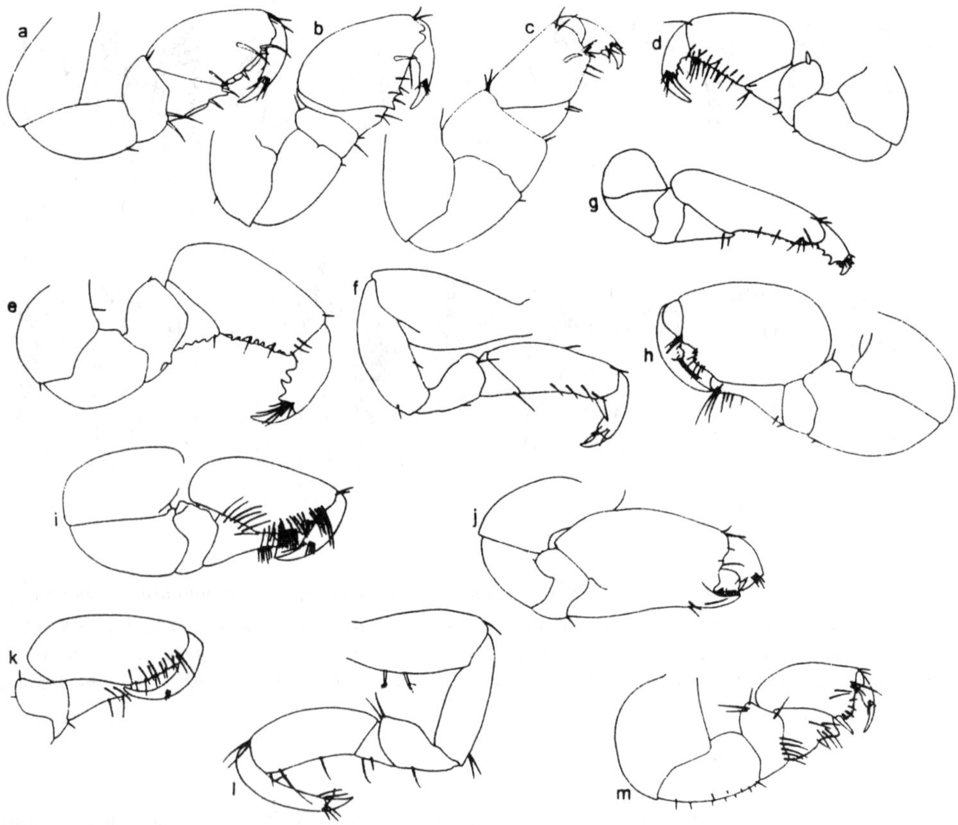

Figure 6. Pereiopod 1. a, *Kupellonura*; b, *Expanathura*; c, *Heptanthura*; d, *Coralanthura*; e, *Rhiganthura*; f, *Eisothistos*; g, *Nemanthura*; h, *Mesanthura*; i, *Apanthura*; j, *Chelanthura*; k, *Haliophasma*; l, *Cetanthura*; m, *Caenanthura*.

it varied between two extremes, cylindrical and about as long as the propodus, and very short, triangular and 'under-riding the propodus.' Intermediate conditions do occur but genera tend to be consistent. An exceptionally large carpus is a characteristic of *Hyssura* and *Quantanthura*. More informative is the distribution of robust setae in a row along the posterior margins of carpus and propodus. A posterodistal robust seta (erroneously called a spine by many authors) is usual in Anthuridea. The disposition of others along the posterior margin is consistent within genus groups. In Antheluridae, Paranthuridae and Leptanthuridae robust setae are numerous while in Anthuridae only a single posterodistal one is seen (except *Quantanthura* with an extra one on the posterior margin of each article) and *Cyathura* (with the carpal one absent). A critical character often previously illustrated but never commented on is the anterodistal transverse row of serrate setae on the propodus of pereiopod 7 of Antheluridae and Expanathuridae. Robust setae on pereiopods are usually cylindrical and flagellate but a unique serrate form is seen in *Heptanthura*.

Pleon (Fig. 11): Pleonites 1-5 are plesiomorphically free, articulating and each almost as long as wide in Hyssuridae. In all other families the pleonites are shortened and the whole pleon at most twice as long as wide and often much shorter. Pleonites are fre-

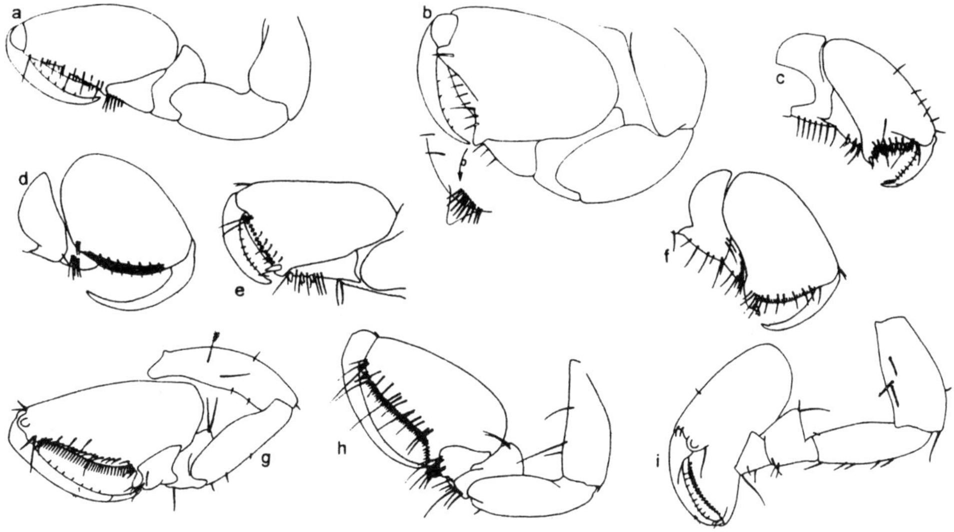

Figure 7. Pereiopod 1. a, *Paranthura*; b, *Colanthura* (with mesial setal row); c, *Bullowanthura*; d, *Pseudanthura*; e, *Ulakanthura*; f, *Aenigmathura*; g, *Cruregens*; h, *Accalathura*; i, *Psittanthura*.

quently fused into a single unit such that only their lateral epimera are visible. In several cases pleonites are fused while cuticular indications of sutures remain dorsally. Authors have reported this sort of pleonite condition as being free (implying articulating) thereby not differentiating the truly articulating condition (Hyssuridae, Leptanthuridae, most Paranthuridae) from fused immobile pleonites showing only a cuticular indication of the individual pleonites. *Apanthura* and *Amakusanthura* are typical anthurid genera with a fused pleon that sometimes show such a pattern. The relative lengths of the pleonites are important within the Expanathuridae. In this family pleonites 5 and 6 are longer than others and fringed with marginal plumose setae.

Pleopods (Fig. 10): In hyssurids all five pairs of pleopods are similar equi-biramous flabelliferan limbs. In all other families pleopod 1 becomes more or less operculiform. In most genera the exopod of pleopod 1 encloses the much smaller endopod and all other pleopods. The extreme condition is seen in highly indurate genera such as *Haliophasma*. The situation is intermediate in most members of Expanathuridae. Here, pleopod 1 does not overlap the others and both exopod and a slightly smaller endopod contribute to the operculum. None of the pleopods reach far along the telson. The exopod is not as indurated as in other families. In the unusual genus *Eisothistos* the two rami of pleopod 1 are fused in the juvenile and pre-spawning female.

Uropod (Figs 10n, o, 11): The uropodal peduncle, triangular in cross-section, lying lateral to the telson and with distal endopod and proximally articulating exopod, is a synapomorphy of the Anthuridea. There appears a tendency for the peduncle to be shorter and the endopod longer in Hyssuridae but this could not be quantified and the character was not used. The exopod ranges from a narrow spike to a broad leaf. Special modifications in armature occur in Expanathuridae particularly.

Pleotelson and statocyst (Fig. 11): As in all isopods pleonite 6 is fused to the telson and the two are relatively immovable. This is in spite of the observation of some authors that

Figure 8. Pereiopod 2. a, *Kupellonura*; b, *Belura*; c, *Xenanthura*; d, *Expanathura*; e, *Rhiganthura*; f, *Heptanthura*; g, *Coralanthura*; h, *Mesanthura*; i, *Notanthura*; j, *Haliophasma*; k, *Curassanthura*; l, *Paranthura*; m, *Bullowanthura*; n, *Ulakanthura*. Pereiopod 3. o, *Xenanthura*; p, *Belura*; q, *Psittanthura*.

they are 'free.' The transverse ridge marking the boundary does not indicate an articulation. In a few genera the ridge is obsolete or absent (e.g. *Coralanthura*, *Pseudanthura*). The evolutionary scenario suggested by the new cladistic analysis proposes that the statocyst is not an apomorphy of the Anthuridea. It was derived at a later stage and is a synapomorphy of the non-hyssurid families. The hypothesis that a basal pair of statocysts evolved into a single medial statocyst that became subsequently lost is well supported (Wägele 1981d). Within these families the change from a pair of statocysts to a single one probably occurred once (Antheluridae + Expanathuridae + Leptanthuridae + Paranthuridae clade). Complete loss occurred twice, in the second and fourth of these families. An alternative scenario is that paired statocysts arose independently in Anthuridae from a single statocyst in Leptanthuridae and Antheluridae. This sequence of events has three evolutionary steps while the first hypothesis has at least four but it conflicts with other character changes and seems less probable. The statocyst is treated as a four-state ordered character in the phylogenetic analysis (Table 3).

The shape of the pleotelson is often consistent within genera.

Figure 9. Pereiopod 7. a, *Hyssura*; b, *Kupellonura*; c, *Xenanthura*; d, *Anthelura*; e, *Expanathura*; f, *Rhiganthura*; g, *Quantanthura*; h, *Cyathura*; i, *Haliophasma*; j, *Pendanthura*; k, *Apanthura*; l, *Accalathura*; m, *Ulakanthura*; n, *Paranthura*; o, *Leptanthura*.

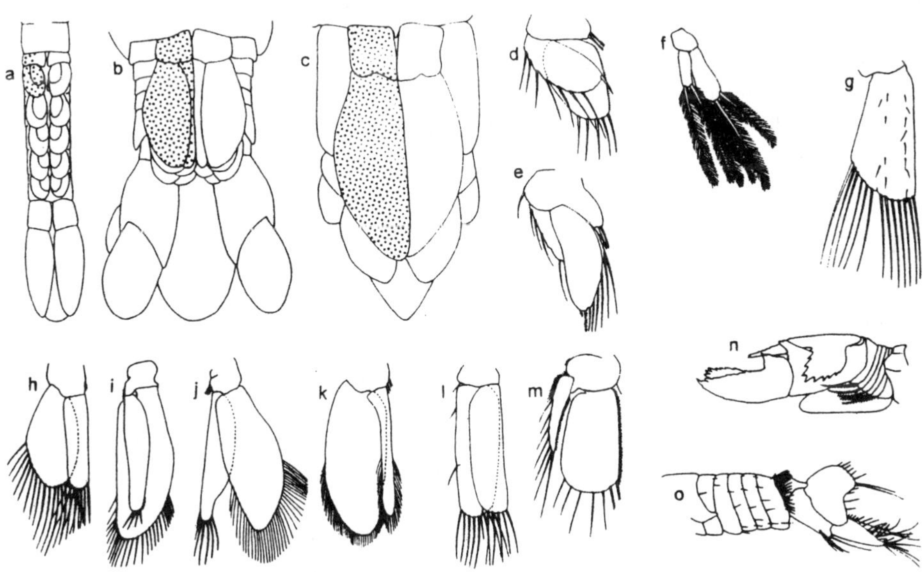

Figure 10. Ventral view of pleon (right pleopod 1 shaded). a, *Kupellonura*; b, *Expanathura*; c, *Haliophasma*. Pleopod 1. d, *Kupellonura*; e, *Belura*; f, *Galziniella*; g, *Eisothistos*; h, *Panathura*; i, *Haliophasma*; j, *Notanthura*; k, *Paranthura*; l, *Ananthura*; m, *Eisothistos*. Lateral view of pleon. n, *Eisothistos*; o, *Expanathura*.

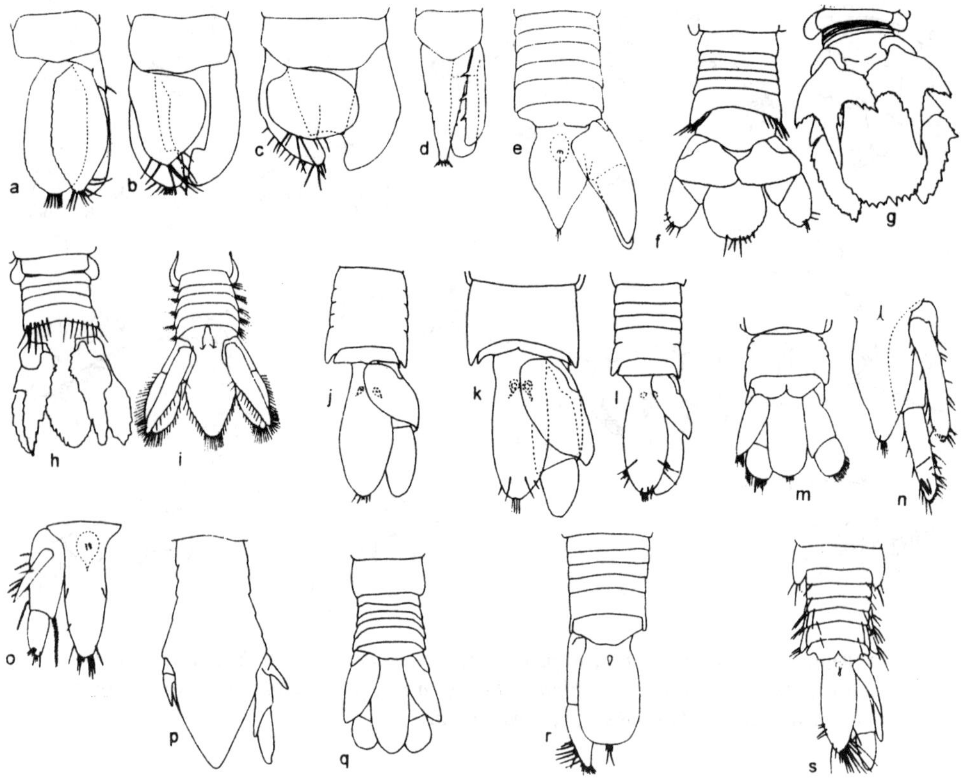

Figure 11. Pleon, telson and uropod(s). a, *Kupellonura* (pleonite 6 and tailfan only); b, *Belura* (pleonite 6 and tail fan only); c, *Xenanthura* (pleonite 6 and tailfan only); d, *Neohyssura* (pleonite 6 and tailfan only); e, *Ananthura*; f, *Expanathura*; g, *Eisothistos*; h, *Heptanthura*; i, *Coralanthura*; j, *Quantanthura*; k, *Cyathura*; l, *Amakusanthura*; m, *Califanthura*; n, *Cruregens* (tailfan only); o, *Bourbonanthura*; p, *Pseudanthura*; q, *Paranthura*; r, *Bullowanthura*; s, *Accalathura*.

3 PHYLOGENY

3.1 *Families*

Analysis of the relationships between families used the six taxa defined in the separate analyses of the four genus-groups. Using 17 characters of which 6 are autapomorphies and only one, the statocyst, was ordered (Tables 2, 3) one tree of length 24 steps was found (Table 1). The Hyssuridae are the sister taxon to all other families and Anthuridae sister taxon to the remainder. Antheluridae and Expanathuridae are sister taxa as are Leptanthuridae and Paranthuridae (Fig. 12). This fully resolved tree was found only if the statocyst (character 17) was treated as an ordered character: absent > paired statocysts > single statocyst > statocyst secondarily lost. If the first and last of these four states were treated as equivalent and the character was unordered the relationships of the three clades, Anthuridae, Antheluridae + Expanathuridae, and Leptanthuridae + Paranthuridae, could not be resolved.

Table 1. Numbers of taxa (excluding hypothetical ancestor) and characters used for cladistic analysis and results of unweighted and self-weighted analysis using HENNIG86 1.5. Weighted analysis of Anthuridea and Hyssuridae did not give greater resolution than unweighted analysis.

Taxon	Taxa	Characters		Unweighted analysis				Weighted analysis			
		Total	Autapomorphies	Trees	Length	CI	RI	Trees	Length	CI	RI
Anthuridea	6	17	6	1	24	95	90	–	–	–	–
Hyssuridae	6	27	11	1	35	88	80	–	–	–	–
Anthuridae	24	38	13	59	89	57	65	25	94	84	82
Antheluridae + Expanathuridae	10	46	15	2	77	71	63	1	77	89	86
Paranthuridae + Leptanthuridae	17	32	6	15	81	58	66	7	81	74	80

Table 2. Anthuridean families. Characters and character states used in cladistic analysis of families. Each character is terminated by a colon and states separated by semicolon, most plesiomorphic condition first.

1. Body: 10-15 times as long as wide, non-males rarely more elongate (pereionite 7 wider than long, much shorter than pereionite 6); slender (pereionite 7 longer than wide, little or not shorter than pereionite 6)
2. Pleonites 1-5: together 3 times as long as wide; not more than twice as long as wide
3. Pleonites 1-5: free and articulating; fused
4. Pleonites: without marginal plumose setae on pleonal epimera or posterior borders of pleonites 4 and 5; with plumose setae on margins of epimera or posterior margins of pleonite 5 and sometimes pleonite 4
5. Antenna 2 flagellum: of fewer than 10 articles, shorter than peduncle; rudimentary (fewer than 5 fused articles), shorter than last article of peduncle; of numerous articles, longer than or shorter than peduncle
6. Mouth parts: not produced anteriorly, mandible with compact and with weakly-toothed transverse incisor; acutely produced anteriorly, mandible with acute anteriorly-directed incisor
7. Maxillipedal endite: reaching palp article 3: absent or obsolete
8. Maxillipedal palp: broad (c. twice as long as wide); narrow and tapering distally to rounded apex (c. 3 times as long as wide); narrow, tapering to acute apex, (c. 3 times as long as wide)
9. Maxillipedal palp: with 5 free articles; with 2 or more articles fused
10. Pereiopod 1 palm: without proximal tooth (palm more or less continuous with that of carpus); defined proximally by angle or proximal tooth
11. Pereiopod 1 palm: marginal setal row irregular if present; with even marginal row of simple or complex robust setae along length
12. Pereiopods 2 and 3 carpus: not or weakly produced posterodistally; strongly produced posterodistally
13. Pereiopods 2 and 3 propodus: with more than 1 robust setae in addition to posterodistal robust seta; with 1 robust seta in addition to posterodistal robust seta; with posterodistal robust seta only (or absent)
14. Pereiopods 4-7 propodus: with more than 1 robust setae in addition to posterodistal robust seta; with posterodistal robust seta only
15. Pereiopod 7 propodus: without anterodistal serrate setae; with anterodistal serrate setae
16. Pleopods: 1-5 similar, none operculiform; pleopod 1 exopod and endopod together weakly operculiform; pleopod 1 exopod operculiform alone
17. Statocyst(s): absent; paired; single; secondarily lost (ordered)

Table 3. Character matrices used in cladistic analyses of families of Anthuridea.

Character number	 12345	1 67890	11111 12345	11 67
Hyssuridae	10000	00000	01000	00
Expanathuridae	01010	00100	00101	13
Antheluridae	01012	00000	00101	22
Anthuridae	01100	00000	00210	21
Paranthuridae	01001	11210	00000	23
Leptanthuridae	01000	10201	10000	22

In the cladogram shown, apomorphies of the Hyssuridae are characters 1 and 12 (Table 4). The body is exceptionally slender, pereionite 7 longer than wide, and the carpus of pereiopods 2 and 3 strongly produced distally. Apomorphies of the sister taxon are characters 2, 16 and 17. Pleonites are shortened and pleopod 1 is operculiform. An elongated pleon is also seen in males of many non-hyssurid genera, a feature that may reflect hyssurid plesiomorphies. Paired statocysts are ancestral in this clade. The most parsimonious analysis suggests that pleopods with broad operculiform exopods and reduced endopods is ancestral and the condition where both rami contribute, as in Expanathuridae, is secondarily derived.

The Anthuridae have three autapomorphies, characters 3, 13 and 14 (Table 4). Pleonites 1-5 are universally fused, a condition seen in some genera in other families. The propodus of pereiopods 2 and 3 has a posterodistal robust seta but never other robust setae on the posterior margin, and in pereiopods 4-7 there is never more than one seta besides the posterodistal one. A pair of statocysts is a retained plesiomorphy.

Clade 9 (including the remaining four families) possesses the synapomorphy of a single statocyst but only two of the four families retain this state.

The sister taxa Antheluridae + Expanathuridae share three unique characters, 4 13 and 15 (Table 4). The posterior borders of at least pleonite 5 and sometimes pleonite 4 are fringed with plumose setae. The propodus of pereiopods 2 and 3 has a posterodistal robust seta and one other robust seta on the posterior margin, a more plesiomorphic condition than in Anthuridae. The propodus of pereiopod 7 has a transverse row of anterodistal setae at the base of the dactylus. Each family has its own apomorphies. Expanathuridae share a narrow tapering maxillipedal palp, weakly developed pleopodal operculum, and have lost the statocyst. Antheluridae have numerous articles in the flagellum of antenna 2 and retain the plesiomorphic single statocyst. The latter condition is also seen in the Leptanthuridae which differ in the piercing mouth parts and other characters.

The sister taxa Leptanthuridae + Paranthuridae (former Paranthuridae) share two synapomorphies: 6 and 8 (Table 4). The acutely produced mouth parts, with acute mandibular incisor and strongly tapering maxillipedal palp, are unique. Leptanthuridae retain a ple-

Table 4. Synapomorphies of families and supporting clades of Anthuridea. Character changes at clades shown in Figure 12 are from state 0 to state 1 unless marked otherwise in brackets (viz., – indicates a reversal from 1 to 0, > a change from a non-zero state to another). Character numbers refer to those in Table 2.

Clade	Taxon	Character change
1.	Hyssuridae	1, 12
2.	Anthuridae	3, 13(2), 14
3.	Expanathuridae	8, 16(2>1), 17(2>3)
4.	Antheluridae	5(2)
5.	Paranthuridae	5, 7, 9, 17(2>3)
6.	Leptanthuridae	10, 11
7.	1-6 (all families)	–
8.	2-6 (non-hyssurid families)	2, 16(2), 17
9.	3-6 (4 families)	17(2)
10.	3-4 (Expanathuridae + Antheluridae)	4, 13, 15
11.	5-6 (Paranthuridae + Leptanthuridae)	6, 8(2)

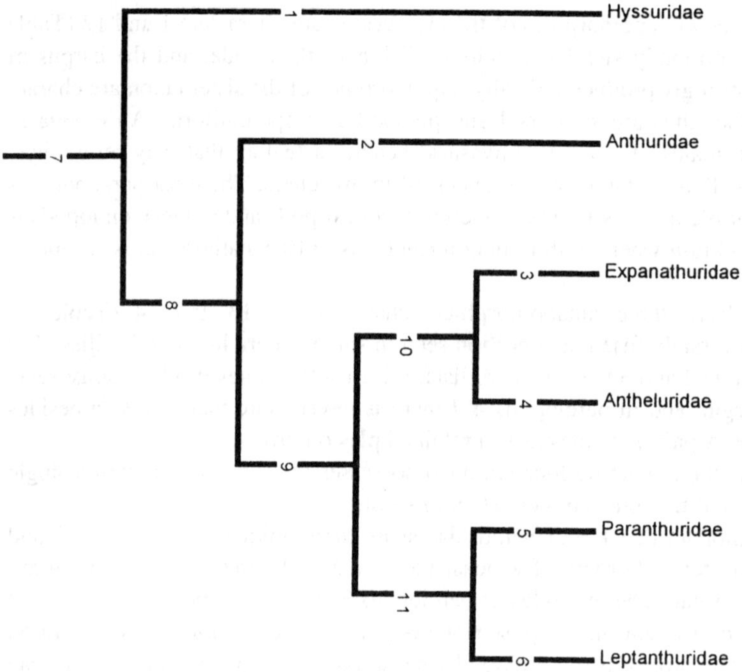

Figure 12. Cladogram of the families of Anthuridea generated by Hennig86 1.5. Characters are listed in Table 2, the data matrix is in Table 3, tree statistics in Table 1. Taxa and clade numbers are those used in Table 4.

siomorphic single statocyst, a condition apparently convergent with that in Antheluridae.Their apomorphies are the palm of pereiopod 1 being defined proximally by an angle or tooth stepped away from the palm of the carpus and a regular row of simple or complex robust setae along the margin. Paranthuridae share four synapomorphies: rudimentary flagellum on antenna 2; absence of the maxillipedal endite; some fusion of maxillipedal palp articles; and loss of the statocyst.

3.2 *Hyssuridae*

Relationships between the six genera of the Hyssuridae were based on 27 characters of which 11 are autapomorphies (Tables 5, 6). A single tree of length 35 steps was found (Fig. 13; Table 1). All genera were supported by 2-5 apomorphies and their relationships clearly defined. Two genera, *Belura* and *Xenanthura*, belong to a clade separated from the rest. This clade 8 has nine synapomorphies (Table 7). The flagellum of antenna 1 is twice as long as wide, the mandibular palp uniarticulate, maxillipedal palp articles 1 and 2 fused, propodus of pereiopods 4-7 without palmar setae, dactylus of pereiopod 7 very short, pleopodal endopods with only one apical seta, uropodal endopod fused to basis and exopod broad, and telson short and excavate apically. Their sister taxa on the other hand (clade 9) share an acutely produced smooth carpus on pereiopods 2 and 3. Refer to Figure 13 and Table 7 for synapomorphies of the remaining clades and genera.

Table 5. Family Hyssuridae. Characters and character states used in cladistic analysis. Each character is terminated by a colon and states separated by semicolon, most plesiomorphic condition first.

1. Eyes: present; absent
2. Antenna 1 flagellum, proximal article: shorter than wide; twice as long as wide
3. Antenna 2 flagellum: of >6 articles, first shorter than others together; of 5 articles, first longer than remaining 4
4. Mandibular molars: articulating cirolanid-type blade; spike-like; blunt; obsolete, with long seta
5. Mandibular palp: of 3 articles; of 1 article with 1 seta
6. Maxilla 1: present; absent
7. Maxillipedal endite: short, triangular, non-indurate lobe; absent
8. Maxillipedal palp: of 5 free articles; articles 1 and 2 fused, 3-5 free; all articles fused
9. Pereiopod 2 carpus: a bluntly produced blade, weakly-strongly crenellate; acutely produced, smooth
10. Pereiopod 2 propodus, palm: axial or slightly oblique; transverse
11. Pereiopod 2 propodus: with robust flagellate setae along cutting edge; without robust setae on cutting edge
12. Pereiopod 3 merus and carpus: not ornamented; with tufts of long marginal setae
13. Pereiopod 3 carpus: not produced; acutely produced posterodistally, different from pereiopod 2
14. Pereiopod 3 carpus: not produced; carpus bluntly produced posterodistally, as in pereiopod 2
15. Pereiopods 4-7 propodus: with at least 3 robust setae along cutting edge; with 1 distal robust seta on cutting edge; without setae on cutting edge
16. Pereiopods 4-7 propodus, transverse row of robust pectinate setae along distal margin: absent; present
17. Pereiopod 7 carpus: triangular, with 1 marginal robust or simple seta; quadrangular, wider than propodus, with 2 or more pectinate setae along free distal margin
18. Pereiopod 7 dactylus: narrow, almost as long as propodus; stout, <two-thirds length of propodus
19. Pleopodal endopods: with several marginal setae; with single apical seta
20. Pleopod 1 endopod: longer and broader than exopod, marginal setae shorter than endopod; shorter and narrower than exopod, marginal setae longer than endopod
21. Uropodal endopod: free; fused to peduncle
22. Uropodal exopod: a narrow leaf articulating proximally; with lateral lobe; ovoid
23. Uropodal exopod mesial margin and lateral margins of telson: unarmed; with cuticular spines
24. Uropodal rami and telson: with entire margins; with dentate margins
25. Telson: ovoid, with rounded apex; with medial projection on excavate apex
26. Telson: ovoid, with rounded apex; triangular
27. Telson: ovoid, with rounded apex; short, excavate apically (with or without medial projection)

Table 6. Character matrices used in cladistic analyses of genera of Hyssuridae.

Character number		1	11111	11112	22222	22
	12345	67890	12345	67890	12345	67
Kupellonura	00020	00010	00011	00000	01010	00
Hyssura	10010	00010	00011	11000	00010	00
Galziniella	10030	11010	10010	00001	00000	00
Neohyssura	00010	00010	00011	00000	00100	10
Xenanthura	01101	00101	11002	10110	12000	01
Belura	11001	00100	10102	00110	12001	01

Table 7. Synapomorphies of genera and supporting clades of Hyssuridae. Character changes at clades shown in Figure 13 are from state 0 to state 1 unless marked otherwise in brackets (viz., – indicates a reversal, > a change from a non-zero state to another). Character numbers refer to those in Table 5.

Clade	Taxon	Character change
1.	*Xenanthura*	3, 10, 12, 16
2.	*Belura*	1, 13, 25
3.	*Galziniella*	1, 4(3), 6, 7, 20
4.	*Neohyssura*	23, 26
5.	*Kupellonura*	4(1 > 2), 22
6.	*Hyssura*	1, 16, 17
7.	1-6 (family)	11
8.	1-2	2, 5, 8, 15(2), 18, 19, 21, 22(2), 27
9.	3-6	9, 14
10.	4-6	4, –11, 15
11.	5-6	24

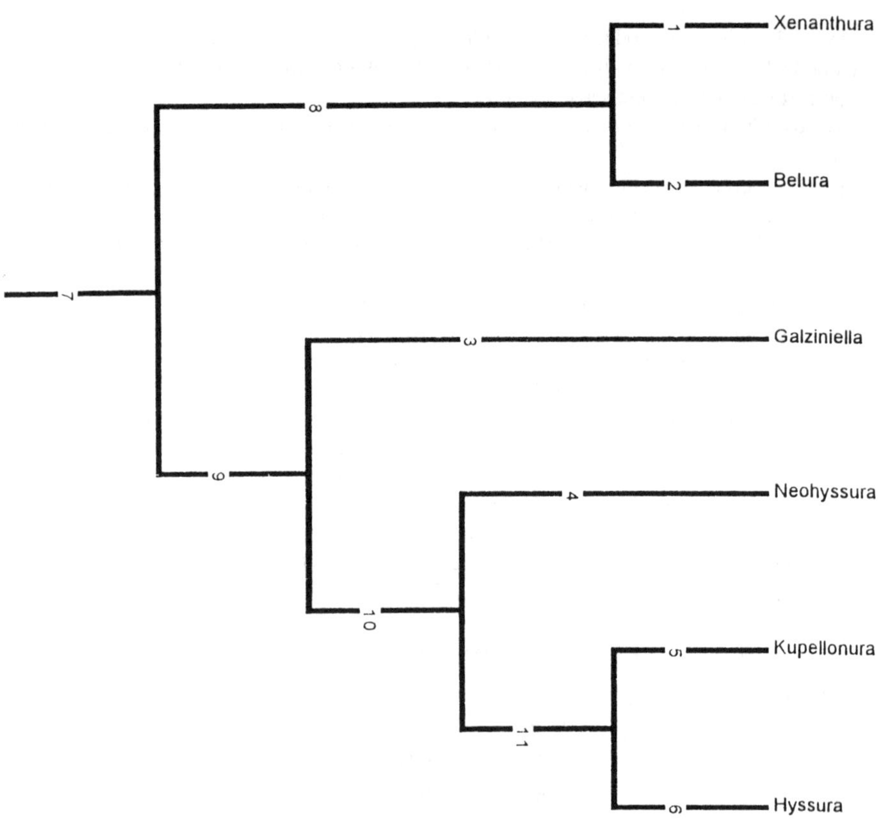

Figure 13. Cladogram of the genera of Hyssuridae generated by Hennig86 1.5. Characters are listed in Table 5, the data matrix is in Table 6, tree statistics in Table 1. Taxa and clade numbers are those used in Table 7.

3.3 *Anthuridae*

Relationships between the 24 genera of the Anthuridae were based on 38 characters of which 13 are autapomorphies (Tables 8-10). Three- and four-state characters 11, 13, 18, 19 and 33 were ordered. The low consistency and retention indices of the unweighted analysis (966 equal-length trees) were improved by automatic weighting (25 trees). Some genera and well-defined clusters in the resulting consensus tree are discussed (Fig. 14; Table 10).

Quantanthura is the primitive sister taxon to all other anthurid genera. It retains several plesiomorphies including a 5-articled maxillipedal palp and marginal robust setae on the posterior margins of the carpus and propodus of pereiopods 4-7. Clade 26 includes all other genera and is defined by four synapomorphies: reduced antenna 1 flagellum, some maxillipedal palp fusion, and no robust setae on the posterior margins of the carpus and propodus of pereiopods 4-7.

Indanthura, Malacanthura and *Pilosanthura* lack the synapomorphies of most anthurid genera. *Malacanthura* is problematic in that it lacks autapomorphies (see discussion in taxonomic section). Clade 27 links three genera (*Haliophasma, Cetanthura* and *Apanthuroides*) with two synapomorphies: pitted body and absence of maxillipedal endite (in two of three genera).

Most anthurid genera are supported by clade 29 the synapomorphies of which are absence of maxillipedal endite and fusion of the maxillipedal palp into two sections (characters 18 and 19). Reversals of both characters are seen in this scenario: return of the endite in three clades and lesser fusion of the palp in clade 38.

Anthura, Ptilanthura and *Exallanthura* that share a completely fused maxillipedal palp are grouped in clade 30. Clade 32, which unites more than half of the genera of the Anthuridae, is characterised by an antenna 2 flagellum with 3-4 articles, shorter than the peduncle and a rectangular carpus on pereiopods 4-7. *Notanthura* and *Cortezura* (clade 34) share a unique triangular mandibular molar, unique setose pereiopods 2-7, and rectangular carpus on pereiopods 4-7 (characters 16, 27 and 29).

Table 8. Family Anthuridae. Characters and character states used in cladistic analysis. Each character is terminated by a colon and states separated by semicolon, most plesiomorphic condition first.

1. Body: of usual proportions; very slender
2. Body total length: more than 4 mm; < 3 mm
3. Body: not pigmented; irregularly pigmented; with dark specific pigment pattern
4. Integument: smooth; covered at least in part with setules (fur)
5. Body: smooth; pitted
6. Head: more or less parallel-sided in both sexes; much wider anteriorly in male
7. Head: without produced chin; with produced chin in males
8. Pleonites 1-5: as long as greatest width; longer than greatest width; shorter than wide; much shorter than wide
9. Pleonites 1-5: fused, sutures visible only laterally; fused (but separated by folds dorsally and laterally except dorsally between 4 and 5); sutures all indicated
10. Pleotelson: with posterior margin of pleonite 6 indicated dorsally, delineated from telson; without indication of posterior margin of pleonite 6
11. Antenna 1 flagellum of male: of 1 basal + c. 20 aesthetasc-bearing articles; of 1 basal + 4-10 aesthetasc-bearing articles; of 1 basal + 1 or 2 short aesthetasc-bearing articles (ordered)

Table 8. Continued.

12.	Antenna 1 flagellum: of c. 7 articles; of 3-4 articles (second longest)
13.	Antenna 2 flagellum: of 4 or 5 articles, longer than article 5 of peduncle; of 3-4 articles, shorter than article 5 of peduncle; of 1-2 articles, very short (ordered)
14.	Eyes: present; absent
15.	Mandibular incisor: directed transversely (exceeded by palp); produced anteriorly (beyond palp)
16.	Mandibular molars: blunt; asymmetrical, left molar with tooth fitting socket of right molar; spike-like on right side, absent on left; triangular blade
17.	Mandibular palp: of 3 articles; of 2 articles, second very setose; of 1-2 short articles; 1 minute article
18.	Maxillipedal endite: present, reaching palp article 3; weak triangular lobe; obsolete or absent (ordered)
19.	Maxillipedal palp articles: 1-5 free; 1-2 fused, 3 free and 4-5 fused or 3-5 fused; 1-3 fused, 4-5 fused; 1-5 fused (ordered)
20.	Maxillipedal palp with articles 4-5: with terminal article (4-5) oblique, without a free mesial margin between its suture and distal group of setae; at least one-third as long as penultimate article (3), almost as long as broad, with a free mesial margin between its suture and distal group of setae; subcircular, transverse
21.	Maxillipedal palp with article 3: barely produced medially beyond articles 4-5 (if at all); produced as lobe beyond distolateral articles 4-5
22.	Pereiopod 1: subchelate, dactylus and terminal unguis closing on axial palm, carpus and propodus separated by suture; chelate, propodus produced posterodistally, dactylus with complexly ridged distal margin, unguis subterminal, carpus and propodus fused
23.	Pereiopod 1 propodus: swollen, subchelate, overlapping merus; cylindrical, not in contact with merus
24.	Pereiopod 1 propodus palm: straight or sinuous, without prominent tooth; with step or prominent tooth (more pronounced in male)
25.	Pereiopod 1 dactylus: smooth, closing on palm; with tubercular teeth along lower margin
26.	Pereiopod 1 palm: without strong proximal seta; with strong proximal seta (about as long as unguis)
27.	Pereiopods 2-7 merus-propodus: bearing few short setae, most on lower margin (merus-carpus posteriorly lobed and setose in some); bearing long setae on upper and lower margins
28.	Pereiopods 2, 3 carpus: without robust seta posterodistally; with robust seta posterodistally
29.	Pereiopods 4-7 with carpus: more or less pentagonal, anterior margin half length of posterior margin; more or less rectangular, anterior (upper) margin nearly as long as posterior (lower) margin, distal margin transverse and without posterodistal lobe
30.	Pereiopods 4-7 carpus: with free anterior margin; triangular, without free anterior margin
31.	Pereiopods 4-7 with carpus: little wider than propodus, without extensive free distal margin; twice width of propodus, rectangular, with obvious free distal margin (usually with row of pectinate setae) defined by stout seta on posterodistal angle
32.	Pereiopods 4-7 carpus: shorter than propodus; longer and wider than propodus
33.	Pereiopods 4-7 carpus: with robust seta on posterior margin and on posterodistal angle; with robust setae on posterodistal angle only; without robust setae on posterior margin or on posterodistal angle (ordered)
34.	Pereiopods 4-7 propodus: with 1 robust setae on posterior margin (in addition to posterodistal robust seta); with 1 posterodistal robust seta only
35.	Pereiopod 7: present on adults; absent on adults
36.	Pleopod 1 endopod: apically rounded; tapering to digitiform apex
37.	Uropodal exopod: leaf-shaped, articulating along peduncle; linear, articulating transversely
38.	Uropodal exopod: leaf-shaped, articulating along peduncle; apically excavate

Table 9. Character matrices used in cladistic analyses of genera of Anthuridae.

Character number		1	11111	11112	22222	22223	33333	333
	12345	67890	12345	67890	12345	67890	12345	678
Quantanthura	00000	00000	10000	00000	00000	00010	10000	000
Indanthura	00000	00000	?0000	00211	00000	00010	00110	000
Malacanthura	00000	00000	01000	00010	00000	00010	00110	000
Haliophasma	00001	00000	01100	00210	00000	00010	00110	000
Nemanthura	10000	00000	?1000	00220	00001	00010	00110	000
Cetanthura	00001	00000	11001	?0210	00100	00000	00111	000
Caenanthura	00000	00000	01200	02220	10000	10000	00110	000
Anthura	00100	00000	01000	00230	00000	00010	00110	000
Ptilanthura	00000	00000	11100	03230	00000	00010	00110	000
Exallanthura	00000	00000	01000	03230	00100	00010	00111	001
Pilosanthura	00010	00020	?1010	00010	00000	00010	00110	000
Apanthuropsis	00000	00000	01100	10210	00010	00000	00110	000
Amakusanthura	00000	01010	01100	00110	00010	00000	00110	000
Apanthuroides	00001	00001	11001	20012	00100	00000	00110	000
Apanthura	00000	01000	11100	00110	00010	00000	00110	000
Skuphonura	00000	11000	21100	00210	00010	00000	00110	000
Mesanthura	00200	00000	11100	00211	00010	00000	00110	000
Chelanthura	00200	00000	11100	00211	01010	00000	00110	000
Cyathura	00110	00100	11100	00221	00010	00000	00210	000
Stygocyathura	00010	00100	11210	00221	00010	00000	00210	010
Pendanthura	01110	00201	11200	03130	00000	00001	00210	000
Sauranthura	01100	00100	?1200	00130	00010	00100	00210	000
Notanthura	00000	00000	11110	30220	00010	01010	01110	100
Cortezura	00000	00000	11200	31120	00010	01010	00110	000

Table 10. Synapomorphies of genera and supporting clades of Anthuridae. Character changes at clades shown in Figure 14 are from state 0 to state 1 unless marked otherwise in brackets (viz., – indicates a reversal, > a change from a non-zero state to another, – no synapomorphies). Character numbers refer to those in Table 8.

Clade	Taxon	Character change
1.	*Quantanthura*	11, 31
2.	*Indanthura*	–12, 18(2), 20
3.	*Malacanthura*	–
4.	*Pilosanthura*	4, 9(2), 14
5.	*Haliophasma*	13
6.	*Cetanthura*	35
7.	*Apanthuroides*	10, 16(2), 18(–2), 20(2)
8.	*Nemanthura*	1, 25
9.	*Anthura*	3
10.	*Ptilanthura*	11, 13
11.	*Exallanthura*	23, 35, 38
12.	*Caenanthura*	13(1>2), 17(2), 21, 26
13.	*Notanthura*	14, 32, 36
14.	*Cortezura*	13(1>2), 17, 18(2>1)
15.	*Stygocyathura*	14, 20, 37

Table 10. Continued.

Clade	Taxon	Character change
16.	*Cyathura*	13(2>1), 20
17.	*Pendanthura*	8(1>2), 16(3), N̄ 24, 30
18.	*Sauranthura*	–4, –11, 28
19.	*Apanthuropsis*	–11, 16
20.	*Mesanthura*	–
21.	*Chelanthura*	22
22.	*Skuphonura*	6, 11(1>2)
23.	*Amakusanthura*	9, –11
24.	*Apanthura*	–
25.	1-24 (Anthuridae)	29
26.	2-24	12, 19, 33, 34
27.	5-7	5, 18(2)
28.	6-7	11, 15, 23, -29
29.	8-24	18(2), 19(1>2)
30.	9-11	19(2>3)
31.	10-11	17(3)
32.	12-24	13, –29
33.	13-24	–7, 11, 24
34.	13-14	15(3), 27, 29
35.	15-18	4, 13(1>2), 33(1>2)
36.	16-18	3
37.	17-18	2, 18(2>1), 19(2>3)
38.	19-24	19(2>1)
39.	20-21	3(2), 20
40.	22-24	7
41.	23-24	18(2>1)

The four genera of clade 35 (four synapomorphies) uniquely possess a shorted pleon, especially so in *Pendanthura*. With the exception of *Sauranthura* all are moderately pilose. The flagellum of antenna 2 is exceptionally short (except in hypogean *Stygocyathura*). Moreover, the four genera are the only ones to lack any robust setae on the posterior margin of the carpus of pereiopods 4-7. There is no synapomorphy to support the division of *Cyathura* into two subgenera, *Cyathura* and *Stygocyathura*. The latter is sister taxon to the other three of this group which share some pigmentation.

The *Apanthura*-group of six genera (clade 38) shares a 3-segmented maxillipedal palp (character 19) appearing as a reversal in this analysis. Clade 39 (*Mesanthura* and *Chelanthura*) has two synapomorphies: specific pigment patterns and enlarged terminal segment of the maxillipedal palp. *Mesanthura* lacks an autapomorphy. Genera of clade 40 possess a produced chin in males.

Refer to Figure 14 and Table 10 for synapomorphies of the remaining clades and genera.

3.4 *Antheluridae and Expanathuridae*

Relationships between the 10 genera of the Antheluridae and Expanathuridae were based

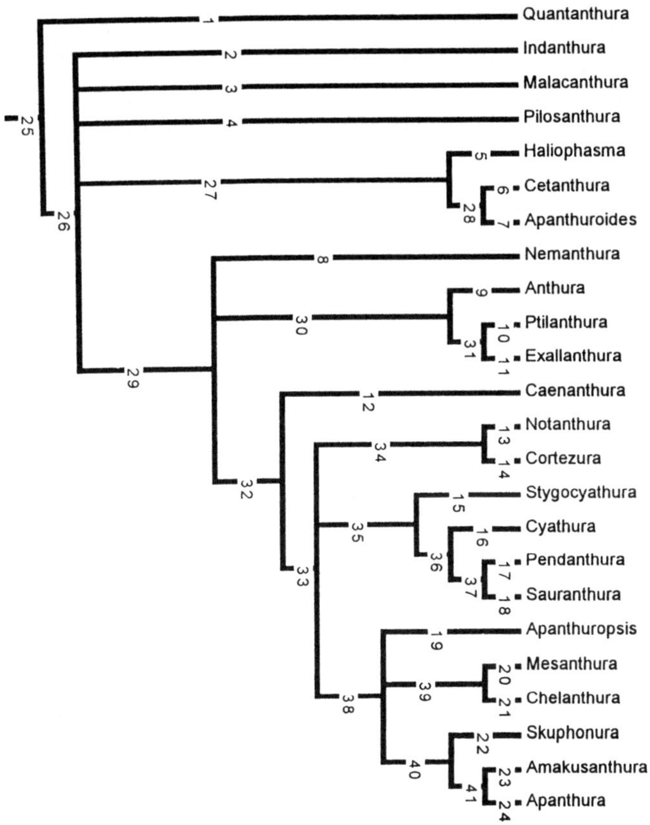

Figure 14. Cladogram of the genera of Anthuridae generated by Hennig86 1.5. Characters are listed in Table 8, the data matrix is in Table 9, tree statistics in Table 1. Taxa and clade numbers are those used in Table 10.

on 46 characters of which 15 are autapomorphies (Tables 11-13). No character was ordered. The consistency and retention indices of the unweighted analysis were improved by automatic weighting, which also resolved the only trichotomy of the unweighted analysis (Table 1). Both supported primary division of the taxa into two clades that are here assigned family rank.

All genera have autapomorphies and in the case of some, many. *Eisothistos* has 18 of which eight are unique (Table 13). Clade 11 which unites the two families has three synapomorphies which are reversed in *Eisothistos* (characters 29 and 31) and *Anthelura* (character 36).

The Antheluridae (clade 12) are united by four synapomorphies: operculiform pleopod 1 with reduced endopod, pleopods 2-5 with hiatus in marginal setal row, and a medial statocyst (characters 34, 36, 38 and 46). The new family Expanathuridae on the other hand has three synapomorphies (characters 28, 31 and 42). The propodus of pereiopods 2 and 3 and the carpus of pereiopods 4-7 have each only one distal robust seta on the cutting edge, and the telson is shortened, rarely more than twice as long.

Table 11. Families Antheluridae and Expanathuridae. Characters and character states used in cladistic analysis. Each character is terminated by a colon and states separated by semicolon, most plesiomorphic condition first.

1. Body: of typical proportions (pereionite 1 longer than wide); female grossly elongate; compact (pereionite 1 wider than long)
2. Body: not pigmented; irregularly pigmented
3. Body setation: body and limbs not covered by scattered long setae; areas of body and bases of some limbs of female covered with scattered long setae
4. Pleonites 1-5: free and articulating; fused
5. Pleonites 1-5: free and articulating; fused; fused in postspawning female
6. Pleonites 5 and 6: not longer than pleonites 1-4; longer than pleonites 1-4
7. Pleonites: pleonites of similar lengths; pleonites 5 and 6 longer than pleonites 1-4; pleonites 1-3 longer than 4 and 5 in male and manca, all of similar lengths in juvenile and female
8. Pleonite 1: as long as pleonite 2; very short, concealed dorsally
9. Pleotelson: with posterior margin of pleonite 6 indicated dorsally, delineated from telson; without indication of posterior margin of pleonite 6
10. Pleonites 1-5: together longer than wide; together wider than long
11. Antenna 1 flagellum of male: of 1 basal article + c. 6-10 aesthetasc-bearing articles; of 1 basal article + c. 20 aesthetasc-bearing articles; of 1 very short basal article bearing numerous aesthetascs and c. 7 longer articles each with 1-2 aesthetascs (or 7 longer articles, second with bunches of aesthetascs)
12. Eyes: present; absent
13. Mouth parts: not produced anteriorly; mandible and maxilla only produced anteriorly
14. Mandible: compact, incisor toothed; linear, incisor a short transverse blade; tapering, incisor simple
15. Mandibular molars: blunt; absent
16. Mandibular palp: of 3 articles; of 1 or 2 articles; of 1 seta; absent
17. Maxillipedal endite: apically rounded; tapering to acute apex (or further reduced)
18. Maxillipedal endite: present; obsolete or absent
19. Maxillipedal palp: broad (1.5-2 times as wide as long), article 4 semicircular; narrow and tapering distally to apex (at least twice as long as wide)
20. Maxillipedal palp: of 5 articles, covering other mouth parts; of 2-5 weakly differentiated narrow articles, much smaller than other mouth parts
21. Maxillipedal palp articles: 1 and 2 separated by suture; 1-2 fused
22. Pereiopods 1-3: differentiated from posterior pereiopods, propodi not 1 swollen; only slightly differentiated from posterior pereiopods, propodus 1 not swollen
23. Pereiopod 1 with mesial setae: in row along length of palm; dominated by 1 much stronger than others
24. Pereiopod 1 merus: narrower than or as wide as swollen propodus; broader than tapering propodus
25. Pereiopod 1 propodus: similar to that of pereiopod 2, cylindrical, palm smooth; as broad as long, triangular, palm dentate proximally
26. Pereiopod 1 palm of dactylus (sometimes carpus and propodus): smooth; tuberculate
27. Pereiopods 1-7 merus: without robust seta on anterior margin; with robust seta on anterior margin
28. Pereiopods 2 and 3 propodi: with 1 distal and 2 more proximal robust setae along cutting edge; with only 1 distal robust seta on cutting edge; without robust setae on cutting edge
29. Pereiopod 7 propodus: without serrate setae mesiodistally; with short transverse row of serrate setae distally (on mesially directed face)

Table 11. (Continued).

30.	Pereiopods 4-7 propodus: with row of robust setae along cutting edge (at least 1 in addition to distal seta); with 1 distal robust seta on cutting edge
31.	Pereiopods 4-7 carpus: with 4 robust setae along cutting edge; with 2 robust setae along cutting edge; with 1 robust seta on cutting edge
32.	Distal palmar robust setae on pereiopods 2-7 propodus and 4-7 carpus: simple or flagellate; bifid and blade-like
33.	Pereiopods 4-7 carpus: about as long as wide, with short anterior margin; longer than wide
34.	Pleopod 1 endopod: as long or longer than exopod and partly exposed; shorter than exopod and totally covered by it
35.	Pleopod 1 rami: similar in both sexes; differentiated between sexes
36.	Pleopod 1: weakly operculiform, with rami fused in juvenile and prespawning female; weakly operculiform in both sexes, with exopod slightly wider than endopod but not enclosing it totally; exopod operculiform alone and enclosing narrower endopod in both sexes
37.	Pleopodal exopods 2-5: little narrower than endopods; much narrower and shorter than endopods in juvenile and female, exopods 1-4 with both rami narrow in male, 5 of male as in female
38.	Pleopods 2-5 exopods: with continuous row of plumose setae around distal and lateral margin; with 1 seta near lateral suture, distinct from other distolateral marginal setae.
39.	Uropodal exopod: a broad leaf articulating longitudinally, with smooth margin; with strong apical spike, with serrate margin; a narrow leaf articulating proximally
40.	Uropodal exopod: without serrate lateral lobe; with serrate lateral lobe between articulation and apical spike
41.	Uropodal endopod and exopod: apices of rami without serrate robust seta; endopod and spike of exopod with apical serrate robust seta
42.	Telson: 2.5 times as long as wide; twice as long as wide
43.	Telson: with smooth distal margin; with dentate distal margin
44.	Telson: without statocyst pore; domed with slit-like pore; flat with small pore
45.	Telson and uropodal rami: apically rounded; tapering to narrow apices
46.	Statocyst: absent; pair present; single and medial

Table 12. Character matrices used in cladistic genera of Antheluridae and Expanathuridae.

Character number		1	11111	11112	22222	22223	33333	33334	44444	4
	12345	67890	12345	67890	12345	67890	12345	67890	12345	6
Anthelura	00000	00000	11000	00000	00000	00010	00110	20100	00021	1
Ananthura	00000	00000	?0000	00000	00000	00010	10110	20100	01001	1
Anthomuda	00000	00000	00000	00000	00000	00010	10010	20100	00010	1
Eisothistos	10101	01001	20111	31111	01000	00101	20001	01011	11100	0
Heptanthura	00010	10000	00020	01110	11110	00111	21000	10010	01100	0
Panathura	00000	10000	00020	00010	00000	00111	20000	10000	01000	0
Minyanthura	01010	10001	00021	21010	00100	00111	20000	10010	01100	0
Expanathura	01000	10000	20021	01010	00101	00210	20010	20000	01000	0
Rhiganthura	00000	10110	00020	10000	10000	10110	20000	10011	01100	0
Coralanthura	01000	00010	00000	00000	00000	11100	20000	10020	01000	0

Table 13. Synapomorphies of genera and supporting clades of Antheluridae and Expanathuridae. Character changes at clades shown in Figure 15 are from state 0 to state 1 unless marked otherwise in brackets (viz., $\bar{N}$ indicates a reversal, > a change from a non-zero state to another). Character numbers refer to those in Table 11.

Clade	Taxon	Character change
1.	*Anthomuda*	44
2.	*Anthelura*	11, 12, –31, 44(2)
3.	*Ananthura*	42
4.	*Rhiganthura*	8, 14(2), 16, 21, 39, 40, 43
5.	*Coralanthura*	27, –29, 39(2)
6.	*Panathura*	
7.	*Expanathura*	11(2), 25, 28(1>2), 34, 36(1>2)
8.	*Minyanthura*	4, 10, 16(2)
9.	*Eisothistos*	3, 5, –6, 7, 10, 11(2), 13, 14(2>1), 16(3), 20, –23, –29, 35, –36, 37, 40, 41
10.	*Heptanthura*	–15, 21, 24, 32
11.	1-10 (Antheluridae + Expanathuridae)	31, 36
12.	1-3 (Antheluridae)	36(1>2), 38, 46
13.	2-3	45
14.	4-10 (Expanathuridae)	31(1>2), 42
15.	4-5	26
16.	6-10	14(2), 19
17.	7-10	17, 23
18.	8-10	39, 43
19.	9-10	22

All the clades within the Expanathuridae are well supported. The primary division is between *Rhiganthura* and *Coralanthura* (clade 15) and the rest (clade 16). Clade 15 is defined by loss of the ridge between pleonite 6 and the telson, and the tuberculate palm of the dactylus of pereiopod 1 (characters 9 and 26). The synapomorphies of clade 16 are a tapering mandible, absence of the mandibular molar (apparently regained in *Heptanthura*), and a narrow and tapering maxillipedal palp (characters 6, 14 and 19).

Eisothistos is anomalous within the Anthuridea and its relationships difficult to determine. I can only conclude that it displays many structural simplifications that obscure its origins and are here apparent as character reversals. Its sister taxon is *Heptanthura* with which it shares two synapomorphies: loss of maxillipedal endite and weak differentiation of pereiopods 1-3 from the rest. There is also a strong resemblance to *Rhiganthura* in the strongly serrate uropods and telson.

Refer to Figure 15 and Table 13 for synapomorphies of the remaining clades and genera.

3.5 *Paranthuridae and Leptanthuridae*

Relationships between the 17 genera of the Paranthuridae and Leptanthuridae were based on 32 characters of which six are autapomorphies (Tables 14-16). No character was ordered. The consistency and retention indices of the unweighted analysis were improved by automatic weighting (Table 1). Both supported primary division of the taxa into two clades but disagreed on the affinities of *Aenigmathura*. The unweighted analysis placed

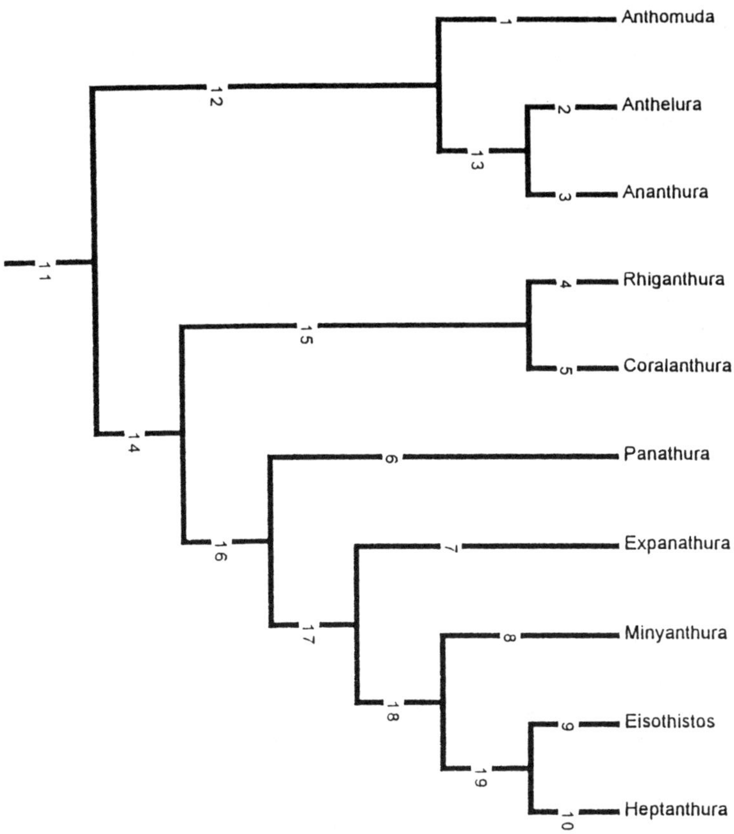

Figure 15. Cladogram of the genera of Antheluridae and Expanathuridae generated by Hennig86 1.5. Characters are listed in Table 11, the data matrix is in Table 12, tree statistics in Table 1. Taxa and clade numbers are those used in Table 13.

this genus within the Paranthuridae s.s. Its affinities seemed intuitively (pereiopod setation, statocyst) more with the leptanthurid clade and these were supported by the weighted analysis which is presented here (Fig. 16, Table 16).

Two genera lack autapomorphies, *Califanthura* and *Bullowanthura*, and only five of the remaining 15 have unique autapomorphies.

The Paranthuridae s.s. comprise six genera (clade 19) that share two synapomorphies (characters 17 and 32). All articles of the maxillipedal palp are fused (with the rare exception of a minute article 5) and there is no statocyst. Equivalent fusion of maxillipedal palp articles is paralleled in two leptanthurid genera. The sister taxon of the Paranthuridae (clade 24) with 11 genera is elevated to family rank, Leptanthuridae. The palm of pereiopod 1 has two synapomorphies (characters 23 and 24), an angle or tooth defining its proximal end and an even marginal row of complex or simple robust setae.

The best-supported genus-group within Paranthuridae is clade 21 with five synapomorphies. This is a group of neotenous genera with shortened pleon, flange-like mandibular molar, mandibular palp absent and pereiopod 7 absent. The hypogean *Cruregens* seems anomalous within this group in possessing a leptanthurid-like palm on pereiopod 1.

Table 14. Families Paranthuridae and Leptanthuridae. Characters and character states used in cladistic analysis. Each character is terminated by a colon and states separated by semicolon, most plesiomorphic condition first.

1. Body: not pigmented; pigmented
2. Pereionite 7: half as long as pereionite 6; one-third as long as pereionite 6; extremely reduced, hidden except dorsally
3. Pleonites: 1-5 free and articulating; 1-5 fused (or sutures immovably indicated); 1-5 fused but indicated by folds dorsally and laterally; 1 free, 2-5 not indicated by folds dorsally
4. Pleonite 1 middorsal length: as long as pleonite 2; longer than pleonite 2, swollen laterally
5. Pleotelson: with posterior margin of pleonite 6 indicated dorsally, delineated from telson; without indication of posterior margin of pleonite 6
6. Pleonites and pleotelson: articulating; fused
7. Eyes: well developed; poorly developed; absent
8. Antenna 2 flagellum: of 5-8 articles, shorter than peduncle; rudimentary (fewer than 5 articles), shorter than last article of peduncle; of 13-20 articles, longer than or equal to peduncle
9. Antenna 2 flagellum articles: articles cylindrical and free; articles short and fused
10. Mandibular molar: absent; flange-like
11. Mandibular incisor: acute; blunt
12. Mandibular palp: of 3 articles; of 1 article; absent
13. Mandibular palp single article: with 1 apical seta; without seta
14. Mandibular palp article 3: with row of 7 or more setae on oblique or longitudinal apex; with fewer than 4 terminal setae
15. Maxillipedal endite: reaching at least article 3; absent or obsolete
16. Maxillipedal palp: almost as long as basis; half as long as basis; third as long as basis
17. Maxillipedal palp: free from basis, of 5 free articles; free from basis, with articles 1-2 fused, 3-4 and 5 free; free from basis, with articles 1-2 fused and 3-5 fused; free from basis, with articles 1-3 fused, 5 minute if visible; free from basis, with articles 1-4 fused and 5 minute if visible; fused to basis and all articles fused (except sometimes for minute distal article)
18. Pereiopods 1-3 propodi: without rolled tessellated margin; with rolled tessellated margin
19. Pereiopod 1 carpus, posterior margin: without short robust setae; with short robust setae (as on propodus)
20. Pereiopod 1 propodus: significantly larger than propodus of pereiopods 2 and 3; narrower than pereiopods 2 and 3
21. Pereiopod 1 propodus: with setae on mesial surface evenly spaced along palm; with row of closely-set setae confined proximally on mesial surface
22. Pereiopod 1 propodus palm: axial or oblique; transverse, its proximal angle produced
23. Pereiopod 1 palm: without proximal tooth (palm more or less continuous with that of carpus); defined proximally by angle distant from palm of carpus; with proximal tooth (defined by clear angle to palm)
24. Pereiopod 1 palm: with marginal setal row irregular if present; with even marginal row of short *complex* robust setae along length; with even marginal row of closely-set *simple* robust setae
25. Pereiopods 2 and 3 carpus margin: without robust setae; with 1-2 robust setae (as in propodus)
26. Pereiopodal: propodi 2 and 3 ovate, 4-7 linear; propodi 2-6 similar, much narrower than long; propodus 2 similar to propodus of pereiopod 1, 3 with transverse palm; propodi 2 and 3 similar to propodus 1 (bases of these limbs much longer and broader than pereiopod 1)
27. Pereiopods 4-7: merus not lobed, dactylus tapering, subapical seta shorter than unguis; merus posterodistally lobed, dactylus not tapering, subapical seta much longer than unguis
28. Pereiopods 4-7 carpus: quadrangular or elongate, with free anterior margin; triangular, enclosed anteriorly by surrounding articles
29. Pereiopod 7: present on adults; absent on adults

Table 14. (Continued).

30.	Pleopodal endopods: with several marginal setae; without marginal setae
31.	Uropodal exopod: narrowly leaf-shaped or broad and more or less notched; linear, tapering, shorter than peduncle; linear, apically rounded
32.	Statocyst: single and medial; absent

Table 15. Character matrices used in cladistic analyses of genera of Paranthuridae and Leptanthuridae.

Character number		1	11111	11112	22222	22223	33
	12345	67890	12345	67890	12345	67890	12
Pseudanthura	00101	12000	00000	04000	00020	00000	11
Paranthura	10000	00110	00001	04000	00000	00000	01
Colanthura	12210	00011	12001	14000	10000	00010	01
Califanthura	12100	00011	12001	14000	10000	00010	01
Cruranthura	11300	00011	12001	15000	10000	00011	01
Cruregens	01001	02011	12001	05000	00120	10011	21
Aenigmathura	00101	00010	00000	02100	00110	00000	10
Curassanthura	01000	0100?	00010	00000	00110	10010	00
Accalathura	10000	00200	00000	01000	00120	00000	00
Calathura	00000	01200	00000	12000	00120	00000	00
Virganthura	00000	00100	00000	01010	00110	00000	00
Bourbonanthura	00000	01000	00000	00000	00110	00000	10
Albanthura	00000	02000	00010	01000	00211	00100	00
Leptanthura	00000	00000	00011	23010	00211	00100	00
Bullowanthura	00000	02000	01001	24010	00111	00100	00
Psittanthura	00000	02000	02001	24001	01110	30100	00
Ulakanthura	00000	02000	01101	23011	00110	31100	00

Table 16. Synapomorphies of genera and supporting clades of Paranthuridae and Leptanthuridae. Character changes at clades shown in Figure 16 are from state 0 to state 1 unless marked otherwise in brackets (viz., – indicates a reversal, > a change from a non-zero state to another, – no apomorphies). Character numbers refer to those in Table 14.

Clade	Taxon	Character change
1.	*Cruregens*	5, 7(2), 17(4>5), 23, 24(2), 26, 30, 31(2)
2.	*Cruranthura*	3(3), 17(4>5), 30
3.	*Colanthura*	3(1>2), 4
4.	*Califanthura*	–
5.	*Paranthura*	1, 8
6.	*Pseudanthura*	3, 5, 6, 7(2), 24(2), 31
7.	*Curassanthura*	2, 7, 14, 26, 29
8.	*Bourbonanthura*	7
9.	*Aenigmathura*	3, 5, 9, 17(2), 18
10.	*Virganthura*	8, 19,
11.	*Calathura*	7, 16, 17(1>2)
12.	*Accalathura*	1
13.	*Albanthura*	7(2), 14, 23(1>2)
14.	*Leptanthura*	14, 17(4>3), 23(1>2)

Table 16. Continued.

Clade	Taxon	Character change
15.	*Bullowanthura*	–
16.	*Psittanthura*	12(1>2), –19, 22
17.	*Ulakanthura*	13, 17(4>3), 27
18.	1-17 (Paranthuridae + Leptanthuridae)	–
19.	1-6 (Paranthuridae)	17(4), 32
20.	1-5	9, 15
21.	1-4	2, 10, 11, 12(2), 29
22.	2-4	1, 16, 21
23.	3-4	2(1>2), 3
24.	7-17 (Leptanthuridae)	23, 24
25.	8-9	31
26.	10-17	17
27.	11-12	8(2), 24(1>2)
28.	13-17	25, 28
29.	14-17	15, 16(2), 19
30.	15-17	7(2), 12, 17(1>4)
31.	16-17	20, –25, 26(3)

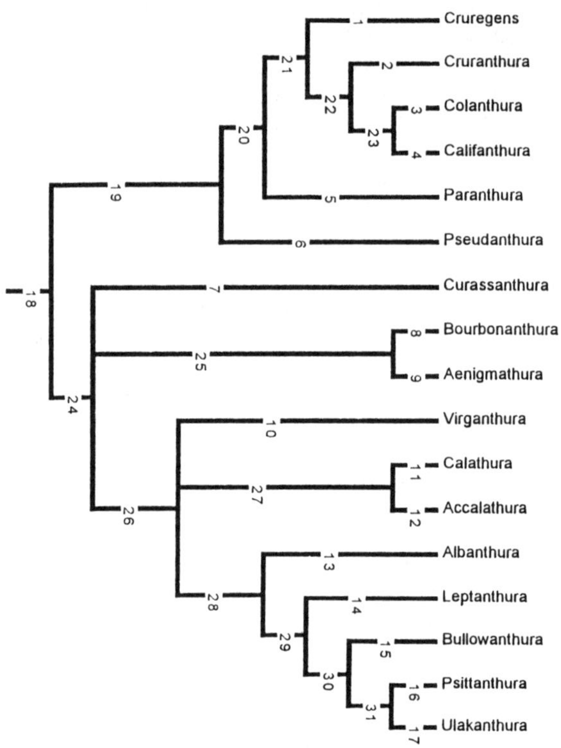

Figure 16. Cladogram of the genera of Paranthuridae and Leptanthuridae generated by Hennig86 1.5. Characters are listed in Table 14, the data matrix is in Table 15, tree statistics in Table 1. Taxa and clade numbers are those used in Table 16.

Basal clades within Leptanthuridae are not fully resolved but the similarity of *Calathura* to *Accalathura* is established (clade 27). The two share a long flagellum on antenna 2 and a row of simple (rather than complex) setae on pereiopod 1. Clade 28 (five genera) has two synapomorphies: robust setae on the carpus of pereiopods 2 and 3, and triangular carpus on pereiopods 4-7.

Refer to Figure 16 and Table 16 for synapomorphies of the remaining clades and genera.

4 A NEW CLASSIFICATION

4.1 *Systematic changes*

The cladistic analysis supports recognition of six families of Anthuridea, two new.

Suborder Anthuridea Monod, 1922
- Family Anthuridae Leach, 1814 (24 genera)
- Family Antheluridae Poore & Lew Ton, 1988b (3 genera)
- Family Expanathuridae fam. nov. (7 genera)
- Family Hyssuridae Wägele, 1981d (6 genera)
- Family Leptanthuridae fam. nov. (11 genera)
- Family Paranthuridae Menzies & Glynn, 1968 (6 genera)

The significant taxonomic changes proposed here are:

1. Erection of the new family Expanathuridae for seven genera previously included in Hyssuridae, Paranthuridae and Anthuridae;
2. Erection of the new family Leptanthuridae for 11 genera previously included in Paranthuridae;
3. Synonymy of *Indonesanthura* Negoescu, 1997b with *Caenanthura* Kensley, 1978a, already senior synonym of *Arabanthura* Kensley & Reid, 1984.
4. Synonymy of *Filanthura* Wägele, 1981d with *Indanthura* Pillai & Eapen, 1966, already senior synonym of *Alloanthura* Kensley, 1980a;
5. Synonymy of *Centranthura* Wägele, 1981d and *Monodanthura* Wägele & Platvoet, 1982 with *Notanthura* Monod, 1927;
6. Synonymy of *Agulanthura* Kensley, 1975a with *Quantanthura* Menzies & George, 1972;
7. Synonymy of *Stellanthura* Wägele, 1979d, *Chalixanthura* Kensley, 1984 and *Licranthura* Kensley & Schotte, 1987 with *Heptanthura* Kensley, 1978a;
8. Synonymy of *Neoanthura* Menzies, 1956 with *Leptanthura* Sars, 1897.

4.2 *Comparison with alternate phylogenies*

4.2.1 *Families*

The new phylogeny does not elucidate the relationships between all the families. It provides good evidence for the Hyssuridae being the sister taxon for all other anthurideans and for the Antheluridae-Expanathuridae and Leptanthuridae-Paranthuridae being family pairs. The position of the Hyssuridae agrees with that proposed in the only other phylogenetic analysis. Wägele (1981d: 67, Fig. 26; and again in 1989: 150, Fig. 73) concluded that a narrow body and atrophied mouth parts in the male are synapomorphies of the Hyssuridae and the presence of two statocysts and operculiform pleopods 1 characterize the remaining anthurideans. He did not mention the produced carpus on pereiopods 2 and 3 of hyssurids. His second apomorphy for this family may be more general than he allows.

While I am inclined to agree with Wägele's well-argued interpretation of the statocysts, the distribution of this most unusual feature may be interpreted in other ways. The organ may have evolved independently as a paired or unpaired structure in only those families in which it occurs, Anthuridae, Antheluridae, and Leptanthuridae. Similarly, not all non-hyssurid anthurideans have equally operculiform pleopods 1. In Expanathuridae in particular, the operculum is weak if it exists at all. While in other families, the exopod of pleopod 1 encloses pleopods 2-4 it is not so enlarged in Expanathuridae. Here both rami of pleopod 1 are exposed and are often not longer or more indurate than the other pleopods. There is not compelling evidence that a well developed operculate pleopod with indurate exopod has not evolved more than once.

Wägele (1981d, 1989) recognized only two families besides Hyssuridae, each with two synapomorphies, but this taxonomy does not correspond with my interpretation.

4.2.2 *Hyssuridae*

Poore & Lew Ton (1988c) discussed differences between the Hyssuridae sensu Wägele (1981d, 1989: 153, Fig. 74), with ten genera and their new restricted group of five genera, now six. Hyssuridae sensu Poore & Lew Ton and as used here is equivalent to Wägele's group A (Wägele 1981d: 74, Fig. 30). Comparison of the phylogenies of Wägele (1981d), Poore & Lew Ton (1988c) and herein must account for additional genera and some synonymies. All recognize the difference of *Xenanthura* (with or without its sister genus *Belura*) from the remainder. Poore & Lew Ton (1988c) could not define a synapomorphy for the remainder united by Wägele in possession of a 4-articled flagellum on antenna 1. In the new analysis, an acutely produced carpus on pereiopods 2 and 3 unites this clade. Both Wägele and Poore & Lew Ton found *Kupellonura* a plesiomorphic genus and *Neohyssura* and *Hyssura* more related. This is not supported by the new analysis that finds that *Kupellonura* and *Hyssura* share the weak character of a dentate margin to the telson (not spiny as in *Neohyssura* or smooth as in *Galziniella*).

4.2.3 *Anthuridae*

Wägele (1981d: 82, Fig. 37) presented a dendrogram of 25 taxa, mostly genera, of Anthuridae but was not able to find affinities for all genera or synapomorphies for all clades. Some of his groups coincided with those in my new analysis of 24 genera. For example, he apparently recognized the synonymy of *Quantanthura* and *Agulanthura*, and grouped *Exallanthura*, *Anthura* and *Ptilanthura* on the basis of a fused maxillipedal palp. He placed *Mesanthura* and *Cyathura* together and (allowing for nomenclatural change) allied these with *Apanthura*, *Amakusanthura* and *Skuphonura*. He placed *Notanthura* and *Centranthura* close together (synonyms in my analysis), and recognized the relationship between *Venezanthura* (=*Cortezura*) and *Notanthura*.

Some of the inconclusive areas of Wägele's dendrogram were resolved in a later version (Wägele 1989: 154, Fig. 75) that is even closer to my new analysis. *Quantanthura* plus *Agulanthura* (which I have synonymized) are sister taxon to all other anthurids. The relationships of *Pendanthura*, *Sauranthura*, *Cyathura* and *Mesanthura* are confirmed. In other areas, there are significant differences. *Skuphonura*, which I believe to be very similar to *Apanthura* and relatives (group I), is excluded from this group. The similarities between *Notanthura* and *Cortezura* are remote while those between *Notanthura* and *Venezanthura* (junior synonym of *Cortezura*) are close. *Cetanthura* is placed with the *Ptilanthura* group (E) while I believe it belongs more closely with *Haliophasma*. Like me, Wägele was ambivalent about the definition of *Haliophasma* and *Malacanthura* but was more reluctant than I to synonymize genera like these (*Alloanthura*, *Filanthura*, *Exanthura*).

4.2.4 *Antheluridae*

Wägele (1981d, 1989) recognized the monophyly (and in the later paper the family name) of the Antheluridae but included them as a clade of 'Paranthuridae' in the broadest sense. In both of his dendrograms, he included most of the seven nominal genera which Poore & Lew Ton (1988b) synonymized into three. What little resolution appeared in Wägele (1981d: 88, Fig. 39) disappeared in the second paper (1989: 160, Fig. 76). Wägele (1989) included *Coralanthura* in this group but this genus is without a single statocyst, a uniting feature of the family, and belongs in Expanathuridae.

4.2.5 *Expanathuridae*

The genera that I have included in the new family Expanathuridae were included in two unrelated groups by Wägele (Wägele 1981d, 1989). He did not recognize the synonymies I have proposed here and partly as a consequence was unable to resolve relationships between *Eisothistos*, *Heptanthura* and *Rhiganthura* that he included as a well-defined clade in Hyssuridae. While these genera share with the Hyssuridae absence of statocyst they are not unusually elongate nor do they have undifferentiated pleopods on long pleonites that are characteristic of Hyssuridae s.s. *Panathura*, *Expanathura* and *Minyanthura* were placed as group C in Wägele's (1989: 160, Fig. 76) dendrogram of Paranthuridae. The seventh genus, *Coralanthura*, was aligned with another clade of Paranthuridae, the group I call Antheluridae. It is not surprising therefore that Wägele did not utilize the character which unites the Expanathuridae and Antheluridae (distal setae row on pereiopod 7 propodus) nor the narrow tapering maxillipedal palp of Expanathuridae. The groups of three he did recognize have no correlates in my analysis.

4.2.6 *Leptanthuridae*

The genera of Leptanthuridae were included by most authors in Paranthuridae sensu Menzies & Glynn (1968), a more restricted family than Wägele's (1981d, 1989) usage. In the dendrograms of the latter author (1981d: 92, Fig. 41: 1989: 160, Fig. 76), the genera are paraphyletic and their synapomorphies (regular row of robust setae along the palm of pereiopod 1, which is defined by a strong angle or tooth) are not recognized. One clade is defined by both Wägele (1989) and me, viz. *Leptanthura* and its allies, *Ulakanthura*, *Bullowanthura*, *Psittanthura* and *Albanthura*. However, the only synapomorphy in common in the two analyses is shortening of the carpus of pereiopods 4-7. Both treatments place *Accalathura* close to *Calathura* but Wägele (1989) linked *Pseudanthura* to these two. Wägele said that *Calathura* and *Pseudanthura* have reduced statocyst. This is true of *Pseudanthura*, which places it in another family, but is not true of *Calathura*. Both cladograms deduce that *Curassanthura* is an early branch from the clade of leptanthurid genera.

The unusual produced mouth parts of *Apanthuroides* suggested to Wägele in 1981 that the genus was possibly a member of Anthuridae but in 1989 he placed the genus in Paranthuridae, at the base of the genera I now put in Leptanthuridae. In my view, this genus has a *Haliophasma*-like ancestor (pitted cuticle, maxilliped) and is a true anthurid.

4.2.7 *Paranthuridae*

The new analysis confines the Paranthuridae to only six genera, those former paranthurids without a statocyst and having at least articles 1-4 of the maxillipedal palp fused. There is general agreement that *Paranthura* is sister taxon to the four neotenous genera discussed by Poore (1984a) and that they share lost maxillipedal endite. Placement of *Pseudanthura* in this family is more contentious, Wägele (1989) placing it with the leptanthurid genera. The genus lacks the peculiar pereiopod 1 palm of leptanthurids.

There is general agreement between the dendrograms of Wägele and those published here in spite of the different methodological approaches. The agreement is surprising more because of the small overlap in character suites used. For example, Wägele relied heavily for many clades in the numbers of articles in antenna 1 and 2 whereas I was not able to define many genera unambiguously using this character. I investigated homologies between articles in the maxilliped and was able to find similarities that were not apparent when simple counts of articles were made. I put a lot of weight on homologies between robust setae on the posterior margins of pereiopodal propodi and carpi, a set of characters that helped to define the new families. Some new discoveries, like a statocyst in *Calathura*, helped place otherwize enigmatic genera. The most radical propositions resulting from this study are diagnosis of a new family, Expanathuridae based on a tapering maxillipedal palp and weakly operculiform pleopod 1, and discovery of its relationship to Antheluridae with which it shares a peculiar setal row on the distal margin of the propodus of pereiopod 7, marginal setae on posterior pleonites and a single marginal seta (in addition to the posterodistal one) on the carpus of pereiopods 4-7. New characters play an important place in this hypothesis.

5 SYSTEMATICS

Anthuridea Monod

Anthuridea Monod; 1922: 136; Wägele 1981d: 67; Negoescu & Wägele 1984: 100; Wägele 1989: 143.

Diagnosis: Elongate, cylindrical flabelliferan Isopoda with uropodal exopod more or less dorsally and proximally displaced; mandibular molar process never triturative, lacinia mobilis and spine row more or less fused into a lamina dentata, incisor process never toothed; maxilla 2 absent; maxillipedal endite a small triangular plate without coupling hooks or absent; pereiopod 1 or 1-3 subchelate.

Key to families of Anthuridea

1. Pleonites 1-5 free and as long as wide; pleopods 1-5 equal (Fig. 10a); slender body form (pereionite 7 longer than wide, little or not shorter than pereionite 6) (Fig. 1a);

pereiopods 2 and 3 with carpus produced posterodistally (Figs 8a-c, o, p) ....Hyssuridae
- Pleonites 1-5 free or fused, some at least shorter than wide; pleopod 1 exopod at least slightly wider than rami of other pleopods (Figs 10b, c); body form not exceptionally slender (at least pereionite 7 wider than long) (Figs 1b-d, 2a-e); pereiopods 2 and 3 with carpus rarely produced posterodistally (Figs 8d-q) 2

2. Mouth parts not strongly and acutely produced (Figs 3d, e); mandible compact, incisor usually transverse and weakly toothed (Figs 4e-q) 3
- Mouth parts strongly and acutely produced (Fig. 3k); mandible tapering, incisor an simple spike (Figs 4r-v)...5

3. Pair of statocysts at base of telson (Figs 11j-l); pleonites 1-5 fused (or weakly indicated even though immobile) (Figs 11j-l); pereiopods 4-7 propodi with 0 or 1 posterodistal robust seta (Figs 9h-k), rarely 1 posterior (Fig. 9g); pereiopod 7 propodus without anterodistal lateral serrate seta (Figs 9g-k) ....Anthuridae
- Single medial (Fig 11e) or no statocyst (Figs f-i) at base of telson; pleonites 1-5 free (rarely fused) (Figs 11e-i); pereiopods 4-7 propodi with at least 1 posterior in addition to posterodistal robust seta (Figs 9d-f); pereiopod 7 propodus usually with anterodistal lateral serrate seta or setae (if not, one of other characters apply) (Figs 9d-f)..................4

4. Telson without statocyst (Figs 11f-i); maxilliped usually with narrow acute endite and narrow tapering palp of 4-5 articles (Figs 5c, d) or single broad article or vestigial (Fig. 5f); uropodal exopod typically leaf-life (Fig. 11f) or with strong dorsal spike (Figs 11g-i) ....Expanathuridae
- Telson with single medial statocyst (Fig. 11e); maxilliped usually with broad short endite and broad palp of 5 articles (Fig. 5e); uropodal exopod typically leaf-life, never with strong dorsal spike (Fig. 11e)..Antheluridae

5. Telson without statocyst (Figs 11m, q); pereiopod 1 palm without tooth or strong proximal angle (except *Cruregens*) and without even row of robust setae (Figs 7a, b, d) ....Paranthuridae
- Telson with 1 medial statocyst at base of telson, usually opening to pore (Figs 11o, r, s); pereiopod 1 palm with tooth or strong proximal angle and with even row of robust simple or complex setae (Figs 7c, e, f, h, i) ....Leptanthuridae

Antheluridae Poore & Lew Ton

Antheluridae Poore & Lew Ton 1988b: 490; Negoescu & Svavarsson 1997: 160-161.

Diagnosis: Body 10-15 times as long as wide, non-males rarely more elongate (pereionite 7 wider than long, much shorter than pereionite 6). Pleonites 1-5 together not more than twice as long as wide, free and articulating; with plumose setae on margins of epimera or posterior margins of pleonite 5 and sometimes pleonite 4.

Antenna 2 flagellum of numerous articles, longer than or shorter than peduncle. Mouth parts not produced anteriorly. Mandible compact and with weakly-toothed transverse incisor. Maxillipedal endite reaching palp article 3; palp broad (c. twice as long as wide), with 5 free articles.

Pereiopods 2 and 3 carpus not or weakly produced posterodistally; propodus with 1 robust seta in addition to posterodistal robust seta. Pereiopods 4-7 propodus with more than 1 robust setae in addition to posterodistal robust seta. Pereiopod 7 propodus with anterodistal serrate setae.

Pleopods: pleopod 1 exopod operculiform alone. Statocyst single.

Remarks: Poore & Lew Ton (1988b) gave a short diagnosis and longer list of descriptive features of their new family. Their characterization does not contradict that presented here but they relied heavily on the 'exceptional width of the maxillipedal endite and palp' as the family's unique apomorphy. This condition does not stand scrutiny when compared with anthurid genera. The three genera are distinguished from members of Anthuridae, which they superficially resemble, by having a single medial statocyst and an anterodistal row of serrate setae on the propodus of pereiopod 7. These setae lie on the mesially-directed face of the leg. Negoescu & Svavarsson (1997) discussed differences between the genera.

The cladistic analysis found *Ananthura* and *Anthelura* to be more similar to each other than to *Anthomuda*. These two are deep-water or cool-temperate or polar while *Anthomuda* is found in warmer and shallower seas.

Key to genera of Antheluridae

1. Pereiopods 4-7 carpus much longer than wide (Fig. 9e); telson with longitudinal ridge and no statocyst pore .... *Anthelura*
– Pereiopods 4-7 carpus about as long as wide; telson domed or flat and with statocyst pore 2
2. Telson domed with slit-like pore (Fig. 11e); antenna 2 flagellum of 5 or more articles .... *Ananthura*
– Telson flat with small pore; antenna 2 flagellum of 2-3 articles .... *Anthomuda*

Genera of Antheluridae

Generic characters with implicit states. Ten characters were required to diagnose the three genera; none is implicit.

Ananthura Barnard

Ananthura Barnard 1925: 136; Menzies 1962: 193; Poore & Lew Ton 1988b: 495.
Bathura Schultz 1966: 12 (type species: *Bathura luna* Schultz 1966, original designation).
Austranthura Kussakin 1967 [1968: 250-251] (type species: *Austranthura elegans* Kussakin 1967, original designation).
Valoranthura Kensley 1978a: 790; Kussakin 1982: 23; (type species: *Anthelura abyssorum* Norman & Stebbing 1886, original designation).

Type species: *Ananthura sulcaticauda* Barnard, 1925, subsequent designation by Menzies (1962).

Diagnosis: Body not pigmented. With plumose setae on dorsolateral and lateral margins of pleonites 4 and 5. Eyes present or absent. Antenna 2 flagellum of 5-11 isometric articles. Pereiopod 1 merus as wide as narrow propodus. Pereiopods 4-7 carpus longer than wide, with 2 robust setae along cutting edge. Telson twice as long as greatest width; uropodal rami tapering to narrow apices. Telson domed with slit-like pore.

Male: Unknown.

Species and distribution: Seven. Temperate to polar seas; 100-4596 m.

Remarks: Species of *Ananthura* are distinguished by the dome over the telsonic statocyst which opens by a slit-like pore. The complex synonymy was discussed by Poore & Lew Ton (1988b). Kensley (1987b) and Negoescu & Svavarsson (1997) redescribed the type species. *Ananthura flexibilis* Pasternak, 1982 is a species of the hyssurid genus *Kupellonura*. The synonymy proposed by Poore & Lew Ton (1988b) has not been disputed. They diagnosed *Ananthura* as not having eyes but illustrated a pair on their species, *A. billarderia*. *Pananthura* [sic.] *ocellata* Nunomura, 1992a can be assigned to this genus on the basis of the long pleon and telson.

Anthelura Norman & Stebbing

Anthelura Norman & Stebbing 1886: 120; Richardson 1905: 68; Barnard 1925: 135; Menzies 1962: 192; Kensley 1978a: 787; Kussakin 1982: 20; Poore & Lew Ton 1988b: 491-495.

Type species: *Anthelura elongata* Norman & Stebbing, 1886, subsequent designation by Menzies (1962).

Diagnosis: Body not pigmented. Pleonites with plumose setae on dorsolateral and lateral margins of pleonites 4 and 5. Eyes absent. Antenna 2 flagellum of 7 isometric articles. Pereiopod 1 merus broader than tapering propodus. Pereiopods 4-7 carpus longer than wide, with 4 robust setae along cutting edge. Telson 2.5 times as long as wide; uropodal rami tapering to narrow apices. Telson without statocyst pore.

Male: Antenna 1 with long flagellum of c. 30 articles reaching to end of pereionite 3. Pereiopod 1 propodus narrow, with numerous mesial setae.

Species and distribution: Monotypic. NE Atlantic Ocean; 1100-1590 m.

Remarks: Poore & Lew Ton (1988b) discussed the fate of the five species previously assigned to this genus. Only one deep water species, still known only from its type material, exists. Others have been transferred to Anthuridae.

Anthomuda Schultz

Anthomuda Schultz 1979: 907-909; Poore & Lew Ton 1988b: 500.
Diaphoranthura Kensley 1980a: 15 (type species: *Diaphoranthura cracens* Kensley 1980a, original designation).

Type species: *Anthomuda stenotelson* Schultz, 1979, original designation.

Diagnosis: Body not pigmented or irregularly pigmented. With plumose setae on dorsolateral and lateral margins of pleonites 4 and 5. Eyes present in both sexes. Antenna 2 flagellum of 3-4 shortened articles. Pereiopod 1 merus as wide as narrow propodus. Pereiopods 4-7 carpus about as long as wide, with short anterior margin, with 2 robust setae along cutting edge. Telson 2.5 times as long as wide; uropodal rami apically rounded. Telson flat with small pore.

Male: Antenna 1 flagellum of male of 1 basal article + c. 10 aesthetasc-bearing articles. Eyes enlarged. Pleon elongated, twice as long as wide. Pleopod 2 with appendix masculina terminally hooked.

Species and distribution: Seven. Tropical to warm temperate; intertidal to 61 m.

Remarks: *Anthomuda* differs from the other two genera in the shorter flagellum of antenna 2 and shorter carpus of pereiopods 4-7. Its species are inhabitants of shallow reef environments whereas the other genera are mostly deep shelf or abyssal. Moore (1996) found in a cladistic analysis that a species from the central Pacific *A. alba* was the most plesiomorphic, those from the western Indian Ocean more derived, and those from Australia and the Caribbean the most derived.

Anthuridae Leach

Anthuridae Leach 1814: 433; Wägele 1981a: 115; Wägele 1981d: 68, 106-108; Kussakin 1982: 15-16.

Diagnosis: Body 10-15 times as long as wide, non-males rarely more elongate (pereionite 7 wider than long, much shorter than pereionite 6). Pleonites 1-5 together not more than twice as long as wide, fused; without marginal plumose setae on pleonal epimera or posterior borders of pleonites 4 and 5. Antenna 2 flagellum of fewer than 10 articles, shorter than peduncle. Mouth parts not produced anteriorly. Mandible compact and with weakly-toothed transverse incisor. Maxillipedal endite reaching palp article 3, or absent or obsolete; palp broad (c. twice as long as wide), with 5 free articles or with 2 or more articles fused. Pereiopods 2 and 3 carpus not or weakly produced posterodistally; propodus with posterodistal robust seta only. Pereiopods 4-7 propodus with posterodistal robust seta only. Pereiopod 7 propodus without anterodistal serrate setae. Pleopods: pleopod 1 exopod operculiform alone. Statocysts paired.

Remarks: The concept of the family presented here is more restricted than that of most recent authors in excluding genera now placed in Expanathuridae. Several nominal genera are synonymized and only 24 remain. Some (*Cyathura, Haliophasma* and *Apanthura* in particular) remain probably polyphyletic or paraphyletic.

Key to genera of Anthuridae

1. Pereiopods 4-7 with carpus twice width of propodus, rectangular, with obvious free distal margin (usually with row of pectinate setae) defined by stout seta on posterodistal angle (Fig. 9g); maxillipedal endite half length of palp, distally rounded; palp articles 1 and 2 separated by suture (4 or 5 articles visible) (Fig. 5h)............. *Quantanthura*
– Pereiopods 4-7 with carpus of similar width to propodus, or if wider triangular, only short free distal margin between base of dactylus and stout seta (Fig. 9i); maxillipedal endite weak or absent, acute if present; palp articles 1-2 and 4-5 fused (maximum of 3 articles visible) 2

2. Pereiopods 2-7 ischium-propodus bearing long setae on upper and lower margins (Fig. 8i); blind or eyes weakly pigmented 3
– Pereiopods 2-7 ischium-propodus bearing few short setae, most on lower margin (merus-carpus posteriorly lobed and setose in some *Apanthura*) (Fig. 8j); eyes usually pigmented 4

3. Carpus of pereiopods 4-7 longer and wider than propodus; mandibular palp of 3 articles .... *Notanthura*
– Carpus of pereiopods 4-7 shorter than propodus; mandibular palp of 2 articles (Fig. 4j) .... *Cortezura*

4. Pereiopods 4-7 with carpus more or less triangular, anterior (upper) margin much shorter than posterior (lower) margin, distal margin oblique and with posterodistal lobe (Fig. 9k) 5
– Pereiopods 4-7 with carpus more or less rectangular, anterior (upper) margin nearly as long as posterior (lower) margin, distal margin transverse and without posterodistal lobe (Fig. 9i) 14

5. Maxillipedal palp with fused articles 4-5 at least one-third as long as penultimate article (3), almost as long as broad, with a free mesial margin between its suture and distal group of setae (Fig. 5m); male antenna 1 flagellum shorter than head; pereiopod 1 with short distal row of mesial setae or no mesial setae; mandibular palp article 3 longer than article 1, usually with longitudinal row of setae (Fig. 4k) 6
– Maxillipedal palp with terminal article (4-5) minute, without a free mesial margin between its suture and distal group of setae (Fig. 5i); male antenna 1 flagellum longer than head; pereiopod 1 with row of mesial setae along palm; mandibular palp article 3 shorter than article 1, with terminal setae only (Fig. 4n) 9

6. Maxillipedal palp with articles 1-2 fused, 3 free and 4-5 fused (Fig. 5m); pereiopods 4-7 carpus with posterodistal stout seta (Fig. 9k); body with dark specific pigment pattern (e.g. Fig. 2b) 7
– Maxillipedal palp with articles 1-3 and 4-5 fused (Fig. 5o); pereiopods 4-7 carpus without posterodistal stout seta (Fig. 9h); body at best irregularly patterned 8

7. Pereiopod 1 chelate, propodus produced posterodistally, dactylus with complexly ridged distal margin, unguis subterminal, carpus and propodus fused (Fig. 6j) .... *Chelanthura*
– Pereiopod 1 subchelate, dactylus and terminal unguis closing on axial palm, carpus and propodus separated by suture (Fig. 6h) .... *Mesanthura*

8. Uropodal exopod leaf-shaped, articulating along peduncle; marine and estuarine .... *Cyathura*
– Uropodal exopod linear, articulating transversely; stygiobiontic in Caribbean .... *Stygocyathura*

9. Maxillipedal palp with fused articles 1-2 longer than broad; mandibles asymmetrical, left molar with tooth fitting socket of right molar; pereiopods 2 and 3 with propodus discoid .... *Apanthuropsis*
– Maxillipedal palp with fused articles 1-2 broader than long; mandibles symmetrical; pereiopods 2 and 3 with propodus linear 10

10. Pleonites 1-5 separated by folds dorsally and laterally except dorsally between 4 and 5 (Fig. 11l) .... *Amakusanthura*
– Pleonites 1-5 not separated by folds dorsally 11

11. Maxillipedal palp with articles 1-2 fused, 3 free and 4-5 fused (Fig. 5i); pereiopods 4-7 carpus with posterodistal stout seta (Fig. 9k) 12

– Maxillipedal palp with articles 1-5 fused (Fig. 5q); pereiopods 4-7 or 4-6 carpus without posterodistal stout seta (Fig. 9j) 13

12. Antenna 2 flagellum with 3-4 articles; male with antenna 1 flagellum of 1 basal + 10 aesthetasc-bearing articles, as long as head (Fig. 3s); head without produced chin except in some males (Fig. 3e) .... *Apanthura*
– Antenna 2 flagellum with 1-2 articles; male with antenna 1 flagellum of 1 basal + 1 or 2 short aesthetasc-bearing articles, never as long as head (Figs 3p, v); head with produced chin in both sexes (Fig. 3e) .... *Skuphonura*

13. Pleon shorter than wide; mandibular palp of 1 article with long seta (Fig. 4e) ....*Pendanthura*
– Pleon longer than wide (Fig. 2c); mandibular palp of 3 articles .... *Sauranthura*

14. Maxillipedal palp of 3 articles of similar length; maxilliped with endite; mandible with spike-like molar on right side, none on left (Fig. 4p); pleonite 6 not separated from telson by transverse ridge ....*Apanthuroides*
– These characters not in combination; mandible with triangular or blunt molar, symmetrical 15

15. Maxillipedal palp with articles 1-2 fused, 3 free, 4-5 fused or 3-5 fused (Fig. 5k); endite triangular (Fig. 5i) or absent (Fig. 5l) 16
– Maxillipedal palp with articles 1-3 fused and 4-5 fused or all fused; endite absent (Fig. 5n) 19

16. Maxillipedal palp with terminal fused articles 4-5 contributing to mesial margin of palp, with setae on mesial margin (Fig. 5k); telson with longitudinal middorsal ridge; uropodal endopod short, oblique ....*Indanthura*
– Maxillipedal palp with terminal fused articles 4-5 minute, triangular or semicircular, not contributing to mesial margin of palp, with setae on distal or distomesial margin (Fig. 5l); telson broadly rounded, rarely with longitudinal middorsal ridge; uropodal endopod at least as long as wide, its suture with peduncle more or less transverse 17

17. Pereiopod 1 merus barely touching cylindrical propodus on anterior margin (Fig. 6l); maxillipedal endite absent (Fig. 5i); male antenna 1 of 10 articles .... *Cetanthura*
– Pereiopod 1 merus cupping swollen subchelate propodus (Fig. 6i); maxillipedal endite present (Fig. 5l); (male antenna 1 of 20 articles) 18

18. Body covered at least in part with setules; pleonites 1-5 with sutures indicated dorsally (Fig. 2d); antenna 2 flagellum of 7 articles; eyes absent ....*Pilosanthura*
– Body smooth, without setules; pleonites 1-5 without sutures indicated dorsally (Fig. 11k); antenna 2 flagellum of 5 articles; eyes present ....*Malacanthura*

19. Pereiopod 1 palm with strong proximal seta (about as long as unguis) (Fig. 6m); maxillipedal palp with fused articles 4-5 small, transverse, lateral to distomesial lobe of fused articles 1-3 (Fig. 5p); mandibular palp of 1 or 2 articles (Fig. 4o) ...*Caenanthura*
– Pereiopod 1 palm without strong proximal seta (Fig. 6k); maxillipedal palp with fused

articles 4-5 oblique (Fig. 5n); mandibular palp of 3 articles (Fig. 4i) 20

20. Maxillipedal palp with articles 1-3 fused and 4-5 fused (Fig. 5n) 21
– Maxillipedal palp with articles 1-5 fused (Fig. 5q) 22

21. Pereiopod 1 propodus swollen, subchelate; dactylus simple and closing on palm (Fig. 6k); body of usual proportions, rarely slender .... *Haliophasma*
– Pereiopod 1 propodus cylindrical; dactylus with teeth along lower margin, not closing on palm (Fig. 6g); body very slender .... *Nemanthura*

22. Mandibular palp of 3 articles (Fig. 4k) .... *Anthura*
– Mandibular palp of 1 article (Fig. 4m) 23

23. Pereiopod 7 present .... *Ptilanthura*
– Pereiopod 7 absent .... *Exallanthura*

Genera of Anthuridae

Generic characters with implicit states: Of the 35 characters used to diagnose anthurid genera 22 have implicit values. Their states follow.

Body of usual proportions, total length of adult more than 4 mm. Head more or less parallel-sided in both sexes (not wider anteriorly), without produced chin. Pleonites 1-5 fused, sutures visible only laterally; pleotelson with posterior margin of pleonite 6 indicated dorsally, delineated from telson.

Antenna 1 flagellum in juvenile and female of 3-4 articles (second longest). Eyes present. Mandibular incisor directed transversely (exceeded by palp); molars blunt; palp of 3 articles; palp article 3 with 1-4 setae on distal margin.

Pereiopod 1 subchelate, dactylus and terminal unguis closing on axial palm, carpus and propodus separated by suture; propodus swollen, subchelate, overlapping merus, palm without strong proximal seta; dactylus smooth, closing on palm. Pereiopods 2-7 merus-propodus setose on posterior margin, group of setae on anterodistal margin of merus. Pereiopods 2, 3 carpus without robust seta posterodistally. Pereiopods 4-7 propodus with 1 posterodistal robust seta only. Pereiopod 7 present on adults.

Pleopod 1 endopod apically rounded. Uropodal exopod leaf-shaped, articulating along peduncle.

Amakusanthura (Nunomura)

Amakusanthura Nunomura 1977: 79-80; Poore & Lew Ton 1988a: 108.
Apanthuretta Wägele 1981d: 85, 112; Fig. 37 (nomen nudum, no type species).
Apanthuretta Wägele 1981a: 134-135 (type species: *Apanthuretta pori* Wägele 1981a, original designation); Poore & Lew Ton 1985a: 136.

Type species: *Amakusanthura longiantennata* Nunomura, 1977, original designation.

Diagnosis: Body not darkly pigmented, smooth. Head with produced chin in males. Pleonites 1-5 together longer than greatest width, fused (but separated by folds dorsally and laterally except dorsally between 4 and 5) or fused, sutures visible only laterally. Antenna 2 flagellum of 3-4 articles, shorter than article 5 of peduncle. Maxillipedal endite present as a triangular lobe, obsolete or absent; palp articles 1-2 fused, 3 free and 4-5 fused; palp article 3 barely produced medially beyond articles 4-5 (if at all); palp terminal articles (4-5) oblique, without a free mesial margin between its suture and distal group of setae. Pereiopod 1 propodus palm with step or prominent tooth (more pronounced in male). Pereiopods 4-7 carpus more or less pentagonal, anterior margin half length of posterior margin, with 1 robust seta on posterodistal angle.

Male: Antenna 1 flagellum of 1 basal + > 10 aesthetasc-bearing articles. Head smaller than in juvenile or female, with broad flattened rostrum. Pereiopod 1 not grossly modified. Telson narrower, with middorsal longitudinal depression.

Species and distribution: 40. Cosmopolitan except polar regions; intertidal to 620 m.

Remarks: Typical species of *Amakusanthura* are distinguishable by absence of the complete dorsal line between pleonites 4 and 5. A line delineates pleonites 1-4 and a suture pleonites 5 and 6. In some species no dorsal lines are present between pleonites (like in *Apanthura*) but the pleon is always more elongate than in this genus and male antenna 1 flagellum much longer. Poore & Lew Ton (1988a) provided a key to Australian species and proposed the synonymy. In the cladistic analysis, *Amakusanthura* falls on the same clade as *Apanthura*, the latter without any autapomorphies. Although the genus may not be justified from a cladistic point of view its species can be distinguished in a practical sense. Poore & Lew Ton (1988a) suggested that other characters (pereiopodal setae and armature) might be useful. They found that *Amakusanthura* is more speciose in tropical Australia while *Apanthura* is more common in temperate seas. There are 81 described species in the two genera which are worthy of more exploration.

Anthura (Leach)

Anthura Leach 1814: 404; Norman & Stebbing 1886: 121-123;. Barnard 1925: 129; Kussakin 1982: 16.

Type species: *Oniscus gracilis* Montagu, 1808, monotypy.

Diagnosis: Body irregularly darkly pigmented, smooth. Pleonites 1-5 together longer than greatest width. Antenna 2 flagellum of 7-9 articles, longer than article 5 of peduncle. Maxillipedal endite obsolete or absent; palp articles 1-5 fused. Pereiopod 1 propodus palm straight or sinuous, without prominent tooth. Pereiopods 4-7 carpus more or less rectangular, anterior (upper) margin nearly as long as posterior (lower) margin, distal margin transverse and without posterodistal lobe, with 1 robust seta on posterodistal angle.

Male: Antenna 1 flagellum of 1 basal + c. 20 aesthetasc-bearing articles.

Species and distribution: Unique. UK, NE Atlantic; intertidal to shallow shelf.

Remarks: Wägele (1980a) redescribed the type species in detail. It is recognisable by the pigmentation, broad telson and single article of the maxillipedal palp. *A. gracilis* belongs to a small clade of four species in three genera possessing completely fused maxillipedal palp.

Apanthura (Stebbing)

Apanthura Stebbing 1900: 621; Barnard 1925: 141; Menzies & Glynn 1968: 28; Kensley 1982c: 98; Kruczynski & Myers 1976: 354; Wägele 1981a: 116; Kussakin 1982: 25; Poore & Lew Ton 1985a: 107; Poore & Lew Ton 1988a: 136.

Diagnosis: Body not darkly pigmented or irregularly darkly pigmented, smooth. Head with produced chin in males. Pleonites 1-5 together as long as greatest width. Antenna 2 flagellum of 3-4 articles, shorter than article 5 of peduncle. Maxillipedal endite present as a triangular lobe; palp articles 1-2 fused, 3 free and 4-5 fused; palp article 3 barely produced medially beyond articles 4-5 (if at all); palp terminal articles (4-5) oblique, without a free mesial margin between its suture and distal group of setae. Pereiopod 1 propodus palm with step or prominent tooth (more pronounced in male). Pereiopods 4-7 carpus more or less pentagonal, anterior margin half length of posterior margin, with 1 robust seta on posterodistal angle.

Male: Antenna 1 flagellum of 1 basal + c. 10 aesthetasc-bearing articles.

Species and distribution: 41. Cosmopolitan except polar regions; intertidal to 850 m.

Remarks: Poore & Lew Ton (1985a) recognized two species groups in southern Australian species of this genus, separated on setation of antenna 2, but these seem not to be apparent in species from other localities. *Apanthura* is best separated from *Amakusanthura* by not having pleonites distinguishable except in males and in the shorter flagellum of antenna 1 in males. Poore & Lew Ton (1988a) provided a key to Australian species. Negoescu (1997b) showed that the number of setae on the mandibular palp of species of *Apanthura* ranged from one to ten; this character does not appear in my diagnosis. In the cladistic analysis *Apanthura* has no autapomorphies (see discussion under *Amakusanthura*).

Apanthuroides (Menzies & Glynn)

Apanthuroides Menzies & Glynn 1968: 29; Kensley 1982c: 106.
Natalanthura Kensley 1978d: 5-6 (type species: *Natalanthura foveolata* Kensley, 1978d, original designation).

Type species: *Apanthuroides millae* Menzies & Glynn 1968, original designation.

Diagnosis: Body not darkly pigmented, pitted. Pleonites 1-5 together as long as greatest width; pleotelson without indication of posterior margin of pleonite 6.

Antenna 2 flagellum of 7-9 articles, longer than article 5 of peduncle. Mandibular incisor produced anteriorly (beyond palp); molars spike-like on right side, reduced or absent on left. Maxillipedal endite present (reaching palp article 3); palp articles 1-2 fused, 3 free and 4-5 fused; palp article 3 well exceeded by distal articles; palp terminal articles (4-5) subcircular, transverse. Pereiopod 1 propodus cylindrical, not in contact with merus, palm straight or sinuous, without prominent tooth. Pereiopods 4-7 carpus more or less pentagonal, anterior margin half length of posterior margin, with 1 robust seta on posterodistal angle. Pereiopod 7 sometimes absent.

Male: Antenna 1 flagellum of 1 basal + 6-7 aesthetascs-bearing articles.

Species and distribution: Six. Mediterranean, Caribbean, South Pacific; 1-850 m.

Remarks: The genus is unique in the possession of mandibles with a spike-like molar on the right side and no molar on the left. Pleonite 6 is not separated from the telson by a transverse ridge. The exoskeleton is strongly pitted as in *Cetanthura* and *Haliophasma*, members of the clade to which it belongs. See *Cetanthura* for further discussion. Kensley (1982c) examined new material of *Natalanthura foveolata* Kensley 1978d and figured a molar and pleon identical to that in *Apanthuroides millae*. These were not shown in Menzies & Glynn's (1968) figure of a male. Kensley (1982c) and Negoescu & Wägele (1984) agreed with this synonymy.

Apanthuropsis Poore & Lew Ton

Apanthuropsis Poore & Lew Ton 1985a: 146.

Type species: *Apanthuropsis richea* Poore & Lew Ton 1985a, original designation.

Diagnosis: Body not darkly pigmented, smooth. Pleonites 1-5 together as long as greatest width. Antenna 2 flagellum of 3-4 articles, shorter than article 5 of peduncle. Mandibular molars asymmetrical, left molar with tooth fitting socket of right molar. Maxillipedal endite obsolete or absent; palp articles 1-2 fused, 3 free and 4-5 fused; palp article 3 barely produced medially beyond articles 4-5 (if at all); palp terminal articles (4-5) oblique, without a free mesial margin between its suture and distal group of setae. Pereiopod 1 propodus palm with step or prominent tooth (more pronounced in male). Pereiopods 4-7 carpus more or less pentagonal, anterior margin half length of posterior margin, with 1 robust seta on posterodistal angle.

Male: Antenna 1 flagellum of 1 basal + c. 20 aesthetasc-bearing articles. Eyes enlarged; head wider anteriorly and rostrum more pronounced and with a dorsal depression. Pereiopod 1 with stronger tooth and palm more setose; all pereiopods more elongate than in female and juvenile.

Species and distribution: Unique. SE Australia; 8-14 m.

Remarks: This monotypic genus is distinguished from *Apanthura* and *Amakusanthura* by the ovoid propodi on pereiopods 2 and 3 and the unique tooth-and-socket arrangement of the mandibles. Negoescu (1997b) reported the manca of a second undescribed species from Indonesia. This specimen has the typical asymmetrical mandible but not the ovoid propodi. *Apanthuropsis richea* should be included in any revision of the two more diverse and weakly differentiated genera.

Caenanthura Kensley

Caenanthura Kensley 1978a: 785.

Arabanthura Kensley & Reid 1984: 674 (type species: *Arabanthura enigmatica* Kensley & Reid, 1984, original designation).

Indonesanthura Negoescu 1997b: 200-201 (type species: *Indonesanthura gutui* Negoescu 1997b, original designation). syn. nov.

Type species: *Cyathura siamensis* Barnard 1925, original designation.

Diagnosis: Body not darkly pigmented, smooth. Pleonites 1-5 together longer than greatest width. Antenna 2 flagellum of 1-2 articles, very short. Mandibular palp of 1 or 2 very short article. Maxillipedal endite obsolete or absent; palp articles 1-3 fused, 4-5 fused; palp article 3 produced as lobe beyond distolateral articles 4-5; palp terminal articles (4-5) oblique, without a free mesial margin between its suture and distal group of setae. Pereiopod 1 propodus palm straight or sinuous, without prominent tooth, palm with strong proximal seta (about as long as unguis). Pereiopods 4-7 carpus more or less pentagonal, anterior margin half length of posterior margin, with 1 robust seta on posterodistal angle.

Male: Antenna 1 flagellum of 1 basal + c. 20 aesthetasc-bearing articles.

Species and distribution: Four. Indian Ocean region; 1-110 m.

Remarks: The type species of these three genera cannot be distinguished at the generic level. The maxillipedal shape (especially the lobe on the distomesial corner of the largest segment; Fig. 5p) is unusual and shared by all species. Pereiopod 1 has a short oblique palm with a row of robust setae of which the most proximal is stronger, and the merus also has a characteristic row of robust setae (Fig. 6m). Pereiopod 3 is particularly setose. *Caenanthura siamensis* has a mandibular palp with one article (Kensley 1978a) while *Arabanthura enigmatica* and *Indonesanthura gutui* have two but the fact that all are reduced suggests a relationship. The monophyly of the group of only four species is supported by their biogeographic affinities. All are from the Indian Ocean region. Wägele (1989) has already suggested the synonymy of *Arabanthura* with *Caenanthura*.

Cetanthura Kensley

Cetanthura Kensley 1982b: 3.

Type species: *Cetanthura foveoderma* Kensley 1982b, original designation.

Diagnosis: Body not darkly pigmented, pitted. Pleonites 1-5 together longer than greatest width. Antenna 2 flagellum of 3-4 articles, shorter than article 5 of peduncle. Mandibular incisor produced anteriorly (beyond palp). Maxillipedal endite obsolete or absent; palp articles 1-2 fused, 3-5 fused. Pereiopod 1 propodus cylindrical, not in contact with merus, palm straight or sinuous, without prominent tooth. Pereiopods 4-7 carpus more or less pentagonal, anterior margin half length of posterior margin, with 1 robust seta on posterodistal angle. Pereiopod 7 absent on adults.

Male: Antenna 1 flagellum of 1 basal + c. 10 aesthetasc-bearing articles.

Species and distribution: Unique. Namibia; 620 m.

Remarks: The pitted telson allies this genus to *Haliophasma* but the produced mandible and elongate pereiopod 1 are shared by its sister taxon, *Apanthuroides*. It differs from this larger genus in the absence of pereiopod 7, delimited pleonite 6 and telson, standard molar and maxilliped.

Chelanthura Poore & Bardsley

Chelanthura Poore & Bardsley 1990: 109.

Type species: *Chelanthura ajuga* Poore & Bardsley 1990, original designation.

Diagnosis: Body with dark species-specific pigment pattern, smooth. Pleonites 1-5 together as long as greatest width. Antenna 2 flagellum of 3-4 articles, shorter than article 5 of peduncle. Mandibular palp article 3 with row of 10-15 setae on distal oblique margin. Maxillipedal endite obsolete or absent; palp articles 1-2 fused, 3 free and 4-5 fused; palp article 3 well exceeded by distal articles; palp terminal articles (4-5) at least one-third as long as penultimate article (3), almost as long as broad, with a free mesial margin between its suture and distal group of setae. Pereiopod 1 chelate, propodus produced posterodistally, dactylus with complexly ridged distal margin, unguis subterminal, carpus and propodus fused; propodus palm with step or prominent tooth (more pronounced in male). Pereiopods 4-7 carpus more or less pentagonal, anterior margin half length of posterior margin, with 1 robust seta on posterodistal angle.

Male: Unknown but assumed to be like *Mesanthura*.

Species and distribution: Four. Southern Australia; intertidal to 21 m.

Remarks: The four species of *Chelanthura* are a derived southern Australian clade from the larger similarly coloured genus *Mesanthura*. They are characterized by a chelate pereiopod 1 and fusion of carpus and propodus in this limb. See *Mesanthura* for further discussion.

Cortezura Schultz

Cortezura Schultz 1977: 842-843.

Venezanthura Kensley 1978a: 782 (type species: *Venezanthura confixa* Kensley, 1978a, original designation). syn. nov.

Type species: *Cortezura penascoensis* Schultz 1977, original designation.

Diagnosis: Body not darkly pigmented, smooth. Pleonites 1-5 together longer than greatest width. Antenna 2 flagellum of 1-2 articles, very short. Eyes weakly pigmented. Mandibular molars triangular blades; palp of 2 articles, second very setose. Maxillipedal endite present as a triangular lobe; palp articles 1-3 fused, 4-5 fused, or 1-5 fused. Pereiopod 1 propodus palm with step or prominent tooth (more pronounced in male). Pereiopods 2-7 merus-propodus bearing long setae on upper and lower margins. Pereiopods 4-7 carpus more or less rectangular, anterior (upper) margin nearly as long as posterior (lower) margin, distal margin transverse and without posterodistal lobe, with 1 robust seta on posterodistal angle.

Male: Antenna 1 flagellum of 1 basal + 1 or 2 short aesthetasc-bearing articles.

Species and distribution: Two. Venezuela, Gulf of California; intertidal to 15 m.

Remarks: Several characters, all confirmed by my own observations, Kensley & Schotte (1989) and by Müller (1992a), support the synonymy of these two genera represented by two central American species. Both have a mandibular palp of only 2 articles, the second with long setae, maxillipedal endite triangular, setose pereiopods 2-7, the male with the antenna 1 flagellum of 1 article bearing numerous aesthetascs, and are blind. The separate pleonites illustrated by Schultz (1977) could not be seen by me but a maxillipedal endite, not reported by him, is present. If the two species belonging to this genus are not synonymous they are sister taxa separated by the Panamanian Isthmus.

Cyathura Norman & Stebbing

Cyathura Norman & Stebbing 1886: 121; Barnard 1925: 139; Kensley 1982c: 113; Kussakin 1982: 28-29; Poore & Lew Ton 1985b: 89-90.

Cyathura (*Cyathura*); Botosaneanu & Stock 1982: 38-39.

Type species: *Anthura carinata* Krøyer 1847, monotypy.

Diagnosis: Body irregularly darkly pigmented, smooth or covered at least in part with setules (fur). Pleonites 1-5 together shorter than wide; pleotelson with posterior margin of pleonite 6 indicated dorsally, delineated from telson, or without indication of posterior margin of pleonite 6. Antenna 2 flagellum of 3-4 articles, shorter than article 5 of peduncle. Eyes present, or absent. Mandibular palp article 3 with 1-4 setae on distal margin, or with row of 10-15 setae on distal oblique margin. Maxillipedal endite obsolete or absent;

palp articles 1-3 fused, 4-5 fused; palp article 3 well exceeded by distal articles; palp terminal articles (4-5) at least one-third as long as penultimate article (3), almost as long as broad, with a free mesial margin between its suture and distal group of setae. Pereiopod 1 propodus palm with step or prominent tooth (more pronounced in male). Pereiopods 4-7 carpus more or less pentagonal, anterior margin half length of posterior margin, without robust setae on posterior margin or on posterodistal angle.

Male: Antenna 1 flagellum of male of 4-5 aesthetasc-bearing articles, retractile. Pereiopods more elongate than in female or juvenile. Pleopod 2 with appendix masculina ending simply.

Species and distribution: 48. Cosmopolitan except polar regions; intertidal to 2707 m.

Remarks: *Cyathura* is a widespread genus with many species invading brackish water. Species are often covered with a dense fur and sometimes pigmented. The terminal article of the maxilliped is usually large and the carpus of pereiopods 4-7 lacks a posterodistal robust seta. Botosaneanu & Stock (1982) erected the subgenus *Stygocyathura* for stygiobiontic species, said to be a monophyletic clade derived from *Cyathura* s.s., a view widely accepted (Poore & Lew Ton 1985b). In the phylogenetic analysis the two subgenera were treated as separate taxa and in the most parsimonious result they were found not to be sister taxa. I treat both as separate genera. The type species was illustrated by Kussakin (1982) and the mouth parts described in detail by Wägele (1979b). The reproductive cycle of *C. carinata* is one of the best understood of all anthurideans (Wägele 1979a).

Exallanthura Kensley

Exallanthura Kensley 1980a: 18-20.

Type species: *Exallanthura sexpes* Kensley 1980a, original designation.

Diagnosis: Body not darkly pigmented, smooth. Pleonites 1-5 together as long as greatest width. Antenna 2 flagellum of 7-9 articles, longer than article 5 of peduncle. Mandibular palp of 1 article. Maxillipedal endite obsolete or absent; palp articles 1-5 fused. Pereiopod 1 propodus cylindrical, not in contact with merus, palm straight or sinuous, without prominent tooth. Pereiopods 4-6 carpus more or less rectangular, anterior (upper) margin nearly as long as posterior (lower) margin, distal margin transverse and without posterodistal lobe, with 1 robust seta on posterodistal angle. Pereiopod 7 absent on adults. Uropodal exopod broad, apically concave.

Male: Unknown.

Species and distribution: Unique. Indian Ocean; intertidal to 49 m.

Remarks: The single species referred to this genus is unique among Anthuridae in lacking pereiopod 7 and pereionite 7 being very short. No ovigerous females or adult males are known and if these are found with pereiopods 7 present it would not be the first occasion a genus has been erroneouslydiagnosed on the basis of this manca feature. Nevertheless, only three other anthurid genera have a uniarticulate mandibular palp. *Exallanthura* is readily distinguished from *Pendanthura* and *Caenanthura* but is sister taxon to *Ptilanthura*. *E. sexpes* has an unusually small propodus on pereiopod 1.

Haliophasma Haswell

Haliophasma Haswell 1881: 476; Barnard 1925: 131; Barnard 1940: 382; Menzies & Barnard 1959:

17; Kensley 1975b: 213; Poore 1975: 504; Kensley 1982c: 116; Poore & Lew Ton 1988d: 85-87.

Exanthura Barnard 1914: 336a; Barnard 1925: 130 (type species: *Exanthura macrura* Barnard 1914, monotypy).

Silophasma Schultz 1977: 840 (type species: *Haliophasma geminatum* Menzies & Barnard 1959, original designation).

Type species: *Haliophasma purpureum* Haswell 1881, monotypy.

Diagnosis: Body not darkly pigmented, pitted. Pleonites 1-5 together as long as greatest width or longer than greatest width. Antenna 2 flagellum of 7-9 articles, longer than article 5 of peduncle. Mandibular palp article 3 with row of 10-15 setae on distal oblique margin, or sometimes fewer. Maxillipedal endite obsolete or absent; palp articles 1-3 fused, 4-5 fused; palp article 3 barely produced medially beyond articles 4-5 (if at all), or well exceeded by distal articles; palp terminal articles (4-5) oblique, without a free mesial margin between its suture and distal group of setae. Pereiopod 1 propodus palm straight or sinuous, without prominent tooth. Pereiopods 4-7 carpus more or less rectangular, anterior (upper) margin nearly as long as posterior (lower) margin, distal margin transverse and without posterodistal lobe, with 1 robust seta on posterodistal angle.

Male: Antenna 1 flagellum of male of 1 basal + c. 20 aesthetasc-bearing articles.

Species and distribution: 39. Cosmopolitan; intertidal to 775 m.

Remarks: Poore & Lew Ton (1988d) listed *Nemanthura* in their synonymy of this genus but it is now regarded as different. Their paper provided a key to Australian species. *Haliophasma* remains a diverse genus and its species have sometimes been placed in *Malacanthura*, especially in South Africa (Kensley 1982c). Typical species have a pitted or sculptured exoskeleton and smoother species with similar maxillipeds and pereiopods may indicate that the genus is polyphyletic. Similar genera are *Cetanthura* and *Apanthuroides*. See further remarks under *Malacanthura*.

Indanthura Pillai & Eapen

Indanthura Pillai & Eapen 1966: 13-14.

Alloanthura Kensley 1980a: 8 (type species: *Alloanthura sculpta* Kensley, 1980a, original designation).

Filanthura Wägele 1981d: 113-114 (type species: *Malacanthura caribbica* Paul & Menzies, 1971, original designation). syn. nov.

Type species: *Indanthura carinata* Pillai & Eapen 1966, original designation.

Diagnosis: Body not darkly pigmented, smooth. Pleonites 1-5 together longer than greatest width, fused, sutures visible only laterally or sutures all indicated dorsally and laterally. Antenna 2 flagellum of 7-9 articles, longer than article 5 of peduncle. Mandibular palp article 3 with row of 10-15 setae on distal oblique margin. Maxillipedal endite obsolete or absent; palp articles 1-2 fused, 3 free and 4-5 fused; palp article 3 well exceeded by distal articles; palp terminal articles (4-5) at least one-third as long as penultimate article (3), almost as long as broad, with a free mesial margin between its suture and distal group of setae. Pereiopod 1 propodus palm straight or sinuous, without prominent tooth. Pereiopods 4-7 carpus more or less rectangular, anterior (upper) margin nearly as long as posterior (lower) margin, distal margin transverse and without posterodistal lobe, with 1 robust seta on posterodistal angle.

Male: Antenna 1 flagellum in juvenile and female of c. 5-7 articles.

Species and distribution: Four. Caribbean, Mediterranean, Indian Ocean; 40-135 m.

Remarks: Although I have been unable to examine type material of the three genera here synonymized there are several characters supporting this action. This synonymy is based on proportions of the maxilliped, the mandibular palp article 3 having a long row of setae, and form of the uropodal rami. The maxilliped is figured by all authors as setulose (which is unusual). In both Paul & Menzies' (1971) figure of *Malacanthura caribbica* and Kensley's (1980a) figure of *Alloanthura sculpta* an unusual oblique terminal maxillipedal suture is drawn. In all taxa the uropodal exopod is especially oblique and distally concave. Negoescu & Wägele (1984) agreed with the synonymy of *Alloanthura* but they placed *Filanthura* in synonymy with *Malacanthura*. *Malacanthura cumanensis* Paul & Menzies 1971 is a probable synonym of *I. cantabrica*. The relationships of *Indanthura* to other anthurids was not clarified by the phylogenetic analysis. It possesses many plesiomorphies.

Malacanthura (Barnard)

Malacanthura Barnard 1925: 133; Kensley 1982b: 12; Kensley 1982c: 128-130.

Type species: *Malacanthura linguicauda* Barnard 1920, monotypy.

Diagnosis: Body not darkly pigmented, smooth. Pleonites 1-5 together longer than greatest width. Antenna 2 flagellum of 3-4 articles, shorter than article 5 of peduncle. Maxillipedal endite present (reaching palp article 3); palp articles 1-2 fused, 3 free and 4-5 fused; palp article 3 barely produced medially beyond articles 4-5 (if at all); palp terminal articles (4-5) oblique, without a free mesial margin between its suture and distal group of setae. Pereiopod 1 propodus palm straight or sinuous, without prominent tooth. Pereiopods 4-7 carpus more or less rectangular, anterior (upper) margin nearly as long as posterior (lower) margin, distal margin transverse and without posterodistal lobe, with 1 robust seta on posterodistal angle.

Male: Antenna 1 flagellum of male of 1 basal + c. 20 aesthetasc-bearing articles.

Species and distribution: Four. Atlantic Ocean; intertidal to 1379 (2870) m.

Remarks: The type species of *Malacanthura* was described from a male and this specimen was reillustrated by Kensley (1982c). It is one of the most confused genera and its species have been placed arbitrarily in this genus or in *Haliophasma*. Both remain confused but I differentiate the two on the basis of a maxillipedal endite (present in *Malacanthura* and absent in *Haliophasma*), antenna 2 flagellum of fewer than 7 articles and shorter than article 5 of the peduncle in *Malacanthura* and more than seven articles and longer than article 5 of the peduncle in *Haliophasma*. Species of *Malacanthura* are not pitted while those of *Haliophasma* usually are. Even this separation may be arbitrary but further work is required to understand these two genera. Kensley (1982b) redefined *Malacanthura* such that the maxillipedal endite may or may not be present but failed to differentiate the genus from *Haliophasma*. He placed many South African species in *Malacanthura* but absence of a maxillipedal endite puts them in *Haliophasma*. In the possession of many plesiomorphies its relationships could not be elucidated by phylogenetic analysis.

Mesanthura Barnard

Mesanthura Barnard 1914: 343a; Barnard 1925: 143; Menzies & Glynn 1968: 25-26; Kensley

1982c: 153; Menzies & Kruczynski 1983: 28; Poore & Lew Ton 1986a: 90-91; Müller 1993a: 19.

Type species: *Anthura catenula* Stimpson 1855, monotypy.

Diagnosis: Body with dark species-specific pigment pattern, smooth. Pleonites 1-5 together as long as greatest width. Antenna 2 flagellum of 3-4 articles, shorter than article 5 of peduncle. Mandibular palp article 3 with row of 10-15 setae on distal oblique margin. Maxillipedal endite obsolete or absent; palp articles 1-2 fused, 3 free and 4-5 fused; palp article 3 well exceeded by distal articles; palp terminal articles (4-5) at least one-third as long as penultimate article (3), almost as long as broad, with a free mesial margin between its suture and distal group of setae. Pereiopod 1 propodus palm with step or prominent tooth (more pronounced in male); dactylus often with complex toothed surface at base of unguis. Pereiopods 4-7 carpus more or less pentagonal, anterior margin half length of posterior margin, with 1 robust seta on posterodistal angle.

Male: Antenna 1 flagellum of male of 1 basal + c. 10 aesthetasc-bearing articles.

Species and distribution: 40. Cosmopolitan except polar regions; intertidal to 140 m (usually very shallow).

Remarks: Species of *Mesanthura* are distinguishable by colour patterns which are supported by morphological differences in proportions (Wägele 1984b, Poore & Lew Ton 1986a). This observation has been supported by the discovery in Australian collections of up to 20 undescribed species all subtly able to be differentiated. The four colour morphs of *M. protei* (Kensley 1980a, Kensley & Poore 1982, Wägele 1984a) probably represent four species. The only other genus with similar colour morphs is *Chelanthura* which is probably an Australian clade derived from the more widely distributed probably paraphyletic genus. *Mesanthura* belongs to a well-defined clade of genera, all with a three-segmented maxillipedal palp. Poore & Lew Ton (1986a) discussed the wide use of the Australian species name *M. maculata*. It is probable that several species are involved.

Nemanthura Wägele

Nemanthura Wägele 1981d: 114.

Type species: *Haliophasma valeriae* Paul & Menzies 1971, original designation.

Diagnosis: Body very slender, not darkly pigmented, smooth. Pleonites 1-5 together longer than greatest width. Antenna 2 flagellum of 7-9 articles, longer than article 5 of peduncle. Maxillipedal endite obsolete or absent; palp articles 1-3 fused, 4-5 fused; palp article 3 well exceeded by distal articles; palp terminal articles (4-5) oblique, without a free mesial margin between its suture and distal group of setae. Pereiopod 1 propodus palm straight or sinuous, without prominent tooth; dactylus with tubercular teeth along lower margin. Pereiopods 4-7 carpus more or less rectangular, anterior (upper) margin nearly as long as posterior (lower) margin, distal margin transverse and without posterodistal lobe, with 1 robust seta on posterodistal angle.

Male: Unknown.

Species and distribution: Unique. Venezuela; 95 m.

Remarks: *Nemanthura* was placed in synonymy with *Haliophasma* by Poore & Lew Ton (1988), a view no longer held. Pereiopod 1 has a cylindrical propodus and a dactylus with teeth along its lower margin. The body is very slender. *Haliophasma irmae* Paul &

Menzies 1971, described at the same time as the type species, is a probable synonym of *N. valeriae*.

Notanthura (Monod)

Notanthura Monod 1927: 201.
Centranthura Wägele 1981d: 113; Kensley 1982c: 109 (type species: *Haliophasma caecus* Kensley 1975b, original designation). syn. nov.
Monodanthura Wägele & Platvoet 1982: 214 (type species: *Monodanthura maroccana* Wägele & Platvoet 1982, original designation). syn. nov.

Type species: Notanthura barnardi Monod 1927 monotypy.

Diagnosis: Body not darkly pigmented, smooth. Pleonites 1-5 together as long as greatest width. Antenna 2 flagellum of 3-4 articles, shorter than article 5 of peduncle. Eyes present, or absent. Mandibular molars triangular blades; palp article 3 with 1-4 setae on distal margin, or with row of 10-15 setae on distal oblique margin. Maxillipedal endite obsolete or absent; palp articles 1-3 fused, 4-5 fused; palp article 3 barely produced medially beyond articles 4-5 (if at all); palp terminal articles (4-5) oblique, without a free mesial margin between its suture and distal group of setae. Pereiopod 1 propodus palm with step or prominent tooth (more pronounced in male). Pereiopods 2-7 merus-propodus bearing long setae on upper and lower margins. Pereiopods 4-7 carpus more or less rectangular, anterior (upper) margin nearly as long as posterior (lower) margin, distal margin transverse and without posterodistal lobe, shorter than propodus, with 1 robust seta on posterodistal angle. Pleopod 1 endopod tapering to digitiform apex.

Male: Antenna 1 flagellum of male of 4-5 aesthetasc-bearing articles, retractile.

Species and distribution: Four. NW and South Africa; intertidal to 68 m.

Remarks: These three nominal genera are represented by only four species, all from the west and south coasts of Africa. All have the eyes absent or weak, a triangular molar process with fine denticles, a maxillipedal palp with all articles fused (Kensley 1975b, 1979b, 1982c) or with obscure sutures between some sutures. Except in the species of *Centranthura* the endopod of pleopod 1 is narrowed distally. The sister taxon of *Notanthura* (West Africa) is *Cortezura* (central America). They share an unusual mandibular molar, maxillipedal palp and setose pereiopods.

Pendanthura Menzies & Glynn

Pendanthura Menzies & Glynn 1968: 31.

Type species: *Pendanthura tanaiformis* Menzies & Glynn 1968, original designation.

Diagnosis: Body total length of adult less than 3 mm, irregularly darkly pigmented, smooth. Pleonites 1-5 together much shorter than wide; pleotelson without indication of posterior margin of pleonite 6. Antenna 2 flagellum of 1-2 articles, very short. Mandibular palp of 1 article. Maxillipedal endite present as a triangular lobe; palp articles 1-5 fused. Pereiopod 1 propodus palm straight or sinuous, without prominent tooth. Pereiopods 2, 3 carpus without robust seta posterodistally. Pereiopods 4-7 carpus triangular, without free anterior margin, without robust setae on posterior margin or on posterodistal angle.

Male: Antenna 1 flagellum of male of 4-5 aesthetasc-bearing articles, retractile.

Species and distribution: Four. Caribbean, South Pacific; intertidal to 260 m.

Remarks: Species of *Pendanthura* are recognizable by the long rostrum, uniarticulate maxillipedal palp, and uniarticulate mandibular palp. Kensley (1982a) reillustrated the type species. Negoescu (1994) provided a key to the species, all of which are tropical. Its sister taxon is another tropical genus, monotypic *Sauranthura*, from which it differs in the uniarticulate mandibular palp. Both are small inhabitants of coral rubble. An undescribed species from Aldabra, Indian Ocean, is unusual among anthurideans in having the endopod of pleopod 1 longer and broader than the exopod, the two together operculiform.

Pilosanthura Wägele

Pilosanthura Wägele 1989: 156, 158.

Type species: *Anthelura fresii* Wägele 1980a, original designation.

Diagnosis: Body not darkly pigmented, covered at least in part with setules (fur). Pleonites 1-5 together longer than greatest width, sutures all indicated dorsally and laterally. Antenna 2 flagellum of 7-9 articles, longer than article 5 of peduncle. Eyes absent. Maxillipedal endite present (reaching palp article 3); palp articles 1-2 fused, 3 free and 4-5 fused; palp article 3 barely produced medially beyond articles 4-5 (if at all); palp terminal articles (4-5) oblique, without a free mesial margin between its suture and distal group of setae. Pereiopod 1 propodus palm straight or sinuous, without prominent tooth. Pereiopods 4-7 carpus more or less rectangular, anterior (upper) margin nearly as long as posterior (lower) margin, distal margin transverse and without posterodistal lobe, with 1 robust seta on posterodistal angle.

Male: Antenna 1 flagellum of 1 basal + 11 aesthetasc-bearing articles.

Species and distribution: Unique. Mediterranean; 0-1739 m.

Remarks: The single species is notable for the covering of fine setules, a characteristic of some species of *Cyathura*. It lacks eyes but is otherwise poorly differentiated from the ill-defined genus *Malacanthura*. Negoescu (1989) described the male and confirmed that the pleonites are free, an unusual feature of the Anthuridae.

Ptilanthura Harger

Ptilanthura Harger 1878: 377; Richardson 1905: 66-67; Barnard 1925: 130; Kussakin 1982: 19-20; Kensley 1996: 763.

Type species: Ptilanthura tenuis Harger 1878, monotypy.

Diagnosis: Body not darkly pigmented, smooth. Pleonites 1-5 together as long as greatest width. Antenna 2 flagellum of 3-4 articles, shorter than article 5 of peduncle. Mandibular palp of 1 article. Maxillipedal endite obsolete or absent; palp articles 1-5 fused. Pereiopod 1 propodus palm straight or sinuous, without prominent tooth. Pereiopods 4-7 carpus more or less rectangular, anterior (upper) margin nearly as long as posterior (lower) margin, distal margin transverse and without posterodistal lobe, with 1 robust seta on posterodistal angle.

Male: Antenna 1 flagellum of primary male (4-5 mm) of 1 basal + c. 10 aesthetasc-bearing articles; of secondary male (6-10 mm) of 1 basal + c. 20 aesthetasc-bearing articles.

Species and distribution: Two. Gulf of Mexico, eastern Atlantic; intertidal to 141 m.

Remarks: Kensley (1996) redescribed the type species in detail, described a second and synonymized another. He provided evidence for maturation as males at two sizes. The genus is noted for the uniarticulate maxillipedal palp (see *Exallanthura*).

Quantanthura Menzies & George

Quantanthura Menzies & George 1972: 9.29; Kensley & Koening 1979: 953; Kensley 1982c: 164; Poore & Lew Ton 1986c: 75-76.

Agulanthura Kensley 1975a: 72 type species: *Agulanthura serenasimus* Kensley 1975a, original designation. syn. nov.

Type species: *Quantanthura globitelson* Menzies & George 1972, original designation.

Diagnosis: Body not darkly pigmented, smooth. Pleonites 1-5 together as long as greatest width or longer than greatest width. Antenna 1 flagellum in juvenile and female of c. 5-7 articles. Antenna 2 flagellum of 7-9 articles, longer than article 5 of peduncle. Eyes present, or absent. Maxillipedal endite present (reaching palp article 3); palp articles 1-5 free, or minute article 5 fused with article 4; palp article 3 barely produced medially beyond articles 4-5 (if at all); palp terminal articles (4-5) oblique, without a free mesial margin between its suture and distal group of setae. Pereiopod 1 propodus palm straight or sinuous, without prominent tooth. Pereiopods 4-7 carpus twice width of propodus, rectangular, with obvious free distal margin (usually with row of pectinate setae) defined by stout seta on posterodistal angle, with robust seta on posterior margin and on posterodistal angle; propodus with 1 robust setae on posterior margin (in addition to posterodistal robust seta).

Male: Antenna 1 flagellum of male of 1 basal + c. 20 aesthetasc-bearing articles.

Species and distribution: 12. Widespread in Southern Hemisphere, Iceland; 1-4383 m.

Remarks: Species of *Quantanthura* are recognized by the broad carpus of pereiopods 4-7 with a row of pectinate setae along the free distal margin. Poore & Lew Ton (1986c) recognized two species-groups among the nine species then known. Negoescu (1994) provided a key to ten species. Examination of material of *Agulanthura serenasimus* confirms that, contrary to Kensley's (1975a) figures, the maxillipedal palp has a distinct basal article and an endite is present as figured by Kensley (1982c). This condition and the broad pereiopodal carpi are exactly as seen in several species of *Quantanthura* and the two genera are here synonymized. *Quantanthura* is the most plesiomorphic of anthurid genera, notably in the 5-articulate maxillipedal palp, maxillipedal endite and robust setae on the margins of carpus and propodus of pereiopods 4-7.

Sauranthura Poore & Kensley

Sauranthura Poore & Kensley 1981: 510.

Type species: *Sauranthura goldmanorum* Poore & Kensley 1981, original designation.

Diagnosis: Body total length of adult less than 3 mm, with dark species-specific pigment pattern, smooth. Pleonites 1-5 together shorter than wide.

Antenna 2 flagellum of 1-2 articles, very short. Maxillipedal endite present as a triangular lobe; palp articles 1-5 fused. Pereiopod 1 propodus palm with step or prominent tooth (more pronounced in male). Pereiopods 2, 3 carpus with robust seta posterodistally. Pereiopods 4-7 carpus more or less pentagonal, anterior margin half length of posterior margin, without robust setae on posterior margin or on posterodistal angle.

Male: Unknown.

Species and distribution: Unique. Great Barrier Reef, Australia; intertidal to 10 m.

Remarks: The only species was described on the basis of a collection of small individuals, all mancas. More recently collected material in the collections of Museum Victoria includes adult females with seventh pereiopods. The small size, 2 mm long, compact habitus, and fused maxillipedal palp articles distinguish the genus (cf. *Pendanthura*).

Skuphonura (Barnard)

Skuphonura Barnard 1925: 145; Kensley 1980c: 731; Menzies & Kruczynski 1983: 30; Brusca & Müller 1991: 593-594.

Type species: *Skuphonura laticeps* Barnard 1925, monotypy.

Diagnosis: Body not darkly pigmented, smooth. Head much wider anteriorly in male, or parallel-sided, with produced chin in males. Pleonites 1-5 together as long as greatest width. Antenna 2 flagellum of 1-2 articles, shorter than article 5 of peduncle. Maxillipedal endite obsolete or absent; palp articles 1-2 fused, 3 free and 4-5 fused; palp article 3 barely produced medially beyond articles 4-5 (if at all); palp terminal articles (4-5) oblique, without a free mesial margin between its suture and distal group of setae. Pereiopod 1 propodus palm with step or prominent tooth (more pronounced in male). Pereiopods 4-7 carpus more or less pentagonal, anterior margin half length of posterior margin, with 1 robust seta on posterodistal angle.

Male: Antenna 1 flagellum of male of 1 basal + 1 or 2 short aesthetasc-bearing articles.

Species and distribution: Five. Caribbean, Gulf of Mexico, Brazil; intertidal to 73 m.

Remarks: The type species was redescribed and three others erected by Kensley (1980c). The genus was reviewed by Brusca & Müller (1991) who provided a key to the species. The genus is closest to the *Apanthura* group of genera but differs from these in the possession of a very short antenna 1 flagellum in the male and only 1 or 2 articles in the flagellum of antenna 2. A ventral chin is always present in both males and females, a feature of some members of the other genera.

Stygocyathura Botosaneanu & Stock

Cyathura (*Stygocyathura*) Botosaneanu & Stock 1982: 38-39.

Type species: *Cyathura curassavica* (Stork 1940), original designation.

Diagnosis: Body not darkly pigmented, smooth. Pleonites 1-5 together shorter than wide; pleotelson without indication of posterior margin of pleonite 6. Antenna 2 flagellum of 1-2 articles, very short. Eyes absent. Mandibular palp article 3 with 1-4 setae on distal margin, or with row of 10-15 setae on distal oblique margin. Maxillipedal endite obsolete or absent; palp articles 1-2 fused, 3 free and 4-5 fused; palp article 3 barely produced medially beyond articles 4-5 (if at all); palp terminal articles (4-5) at least one-third as long as penultimate article (3), almost as long as broad, with a free mesial margin between its suture and distal group of setae. Pereiopod 1 propodus palm with step or prominent tooth (more pronounced in male). Pereiopods 4-7 carpus more or less pentagonal, anterior margin half length of posterior margin, without robust setae on posterior margin or on posterodistal angle. Uropodal exopod linear, articulating transversely.

Male: Antenna 1 flagellum of male of 1 basal + c. 10 aesthetasc-bearing articles.

Species and distribution: 20. Caribbean and South Pacific Ocean islands, Oman; freshwater stygiobiont.

Remarks: *Stygocyathura* was erected as a monophyletic subgenus of *Cyathura* for stygiobiont species of *Cyathura* from the Caribbean Sea. The characters differentiating the species from species of *Cyathura* are the narrow uropodal exopod, never lying against the telson like an 'elephant's ear,' eyes lacking, pleonite 6 and telson completely fused, pleopod 1 endopod reduced, and smooth pereiopod 1 palm. Wägele et al. (1987) and Wägele (1985a) agreed that the subgenus is a monophyletic taxon adding that the telson is generally broader than in *Cyathura*. They concluded that the group had invaded freshwater at the margins of the Tethys Sea. The apex of the appendix masculina of species of *Stygocyathura* is more complex than the plesiomorphic simple condition in most species of *Cyathura* (Wägele 1985a). However, some species of *Cyathura* (e.g. *C. polita*, *C. carinata*) are complex too. Phylogenetic analysis found that *Stygocyathura* is not a sister taxon to *Cyathura* but to a larger group of genera.

Expanathuridae fam. nov.

Type genus: *Expanathura* Wägele 1981d.

Diagnosis: Body 10-15 times as long as wide, non-males rarely more elongate (pereionite 7 wider than long, much shorter than pereionite 6). Pleonites 1-5 together not more than twice as long as wide, free and articulating or fused; with plumose setae on margins of epimera or posterior margins of pleonite 5 and sometimes pleonite 4. Antenna 2 flagellum of fewer than 10 articles, shorter than peduncle, or of numerous articles, longer than or shorter than peduncle. Mouth parts not produced anteriorly. Mandible tapering and incisor simple, or compact and with weakly-toothed transverse incisor (rarely). Maxillipedal endite reaching palp article 3, or absent or obsolete; palp narrow and tapering distally to rounded apex (c. 3 times as long as wide) or reduced, digitiform (*Eisothistos*), with 5 free articles or with 2 or more articles fused. Pereiopods 2 and 3 carpus not or weakly produced posterodistally; propodus with 1 robust seta in addition to posterodistal robust seta, or with posterodistal robust seta only, or without posterodistal robust seta. Pereiopods 4-7 propodus with more than 1 robust setae in addition to posterodistal robust seta, or with posterodistal robust seta only. Pereiopod 7 propodus with anterodistal serrate setae. Pleopods: pleopod 1 exopod and endopod together weakly operculiform, or pleopod 1 exopod operculiform alone (*Expanathura* only). Statocysts absent.

Remarks: The Expanathuridae are on first impression a heterogeneous family but this is less apparent when the highly derived genus *Eisothistos* is ignored. Characters which unite the remaining genera were discussed in more detail in the *Phylogeny* section but include: absence of statocysts, presence of a maxillipedal endite and palp of five articles, anterodistal serrate setae on the propodus of pereiopod 7, and the weakly operculiform pleopod 1. Some genera have highly ornamented telson and uropodal margins (not seen in other families) and it is this condition which is shared with *Eisothistos* and supports its placement in the family.

Expanathurids are distinguished from anthurids, which they superficially resemble, by the absence of statocysts, usually free pleonites, and by the primitive condition of the maxilliped. The family is most closely related to Antheluridae with which they share several plesiomorphies but differ in the absence of statocyst, having a more acute maxilliped and less developed operculiform pleopods 1.

Key to genera of Expanathuridae

1. Pereiopods 1-3 only slightly differentiated from posterior pereiopods, propodi not swollen (Figs 1b, c, 6f); pleopod 1 rami fused in juveniles (Fig. 10g); mandible with molar process and palp absent (Fig. 4g); maxillipedal endite absent, palp of 2-5 weakly differentiated elongate articles (Fig. 5f); female pereion and pereiopods covered with long setae, body becoming extremely elongate after spawning .... *Eisothistos*
– Pereiopods 1-3 well differentiated from posterior pereiopods, propodi swollen (Figs 1d, 6b-e); pleopod 1 rami free (Fig. 10h); mandible with molar process and palp present (Fig. 4f); maxillipedal endite present, palp of well differentiated articles (Fig. 5g); female pereion and pereiopods not covered with long setae, never elongate 2

2. Pleonites 5 and 6 not longer than pleonites 1-4; pleonite 6 clearly fused dorsally to telson; uropodal exopod a simple narrow leaf (Fig. 11i) .... *Coralanthura*
– Pleonites 5 and 6 longer than pleonites 1-4 or all pleonites fused; pleonite 6 usually marked off dorsally from telson; uropodal exopod broad or complex (Fig. 11f) 3

3. Uropodal exopod with spike-like dorsal lobe, rami and telson with dentate margins (Fig. 11h) 4
– Uropodal exopod with obsolete dorsal lobe, rami and telson with entire or dentate margins (Fig. 11f) 6

4. Uropodal exopod spike shorter than proximal plate (Fig. 11h); mandibular palp with 3 articles; maxillipedal endite absent .... *Heptanthura*
– Uropodal exopod spike shorter or longer than proximal plate; mandibular palp with fewer than 3 articles; maxillipedal endite present 5

5. Telsonic apex tapering to point from midpoint; mandibular palp of 2 articles (Fig. 4e); maxillipedal palp broad (Fig. 5d) .... *Rhiganthura*
– Telsonic apex apically rounded; mandibular palp of 1 seta only; maxillipedal palp acute ....*Minyanthura*

6. Telson 1.5 times as long as wide (Fig. 11f); pereiopod 1 propodus broader than that of pereiopod 2, strongly tapering, obscurely dentate proximally (Figs 6b, d); male antenna 1 flagellum with basal article short and with numerous aesthetascs (Fig. 3q) ....*Expanathura*
– Telson twice as long as wide; pereiopod 1 propodus similar to that of pereiopod 2, scarcely tapering, palm smooth (Fig. 6d); male antenna 1 flagellum with rows of aesthetascs on all articles (Fig. 3r) .... *Panathura*

Genera of Expanathuridae

Generic characters with implicit states. Of 37 characters used to diagnose expanathurid genera nine have implicit values, that is, apply to six of the seven genera. Their states follow. Body and limbs not covered by scattered long setae. Mouth parts not produced anteriorly. Pereiopod 1 propodus cylindrical, palm smooth; merus narrower than swollen propodus. Pereiopods 1-7 merus without robust seta on anterior margin. Distal palmar robust setae on pereiopods 2-7 propodus and 4-7 carpus simple or flagellate. Pleopodal 2-5 exopods little narrower than endopods, with continuous row of plumose setae around distal and lateral margin. Apices of rami without serrate robust seta.

Coralanthura Poore & Kensley

Coralanthura Poore & Kensley 1981: 503-506; Negoescu & Wägele 1984: 113.

Type species: *Coralanthura endeavourae* Poore & Kensley 1981, original designation.

Diagnosis: Body typically elongate (pereionite 1 longer than wide), irregularly pigmented; body and limbs not covered by scattered long setae. Pleonites 1-5 free and articulating, together longer than wide in juvenile; pleonites of similar lengths; with plumose setae on ventral margins of all epimera. Pleotelson without indication of posterior margin of pleonite 6. Eyes present in both sexes. Mouth parts not produced anteriorly. Mandible compact, incisor toothed; molar blunt; palp of 3 articles. Maxillipedal endite apically rounded, reaching to palp article 3; palp broad (1.5-2 times as wide as long), article 4 semicircular; palp articles 1 and 2 separated by suture. Pereiopods 1-3 differentiated from posterior pereiopods, propodus 1 swollen. Pereiopod 1 palm of dactylus tuberculate; propodus cylindrical, palm smooth, with mesial setae few, in row along length of palm; merus narrower than swollen propodus. Pereiopods 1-7 merus with robust seta on anterior margin. Pereiopods 2 and 3 propodus with 1 distal robust seta on cutting edge. Pereiopods 4-7 propodus with 1 distal robust seta on cutting edge; carpus about as long as wide, with short anterior margin. Distal palmar robust setae on pereiopods 2-7 propodus and 4-7 carpus simple or flagellate. Pereiopod 7 propodus without serrate setae mesiodistally. Pleopodal rami similar in both sexes, pleopod 1 weakly operculiform in both sexes, with exopod slightly wider than endopod but not enclosing it totally, pleopod 1 endopod as long or longer than exopod and partly exposed. Pleopodal 2-5 exopods little narrower than endopods, with continuous row of plumose setae around distal and lateral margin. Uropodal exopod a narrow leaf articulating proximally; apices of rami without serrate robust seta. Telson with smooth distal margin.

Male: Antenna 1 flagellum of male of 6-7 aesthetasc-bearing articles.

Species and distribution: Monotypic. Tropical Australia and Coral Sea; coral reefs; 0-21 m.

Remarks: The single species, *C. endeavourae* is placed within Expanathuridae on the basis of biting mandible, maxillipedal endite and 5-articled palp, axial pereiopodal palm, no statocyst, weakly operculiform pleopod 1. It differs from the other genera in that pleonites 1-4 are not more shortened than pleonite 5 and the mandibular incisor is not produced. It does not possess the anterodistal serrate setae characteristic of pereiopod 7 of other expanathurids (except the specialised *Eisothistos*) but has instead simple setae, extending along the anterior margin of adult females. These may be homologous. The species is superficially similar to *Expanathura amstelodami* and *E. ardea* in the lanceolate uropodal exopod and tapering telson but the mouth parts and pleopod 1 are quite different.

The species is unusual among anthurideans in the strong spiniform seta on the anterior margin of the pereiopodal merus. The genus has not been discussed by other authors. Poore & Kensley (1981) included a second species in their original paper; this, *C. ardea*, is now a species of *Expanathura*.

Eisothistos Haswell

Eisothistos Haswell 1884: 677-679; Barnard 1925: 133; Wägele 1979c: 2-19; Wägele 1981d: 72-75; Negoescu & Wägele 1984: 117.

Heteranthura Kensley 1980a: 24; Shyamasundari et al. 1991: 262 (type species: *Heteranthura anomala* Kensley 1980a, original designation).

Type species: *Eisothistos vermiformis* Haswell 1884, monotypy.

Diagnosis: Body of post-spawning female grossly elongate; of juvenile and male of usual proportions, not pigmented; areas of body and bases of some limbs of female covered with scattered long setae. Pleonites 1-5 free and articulating and fused in post-spawning female, together longer than wide in male and much wider than long in juvenile and pre-spawning female; pleonites 1-3 longer than 4 and 5 in male and manca, all of similar lengths in juvenile and female; with plumose setae on ventral margins of all epimera. Pleotelson with posterior margin of pleonite 6 indicated dorsally, delineated from telson, or without indication of posterior margin of pleonite 6. Eyes present in both sexes, or absent and grossly enlarged in male. Mandible and maxilla produced anteriorly beyond rostrum. Mandible linear, incisor a short transverse blade; molar absent; palp absent. Maxillipedal endite obsolete or absent; palp of 1-5 weakly differentiated narrow articles. Pereiopods 1-3 only slightly differentiated from posterior pereiopods, propodi not more swollen than merus. Pereiopod 1 palm of dactylus smooth; propodus cylindrical, palm smooth, with mesial setae few, in row along length of palm; merus as wide as narrow propodus. Pereiopods 1-7 merus without robust seta on anterior margin. Pereiopods 2 and 3 propodus with 1 distal robust seta on cutting edge. Pereiopods 4-7 propodus with 1 distal robust seta on cutting edge; carpus about as long as wide, with short anterior margin. Distal palmar robust setae on pereiopods 2-7 propodus and 4-7 carpus simple or flagellate. Pereiopod 7 propodus without serrate setae mesiodistally. Pleopodal rami differentiated between sexes, pleopod 1 weakly operculiform, with rami fused in juvenile and prespawning female. Pleopodal 2-5 exopods much narrower and shorter than endopods in juvenile and female, exopods 1-4 with both rami narrow in male, 5 of male as in female. Uropodal exopod with strong apical spike, with serrate margin; with serrate lateral lobe between articulation and apical spike; endopod and spike of exopod with apical serrate robust seta. Telson with dentate distal margin.

Male: Pleonites 1-3 elongate. Eyes grossly enlarged. Antenna 1 flagellum of male of 1 very short basal article bearing numerous aesthetascs and c. 7 longer articles each with 1-2 aesthetascs (or 7 longer articles, second with bunches of aesthetascs). Mouth parts reduced. Pereiopods more slender than in juvenile or undifferentiated female. Pleopods 1-3 with elongate peduncle and free rami, with long plumose setae.

Post-spawning female: All pleonites very short. Pereion and pereiopods covered with long setae. Body becoming moderately or extremely elongate after spawning.

Species and distribution: 19. Cosmopolitan except Antarctica but rare; intertidal and shallow subtidal to 385 m.

Remarks: In his detailed review of the genus and descriptions of three new species Wägele (1979c) described the extreme differences between juveniles, adult females, post-spawning females and adult males of species of *Eisothistos*. Juveniles and undifferentiated females have fused rami on pleopod 1 which is barely if at all operculiform. The exopods on pleopods 2-5 are reduced. Although the mouth parts are in some ways reduced (linear maxilliped without endite, molar and mandibular palp absent) the curved teeth on the outer lobe of the maxilla are not as obsolete as in other anthurideans. The male is small and possesses pleonites 1-3 enlarged relative to the shorter pleonites 4 and 5. Its pleopods 1-3 are all biramous and have elongate peduncle and rami; the appendix masculina on pleopod 2 is simple; and pleopods 4 and 5 are as in the juvenile. The antenna 1 has dense rows of proximal aesthetascs, or callynophore (Lowry 1986), on the first article of the flagellum while the remaining flagellar articles are as in the female. The eyes are grossly enlarged and occupy most of the side of the head. Pereiopod 1 has a linear row of spiniform setae along the elongate carpus and propodus. The mouth parts are atrophied.

It appears that the female passes through several instars but this is not clear. At first the mouth parts (except the maxilliped) are reduced to swollen lobes with reduced spines. Pereiopod 1, including the coxal plates attached to the side of the body, and later pereiopods 2-7, become covered with long setae. The presumed terminal vermiform phase is rarely seen. No stage with oostegites has been reported.

A unique feature of the genus is the presence of a strong setulate spiniform seta at the apex of the spike of the uropodal exopod and on the uropodal endopod.

Kensley (1980a) erected *Heteranthura* for one new species, *H. anomala*, without comparing it to other anthuridean genera. The illustrations of the holotype are clearly that of a juvenile *Eisothistos*, an observation made by Negoescu & Wägele (1984: 103) who synonymised the two genera.

The type species and others were taken from the tube of a serpulid polychaete and Wägele (1981d) recorded that these worms were eaten by *Eisothistos*.

Expanathura Wägele

Expanathura Wägele 1981d: 89; 121-122. Negoescu & Wägele 1984: 118, 136.

Type species: *Panathura amstelodami* Kensley 1976, original designation.

Diagnosis: Body typically elongate (pereionite 1 longer than wide), irregularly pigmented; body and limbs not covered by scattered long setae. Pleonites 1-5 free and articulating, together longer than wide in juvenile; pleonites 5 and 6 longer than pleonites 1-4; with plumose setae on dorsal and lateral margins of pleonite 5 (at least in some life stages). Pleotelson with posterior margin of pleonite 6 indicated dorsally, delineated from telson. Eyes present in both sexes and grossly enlarged in male. Mouth parts not produced anteriorly. Mandible tapering, incisor simple; molar absent; palp of 3 articles. Maxillipedal endite tapering to acute apex, reaching to at least palp article 4; palp narrow and tapering distally to apex (at least twice as long as wide); palp articles 1 and 2 separated by suture. Pereiopods 1-3 differentiated from posterior pereiopods, propodus 1 swollen. Pereiopod 1 palm of dactylus smooth; propodus as broad as long, triangular, palm dentate proximally, with mesial setae dominated by 1 much stronger than others; merus narrower than swollen propodus. Pereiopods 1-7 merus without robust seta on anterior margin. Pereiopods 2 and 3 propodus without robust setae on cutting edge. Pereiopods 4-7 propodus with row of robust setae along cutting edge (at least 1 in addition to distal seta); carpus about as long as wide, with short anterior margin. Distal palmar robust setae on pereiopods 2-7 propodus and 4-7 carpus simple or flagellate. Pereiopod 7 propodus with short transverse row of serrate setae distally (on mesially directed face). Pleopodal rami similar in both sexes, pleopod 1 exopod operculiform alone and enclosing narrower endopod in both sexes, pleopod 1 endopod shorter than exopod and totally covered by it. Pleopodal 2-5 exopods little narrower than endopods, with continuous row of plumose setae around distal and lateral margin. Uropodal exopod a broad leaf articulating longitudinally, with smooth margin; without serrate lateral lobe; apices of rami without serrate robust seta. Telson with smooth distal margin.

Male: Head with rostrum more strongly developed, anterolateral lobes reduced, eyes enlarged. Antenna 1 flagellum of about 8 articles, basal article short and broad and bearing aesthetascs around its proximal margin; distal articles without aesthetascs; terminal article with 2 long setae. Pereiopods 1-3 more elongate than in female or juvenile; pereiopod 1 propodus without dense mesial setae. Pleon elongate; pleonites 5 not markedly longer than anterior ones.

Species and distribution: Six. Indian Ocean (African and Australian coasts), and tropical and subtropical south-western Pacific (islands and northern Australia).

Remarks: Wägele (1981d) recognised that some species of *Panathura* (*P. amstelodami*, *P. collaris* (Kensley) and *P. macronesia* (Kensley) have different mouth parts from the type species (*P. serricauda* Barnard). He erected *Expanathura* for them on the basis of their long, apically acute maxillipedal endite and elongate mandible with reduced molar (Negoescu & Wägele 1984). Other characters which differentiate *Expanathura* from *Panathura* are the form of pereiopods 1-3, the shortened pleopod 1 endopod, and the unique form of male antenna 1.

The genus can be divided into two groups: the first including *Expanathura amstelodami*, *E. ardea* (Poore & Kensley) and *E. mooreae* (Müller); and the second, *E. collaris*, *E. haddae* (Kensley & Poore) and *E. macronesia*. The first group is characterized by narrow uropodal rami, tapering telson, and short pleopod 1 endopod. Species of the second group are morphologically very similar to each other; they have very broad uropodal rami, broad telson, and moderately shortened pleopod 1 endopod. Many specimens belonging to this group have brown pigmentation dorsally and laterally. They can be distinguished using the shape of the uropod and telson.

Heptanthura Kensley

Heptanthura Kensley 1978a: 775-776; Wägele 1981d: 72, 74; Negoescu & Wägele 1984: 120.

Stellanthura Wägele 1979d: 21; Wägele 1981d: 72, 74; Negoescu & Wägele 1984: 134 (type species: *Stellanthura cryptobia* Wägele, 1981d, original designation). syn. nov.

Chalixanthura Kensley 1984: 2 (type species: *Chalixanthura scopulosa* Kensley, 1984, original designation). syn. nov.

Licranthura Kensley & Schotte 1987: 216-217 (type species: *Licranthura amyle* Kensley & Schotte 1987, monotypy). syn. nov.

Type species: *Heptanthura novaezealandiae* Kensley 1978a, original designation.

Diagnosis: Body typically elongate (pereionite 1 longer than wide), not pigmented or irregularly pigmented; body and limbs not covered by scattered long setae. Pleonites 1-5 free and articulating or fused, together longer than wide in juvenile; pleonites 5 and 6 longer than pleonites 1-4; with plumose setae on dorsal and lateral margins of pleonite 5 (at least in some life stages). Pleotelson with posterior margin of pleonite 6 indicated dorsally, delineated from telson. Eyes present in both sexes. Mouth parts not produced anteriorly. Mandible tapering, incisor simple; molar blunt; palp of 3 articles, or absent. Maxillipedal endite obsolete or absent; palp narrow and tapering distally to apex (at least twice as long as wide); palp articles 1-2 fused. Pereiopods 1-3 only slightly differentiated from posterior pereiopods, propodi not more swollen than merus. Pereiopod 1 palm of dactylus smooth; propodus cylindrical, palm smooth, with mesial setae dominated by 1 much stronger than others; merus narrower than swollen propodus. Pereiopods 1-7 merus without robust seta on anterior margin. Pereiopods 2 and 3 propodus with 1 distal robust seta on cutting edge. Pereiopods 4-7 propodus with 1 distal robust seta on cutting edge; carpus about as long as wide, with short anterior margin. Distal palmar robust setae on pereiopods 2-7 propodus and 4-7 carpus bifid and blade-like. Pereiopod 7 propodus with short transverse row of serrate setae distally (on mesially directed face). Pleopodal rami similar in both sexes, pleopod 1 weakly operculiform in both sexes, with exopod slightly wider than endopod but not enclosing it totally, pleopod 1 endopod as long or longer than exopod and partly exposed. Pleopodal 2-5 exopods little narrower than endopods, with continuous row of plumose setae around distal and lateral margin. Uropodal exopod with strong apical spike, with serrate margin; without serrate lateral lobe; apices of rami without serrate robust seta. Telson with dentate distal margin.

Male: Body and all appendages more elongate than in female and juvenile. Eyes grossly enlarged. Antenna 1 flagellum of 11 aesthetasc-bearing articles. Mouth parts reduced: mandibular incisor, lamina dentata and molar obsolete; maxilliped narrow. Pereiopod 1 propodus with few marginal and mesial setae. Pereiopod 7 without pectinate setae on anterodistal margin. Pleopod 2 with simple appendix masculina, 2-articled exopod. Pleon elongate; pleonite 5 not markedly longer than anterior pleonites.

Species and distribution: Eight. Caribbean, New Zealand, Australia; intertidal to 101 m.

Remarks: Synonymy of *Stellanthura* with *Heptanthura* is based on comparison of Wägele's (1979d) detailed drawings of the type species, *S. cryptobia*, with Kensley's (1978a) figures of *H. novaezealandiae*, a similar species from Australia (Kensley & Poore 1982) and an undescribed Australian species. The broad dentate palmar setae of pereiopods 4-7 of all species is unique. Wägele also figured a bicuspid seta at the base of the unguis of pereiopod 1, identical to that of *Heptanthura*. Kensley illustrated the maxilliped with the usual first palp article free. Wägele's figures show it fused to article 2 as in the similar genus *Rhiganthura* and this condition is confirmed in *Heptanthura*.

Chalixanthura scopulosa resembles species of *Heptanthura* in several characters. General body form, pigmentation, pereiopod shape and setation, maxilliped, pleopods, and shape of the telson and uropodal rami are very similar. Notable are the broad dentate palmar setae of pereiopods 4-7 and the short proximal spines on the dactylus of pereiopod 2. The maxillipedal endite of *C. scopulosa* is shorter than in any other species assigned to *Heptanthura*. *C. columbiana* is very similar to *C. scopulosa*, especially in uropodal exopod but both can be accommodated in *Heptanthura*.

Chalixanthura lewisi, on the other hand, is very different and may warrant a separate genus. It lacks the typical maxilliped, uropod, pleopod 1 and pereiopod setation of *Heptanthura* and in its free pleonites, long acute maxillipedal endite and palp, pleopod 1, telson, uropodal exopod and pereiopodal setation is similar to *Expanathura*. The shape of the pereiopods 1-3, with cylindrical propodi, is more like species of *Heptanthura* than of *Expanathura* which are very characteristic. Kensley (1984) placed *Chalixanthura* in Hyssuridae and compared it with *Coralanthura* and *Panathura*; Kensley & Snelgrove (1987) moved it without comment to Anthuridae.

Licranthura amyle Kensley & Schotte 1987 differs only slightly from those already mentioned: non-swollen propodi of pereiopods 1 and 2, the unusual complex distal palmar seta of the pereiopods, serrate margins of the telson and uropodal exopod and the general shape of the latter. The unique toothed article 3 of antenna 2 and the absence of a mandibular molar and palp are viewed as specific characters.

Minyanthura (Kensley)

Minyanthura Kensley 1982a: 341-343.

Type species: *Minyanthura corallicola* Kensley 1982a, original designation.

Diagnosis: Body compact (pereionite 1 shorter than wide) and total length less than 4 mm, irregularly pigmented; body and limbs not covered by scattered long setae. Pleonites 1-5 fused, together wider than long in both sexes; with plumose setae on dorsolateral and lateral margins of pleonites 4 and 5. Pleotelson without indication of posterior margin of pleonite 6. Eyes present in both sexes. Mouth parts not produced anteriorly. Mandible tapering, incisor simple; molar absent; palp of 1 seta. Maxillipedal endite tapering to acute apex, reaching to at least palp article 4; palp narrow and tapering distally to apex (at least twice as long as wide); palp articles 1 and 2 separated by suture. Pereiopods 1-3 differentiated from posterior pereiopods, propodus 1 swollen. Pereiopod 1 palm of dactylus smooth; propodus cylindrical, palm smooth, with mesial setae dominated by 1 much stronger than others; merus narrower than swollen propodus. Pereiopods 1-7 merus without robust seta on anterior margin. Pereiopods 2 and 3 propodus with 1 distal robust seta on cutting edge. Pereiopods 4-7 propodus with 1 distal robust seta on cutting edge; carpus about as long as wide, with short anterior margin. Distal palmar robust setae on pereiopods 2-7 propodus and 4-7 carpus simple or flagellate. Pereiopod 7 propodus with short transverse row of serrate setae distally (on mesially directed face). Pleopodal rami similar in both sexes, pleopod 1 weakly operculiform in both sexes, with exopod slightly wider than endopod but not enclosing it totally, pleopod 1 endopod as long or longer than exopod and partly exposed. Pleopodal 2-5 exopods little narrower than endopods, with continuous row of plumose setae around distal and lateral margin. Uropodal exopod with strong apical spike, with serrate margin; without serrate lateral lobe; apices of rami without serrate robust seta. Telson with dentate distal margin.

Male: Body more elongate than in juvenile. Pleonites 1-5 free. Eyes grossly enlarged in male. Antenna 1 flagellum of male of 6 articles 4 with aesthetascs.

Species and distribution: Three. Caribbean, Réunion I., Society Is; 0-24 m.

Remarks: *Minyanthura* is placed in Expanathuridae on the basis of a spike on the uropodal endopod, serrate margins to uropodal rami and telson, pigmentation, the serrate seta(e) on the anterodistal margin of the propodus of pereiopod 7, the non-overlapping rami of pleopod 1, mandible, maxilliped and marginal plumose setae on pleonites 4 and 5.

It is close to *Heptanthura* but differs in the reduced mandibular palp and absence of maxillipedal endite. *Rhiganthura* also has reduced mouth parts but differs in the structure of the telson and uropods.

Juveniles of *Minyanthura* have fused pleonites 1-5, a character state shared only with an undescribed species of *Heptanthura* in this family and with members of Anthuridae and Antheluridae. The pleonites are free in the male (Müller 1990a, 1991a). Like members of *Expanathura*, *Eisothistos* and *Heptanthura*, the largest species is very small at 1.7 mm total length.

Kensley (1982a) reported two statocysts at the base of the telson of the type species but Müller (1990c) interpreted these as 'pits' in a juvenile female figured by him. Müller (1991a) observed 'no statocysts' in a male. His illustrations may be of another species but absence of statocysts would be consistent with placement of *Minyanthura* in Expanathuridae. There are no statocysts in the male of *M. firingae* (Müller 1990a).

Moore (1996) found that *M. corallicola* and *M. firingae* were more closely related to each other than to *M. pacifica*.

Panathura Barnard

Panathura Barnard 1925: 143; Kensley 1982c: 159-160; Negoescu & Wägele 1984: 128-129.

Type species: *Apanthura serricauda* Barnard 1920, monotypy.

Diagnosis: Body typically elongate (pereionite 1 longer than wide), not pigmented or irregularly pigmented; body and limbs not covered by scattered long setae. Pleonites 1-5 free and articulating, together longer than wide in juvenile; pleonites 5 and 6 longer than pleonites 1-4; with plumose setae on dorsolateral and lateral margins of pleonites 4 and 5. Pleotelson with posterior margin of pleonite 6 indicated dorsally, delineated from telson. Eyes absent, or present in both sexes. Mouth parts not produced anteriorly. Mandible tapering, incisor simple; molar blunt; palp of 3 articles. Maxillipedal endite apically rounded, reaching to palp article 3; palp narrow and tapering distally to apex (at least twice as long as wide); palp articles 1 and 2 separated by suture. Pereiopods 1-3 differentiated from posterior pereiopods, propodus 1 swollen. Pereiopod 1 palm of dactylus smooth; propodus cylindrical, palm smooth, with mesial setae few, in row along length of palm; merus as wide as narrow propodus. Pereiopods 1-7 merus without robust seta on anterior margin. Pereiopods 2 and 3 propodus with 1 distal robust seta on cutting edge. Pereiopods 4-7 propodus with 1 distal robust seta on cutting edge; carpus about as long as wide, with short anterior margin. Distal palmar robust setae on pereiopods 2-7 propodus and 4-7 carpus simple or flagellate. Pereiopod 7 propodus with short transverse row of serrate setae distally (on mesially directed face). Pleopodal rami similar in both sexes, pleopod 1 weakly operculiform in both sexes, with exopod slightly wider than endopod but not enclosing it totally, pleopod 1 endopod as long or longer than exopod and partly exposed. Pleopodal 2-5 exopods little narrower than endopods, with continuous row of plumose setae around distal and lateral margin. Uropodal exopod a broad leaf articulating longitudinally, with smooth margin; without serrate lateral lobe; apices of rami without serrate robust seta. Telson with smooth distal margin.

Male: Head with rostrum more strongly developed, anterolateral lobes reduced, eyes grossly enlarged. Antenna 1 flagellum of 7 or more articles, all with aesthetascs on lateral margins only; terminal article with 2 long setae. Pereiopods 1-7 more elongate than in female or juvenile; pereiopod 1 propodus without dense mesial setae. Pleon elongate; pleonites 5 not markedly longer than anterior ones. Appendix masculina longer than pleopod 2 rami.

Species and distribution: Monotypic (except for undescribed Australian species). South Africa, Amsterdam and St Paul I.; intertidal to 90 m.

Remarks: The type species was described and illustrated by Barnard (1920), Kensley (1976, 1978b, 1982c) and Wägele (1981d). Undescribed species exist in Australia to a depth of 1000 m.

Rhiganthura Kensley

Rhiganthura Kensley 1978a: 779; Wägele 1981d: 72, 74.

Type species: *Rhiganthura spinosa* Kensley 1978a, original designation.

Diagnosis: Body compact (pereionite 1 shorter than wide) and total length less than 4 mm, not pigmented; body and limbs not covered by scattered long setae. Pleonites 1-5 free and articulating, together wider than long in both sexes; pleonites 5 and 6 longer than pleonites 1-4 and pleonite 1 very short, concealed dorsally; with plumose setae on dorsal and lateral margins of pleonite 5 (at least in some life stages). Pleotelson without indication of posterior margin of pleonite 6. Eyes present in both sexes. Mouth parts not produced anteriorly. Mandible tapering, incisor simple; molar blunt; palp of 1 or 2 articles. Maxillipedal endite apically rounded, reaching to palp article 3; palp broad (1.5-2 times as wide as long), article 4 semicircular; palp articles 1-2 fused. Pereiopods 1-3 differentiated from posterior pereiopods, propodus 1 swollen. Pereiopod 1 palm of dactylus tuberculate and sometimes carpus and propodus tuberculate; propodus cylindrical, palm smooth, with mesial setae few, in row along length of palm; merus narrower than swollen propodus. Pereiopods 1-7 merus without robust seta on anterior margin. Pereiopods 2 and 3 propodus with 1 distal robust seta on cutting edge. Pereiopods 4-7 propodus with row of robust setae along cutting edge (at least 1 in addition to distal seta); carpus about as long as wide, with short anterior margin. Distal palmar robust setae on pereiopods 2-7 propodus and 4-7 carpus simple or flagellate. Pereiopod 7 propodus with short transverse row of serrate setae distally (on mesially directed face). Pleopodal rami similar in both sexes, pleopod 1 weakly operculiform in both sexes, with exopod slightly wider than endopod but not enclosing it totally, pleopod 1 endopod as long or longer than exopod and partly exposed. Pleopodal 2-5 exopods little narrower than endopods, with continuous row of plumose setae around distal and lateral margin. Uropodal exopod with strong apical spike, with serrate margin; with serrate lateral lobe between articulation and apical spike; apices of rami without serrate robust seta. Telson with dentate distal margin.

Male: Unknown.

Species and distribution: Monotypic (except for undescribed Australian species). New Zealand; 101 m.

Remarks: *Rhiganthura* is similar to *Heptanthura* and *Minyanthura* with which it shares dentate margins to the uropodal rami and telson, a strong spike-shaped dorsal lobe on the uropodal exopod. It differs in the form of the mandibular palp and fusion of article 5 of the maxillipedal palp. The phylogenetic analysis suggests that these similarities are retained plesiomorphies or convergences; the genus is sister taxon to *Coralanthura*, sharing loss of the ridge between pleonite 6 and the telson and a tuberculate palm on pereiopod 1. It shares short pereionites and pleonites with *Minyanthura*. An undescribed species exists in the intertidal of eastern Australia. It differs from the type in possession of a 1-articulate mandibular palp (2 articles in the type species), and a visible terminal article of the maxillipedal palp.

Hyssuridae Wägele

Hyssuridae (genus-group A) Wägele 1981d: 68, 103; Wägele 1981b: 48.
Hyssuridae; Kussakin 1982: 11; Poore & Lew Ton 1988c: 172.

Diagnosis: Body slender (pereionite 7 longer than wide, little or not shorter than pereionite 6). Pleonites 1-5 together 3 times as long as wide, free and articulating; without marginal plumose setae on pleonal epimera or posterior borders of pleonites 4 and 5. Antenna 2 flagellum of fewer than 10 articles, shorter than peduncle. Mouth parts not produced anteriorly. Mandible compact and with weakly-toothed transverse incisor. Maxillipedal endite reaching palp article 3, or absent or obsolete; palp broad (c. twice as long as wide), with 5 free articles or with all articles fused. Pereiopods 2 and 3 carpus strongly produced posterodistally; propodus with more than 1 robust setae in addition to posterodistal robust seta, or without posterodistal robust seta. Pereiopods 4-7 propodus with more than 1 robust setae in addition to posterodistal robust seta, or with posterodistal robust seta only. Pereiopod 7 propodus without anterodistal serrate setae. Pleopods: 1-5 similar, none operculiform. Statocysts absent.

Remarks: Poore & Lew Ton (1988c) summarized the history of the family Hyssuridae confining it to genus-group A of Wägele (1981d). They synonymized some genera and erected another. All genera were diagnosed and species then known listed. Poore & Lew Ton (1988c) presented a phylogeny and a key for identification of five genera and eight Australian species. More recently a sixth genus has been described (Müller 1991c).

Hyssurids are recognized by their extreme elongate body form and the similarity of all the pleopods (absence of operculiform pleopods 1). All are very small (less than 7 mm) and colorless.

Key to genera of Hyssuridae

1. Uropodal exopod as broad or broader than long, endopod fused to peduncle (Fig. 1b); pleopodal endopods with single apical seta (Fig. 10e); mandibular palp of 1 article with seta (Fig. 4c) 2
– Uropodal exopod longer than wide, endopod free (Fig. 11a); pleopodal endopods with several marginal setae (Fig. 10d); mandibular palp of 3 articles (Fig. 4a) 3

2. Pereiopod 3 merus and carpus with tufts of long setae (Fig. 8o); telson short, excavate apically (Fig. 11c); eyes present .... *Xenanthura*
– Pereiopod 3 merus and carpus acutely produced posterodistally (Fig. 8b); telson with apical projection (Fig. 11b); eyes absent .... *Belura*

3. Uropodal exopod linear, its mesial margin and lateral margins of telson with cuticular spines (Fig. 11d) .... *Neohyssura*
– Uropodal exopod linear or with lateral lobe, unarmed; telson unarmed (Fig. 11a) 4

4. Pereiopods 4-7 with carpus quadrangular, with 2-3 marginal spiniform setae (Fig. 9a); mandibular molar acute .... *Hyssura*
– Pereiopods 4-7 with carpus triangular, with 1 marginal spiniform seta (Fig. 9b); mandibular molar blunt (Fig. 4a) 5

5. Pleopod 1 endopod longer and broader than exopod (Fig. 10d); maxillipedal endite prominent (Fig. 5a) .... *Kupellonura*
– Pleopod 1 endopod shorter and narrower than exopod (Fig. 10f); maxillipedal enditeabsent .... *Galziniella*

Genera of Hyssuridae

Generic characters with implicit states. Of the 23 characters used to diagnose hyssurid genera six have implicit values, that is, apply to five of the six genera. Their states follow: Antenna 1 flagellum article 1 shorter than wide; maxilla 1 present; pereiopod 2 propodal palm axial or slightly oblique; pereiopod 3 merus and carpus without tufts of long marginal setae; pereiopods 4-7 carpus triangular, with 1 marginal robust or simple seta; pleopod 1 endopod longer and broader than exopod, marginal setae shorter than endopod length.

Belura Poore & Lew Ton

Belura Poore & Lew Ton 1988c: 172.

Type species: *Belura pillara* Poore & Lew Ton 1988c, original designation.

Diagnosis: Eyes absent. Antenna 2 flagellum of 5 articles, first longer than remaining 4. Mandibular molar an articulating cirolanid-type blade; palp of 1 article with 1 seta or of seta alone. Maxillipedal endite a short triangular lobe with distal seta(e); palp of single fused article. Pereiopod 2 carpus a bluntly produced blade, weakly-strongly crenellate; propodus without robust setae on cutting edge. Pereiopod 3 carpus acutely produced posterodistally, different from pereiopod 2. Pereiopods 4-7 propodi with 1 distal robust seta on cutting edge, without pectinate setae along mesiodistal margin. Pereiopod 7 dactylus stout, less than two-thirds length of propodus. Pleopodal endopods with single apical seta. Uropodal endopod fused to peduncle; exopod ovoid. Uropodal rami and telson margins with cuticular spines (especially exopod mesial margin). Telson with medial projection on excavate apex.

Male: Antenna 1 flagellum of 5 articles each bearing about 6 aesthetascs. Pereiopod 2 carpus more strongly produced and propodal palm more transverse than in juvenile. Pleopod 2 endopod narrow, with 2 apical setae, appendix masculina exceeding endopod, broad, apically curved.

Species and distribution: Two. Southern cool temperate seas, SE Australia, Falkland Is; 10-841 m.

Remarks: *Belura* is similar to *Xenanthura*, differing in the presence of eyes, structure of pereiopod 3 and the telson. The largest specimen known is 3.4 mm.

Galziniella Müller

Galziniella Müller 1991c: 289-290

Type species: *Galziniella polynesica* Müller 1991c, original designation.

Diagnosis: Eyes absent. Antenna 2 flagellum of more than 6 articles, first shorter than others together. Mandibular molar obsolete, with strong seta reaching near tip of incisor; palp of 3 articles. Maxilla 1 absent. Maxillipedal endite absent; palp with articles 1 and 2 fused, 3-5 free. Pereiopod 2 carpus acutely produced, smooth; propodus without robust setae on cutting edge. Pereiopod 3 carpus bluntly produced posterodistally, as in pereiopod 2. Pereiopods 4-7 propodi with at least 3 robust setae along cutting edge, without pectinate setae along mesiodistal margin. Pereiopod 7 dactylus narrow, almost as long as propodus. Pleopodal endopods with several marginal setae. Pleopod 1 endopod shorter and narrower than exopod, marginal setae longer than endopod length. Uropodal endopod free; exopod narrowly leaf-shaped. Uropodal rami and telson margins entire. Telson ovoid, with rounded apex.

Male: Unknown.

Species and distribution: Monotypic. Tropical reef; Society Is; intertidal.

Remarks: *Galziniella* is most similar to *Kupellonura*, differing in the absence of maxilla 1 and maxillipedal endite and in the differentiation of pleopod 1 from the others. The genus is known from a single specimen, 1 mm long.

Hyssura Norman & Stebbing

Hyssura Norman & Stebbing 1886: 128; Barnard 1925: 137; Wägele 1981d: 70-75; Wägele 1981b: 49; Kussakin 1982: 11; Poore & Lew Ton 1988c: 176.

Ocsanthura Kensley 1978c: 558; Kussakin 1982: 13 (type species: *Ocsanthura vimsae* Kensley, 1978c, original designation).

Oscanthura Kensley Wägele 1981d: 70, 72, 74, 103; Wägele 1981b: 49 (lapsus calami).

Type species: *Hyssura producta* Norman & Stebbing 1886, monotypy.

Diagnosis: Eyes absent. Antenna 2 flagellum of more than 6 articles, first shorter than others together. Mandibular molar spike-like; palp of 3 articles. Maxillipedal endite a short triangular lobe with distal seta(e); palp of 5 free articles. Pereiopod 2 carpus acutely produced, smooth; propodus with robust flagellate setae along cutting edge. Pereiopod 3 carpus bluntly produced posterodistally, as in pereiopod 2. Pereiopods 4-7 propodi with 1 distal robust seta on cutting edge, with short transverse row of 1-3 pectinate setae along mesiodistal margin; carpus quadrangular, wider than propodus, with or without 2 or more pectinate setae along free distal margin. Pereiopod 7 dactylus narrow, almost as long as propodus. Pleopodal endopods with several marginal setae. Uropodal endopod free; exopod with lateral lobe. Uropodal rami and telson margins entire, or dentate. Telson triangular, or ovoid, with rounded apex.

Male: Antenna 1 flagellum of 12 aesthetasc-bearing articles.

Species and distribution: Six. North Atlantic, Mediterranean; 350-2900 m.

Remarks: *Hyssura* is most similar to *Kupellonura*, being unique within the family in the possession of a broad quadrate carpus on pereiopods 4-7. Poore & Lew Ton (1988c) discussed some of the species. Maximum length seems to be about 5 mm.

Kupellonura Barnard

Kupellonura Barnard, 1925: 146; Wägele 1981d: 71-75; Wägele 1981b: 55; Kensley 1982c: 126; Poore & Lew Ton 1988c: 176.

Kensleyanthura Wägele 1981d: 106 (type species: *Kupellonura capensis* Kensley 1975a, original designation).

Belizanthura Kensley 1982a: 338 (type species: *Belizanthura inswae* Kensley 1982a, original designation).

Type species: *Kupellonura mediterranea* Barnard 1925, monotypy.

Diagnosis: Eyes present. Antenna 2 flagellum of more than 6 articles, first shorter than others together. Mandibular molar blunt; palp of 3 articles. Maxillipedal endite a short triangular lobe with distal seta(e); palp of 5 free articles. Pereiopod 2 carpus acutely produced, smooth; propodus with robust flagellate setae along cutting edge. Pereiopod 3 carpus bluntly produced posterodistally, as in pereiopod 2. Pereiopods 4-7 propodi with 1 distal robust seta on cutting edge, without pectinate setae along mesiodistal margin. Pereiopod 7 dactylus narrow, almost as long as propodus. Pleopodal endopods with several marginal setae. Uropodal endopod free; exopod narrowly leaf-shaped, or with lateral lobe. Uropodal rami and telson margins entire, or dentate. Telson ovoid, with rounded apex.

Male: Head narrower than in juvenile, eyes large. Antenna 1 with 5 aesthetasc-bearing articles plus basal and terminal articles. Maxillipedal endite absent. Pereiopod 1 propodus narrower, palm with longer mesial and marginal setae. Pleopods, uropods and telson more elongate. Pleopod 2 with hooked appendix masculina. Pleon without epimeral setae.

Species and distribution: 15. Tropical to cool temperate seas; coral rubble, sandy benthos; intertidal to 2829 m.

Remarks: The complex synonymy has been discussed earlier by Kensley (1982c), Negoescu & Wägele (1984) and Poore & Lew Ton (1988c). Some species were mapped by George & Negoescu (1985). The genus is most easily recognized by the uropodal exopods, held obliquely erect in preserved material and usually with a broadly based lateral lobe. The largest species is 6.5 mm, most are 3-5 mm.

Neohyssura (Amar)

Neohyssura Amar 1953: 353-354; Wägele 1981d: 70, 71, 72, 75; Kensley 1982c: 159; Wägele 1981b: 72-73; Poore & Lew Ton 1988c: 184-185.

Horolanthura Menzies & Frankenberg 1966: 41-42 (type species: *Horolanthura irpex* Menzies & Frankenberg 1966, original designation).

Type species: *Hyssura spinicauda* Walker 1901, monotypy.

Diagnosis: Eyes absent. Antenna 2 flagellum of more than 6 articles, first shorter than others together. Mandibular molar spike-like; palp of 3 articles. Maxillipedal endite a short triangular lobe with distal seta(e), or absent; palp of 5 free articles. Pereiopod 2 carpus acutely produced, smooth; propodus with robust flagellate setae along cutting edge. Pereiopod 3 carpus bluntly produced posterodistally, as in pereiopod 2. Pereiopods 4-7 propodi with 1 distal robust seta on cutting edge, without pectinate setae along mesiodistal margin. Pereiopod 7 dactylus narrow, almost as long as propodus. Pleopodal endopods with several marginal setae. Uropodal endopod free; exopod narrowly leaf-shaped. Uropodal rami and telson margins with cuticular spines (especially exopod mesial margin). Telson triangular.

Male: Unknown.

Species and distribution: Five. Subtropical and temperate seas; North Atlantic, Mediterranean, Indian Ocean, Australia; intertidal to 850 m.

Remarks: *Neohyssura* can be differentiated from *Kupellonura* using the evenly tapering telson and uropodal and telsonic spines. Specimens reach little over 4 mm.

Xenanthura Barnard

Xenanthura Barnard 1925: 138; Wägele 1981d: 69-75; Wägele 1981b: 84; Poore & Lew Ton 1988c: 186-189.

Type species: *Xenanthura brevitelson* Barnard 1925, monotypy.

Diagnosis: Eyes present. Antenna 1 flagellum article 1 twice as long as wide. Antenna 2 flagellum of 5 articles, first longer than remaining 4. Mandibular molar an articulating cirolanid-type blade; palp of 1 article with 1 seta or of seta alone. Maxillipedal endite a short triangular lobe with distal seta(e), or absent; palp of single fused article. Pereiopod 2 carpus a bluntly produced blade, weakly-strongly crenellate; propodal palm transverse, crenellate, propodus without robust setae on cutting edge. Pereiopod 3 merus and carpus with tufts of long marginal setae; carpus not produced. Pereiopods 4-7 propodi without robust setae on cutting edge (few tubercles on merus-propodus instead), with short transverse row of 1-3 pectinate setae along mesiodistal margin. Pereiopod 7 dactylus stout, less than two-thirds length of propodus. Pleopodal endopods with single apical seta. Uropodal endopod fused to peduncle; exopod ovoid. Uropodal rami and telson margins entire. Telson short, excavate apically (without medial projection).

Male: Unknown.

Species and distribution: Seven. Tropical and warm temperate; Caribbean, Brazil, India, Australia, Sinai; intertidal to 145 m.

Remarks: *Xenanthura* shares with *Belura* similar antennae, pereiopods 4-7, pleopods and uropods. Its pereiopod 3 merus and carpus with tufts of long setae and short telson, excavate apically differentiate the genus. George & Negoescu (1985) mapped five species. The largest individual is 4.4 mm.

Leptanthuridae fam. nov.

Type genus: *Leptanthura* Sars 1897.

Diagnosis: Body 10-15 times as long as wide, non-males rarely more elongate (pereionite 7 wider than long, much shorter than pereionite 6). Pleonites 1-5 together not more than twice as long as wide, free and articulating or fused; without marginal plumose setae on pleonal epimera or posterior borders of pleonites 4 and 5. Antenna 2 flagellum of fewer than 10 articles, shorter than peduncle, or of numerous articles, longer than or shorter than peduncle. Mouth parts acutely produced anteriorly. Mandible with acute anteriorly-directed incisor. Maxillipedal endite reaching palp article 3, or absent or obsolete; palp narrow, tapering to acute apex, of fewer than 4 articles (c. 3 times as long as wide), with 5 free articles (rarely) or with 2 or more articles fused. Pereiopod 1 palm without proximal tooth (palm more or less continuous with that of carpus), or defined proximally by angle distant from palm of carpus, or with proximal tooth (defined by definite angle to palm); with even marginal row of short complex robust setae along length, or with even marginal row of closely-set simple robust setae. Pereiopods 2 and 3 carpus not or weakly produced posterodistally; propodus with more than 1 robust setae in addition to posterodistal robust seta. Pereiopods 4-7 propodus with more than 1 robust setae in addition to posterodistal robust seta. Pereiopod 7 propodus without anterodistal serrate setae. Pleopods: pleopod 1 exopod operculiform alone. Statocyst single.

Remarks: The new family includes 12 valid genera all previously included in Paranthuridae. All retain a single statocyst, which may be viewed as plesiomorphic if the loss of statocyst in its sister taxon, Paranthuridae, is viewed as apomorphic. All have an even row of robust setae, often complex, along the palm of pereiopod 1 which is usually defined proximally by a square angle or a protruding tooth.

Key to genera of Leptanthuridae

1. Pereiopod 7 absent on adults; pereiopodal propodi 2-6 similar, much narrower than long (Fig. 8k) .... *Curassanthura*
– Pereiopod 7 present on adults; pereiopodal propodi 2 and 3 differentiated from 4-7, even slightly (Figs. 7h, 9l) 2

2. Maxillipedal endite reaching at least end of first palp article (Fig. 5s) 3
– Maxillipedal endite absent or obsolete (Fig. 5t) 8

3. Pleonites 1-5 fused; pleonite 6 not distinguishable from telson; pereiopods 1-3 similar, propodus with rolled tessellated palm and strong proximal thumb (Fig. 7f) .... *Aenigmathura*
– Pleonites free and articulating (rarely 1-5 fused) (Fig. 11s); pleonite 6 marked off dorsally from telson; pereiopods 2 and 3 or 3 only differentiated from pereiopod 1, none with rolled tessellated propodal palm 4

4. Antennae 1 and 2 flagella of >12 articles, longer than or equal to peduncles (Fig. 3k) .... *Accalathura*
– Antennae 1 and 2 flagella of <12 articles, shorter than peduncles (Fig. 3l) 5

5. Pereiopods 4-7 carpus triangular, enclosed anteriorly by surrounding articles (Fig. 9o) .... *Albanthura*
– Pereiopods 4-7 carpus quadrangular or elongate, with free anterior margin (Fig. 9l) 6

6. Maxillipedal palp of 5 articles (Fig. 5r); pereiopod 2 much more elongate than pereiopod 1 .... *Bourbonanthura*
– Maxillipedal palp of 3 articles; pereiopod 2 similar to but smaller than pereiopod 1......7

7. Eyes poorly developed; pereiopod 1 carpus without robust setae on posterior margin; propodal palm with row of similar simple robust setae .... *Calathura*
– Eyes well developed; pereiopod 1 carpus with robust setae on posterior margin; propodal palm with row of complex robust setae and distinctive setae on tooth (Fig. 7e) .... *Virganthura*

8. Mandibular palp of 3 articles (Fig. 4s) .... *Leptanthura*

– Mandibular palp of 1 article (Fig. 4v) or absent 9

9. Pereiopod 1 propodus much larger than that of pereiopod 2 (Figs 7c, 8m); mandibular palp of 1 article with 1 seta .... *Bullowanthura*
– Pereiopods 1 and 2 or 1-3 similar (Figs 7e, 7n); mandibular palp of 1 article without seta or absent (Fig. 4v) 10

10. Pereiopods 1 and 2 similar, enlarged, propodus grossly produced at proximal end of palm so that palm is transverse (Figs 7i, 7q); pereiopods 4-7 with merus not postero-distally lobed; mandibular palp absent .... *Psittanthura*
– Pereiopods 1-3 similar, propodus with small thumb at proximal end of axial palm (Figs 7e, 8n); pereiopods 4-7 with merus lobed (Fig. 8m); mandibular palp a simple article (Fig. 4v) .... *Ulakanthura*

Genera of Leptanthuridae

Generic characters with implicit states. Of the 28 characters used to diagnose leptanthurid genera 10 have implicit values, that is, apply to 11 of the 12 genera. Their states follow. Body not pigmented. Pereionite 7 half as long as pereionite 6. Pleonite 1 middorsal length as long as pleonite 2. Pleotelson with posterior margin of pleonite 6 indicated dorsally, delineated from telson. Pleonites and pleotelson articulating. Antenna 2 flagellum articles cylindrical and free. Pereiopod 1 propodus significantly larger than propodus of pereiopods 2 and 3; propodus palm axial or oblique. Pereiopods 4-7 merus not lobed; dactylus tapering, subapical seta shorter than unguis. Pereiopod 7 present on adults. Pleopodal endopods with several marginal setae.

Accalathura (Barnard)

Accalathura Barnard 1925: 147; Menzies & Glynn 1968: 33; Poore 1980: 58-59; Poore 1981b: 57-58; Kussakin 1982: 39; Menzies & Kruczynski 1983: 35; Kensley 1982c: 169; Poore & Lew Ton 1990: 381.
Katanthura Nierstrasz 1941: 243 (type species: *Katanthura barnardi* Nierstrasz 1941 monotypy).
Metanthura Nierstrasz 1941: 247 (nomen nudum, no type species indicated).
Zulanthura Poore 1980: 65 (type species: *Katanthura laevitelson* Kensley 1975a, original designation).

Type species: *Calathura crenulata* Richardson 1901, subsequent designation by Menzies & Glynn (1968).

Diagnosis: Body not pigmented, or pigmented. Pleonites 1-5 free and articulating. Pleotelson with posterior margin of pleonite 6 indicated dorsally, delineated from telson. Pleonites and pleotelson articulating. Eyes well developed. Antenna 2 flagellum of 13-20 articles, longer than or slightly shorter than peduncle, articles cylindrical and free. Mandibular palp of 3 articles, palp article 3 with row of 7 or more setae on oblique apex. Maxillipedal endite reaching at least article 3; palp almost as long as basis, palp free from basis, with articles 1-2 fused, 3-4 and 5 free or free from basis, with articles 1-2 fused and 3-5 fused. Pereiopod 1 carpus, posterior margin without short robust setae; palm with proximal tooth (defined by clear angle to palm), palm with even marginal row of closely-set simple robust setae, palm without distinctive robust setae on proximal tooth. Pereiopods 2 and 3 carpus margin without robust setae. Pereiopodal propodi 2 and 3 ovate, 4-7 linear. Pereiopods 4-7 carpus quadrangular or elongate, with free anterior margin. Uropodal exopod narrowly leaf-shaped, or broad and more or less notched.

Male: Pleon, uropods and telson more elongate than in juveniles. Antenna 1 flagellum of proximal discoid articles bearing dense whorls of aesthetascs and distal articles without aesthetascs. Pereiopod 1 elongate and setose; pereiopods 2 and 3 often with strong teeth.

Species and distribution: 24. Indo-west Pacific, North Pacific, Southern Ocean, Caribbean; intertidal to 655 m.

Remarks: Poore (1980) synonymized Nierstrasz's (1941) two genera on the basis of published illustrations. It is now realized that one name is a nomen nudum. Types of the three species involved have subsequently been examined by me and the synonymy is confirmed. Nierstrasz stated that species of '*Metanthura*' lack a statocyst but this is not so. *Zulanthura* Poore is based on a manca and is clearly a junior synonym of *Accalathura*.

Accalathura is a relatively easy genus to recognize because of the long multi-articulate antennal flagella. Many descriptions of species of *Accalathura* state that eyes are not present. Dissection has shown that in the case of *A. gigas* and *A. gigantissima* non-pigmented eyes are present in preserved material (Poore 1981b). Negoescu (1994) noted the presence of fields of 'sucker-like cuticular structures on the ventral face of articles 3-5 of antennal peduncle' in her new species *A. singularia*. Such structures have not been reported in any other anthuridean. *Accalathura* is sister genus to the monotypic *Calathura* which differs in the shorter maxillipedal palp with fewer articles.

The greatest concentration of species known is in the tropical southwestern Pacific. Poore (1990) discussed the validity of specific names used in the Indian Ocean. One western Atlantic species, *A. crenulata* (Richardson 1901), is especially frequently recorded (Müller 1991a). The genus includes one of the largest anthurideans known, *A. gigantissima*, at 51 mm long.

***Aenigmathura* Thomson**

Aenigmathura Thomson 1951: 5; Poore 1980: 59-60; Poore 1981b: 66; Poore & Lew Ton 1986b: 60.

Type species: *Aenigmathura lactanea* Thomson 1951, monotypy.

Diagnosis: Pleonites 1-5 fused (or sutures immovably indicated). Pleotelson without indication of posterior margin of pleonite 6. Pleonites and pleotelson articulating. Eyes well developed. Antenna 2 flagellum of 5-8 articles, shorter than peduncle, articles short and fused. Mandibular palp of 3 articles, palp article 3 with row of 7 or more setae on oblique apex. Maxillipedal endite reaching at least article 3; palp almost as long as basis, palp free from basis, with articles 1-2 fused and 3-5 fused. Pereiopod 1 carpus, posterior margin without short robust setae; propodus palm axial or oblique, palm with proximal tooth (defined by clear angle to palm), palm with even marginal row of short complex robust setae along length, palm with 1-2 distinctive robust setae on tooth. Pereiopods 2 and 3 carpus margin without robust setae. Pereiopods 4-7 carpus quadrangular or elongate, with free anterior margin.

Uropodal exopod linear, tapering, as long as or shorter than peduncle.

Male: Antenna 1 flagellum of about 12 aesthetasc-bearing articles. Pereiopods 1-3 more setose; all pereiopods more elongate. Appendix masculina a simple elongate rod.

Species and distribution: Four. Australia, Coral Sea, Society Is.; intertidal to 520 m.

Remarks: Poore (1981b) gave an extended diagnosis of this genus recognized by the tessellated rolled palm of pereiopod 1 and lack of distinction between pleonite 6 and telson. Negoescu (1994) provided a key to the species. Her new species is from a slope environment while all others are from shelf and coral reef. The genus shares with *Bourbonanthura* a narrow uropodal exopod but has a more reduced maxillipedal palp.

Albanthura (Wägele)

Albanthura Wägele 1985b: 366.

Type species: *Albanthura stenodactyla* Wägele 1985b, original designation.

Diagnosis: Pleonites 1-5 free and articulating. Pleotelson with posterior margin of pleonite 6 indicated dorsally, delineated from telson. Pleonites and pleotelson articulating. Eyes absent. Antenna 2 flagellum of 5-8 articles, shorter than peduncle, articles cylindrical and free. Mandibular palp of 3 articles, palp article 3 with fewer than 4 terminal setae. Maxillipedal endite reaching at least article 3; palp almost as long as basis, palp free from basis, with articles 1-2 fused, 3-4 and 5 free. Pereiopod 1 carpus, posterior margin without short robust setae; palm defined proximally by angle distant from palm of carpus, palm with even marginal row of short complex robust setae along length, palm with 1 larger simple robust seta on proximal angle. Pereiopods 2 and 3 carpus margin with 1-2 robust setae (as in propodus). Pereiopodal propodi 2 and 3 ovate, 4-7 linear. Pereiopods 4-7 carpus triangular, enclosed anteriorly by surrounding articles. Uropodal exopod broad and more or less notched.

Male: Antenna 1 flagellum with 13 aesthetasc-bearing articles.

Species and distribution: Two. New Zealand; 31-808 m.

Remarks: Wägele's (1985b) original species are the only known. They differ from *Leptanthura* in the absence of eyes, the tooth on the palm of pereiopod 1 and the presence of a maxillipedal endite. Wägele suggested that *Leptanthura sculpta* Pasternak 1982 might belong to *Albanthura* but this species lacks a maxillipedal endite. Eyes are not often reported for *Leptanthura* but they are present in living material of some species and quickly fade in preservative (own observation). The same might be true of *Albanthura*.

Bourbonanthura Müller

Bourbonanthura Müller 1990a: 376.

Type species: *Bourbonanthura littoralis* Müller 1990a, original designation.

Diagnosis: Pleonites 1-5 free and articulating. Pleotelson with posterior margin of pleonite 6 indicated dorsally, delineated from telson. Pleonites and pleotelson articulating. Eyes poorly developed, or absent. Antenna 2 flagellum of 5-8 articles, shorter than peduncle, articles cylindrical and free. Mandibular palp of 3 articles, palp article 3 with row of 7 or more setae on oblique apex. Maxillipedal endite reaching at least article 3; palp almost as long as basis, palp free from basis, of 5 free articles. Pereiopod 1 carpus, posterior margin without short robust setae; palm with proximal tooth (defined by clear angle to palm), palm with even marginal row of short complex robust setae along length, palm without distinctive robust setae on proximal tooth. Pereiopods 2 and 3 carpus margin without robust setae. Pereiopodal propodi 2 and 3 ovate, 4-7 linear. Pereiopods 4-7 carpus quadrangular or elongate, with free anterior margin. Uropodal exopod linear, tapering, as long as or shorter than peduncle.

Male: Antenna 1 flagellum of 10 articles, 6 bearing whorls of aesthetascs; pereiopods elongate (Müller 1993b).

Species and distribution: Two. Réunion and Society Is; intertidal.

Remarks: The genus is unusual in the possession of five maxillipedal palp articles. It is known from only two species from widely separated tropical islands, one in the Indian and the other in the South Pacific Ocean.

Bullowanthura (Poore)

Bullowanthura Poore 1978: 144-146; Poore 1980: 60.

Type species: *Bullowanthura pambula* Poore 1978, original designation.

Diagnosis: Pleonites 1-5 free and articulating. Pleotelson with posterior margin of pleonite 6 indicated dorsally, delineated from telson. Pleonites and pleotelson articulating. Eyes absent. Antenna 2 flagellum rudimentary (fewer than 5 articles), shorter than last article of peduncle, articles cylindrical and free. Mandibular palp of 1 article, palp with 1 apical seta. Maxillipedal endite absent or obsolete; palp one-third as long as basis, palp free from basis, with articles 1-2 fused and 3-5 fused. Pereiopod 1 carpus, posterior margin with short robust setae (as on propodus); palm with proximal tooth (defined by clear angle to palm), palm with even marginal row of short complex robust setae along length, palm with 1-2 distinctive robust setae on tooth. Pereiopods 2 and 3 carpus margin with 1-2 robust setae (as in propodus). Pereiopodal propodi 2 and 3 ovate, 4-7 linear. Pereiopods 4-7 carpus triangular, enclosed anteriorly by surrounding articles. Uropodal exopod broad and more or less notched.

Male: Antenna 1 flagellum of 10-12 aesthetasc-bearing articles; pereiopods more elongate than in juvenile and female; pereiopod 1 with setose mesial palm.

Species and distribution: Four. Australia, New Zealand, New Caledonia, Bay of Biscay; 8-860 m.

Remarks: *Bullowanthura* is characterised by the uniarticulate mandibular palp. Three of the species inhabit the coasts of the Tasman Sea while the fourth is from the North Atlantic. Negoescu (1994) provided a key to the species.

Calathura Norman & Stebbing

Calathura Norman & Stebbing 1886: 122; Sars 1897: 44-45; Barnard 1925: 152; Poore 1980: 60; Kussakin 1982: 46; Brandt & Negoescu 1997: 426.

Type species: *Anthura brachiata* Stimpson 1853, monotypy.

Diagnosis: Pleonites 1-5 free and articulating. Pleotelson with posterior margin of pleonite 6 indicated dorsally, delineated from telson. Pleonites and pleotelson articulating. Eyes poorly developed. Antenna 2 flagellum of 11 articles, shorter than peduncle, articles cylindrical and free. Mandibular palp of 3 articles, palp article 3 with row of 7 or more setae on oblique apex. Maxillipedal endite reaching at least article 3; palp half as long as basis, palp free from basis, with articles 1-2 fused and 3-5 fused. Pereiopod 1 carpus, posterior margin without short robust setae; palm with proximal tooth (defined by clear angle to palm), palm with even marginal row of closely-set simple robust setae, palm without distinctive robust setae on proximal tooth. Pereiopods 2 and 3 carpus margin without robust setae. Pereiopodal propodi 2 and 3 ovate, 4-7 linear. Pereiopods 4-7 carpus quadrangular or elongate, with free anterior margin. Uropodal exopod broad and more or less notched.

Male: Antenna 1 with 16 aesthetasc-bearing articles; ventral protuberances on pereionite 6, small penes on pereionite 7; pereiopods and pleotelson more elongate than in female and juveniles.

Species and distribution: Unique. North Atlantic, North Pacific, Arctic Oceans; 9-2488 m.

Remarks: Only the type species is known and it has been well described (Kussakin 1982, Brandt & Negoescu 1997). These last authors presented detailed information on distribution and synonymy. Most descriptions of *Calathura brachiata* deny the existence of a statocyst and Barnard (1925) went to great lengths to dismiss Hansen's (1916) illustration and description of a medial statocyst. Barnard believed Hansen had a specimen of *Leptanthura*. I have seen a large medial statocyst (under a long slit-like dorsal pore) in three specimens cleared with lactic acid (including in material described by Brandt & Negoescu 1997).

Curassanthura Kensley

Curassanthura Kensley 1981: 131; Kensley 1992: 187-192.

Type species: *Curassanthura halma* Kensley 1981, original designation.

Diagnosis: Pereionite 7 one-third as long as pereionite 6. Pleonites 1-5 free and articulating. Pleotelson with posterior margin of pleonite 6 indicated dorsally, delineated from telson. Pleonites and pleotelson articulating. Eyes poorly developed. Antenna 2 flagellum of 5-8 articles, shorter than peduncle, articles cylindrical and free. Mandibular palp of 3 articles, palp article 3 with fewer than 4 terminal setae. Maxillipedal endite reaching at least article 3; palp almost as long as basis, palp free from basis, of 5 free articles. Pereiopod 1 carpus, posterior margin without short robust setae; palm with proximal tooth (defined by clear angle to palm), palm with even marginal row of short complex robust setae along length, palm without distinctive robust setae on proximal tooth or with 1-2 distinctive robust setae on tooth. Pereiopods 2 and 3 carpus margin without robust setae. Pereiopodal propodi 2-6 similar, much narrower than long. Pereiopods 4-6 carpus quadrangular or elongate, with free anterior margin. Pereiopod 7 absent on adults. Uropodal exopod linear, tapering, as long as or shorter than peduncle.

Male: Antenna 1 flagellum of 12 articles, with whorl of aesthetascs on articles 2-7; pereiopod 1 with mesial palmar setae..

Species and distribution: Four. Caribbean islands, Canary I.; hypogean in beach sand.

Remarks: Wägele (1982a), Wägele & Brandt (1985) and Kensley (1992) redescribed the type species. Wägele (1985a) discussed the zoogeography of this hypogean shoreline genus.

Leptanthura (Sars)

Leptanthura Sars 1897: 47-48; Barnard 1925: 149-150; Wolff 1956: 137; Menzies 1962: 193; Birstein 1963: 138; Poore 1978: 136; Poore 1980: 61; Kensley 1982c: 174; Kensley 1982c: 174; Kussakin 1982: 41; Müller 1992d: 180.

Neoanthura Menzies 1956: 5; Poore 1980: 62 (type species: *Neoanthura coeca* Menzies, 1956, original designation). syn. nov.

Type species: *Paranthura tenuis* Sars 1873, monotypy.

Diagnosis: Pleonites 1-5 free and articulating. Pleotelson with posterior margin of pleonite 6 indicated dorsally, delineated from telson. Pleonites and pleotelson articulating. Eyes poorly developed, or absent. Antenna 2 flagellum rudimentary (fewer than 5 articles), shorter than last article of peduncle, articles cylindrical and free. Mandibular palp of 3 articles, palp article 3 with fewer than 4 terminal setae (or palp absent *L. apalpata*). Maxillipedal endite absent or obsolete; palp one-third as long as basis, palp free from basis, with articles 1-3 fused, 5 minute if visible or free from basis, with articles 1-4 fused and 5 minute if visible. Pereiopod 1 carpus, posterior margin with short robust setae (as on propodus); palm defined proximally by angle distant from palm of carpus, palm with even marginal row of short complex robust setae along length, palm with 1 larger simple robust seta on proximal angle. Pereiopods 2 and 3 carpus margin with 1-2 robust setae (as in propodus). Pereiopodal propodi 2 and 3 ovate, 4-7 linear. Pereiopods 4-7 carpus triangular, enclosed anteriorly by surrounding articles. Uropodal exopod narrowly leaf-shaped, or broad and more or less notched, or linear, tapering, as long as or shorter than peduncle.

Male: Antenna 1 flagellum with multiarticulate aesthetasc-bearing flagellum. Pereiopods and body more elongate than in female and juvenile. Pleopod 2 with appendix masculina ending in flagellum.

Species and distribution: 39. Cosmopolitan; intertidal to 4392 m.

Remarks: *Leptanthura* is a large diverse genus as currently comprised. There is considerable variability in the nature of the pereiopods and there is a good chance that the genus is polyphyletic. One species lacks a mandibular palp (Wägele 1981c). Müller (1992d) was the most recent to list most of the species with complete synonymies. To this is added the only representative of the monotypic genus *Neoanthura*, *N. coeca* Menzies 1956. I have examined the holotype of this poorly illustrated and described species which conforms with *Leptanthura* in all regards.

Psittanthura Wägele

Psittanthura Wägele 1985b: 371.

Type species: *Psittanthura egregia* (Wägele 1985b), original designation.

Diagnosis: Pleonites 1-5 free and articulating. Pleotelson with posterior margin of pleonite 6 indicated dorsally, delineated from telson. Pleonites and pleotelson articulating. Eyes absent. Antenna 2 flagellum of 5-8 articles, shorter than peduncle, articles cylindrical and free. Mandibular palp absent. Maxillipedal endite absent or obsolete; palp one-third as long as basis, palp free from basis, with articles 1-4 fused and 5 minute if visible. Pereiopod 1 carpus, posterior margin without short robust setae; propodus narrower than on similarly-shaped pereiopod 2; propodus palm transverse, its proximal angle strongly produced away from carpus, palm defined proximally by angle distant from palm of carpus, palm with even marginal row of short complex robust setae along length, palm with 1 larger simple robust seta on proximal angle. Pereiopods 2 and 3 carpus margin without robust setae. Pereiopodal propodus 2 similar to propodus of pereiopod 1, propodus 3 smaller, with transverse palm. Pereiopods 4-7 carpus triangular, enclosed anteriorly by surrounding articles. Uropodal exopod broad and more or less notched.

Male: Unknown.

Species and distribution: Unique. New Zealand; 560 m.

Remarks: The single described species is immediately recognizable by its almost chelate pereiopods 1 and 2. The same or a similar species occurs on the southern Australian shelf (Museum Victoria collections).

Ulakanthura Poore

Ulakanthura Poore 1978: 147-150; Poore 1980: 64.

Type species: *Paranthura crassicornis* Haswell 1881, original designation.

Diagnosis: Pleonites 1-5 free and articulating. Pleotelson with posterior margin of pleonite 6 indicated dorsally, delineated from telson. Pleonites and pleotelson articulating. Eyes absent. Antenna 2 flagellum rudimentary (fewer than 5 articles), shorter than last article of peduncle, articles cylindrical and free. Mandibular palp of 1 article, palp without seta. Maxillipedal endite absent or obsolete; palp one-third as long as basis, palp free from basis, with articles 1-3 fused, 5 minute if visible. Pereiopod 1 carpus, posterior margin with short robust setae (as on propodus); propodus narrower than on similarly-shaped pereiopod 2; palm defined proximally by angle distant from palm of carpus, palm with even marginal row of short complex robust setae along length, palm with 1 larger simple robust seta on proximal angle. Pereiopods 2 and 3 carpus margin without robust setae. Pereiopodal propodi 2 and 3 similar to propodus 1 (bases of these limbs much longer and broader than in pereiopod 1). Pereiopods 4-7 merus strongly posterodistally lobed, enclosing part of carpus; dactylus not tapering, subapical seta much longer than unguis; carpus triangular, enclosed anteriorly by surrounding articles. Uropodal exopod broad and more or less notched.

Male: Male antenna 1 flagellum of 8-14 aesthetasc-bearing articles. Pereiopod 1 palm setose; all pereiopods with longer dactyli than in juveniles and females. Appendix masculinis of pleopod 2 simple.

Species and distribution: Seven. Eastern Australian shelf; 2-31 m.

Remarks: *Ulakanthura* is a distinctive genus with characteristic pereiopods known only from seven species on the shelf of eastern Australia where it is very common.

Virganthura Kensley

Virganthura Kensley 1987b: 139.

Type species: *Accalathura crassa* Barnard 1925, original designation.

Diagnosis: Pleonites 1-5 free and articulating. Pleotelson with posterior margin of pleonite 6 indicated dorsally, delineated from telson. Pleonites and pleotelson articulating. Eyes well developed. Antenna 2 flagellum of 5-8 articles, shorter than peduncle, articles cylindrical and free. Mandibular palp of 3 articles, palp article 3 with c. 5 setae in longitudinal row. Maxillipedal endite reaching at least article 3; palp almost as long as basis, palp free from basis, with articles 1-2 fused, 3-4 and 5 free. Pereiopod 1 carpus, posterior margin with short robust setae (as on propodus); palm with proximal tooth (defined by clear angle to palm), palm with even marginal row of short complex robust setae along length, palm with 1-2 distinctive robust setae on tooth. Pereiopods 2 and 3 carpus margin without robust setae. Pereiopodal propodi 2 and 3 ovate, 4-7 linear. Pereiopods 4-7 carpus quadrangular or elongate, with free anterior margin. Uropodal exopod broad and more or less notched.

Male: Unknown

Species and distribution: Unique. Virgin Is; 30 m.

Remarks: The only species of *Virganthura* looks similar to species of *Leptanthura* but differs in possession of a maxillipedal endite.

Paranthuridae Menzies & Glynn

Paranthuridae (part) Menzies & Glynn 1968: 32-33; Poore 1980: 57-58; Kussakin 1982: 38-39; Wägele 1981c: 212-213.

Diagnosis: Body 10-15 times as long as wide, non-males rarely more elongate (pereionite 7 wider than long, much shorter than pereionite 6). Pleonites 1-5 together not more than twice as long as wide, free and articulating or fused; without marginal plumose setae on pleonal epimera or posterior borders of pleonites 4 and 5. Antenna 2 flagellum rudimentary (fewer than 5 fused articles), shorter than last article of peduncle. Mouth parts acutely produced anteriorly. Mandible with acute anteriorly-directed incisor (often secondarily blunt). Maxillipedal endite absent or obsolete; palp narrow, tapering to acute apex, of fewer than 4 articles (c. 3 times as long as wide), with 2 or more articles fused. Pereiopod 1 palm defined proximally by angle distant from palm of carpus; marginal setal row irregular if present, or with even marginal row of closely-set simple robust setae. Pereiopods 2 and 3 carpus not or weakly produced posterodistally; propodus with more than 1 robust setae in addition to posterodistal robust seta. Pereiopods 4-7 propodus with more than 1 robust setae in addition to posterodistal robust seta. Pereiopod 7 propodus without anterodistal serrate setae. Pleopods: pleopod 1 exopod operculiform alone. Statocysts absent.

Remarks: The concept of the family until now included what is treated here as Paranthuridae and Leptanthuridae. Justification for its restriction was presented in the *Phylogeny* section above. All paranthurids lack statocysts, do not have a row of complex robust setae along the palm of pereiopod 1 and most are darkly colored. *Paranthura* and *Pseudanthura* have a full compliment of seven pairs of pereiopods but the others lack pereiopod 7.

Key to genera of Paranthuridae

1. Pereiopod 7 present in adults; mandible with 3-articled palp (Fig. 4s); pereiopod 1 propodus without proximal row of closely-set setae on mesial surface (Fig. 7a) 2
– Pereiopod 7 absent in adults; mandible without palp (Fig. 4u); pereiopod 1 propodus with row of closely-set setae on mesial surface (Fig. 7b) 3

2. Pleonites free (Fig. 11q) or fused; pleonite 6 articulating with pleon and usually distinct from telson; uropodal exopod linear or leaf-shaped; antenna 2 flagellum plate-like, of short fused articles (Fig. 3j) .... *Paranthura*
– Pleon and pleotelson fused into single unit (Fig. 11p); uropodal exopod extremely reduced; antenna 2 flagellum of about 10 articles (Fig. 3l) .... *Pseudanthura*

3. Pereionite 7 very short, not visible laterally (Fig. 11m); head as long as wide 4
– Pereionite 7 one-third as long as pereionite 6; head longer than wide 5

4. Pleonites 1-5 separated by integumental folds dorsally; pleonite 1 twice as long as pleonite 2 .... *Colanthura*
– Pleonites 1-5 fused, rarely with integumental folds dorsally; pleonites of similar lengths .... *Califanthura*

5. Pleonites 2-5 only fused; marine or estuarine; eyes present .... *Cruranthura*
– Pleonites all free (Fig. 11n); hypogean or freshwater; eyes absent .... *Cruregens*

Genera of Paranthuridae

Generic characters with implicit states. Of the 21 characters used to diagnose paranthurid genera eight have implict values, that is, apply to five of the six genera. Their states follow. Pleonite 1 middorsal length as long as pleonite 2. Antenna 2 flagellum articles short and fused. Mandibular molar flange-like; incisor blunt. Maxillipedal palp fused to basis and fused articles (except sometimes for minute distal article). Pereiopod 1 palm without proximal tooth (palm more or less continuous with that of carpus). Pereiopodal propodi 2 and 3 ovate, 4-7 linear.

Califanthura Schultz

Califanthura Schultz 1977: 839-840; Poore 1984a: 699-701.
Tristanthura Sivertsen & Holthuis 1980: 20 (type species: *Tristanthura barnardi* Sivertsen & Holthuis, 1980, original designation).
Colanthura (part) Poore 1980: 66.

Type species: *Colanthura squamosissima* Menzies 1951, original designation.

Diagnosis: Body pigmented. Pereionite 7 extremely reduced, hidden except dorsally. Pleonites 1-5 fused (or sutures immoveably indicated). Pleotelson with posterior margin of pleonite 6 indicated dorsally, delineated from telson. Pleonites and pleotelson articulating. Eyes well developed. Mandibular palp absent. Maxillipedal palp half as long as basis. Pereiopod 1 propodus with row of closely-set setae confined proximally on mesial surface; palm without marginal short complex robust setae. Pereiopod 7 absent on adults. Pleopodal endopods with several marginal setae. Uropodal exopod narrowly leaf-shaped.

Male: Antenna 1 flagellum with about 5 aesthetasc-bearing articles.

Species and distribution: Eight. Southern Hemisphere cool temperate islands, California, West Indies; intertidal to 490 m.

Remarks: Poore's (1980) synonymy of this genus with *Colanthura* was wrong. He (1984a) defined the genus correctly and diagnosed all the species then known, including in an addendum two species described in *Colanthura*, *C. anophthalma* and *C. latimana* (Kussakin & Vasina 1982). Although the descriptions of these species are brief the authors' comments on the pleon and biogeographic considerations suggest *Califanthura* rather than *Colanthura*. *Califanthura* is without autapomorphies in the phylogenetic analysis.

Colanthura Richardson

Colanthura Richardson 1902: 287; Richardson 1905: 79; Barnard 1925: 158; Kensley 1982c: 171; Poore 1980: 60; Kensley 1982c: 171; Kussakin 1982: 57; Poore 1984a: 704.

Type species: *Colanthura tenuis* Richardson 1902, original designation.

Diagnosis: Body pigmented. Pereionite 7 extremely reduced, hidden except dorsally. Pleonites 1-5 fused but indicated by folds dorsally and laterally. Pleonite 1 middorsal length longer than pleonite 2, swollen laterally. Pleotelson with posterior margin of pleonite 6 indicated dorsally, delineated from telson. Pleonites and pleotelson articulating. Eyes well developed. Mandibular palp absent. Maxillipedal palp half as long as basis. Pereiopod 1 propodus with row of closely-set setae confined proximally on mesial surface; palm without marginal short complex robust setae. Pereiopod 7 absent on adults. Pleopodal endopods with several marginal setae. Uropodal exopod narrowly leaf-shaped.

Male: Antenna 1 flagellum with about 5 short aesthetasc-bearing articles.

Species and distribution: Nine. Central America, Indo-west Pacific; intertidal to 775 m.

Remarks: Poore (1980) diagnosed the genus and all the species then known. The genus is distinguished from *Califanthura* by the elongated pleonite 1 but not all authors have noted this in species descriptions.

Cruranthura Thomson

Cruranthura Thomson 1946: 46-49; Poore 1981b: 70; Poore 1984a: 710-711.
Colanthura (part) Poore 1980: 66.

Type species: *Cruranthura simplicia* Thomson 1946, monotypy.

Diagnosis: Body pigmented. Pereionite 7 one-third as long as pereionite 6. Pleonites 1 free, 2-5 not indicated by folds dorsally. Pleotelson with posterior margin of pleonite 6 indicated dorsally, delineated from telson. Pleonites and pleotelson articulating. Eyes well developed. Mandibular palp absent. Maxillipedal palp half as long as basis. Pereiopod 1 propodus with row of closely-set setae confined proximally on mesial surface; palm without marginal short complex robust setae. Pereiopod 7 absent on adults. Pleopodal endopods without marginal setae. Uropodal exopod narrowly leaf-shaped.

Male: Antenna 1 flagellum with about 5 short aesthetasc-bearing articles.

Species and distribution: Four. Southern Australia; intertidal to 30 m.

Remarks: Poore (1980) diagnosed the genus and all the species. The genus is characterised by a free first pleonite and long pereionite 7.

Cruregens Chilton

Cruregens Chilton 1882b: 175; Chilton 1894: 209-211; Barnard 1925: 157; Wägele 1982a: 52-55; Poore 1980: 61; Poore 1984a: 713.

Type species: *Cruregens fontanus* Chilton 1882b, monotypy.

Diagnosis: Body not pigmented. Pereionite 7 one-third as long as pereionite 6. Pleonites 1-5 free and articulating. Pleotelson without indication of posterior margin of pleonite 6. Pleonites and pleotelson fused. Eyes absent. Mandibular palp absent. Maxillipedal palp almost as long as basis. Pereiopod 1 propodus with setae on mesial surface evenly spaced along palm; palm with proximal tooth (defined by clear angle to palm), palm with even marginal row of closely-set simple setae. Pereiopodal propodi 2-6 similar, much narrower than long. Pereiopod 7 absent on adults. Pleopodal endopods without marginal setae. Uropodal exopod linear, apically rounded.

Male: Antenna 1 flagellum with about 5 aesthetasc-bearing articles.

Species and distribution: Monotypic. Hypogean freshwater, New Zealand.

Remarks: The only species lives in an unusual habitat and is well described (Wägele 1982a).

Paranthura Bate & Westwood

Paranthura Bate & Westwood 1866: 163 [see Holthuis (1977) for verification of date of publication]; Barnard 1925: 152; Poore 1980: 63; Kensley 1982c: 185-187; Kensley 1982c: 185; Kussakin 1982: 50; Menzies & Kruczynski 1983: 33; Poore 1984b: 36; Negoescu 1997b: 246-248.
Calamura Boone 1920: 25-26 (type species: *Calamura porteri* Boone, 1920, monotypy).
Edanthura Boone 1923: 148 (type species: *Edanthura linearis* Boone 1923, original designation).

Type species: *Paranthura costana* Bate & Westwood 1866, monotypy.

Diagnosis: Body pigmented. Pereionite 7 half as long as pereionite 6. Pleonites 1-5 free and articulating, or 1-5 fused (or sutures immoveably indicated), or 1 free, 2-5 not indicated by folds dorsally. Pleotelson with posterior margin of pleonite 6 indicated dorsally, delineated from telson. Pleonites and pleotelson articulating. Eyes well developed. Mandibular molar absent; incisor acute; palp of 3 articles, palp article 3 with row of 7 or more setae on longitudinal apex or with fewer than 4 terminal setae. Maxillipedal palp almost as long as basis, palp free from basis, with articles 1-4 fused and 5 minute if visible. Pereiopod 1 propodus with setae on mesial surface evenly spaced along palm; palm without marginal short complex robust setae. Pereiopod 7 present on adults. Pleopodal endopods with several marginal setae. Uropodal exopod narrowly leaf-shaped, or broad and more or less notched.

Male: Body elongated or not. Pleon usually shorter and sometimes extremely shorter than in juvenile and female. Antenna 1 flagellum articles with 2 or many aesthetascs. Pereiopods elongated or with basal articles much broader than in juvenile. Pleopod 2 appendix masculina stylet-like or exceptionally broad. Uropod broader and setose or more elongate than in juvenile.

Species and distribution: 58. Cosmopolitan; intertidal to 2707 m.

Remarks: The type species was described by Wägele (1982b). Poore (1984b) discussed variation in this large, diverse and cosmopolitan genus. It seems clear that division into smaller monophyletic genera on the basis of male morphology may be warranted. Negoescu (1997b) tabulated variation in the mandibular palp, fusion in maxillipedal palp articles, number of free pleonites, length of the pleon, and structure of the appendix masculina. She concluded also that *Paranthura* is a heterogeneous genus but was unable to correlate variation in one character with that in another. *Paranthura* is sister taxon to the group of four genera lacking pereiopod 7. Its own apomorphy is the abbreviated antenna 2 flagellum.

Pseudanthura (Richardson)

Pseudanthura Richardson, 1911: 523-524; Barnard 1925: 157; Kensley 1978[e]: 222-224; Kensley 1982c: 191; Poore 1980: 64; Poore & Lew Ton 1986b: 65-66.

Type species: *Pseudanthura lateralis* Richardson 1911, original designation.

Diagnosis: Body not pigmented. Pereionite 7 third-half as long as pereionite 6. Pleonites 1-5 fused. Pleotelson without indication of posterior margin of pleonite 6. Pleonites and pleotelson fused. Eyes absent. Antenna 2 flagellum articles cylindrical and free. Mandibular molar absent; incisor acute; palp of 3 articles, palp article 3 with row of 7 or more setae on longitudinal apex. Maxillipedal palp almost as long as basis, palp free from basis, with articles 1-3 fused, 5 minute if visible or free from basis, with articles 1-4 fused and 5 minute if visible. Pereiopod 1 propodus with setae on mesial surface evenly spaced along palm; palm without proximal tooth (palm more or less continuous with that of carpus), palm with even marginal row of closely-set simple setae. Pereiopodal propodi 2 and 3 ovate, 4-7 linear. Pereiopod 7 present on adults. Pleopodal endopods with several marginal setae. Uropodal exopod linear, tapering, as long as or shorter than peduncle.

Male: Antenna 1 flagellum of 8-12 articles, proximal articles only with whorls of aesthetascs. Pereiopod 1 with dense setae on mesial faces of merus-propodus. Pereiopods 2-7 more slender than in juvenile. Pleopod 2 with appendix masculina extending well beyond endopod (Poore & Lew Ton 1986b).

Species and distribution: Five. Indonesia, SE Australia, South Atlantic; 79-3200 m.

Remarks: *Pseudanthura* is easily recognised from the fusion of the pleon and pleotelson and narrow uropodal exopod. Poore & Lew Ton (1986b) provided a key to the five species among which they discerned two groups. In two species pereionite 6 is as long as wide, pereionite 7 is less than one-third as long as wide, the pleon is swollen anteriorly, and pleopod 1 has an endopod almost as long as the exopod. In three species pereionite 6 is greater than 1.5 times as long as wide, pereionite 7 is more than half as long as wide, the pleon is narrowest anteriorly, and pleopod 1 has an endopod at most half as long as the exopod.

6 BIOGEOGRAPHY

Negoescu (1997a) presented 'preliminary zoogeographical considerations' on the suborder. She reviewed the distribution of species in terms of salinity, temperature, substratum and depth. Her zoogeographical analysis was based on maps of distribution of families and some genera and species. While she acknowledged that anthurideans are poor dispersers and that the distribution of most species is incompletely known, she concluded that the 'dispersal center of Anthuridea may be situated probably in the tropical warm waters of south west Pacific Ocean.' Her deduction that the evolution of the group 'took place probably in the southern hemisphere' cannot be supported by the facts. Her correct observation that most species occur in the tropical shallow water merely reflects the truth that this is the most extensive marine habitat (outside the deep sea) on earth.

A more profitable explanation for the radiation of the taxon is to seek evolutionary divergences and speciation events in phyletic analysis. Given that anthurideans, like all peracaridans which brood young, are poor dispersers vicariant events are more likely to explain most present-day distributions than dispersal from a center of origin.

Anthurideans, more than 500 species, have been found in most marine environments investigated and some estuarine and freshwater ones too. The only places where they seem to be absent are the shelf of southern South America (R. Bastida, pers. comm.) and much of the cool temperate environments of western North and South America. No species were found by Gomez Simes (1979) in an estuary in Argentina. There are few deep-sea and Antarctic species (compared to asellote isopods, for example) and only three species in the Arctic Ocean.

My phyletic analysis was unable to resolve unambiguously relationships between the families except to agree with Wägele (1981d) that hyssurids are the most plesiomorphic family and sister taxon to the others. Further, all the families are more or less equally widely distributed globally and ecologically and therefore biogeographically uninformative. This suggests that the families are ancient. The hypothesized phylogenies of genera within families are marginally more informative.

Hyssuridae are absent from polar regions; Negoescu's statement (1997a) that they are apparently absent from the Indo-west Pacific is contradicted by Müller (1991c) and B. Kensley's records of undescribed species from the Indian Ocean (pers. comm.). *Hyssura* (six species) is confined to the North Atlantic and Mediterranean. Sister genera *Belura* (two species in southern seas) and *Xenanthura* (seven species in tropics) seem biogeographically separate (Poore & Lew Ton 1988c).

The Anthuridae are the largest and most widespread family. Its largest genera are also cosmopolitan. We will remain ignorant of their biogeography until a reliable phylogeny exists and their apparent monophyly is confirmed. Two of the largest, *Apanthura* and *Amakusanthura*, are certainly closely related and overlap in the Indo-west Pacific but the former appears to replace the latter (in terms of the numbers of species) in southern Australia (Poore & Lew Ton 1988a). Another genus closely related to these two is *Skuphonura* whose four species are confined to the Caribbean and Ecuador. Four small genera whose species are also geographically confined are *Caenanthura* (four species in Indian Ocean-Indonesia), *Chelanthura* (four species which diverged from *Mesanthura* in southern Australia), *Notanthura* (four species on the Atlantic coast of Africa) and *Ptilanthura* (two species in the northwestern Atlantic).

Two small anthurid genera which have radiated largely in the tropics, especially in the Indo-west Pacific are *Pendanthura* and *Apanthuroides*. *Sauranthura* with one species in tropical Australia is sister genus to *Pendanthura*. *Cyathura* is a coastal genus with many representatives confined to estuaries. A deepwater species (*C. profunda*) seems anomalous. Closely related, a subgenus in other taxonomies, is the hypogean *Stygocyathura* whose zoogeography was dealt with by Wägele (1985a). Although the 20 species in this genus are widespread through the tropical Pacific and Caribbean Wägele rightly concluded that they share a common ancestor and invaded this special habitat in the Tethyan Sea. There is evidence therefore in several anthurid genera for relatively recent divergence and radiation.

The two clades which comprise the Antheluridae are ecologically and geographically distinct. *Anthomuda* (seven species) is essentially a shallow-water Indo-west Pacific genus with no records greater than 110 m depth. Its sister taxon, *Ananthura* (eight species) and *Anthelura* (one species), is largely a cold- or deep-water group with only one species recorded at shelf depths outside the Antarctic. The exception is *Ananthura ovalis* from the Mediterranean shelf.

The Expanathuridae are largely a tropical shallow-shelf family of seven genera. Some genera (*Eisothistos*, *Expanathura*, *Heptanthura*) are especially diverse in coral reef environments and other smaller ones live there too (*Coralanthura*, *Rhiganthura*, *Minyanthura*). The exception is *Panathura* (one described species in South Africa and four undescribed species in southern Australia) which is southern temperate. Only the highly derived *Eisothistos* is more widespread with species in southern temperate and Antarctic seas. The family is not recorded in Atlantic or eastern Pacific Oceans except in areas with Tethyan origins (Mediterranean and Caribbean Seas).

Plesiomorphic genera of Leptanthuridae are small and specialised. *Curassanthura* (four species) is a hypogean genus of Tethyan origin whose present distribution results from the opening of the Atlantic (Wägele 1985a). The two species of *Bourbonanthura* are widely separated in the Indo-west Pacific and *Aenigmathura* is similar in being essentially a northern shallow-water Australian genus with a representative in the deep South Pacific islands. Of the two major clades of Leptanthuridae one centres on *Accalathura* and the other on *Leptanthura*. *Accalathura* is highly diverse in the Indo-west Pacific, two representatives only being found in the Caribbean and central eastern American shelf, and one in Antarctica. Its sister genus, *Calathura* with one species, has invaded the North Atlantic Ocean. *Leptanthura* is one of the largest and most widespread of anthuridean genera but is especially diverse in the Indo-west Pacific. There is a chance that the genus is polyphyletic. Three genera derived from it are smaller, *Albanthura* with two species in New Zealand, *Psittanthura* with one species in New Zealand (and another undescribed in southern Australia), *Bullowanthura* with three species in Australia-New Zealand, and the diverse *Ulakanthura* in eastern Australia. *Bullowanthura aquitania* from the northeastern Atlantic is anomalous. Nevertheless, biogeography correlates well with the cladistic analysis of the family.

Of the Paranthuridae *Pseudanthura* is a deep-water widespread genus derived early from this clade. *Paranthura* is possibly polyphyletic; it is the largest and most cosmopolitan of anthuridean genera. Of its much smaller derived taxa, *Colanthura* and *Califanthura*, are also widely scattered while *Cruranthura* is an west Pacific genus. Unlike many other genera, the ancestry of the freshwater hypogean *Cruregens fontanus* in New Zealand is probably very recent as the gravel plains it inhabits are of late Pleistocene origin.

ACKNOWLEDGEMENTS

My work on anthuridean isopods started in the 1970s and I thank the then Fisheries and Wildlife Department, Victoria, for facilitating collection of the specimens. At Museum Victoria I have been assisted by Australian Biological Resources Study grants which allowed collection in diverse habitats throughout Australia. I particularly thank Helen Lew Ton for help in the field and laboratory, sorting and identifications, preparing numerous manuscripts and for stimulating the ideas which have lead to this synthesis. I thank Kate Nolan for inking the figures. I am especially grateful to Brian Kensley and Marilyn Schotte for allowing me to use their published list of isopod species to prepare the appendix .

APPENDIX ALPHABETICAL LIST OF FAMILIES, GENERA AND SPECIES

Each species is given with its authority, geographical locality, depth range in meters, and the genus of the original combination in parenthesis if appropriate.

Antheluridae **Poore & Lew Ton 1988b**

Ananthura Barnard 1925

A. abyssorum (Norman & Stebbing 1886) Davis Strait, Canada 3500-4596 (*Anthelura*)
A. billarderia Poore & Lew Ton 1988b New South Wales, Australia 750
A. elegans Kussakin 1967 Antarctic Ocean 100-14 (*Austranthura*)
A. luna (Schultz 1966) Tanner Canyon, California, USA 783-1298 (*Bathura*)
A. ocellata (Nunomura 1992a) Ryukyu I., Japan (*Panathura*)
A. ovalis Barnard 1925 Italy, 18-150
A. rigida Nunomura 1976a Japan 1008-1050
A. sulcaticauda Barnard 1925 North Atlantic Ocean 568-800

Anthelura Norman & Stebbing 1886
A. elongata Norman & Stebbing 1886 Northeast Atlantic Ocean 1100-1590

Anthomuda Schultz 1979
A. affinis (Richardson 1902) Bermuda 9 (*Anthelura*)
A. alba Müller 1990d Society Is. 0.5-1
A. chorizema Poore & Lew Ton 1988b New South Wales, Australia 10-29
A. cracens (Kensley 1980a) Mozambique; Somalia 42-61 (*Diaphoranthura*)
A. hapla (Kensley 1980a) Mauritius 110 (*Diaphoranthura*)
A. hovea Poore & Lew Ton 1988b Queensland, Australia 1-18
A. poorei Müller 1990a Réunion I. 0.5-1

Anthuridae Leach 1814

Amakusanthura Nunomura 1977
A. africana (Barnard 1914) South Africa intertidal-200 (*Apanthura*)
A. agonis Poore & Lew Ton 1988a Queensland, Australia 5-42
A. angophora Poore & Lew Ton 1988a Queensland, Australia 3-11
A. brachyscome Poore & Lew Ton 1985a Queensland, Australia 3
A. californiensis (Schultz 1964) California, USA 80 (*Apanthura*)
A. coppingeri (Barnard 1925) Northern Territory, Australia 14-40 (*Apanthura*)
A. correa (Poore & Lew Ton 1988a) Queensland, Australia shelf (*Apanthuretta*)
A. dodonaea Poore & Lew Ton 1988a Queensland, Australia 0-15
A. dubia (Barnard 1914) South Africa intertidal-200 (*Apanthura*)
A. eugenia Poore & Lew Ton 1988a Queensland, Australia shallow
A. geminsula (Kensley 1982a) Belize intertidal-1.5 (*Apanthura*)
A. goodenia Poore & Lew Ton 1988a Queensland, Australia 3-10
A. hibbertia Poore & Lew Ton 1985a Queensland, Australia 43-52
A. iberica Reboreda & Wägele 1992 Spain; North Atlantic Ocean 13

A. inornata (Miller & Menzies 1952) Hawaii, USA shelf (*Apanthura*)
A. kingia Poore & Lew Ton 1988a Queensland, Australia intertidal-24
A. koonyumae Bamber 1997 Hong Kong 2-3
A. lathridia (Wägele 1982d) Cuba intertidal (*Apanthuretta*)
A. lechenaultia Poore & Lew Ton 1988a Queensland, Australia shallow
A. libyana (Negoescu 1980a) Libya 26-100 (*Apanthura*)
A. longiantennata Nunomura 1977 Japan 3-4
A. magnifica (Menzies & Frankenberg 1966) Georgia, USA 7-137 (*Apanthura*)
A. mana (Kensley 1979b) Fiji 1 (*Apanthura*)
A. melaleuca Poore & Lew Ton 1988a Queensland, Australia intertidal-12
A. moragallae Müller 1991b Sri Lanka intertidal
A. motasi (Negoescu 1980a) Gulf of Aden 24 (*Apanthura*)
A. olearia Poore & Lew Ton 1985a Victoria, Australia intertidal-56
A. pimelia Poore & Lew Ton 1985a Queensland, Australia shallow
A. pandorea Poore & Lew Ton 1988a Queensland, Australia 12
A. vermiformis Müller 1992a Caribbean Colombia 0.5-30
A. paramagnifica Müller 1992a Caribbean Colombia 1-2
A. pori (Wägele 1981a) Gulf of Elat 215-620 (*Apanthuretta*)
A. signata (Menzies & Glynn 1968) Puerto Rico intertidal-24 (*Apanthura*)
A. stocki (Müller 1991b) Sri Lanka intertidal (*Apanthura*)
A. significa (Paul & Menzies 1971) Venezuela 95 (*Apanthura*)
A. tengo Müller 1992a Caribbean Colombia intertidal-13
A. toyamaensis Nunomura 1992b Toyama, Japan intertidal
A. tristania Poore & Lew Ton 1988a Queensland, Australia 2
A. vermiformis Müller 1992a Caribbean Colombia 0.5-30
A. wahlenbergia Poore & Lew Ton 1988a Queensland, Australia 9

Anthura Leach 1814
A. gracilis (Montagu 1808) UK, Northeast Atlantic Ocean intertidal, shelf (*Oniscus*)

Apanthura Stebbing 1900
A. addui Wägele 1981a Maldive Is. intertidal-15
A. banksia Poore & Lew Ton 1985a New South Wales, Australia 1.5
A. bruscai Müller 1992b Malaysia intertidal-1
A. callitris Poore & Lew Ton 1985a Queensland, Australia shallow
A. childi (Kensley 1979a) Fiji 1 (*Mesanthura*)
A. corsica Amar 1953 Corsica, France, Mediterranean Sea 3-4
A. cracenta Kensley 1984 Belize intertidal-2
A. dimorpha (Kensley 1982c) South Africa 84-90 (*Mesanthura*)
A. drosera Poore & Lew Ton 1985a Bass Strait, Australia 59-127
A. excavata Mezhov 1976 Kurile I., Japan 30
A. fusei Nunomura 1993 Japan
A. harringtoniensis Wägele 1981d Bermuda shelf
A. honshuensis Wägele 1984a Japan intertidal
A. indonesiensis Negoescu 1997b Indonesia 4.5
A. insignifica Kensley 1978d South Africa 550-850
A. isotoma Poore & Lew Ton 1985a Victoria, Australia 4-16

A. kennedia Poore & Lew Ton 1988a Queensland, Australia 21-23
A. lambertia Poore & Lew Ton 1985a Queensland, Australia shallow
A. microps Kensley 1980a Arabian Sea; Somalia; Mozambique 42-95
A. mirbelia Poore & Lew Ton 1985a Bass Strait, Australia 33-120
A. pallida Wägele 1981a Gulf of Elat 42-45
A. pariensis Negoescu 1997b Java Sea 1-1.5
A. pultenaea Poore & Lew Ton 1988a Queensland, Australia 11-82
A. restio Poore & Lew Ton 1988a Queensland, Australia 1-15
A. robertiana (Monod 1925) Morocco 25-28 (*Cyathura*)
A. sandalensis Stebbing 1900 Loyalty Is. shelf
A. senegalensis Barnard 1925 Senegal 10
A. shikokuensis Nunomura 1993 Japan
A. stanjeki Wägele 1981a Gulf of Suez intertidal-11
A. stipa Poore & Lew Ton 1988a Queensland, Australia shallow
A. stocki (Müller 1991b) Sri Lanka intertidal (*Amakusanthura*)
A. styphelia Poore & Lew Ton 1985a New South Wales, Australia 2-31
A. thryptomene Poore & Lew Ton 1985a Victoria, Australia 8-18
A. tiomanae Müller 1992b Malaysia 1-3
A. trioculata Nunomura 1993 Japan
A. tyrrhenica Wägele 1980a Italy 33-330
A. wudu Müller 1990a Réunion I. intertidal-1.5
A. xanthorrhoea Poore & Lew Ton 1985a Victoria, Australia intertidal-33
A. xenocheir Stebbing 1910 Egmont Reef, Indian Ocean
A. zeewykae Kensley & Poore 1982 Houtman Abrolhos I., Western Australia 1.3

Apanthuroides Menzies & Glynn 1968
A. coralliophilus Müller 1993b Society Is. 2
A. fijiensis (Kensley 1979a) Fiji 1 (*Natalanthura*)
A. foveolata (Kensley 1978d) South Africa 550-850 (*Natalanthura*)
A. mediterranea (Negoescu 1981) Libya 70 (*Natalanthura*)
A. millae Menzies & Glynn 1968 Puerto Rico intertidal-30
A. spathulicauda (Wägele 1981a) Mediterranean Sea 70 (*Natalanthura*)

Apanthuropsis Poore & Lew Ton 1985a
A. richea Poore & Lew Ton 1985a Victoria, Australia 8-14

Caenanthura Kensley 1978a
C. enigmatica (Kensley & Reid 1984) Arabian Gulf 1-3 (*Arabanthura*)
C. indica Negoescu 1980b Dijbouti; Gulf of Oman, 19-110
C. gutui (Negoescu 1997b) Java Sea 1-1.5 (*Indonesanthura*)
C. siamensis (Barnard 1925) Gulf of Thailand 6-10 (*Cyathura*)

Cetanthura Kensley 1982b
C. foveoderma Kensley 1982b Namibia 620

Chelanthura Poore & Bardsley 1990
C. ajuga Poore & Bardsley 1990 Western Australia intertidal-3

C. calaena (Poore & Lew Ton 1986a) Victoria, Australia intertidal (*Mesanthura*)
C. salvia Poore & Bardsley 1990 New South Wales, Australia 5
C. westringia Poore & Bardsley 1990 South Australia 21

Cortezura Schultz 1977
C. confixa (Kensley 1978a) Venezuela 1.5-10 (*Venezanthura*)
C. penascoensis Schultz 1977 Gulf of California, Mexico intertidal

Cyathura Norman & Stebbing 1886
C. aegiceras Poore & Lew Ton 1985b Queensland, Australia estuarine
C. bentotae Müller 1991b Sri Lanka intertidal
C. bruguiera Poore & Lew Ton 1985b Queensland, Australia estuarine
C. burbancki Frankenberg 1965 Georgia, USA estuarine,16-77
C. carinata Krøyer 1847 Greenland intertidal-5 (*Anthura*)
C. cubana Negoescu 1979a Cuba estuarine
C. eremophila Monod 1925 Mauritania fishgut
C. esquivel Kensley et al. 1997 Cuba 2
C. estuaria Barnard 1914 South Africa estuarine
C. francispori Negoescu 1981 Suez Canal, Egypt intertidal
C. furcata Nunomura & Hagino 2000 Ibaraki, Japan lake
C. guaroensis Brusca & Iverson 1985 Pacific Costa Rica
C. hakea Poore & Lew Ton 1985b New South Wales, Australia estuarine
C. higoensis Nunomura 1977 Japan estuarine
C. indica Barnard 1925 Sri Lanka 1-100
C. kikuchii Nunomura 1977 Japan estuarine, intertidal-4
C. madelinae Hackney & Ganucheau 1989 North Carolina, USA
C. munda Menzies 1951 California, USA, 18-55
C. muromiensis Nunomura 1974) Japan estuarine
C. omorii Nunomura 1992a Okinawa, Japan estuarine
C. polita (Stimpson 1855) Northwestern Atlantic Ocean intertidal-35 (*Anthura*)
C. profunda Kensley 1982b Argentine Basin, South Atlantic Ocean 508-2707
C. pusilla Stebbing 1904 Sri Lanka intertidal
C. rudloei Kensley 1980a Thailand; Madagascar intertidal-73
C. siamensis Barnard 1925 Thailand 6-10
C. terryae Müller 1990a Réunion I. intertidal-1
C. tridentata Wagner 1990 Dominican Republic intertidal

Exallanthura Kensley 1980a
E. sexpes Kensley 1980a Somalia; Singapore intertidal-49

Haliophasma Haswell 1881
H. adinae (Negoescu 1980b) Gulf of Aden 24-110 (*Exanthura*)
H. alaticauda Amar 1966 Mediterranean Sea 470-500
H. antarctica (Kensley 1982b) Weddell Sea, Antarctic Ocean 108-513 (*Malacanthura*)
H. austroafricanum Kensley 1982c South Africa, 18-620
H. beaufortia Poore & Lew Ton 1988d Western Australia 31-134
H. blandfordia Poore & Lew Ton 1988d Western Australia 6-329

H. canale Poore 1975 Victoria, Australia, 19
H. caprii Wägele 1981a Italy 70
H. coronicauda Barnard 1925 South Africa 7-174
H. cribense Poore 1975 Victoria, Australia 7-101
H. curri Paul & Menzies 1971 Venezuela intertidal-95
H. cycneum Poore 1975 Victoria, Australia 1-124
H. dakarensis Barnard 1925 Senegal 25-29
H. darwinia Poore & Lew Ton 1988d Queensland, Australia 24-37
H. dillwynia Poore & Lew Ton 1988d Western Australia 122-134
H. elongatum Poore 1975 New South Wales, Australia 21-72
H. falcatum Poore 1975 Victoria, Australia 7-18
H. foveolata Barnard 1940 South Africa 120-200
H. geminata Menzies & Barnard 1959 California, USA 9-512
H. hermani Barnard 1940 South Africa 0-15
H. macrurum (Barnard 1914) South Africa intertidal (*Exanthura*)
H. mjoelniri Negoescu & Svavarsson 1997 Iceland 340-508
H. mombasa (Kensley 1980a) Kenya 3.5 (*Malacanthura*)
H. novaezelandiae Wägele 1985b New Zealand 167
H. palmatum Wägele 1981a) Gulf of Elat intertidal-1
H. pinnatum Poore 1975) New South Wales, Australia 28-66
H. platytelson Wägele 1985b) New Zealand 202
H. poorei Kensley 1980a) India; Somalia intertidal-37
H. profunda Negoescu 1994) New Caledonia 1040
H. pseudocarinata Barnard 1940) South Africa intertidal-4
H. pugnatum Poore 1975 Victoria, Australia intertidal-16
H. purpurea Haswell 1881 New South Wales, Australia shelf
H. schotteae (Kensley 1982c) South Africa 80-775 (*Malacanthura*)
H. swainsonia Poore & Lew Ton 1988d New South Wales, Australia 5-20
H. syrtis Poore 1975 Queensland, Australia 12-80
H. templetonia Poore & Lew Ton 1988d New South Wales, Australia intertidal-9
H. transkei (Kensley 1982c) South Africa 710-775 (*Malacanthura*)
H. tricarinata Barnard 1925 South Africa 48-100
H. yarra Poore 1975 Victoria, Australia 3-400

Indanthura Pillai & Eapen 1966
I. caribbica (Paul & Menzies 1971) Venezuela 42-95 (*Malacanthura*)
I. carinata Pillai & Eapen 1966 Southern India 40
I. larwoodi (Wägele 1981a) Mediterranean Sea 100 (*Alloanthura*)
I. sculpta (Kensley 1980a) Mozambique 135 (*Alloanthura*)

Malacanthura Barnard 1925
M. cantabrica Kensley 1982b Bay of Biscay 1739
M. linguicauda (Barnard 1920) South Africa 90-190 (*Anthura*)
M. ornata (Barnard 1957) South Africa intertidal-15 (*Haliophasma*)
M. truncata (Hansen 1916) Guiana Basin, Atlantic Ocean 520-2870 (*Cyathura*)

Mesanthura Barnard 1914
M. adrianae Negoescu 1999 Fiji coral reef
M. affinis (Chilton 1882) New Zealand (*Anthura*)
M. albinotata Thomson 1951 Western Australia shelf
M. albolineata Barnard 1925 Singapore intertidal-2
M. asiatica Müller 1993a Malaysia 1-2
M. astelia Poore & Lew Ton 1986a New South Wales, Australia 1-68
M. atrata Nunomura 1985 Japan 1
M. bermudensis Kensley 1994 Bermuda intertidal
M. bipunctata Thomson 1951 Western Australia shelf
M. bivittata Kensley 1987a Belize 1-2
M. brasiliensis Koening 1980 Brazil 25
M. callicera Pires 1981 Brazil 4
M. catenula (Stimpson 1855) South Africa intertidal-4 (*Anthura*)
M. crucis (Barnard 1925) St Croix 1-8 (*Cyathura*)
M. dianella Poore & Lew Ton 1986a New South Wales, Australia 3-66
M. excelsa Pires 1981 Brazil intertidal
M. fasciata Kensley 1982a Belize 0-24
M. floridensis Menzies & Kruczynski 1983 Gulf of Mexico , 18-37
M. frances Kensley et al. 1997 Cuba 2
M. gerlachi Wägele 1981a Maldive Is. intertidal-4
M. hieroglyphica Miller & Menzies 1952 Hawaii, USA shelf
M. hopkinsi Hooker 1985 Florida, USA intertidal-55
M. javensis Wägele 1984a Java, Indonesia intertidal
M. kiliani Müller 1993a Malaysia 1-3
M. libertia Poore & Lew Ton 1986a Bass Strait, Australia 130-140
M. looensis Kensley & Schotte 1987 Florida, USA 1
M. miersi (Haswell 1885) New South Wales, Australia shelf (*Anthura*)
M. miyakoensis Nunomura 1979a Japan 3
M. moroea Poore & Lew Ton 1986a Victoria, Australia intertidal-2
M. nigra Müller 1993a Malaysia intertidal
M. nigrodorsalis Nunomura 1977 Japan intertidal
M. nubifera Wägele 1984c Baja California, Mexico intertidal
M. occidentalis Menzies & Barnard 1959 California, USA 0-55
M. ocellata Barnard 1925 Gulf of Thailand 2
M. paucidens Menzies & Glynn 1968 Puerto Rico intertidal-15
M. protei Kensley 1980a Madagascar 0-37
M. pulchra Barnard 1925 West Indies 2-36
M. punctillata Kensley 1982a Belize intertidal-20
M. reticulata Kensley 1982a Belize 10-24
M. romulea Poore & Lew Ton 1986a New South Wales, Australia intertidal-4
M. spongicola Schotte et al. 1991 Turks & Caicos Is. 12-28
M. stypandra Poore & Lew Ton 1986a Victoria, Australia intertidal-17

Nemanthura Wägele 1981d
N. valeriae (Paul & Menzies 1971) Venezuela 95 (*Haliophasma*)

Notanthura Monod 1927
N. barnardi Monod 1927 West Africa estuarine
N. caeca (Kensley 1975b) South Africa 5-68 (*Haliophasma*)
N. liouvillei (Monod 1925) Morocco 21-40 *(Cyathura)*
N. maroccana (Wägele & Platvoet 1982) Morocco intertidal (*Monodanthura*)

Pendanthura Menzies & Glynn 1968
P. anophthalma Negoescu 1994 Loyalty Is. 261
P. hendleri Kensley 1984 Belize 0-30
P. rarotonga Kensley 1979a Cook Is. 1
P. tanaiformis Menzies & Glynn 1968 Puerto Rico intertidal-1

Pilosanthura Wägele 1989
P. fresii (Wägele 1980a) Corsica, France, Mediterranean Sea 112-1739 (*Anthelura*)

Ptilanthura Harger 1878
P. colpos Kensley 1996 Gulf of Mexico 14-88
P. tenuis Harger 1878 Eastern North America 0-141

Quantanthura Menzies & George 1972
Q. brasiliensis Kensley & Koening 1979 Brazil 21-166
Q. caledonensis Negoescu 1994 New Caledonia 500
Q. erica Poore & Lew Ton 1986c Victoria, Australia, 19
Q. frema Poore & Lew Ton 1986c Queensland, Australia shelf
Q. globitelson Menzies & George 1972 Peru-Chile Trench, Pacific Ocean 2504-4383
Q. menziesi Kensley & Koening 1979 Brazil 1-96
Q. pacifica Wägele 1985b New Zealand 134
Q. raoulia Poore & Lew Ton 1986c New Zealand, 19
Q. remipes (Barnard 1914) South Africa 78-312 (*Anthelura*)
Q. serenasinus (Kensley 1975a) South Africa 5-125 (*Agulanthura*)
Q. sinuata Kensley 1982b Guiana Basin, Atlantic Ocean 520-550
Q. tyri Negoescu & Svavarsson 1997 Iceland 304-1212

Sauranthura (Poore & Kensley 1981)
S. goldmanorum (Poore & Kensley 1981) Queensland, Australia 0-10

Stygocyathura (Botosaneanu & Stock) 1982
S. beroni (Andreev 1982b) Papua New Guinea fresh water, caves (*Cyathura*)
S. broodbakkeri (Wagner 1990) Dominican Republic fresh water (*Cyathura*)
S. chapmani (Andreev 1982a) Kalimantan, Indonesia fresh water, caves (*Cyathura*)
S. cuborientalis (Botosaneanu & Stock 1982) Cuba fresh water (*Cyathura*)
S. curassavica (Stork 1940) Curacao fresh water (*Cyathura*)
S. fijiensis (Wägele et al. 1987) Fiji fresh water (*Cyathura*)
S. filipinica (Botosaneanu & Sket 1999) Bohol I., Philipinnes freshwater, caves (*Cyathura*)
S. hummelincki (Botosaneanu & Stock 1982) Aruba fresh water (*Cyathura*)
S. milloti (Chappuis et al. 1956) Réunion I. (*Cyathura*)

S. motasi (Botosaneanu & Stock 1982) Haiti fresh water (*Cyathura*)
S. numeae (Wägele 1982c) New Caledonia interstitial (*Cyathura*)
S. omorii (Nunomura 1992a) Japan (*Cyathura*)
S. orighani (Negoescu Vladescu 1983) Cuba fresh water, caves (*Cyathura*)
S. papuae (Wägele et al. 1987) Papua New Guinea fresh water (*Cyathura*)
S. parapotamica (Botosaneanu & Stock 1982) Jamaica fresh water (*Cyathura*)
S. rapanuia (Botosaneanu 1987) Easter & Pitcairn Is. intertidal, interstitial (*Cyathura*)
S. salpiscinalis (Botosaneanu & Stock 1982) Haiti fresh water (*Cyathura*)
S. sbordonii (Argano 1971) Mexico fresh water, caves (*Cyathura*)
S. specus (Bowman 1965) Cuba fresh water, caves (*Cyathura*)
S. univam (Botosaneanu 1983) Venezuela caves (*Cyathura*)
S. wadincola (Botosaneanu & Stock 1997) Oman fresh water (*Cyathura*)

Skuphonura Barnard 1925
S. ecuadorensis Kensley 1980c Ecuador 0-15
S. itapuca Kensley 1980c Brazil 0-15
S. kensleyi Brusca & Müller 1991 Caribbean Colombia intertidal-6
S. laticeps Barnard 1925 St Thomas, West Indies 7-40
S. lindae Menzies & Kruczynski 1983 Gulf of Mexico 6-73

Expanathuridae fam. nov.

Coralanthura (Poore & Kensley 1981)
C. endeavourae (Poore & Kensley 1981) Australia; Coral Sea 1-21

Eisothistos Haswell 1884
E. anomalus (Kensley, 1980a) Madagascar 1-1.5 (*Heteranthura*)
E. antarcticus Vanhöffen 1914 Antarctic Ocean 46-385
E. atlanticus Vanhöffen 1914 Cape Verde I.
E. bataviae Kensley & Poore 1982 Houtman Abrolhos I. 1.3
E. besar Müller 1992c Malaysia 1-2
E. crateris Kensley 1976 St Paul I. shallow
E. macrurus Wägele 1979c Mediterranean Sea 3-35
E. maledivensis Wägele 1979c Maldive Is. 35
E. minutus Sivertsen & Holthuis 1980 Tristan da Cunha, South Atlantic intertidal-5
E. moreirai Pires 1981 Brazil 3
E. neoanomalus (Shyamasundari et al. 1991) Bengal, India intertidal (*Heteranthura*)
E. nipponicus Nunomura 1984 Japan 0.4
E. petrensis Kensley 1984 Belize intertidal-36
E. pumilus Wägele 1979c Mediterranean Sea 6
E. societensis Müller 1990d Society Is. 1-2
E. tayronae Müller 1990b Caribbean Colombia 2-3
E. teri Kensley & Snelgrove 1987 Barbados 9-15
E. vermiformis Haswell 1884 New South Wales, Australia intertidal

Expanathura Wägele 1981d
E. ardea (Poore & Kensley 1981) Queensland, Australia intertidal-5 (*Coralanthura*)

E. amstelodami (Kensley 1976) Amsterdam I. shelf-80 (*Panathura*)
E. collaris (Kensley 1979a) Fiji 1-3 (*Panathura*)
E. haddae (Kensley & Poore 1982) Western Australia intertidal (*Panathura*)
E. macronesia (Kensley 1980a) Mauritius; Madagascar (*Panathura*)
E. mooreae Müller 1993b Society Is. 0.5

Heptanthura Kensley 1978a
H. amyle (Kensley & Schotte 1987) Belize 0-25 (*Licranthura*)
H. caribbica (Müller 1990b) Caribbean Colombia 0-6 (*Stellanthura*)
H. colombiana (Müller 1990b) Caribbean Colombia 52 (*Chalixanthura*)
H. cryptobia (Wägele 1979d) Mediterranean Sea 0-50 (*Stellanthura*)
H. lewisi (Kensley & Snelgrove 1987) Barbados 9-15 (*Chalixanthura*)
H. novaezealandiae Kensley 1978a New Zealand 101
H. scopulosa (Kensley 1984) Belize intertidal (*Chalixanthura*)
H. tuberculata (Kensley & Heard 1991) Turks & Caicos Is. 37-40 (*Licranthura*)

Minyanthura Kensley 1982a
M. corallicola Kensley 1982a Belize 6-24
M. firingae Müller 1990a Réunion I. 0.5-1
M. pacifica Müller 1993b Society Is. 0.5

Panathura Barnard 1925
P. serricauda (Barnard 1920) South Africa intertidal-90 (*Apanthura*)

Rhiganthura Kensley 1978a
R. spinosa Kensley 1978a New Zealand 101

Hyssuridae (Wägele 1981d)

Belura Poore & Lew Ton 1988c
B. pillara Poore & Lew Ton 1988c Bass Strait, Australia 10-841
B. acuticauda (George & Negoescu 1985) Falkland Is. 108 (*Xenanthura*)

Galziniella Müller 1991c
G. polynesica Müller 1991c Society Is. intertidal

Hyssura Norman & Stebbing 1886
H. bacescui (George & Negoescu 1982) North Carolina, USA 445 (*Ocsanthura*)
H. gracilis (Kensley 1982b) Bermuda Rise, Atlantic Ocean 1500-1700 (*Ocsanthura*)
H. ligurica Wägele 1981b Mediterranean Sea 600-690
H. producta Norman & Stebbing 1886 North Atlantic Ocean 2651-2900
H. profunda Barnard 1925 Mediterranean Sea 2018
H. vimsae (Kensley 1978c) New Jersey, Virginia, USA 350-1934 (*Ocsanthura*)

Kupellonura Barnard 1925
K. afareaitu Müller 1991c Society Is. 1-2
K. biriwa Poore & Lew Ton 1988c Victoria, Australia 9-24

K. capensis (Kensley 1975a) South Africa 128-400 (*Horoloanthura*)
K. caudoserrata Negoescu 1994 New Caledonia 310-315
K. currawan Poore & Lew Ton 1988c New South Wales, Australia 777
K. flexibilis (Pasternak 1982) Mediterranean Sea 2827-2829 (*Ananthura*)
K. formosa (Menzies & Frankenberg 1966) Georgia, USA 84-97 (*Panathura*)
K. gidgee Poore & Lew Ton 1988c Queensland, Australia intertidal-14
K. imswe (Kensley 1982a) Belize intertidal-6 (*Belizanthura*)
K. marrongie Poore & Lew Ton 1988c Queensland, Australia 1-4
K. mediterranea Barnard 1925 Mediterranean Sea 70-880
K. proberti Wägele 1985b New Zealand 134
K. racovitzai George & Negoescu 1985 Falkland Is. 108
K. serritelson Wägele 1981b Mediterranean Sea intertidal-33
K. werawera Poore & Lew Ton 1988c Queensland, Australia 9

Neohyssura Amar 1953
N. atlantica Wägele 1987 Cape Verde I. intertidal
N. bilara Poore & Lew Ton 1988c Queensland, Australia intertidal
N. irpex (Menzies & Frankenberg 1966) Georgia, USA 87 (*Horoloanthura*)
N. skolops Kensley 1978d South Africa 90-850
N. spinicauda (Walker 1901) Mediterranean Sea 1-40 (*Hyssura*)

Xenanthura Barnard 1925
X. bacescui Negoescu 1979b Brazil shelf, estuarine
X. brevitelson Barnard 1925 St Thomas, West Indies 8-145
X. conchae Müller 1990b Caribbean Colombia 0.5-4
X. linearis Pillai 1954 India shelf
X. orientalis Barnard 1935 India shelf
X. sinaica Wägele 1981b Sinai 1-35
X. ulawa Poore & Lew Ton 1988c Queensland, Australia intertidal-134

Leptanthuridae fam. nov.

Accalathura Barnard 1925
A. avena Poore & Lew Ton 1990 Queensland, Australia 7-18
A. barnardi (Nierstrasz 1941) Indonesia, 19-113 (*Katanthura*)
A. bassi Poore 1981b Victoria, Australia 8-19
A. borradailei (Stebbing 1904) Maldive Is. 45 (*Calathura*)
A. crenulata (Richardson 1901) Bahamas, Mexico, Brazil intertidal-55 (*Calathura*)
A. dimeria Wägele & Walter 1990 Queensland, Australia 3-13
A. eulalia Wägele & Walter 1990 Northern Territory, Australia 8-14
A. gigantissima Kussakin 1967 Antarctic Ocean 123-655
A. gigas (Whitelegge 1901) New South Wales, Australia 8-83 (*Calathura*)
A. indica (Nierstrasz 1941) Java Sea 73 ('*Metanthura*')
A. laevitelson (Kensley 1975a) South Africa (*Katanthura*)
A. maculata (Haswell 1881) New South Wales, Australia (*Haliophasma*)
A. normani (Nierstrasz, 1941) Indonesia 16-23 ('*Metanthura*')
A. ochotensis Nunomura 1976b Okhotsk Sea 390

A. oryza Poore & Lew Ton 1990 Queensland, Australia 1-9
A. poa Poore & Lew Ton 1990 Queensland, Australia intertidal
A. sehima Poore & Lew Ton 1990 Western Australia 42-122
A. setosa Kensley 1984 Belize; Society Is. intertidal
A. singularia Negoescu 1994 New Caledonia 305-310
A. sladeni Stebbing 1910 Cargados Carajos Is. 48-51 (*Calathura*)
A. spathia Poore & Lew Ton 1990 Western Australia 134-182
A. themeda Poore & Lew Ton 1990 Coral Sea; Australia 1-20
A. triodea Poore & Lew Ton 1990 Western Australia 30-97
A. vulpia Poore & Lew Ton 1990 Queensland, Australia intertidal
A. zoisia Poore & Lew Ton 1990 Western Australia 57-97

Aenigmathura Thomson 1951
A. calliandra Poore & Lew Ton 1986b Coral Sea; Australia 2-15
A. helicia Poore & Lew Ton 1986b Queensland, Australia 1-8
A. lactanea Thomson 1951 Western Australia intertidal-24
A. lifouensis Negoescu 1994 Loyalty Is. 510-520

Albanthura Wägele 1985b
A. rotunduropus Wägele 1985b New Zealand 31
A. stenodactyla Wägele 1985b New Zealand 808

Bourbonanthura Müller 1990a
B. littoralis Müller 1990a Réunion I. 0.5
B. vaitapensis Müller 1993b Society Is. 1

Bullowanthura Poore 1978
B. aquitanica Kensley 1982b Bay of Biscay 641-860
B. crebrui Wägele 1985b New Zealand 520
B. furcillata Negoescu 1994 New Caledonia 760-790
B. pambula Poore 1978 New South Wales, Australia 8-69

Calathura Norman & Stebbing 1886
C. brachiata (Stimpson 1853) North Atlantic Ocean 9-2500 (*Anthura*)

Curassanthura Kensley 1981
C. bermudensis Wägele & Brandt 1985 Bermuda caves
C. canariensis Wägele 1985a Canary Is. caves
C. halma Kensley 1981 Curacao intertidal
C. jamaicensis Kensley 1992 Jamaica caves

Leptanthura Sars 1897
L. affinis (Bonnier 1896) North Atlantic Ocean 516-2185 (*Calathura*)
L. agulhasensis Kensley 1975a South Africa 26-320
L. antarctica Kussakin 1967 Davis Sea, Antarctic Ocean 206-540
L. apalpata Wägele 1981c Mediterranean Sea 300-800
L. argentinae Kensley 1982b Argentine Basin, South Atlantic Ocean 1002-4392

L. baliensis Negoescu 1992a Bali, Indonesia 3-4
L. boweni Poore 1981b Victoria, Australia intertidal-329
L. chardyi Negoescu 1992b France, North Atlantic Ocean 850-2820
L. chiltoni Beddard 1886 New Zealand 1274
L. coeca Menzies 1956 Jamaica 1237
L. coralliophila Müller 1992d Malaysia intertidal-2
L. diemenensis Haswell 1885 Tasmania, Australia 0-192
L. elegans Birstein 1963 Kurile I., Russia 1226
L. exilis Wägele 1985b New Zealand 506
L. flindersi Poore 1981b Tasmania, Australia 12-24
L. geocostarioi Negoescu 1991 Brazil 65
L. glacialis Hodgson 1910 Ross Sea, Antarctic Ocean 50-5216
L. guianae Kensley 1982b Guiana Basin, Atlantic Ocean 516-1487
L. hendili Wolff 1956 Banda Trench 6490-6650
L. kapala Poore 1978 New South Wales, Australia 1200
L. laevigata (Stimpson 1855) South Africa 15-1360 (*Anthura*)
L. micrura Kensley 1982b Brazil 520-2494
L. minuta Kensley 1978d South Africa 550-850
L. monnioti Negoescu 1994 New Caledonia 760-790
L. muelleri Negoescu 1980a Libya 62-63
L. murrayi Poore 1981b Victoria, Australia 12-24
L. natalensis Kensley 1978d South Africa 690-850
L. nunana Poore 1978 Victoria, Australia 8-22
L. orientalis Barnard 1925 Singapore
L. profundicola Wägele 1985b New Zealand 1120
L. sculpta Pasternak 1982 Mediterranean Sea 1113
L. segonzaci Negoescu 1994 New Caledonia 975
L. tenuis (Sars 1873) Norway 0-1500 (*Paranthura*)
L. thalassae Negoescu 1980b Gulf of Aden 24-110
L. thori Barnard 1925 North Atlantic Ocean 957
L. truncata Richardson 1911 Mauritania 888
L. truncatitelson Wägele 1985b New Zealand 560
L. urospinosa Kensley 1975a South Africa 5-200
L. victori Negoescu 1985 North Atlantic Ocean, 1894-1950
L. vitilevui Negoescu 1999 Fiji 2

Psittanthura Wägele 1985b
P. egregia Wägele 1985b New Zealand 560

Ulakanthura Poore 1978
U. colac Poore, 1978 Queensland, Victoria, Australia 5-22
U. cooma Poore 1978 Queensland, Australia 9-10
U. crassicornis (Haswell 1881) New South Wales, Australia 22-31 (*Paranthura*)
U. lara Poore 1978 Victoria, Australia 2-31
U. marlee Poore 1981b Victoria, , New South Wales, Australia 15-20
U. namoo Poore 1978 Queensland, New South Wales, Australia 9-23
U. wanda Poore 1978 Queensland, Australia 9-10

Virganthura Kensley 1987b
V. crassa (Barnard 1925) Virgin Is. 30 (*Accalathura*)

Paranthuridae Menzies & Glynn 1968

Califanthura Schultz 1977
C. anophthalma (Kussakin & Vasina 1982) Kerguelen I. 3-15 (*Colanthura*)
C. barnardi (Sivertsen & Holthuis 1980) Tristan da Cunha, South Atlantic intertidal-135 (*Tristanthura*)
C. latimana (Kussakin & Vasina 1982) Kerguelen I. 268 (*Colanthura*)
C. lowryi Poore 1984a Macquarie I. 6-13
C. minuta Kensley & Heard 1991 Turks & Caicos Is. 1.5-5
C. pingouin (Kensley 1980b) Crozet I. 142-490 (*Colanthura*)
C. rima (Poore 1981a) New Zealand intertidal-22 (*Colanthura*)
C. squamosissima (Menzies 1951) California, USA intertidal-142 (*Colanthura*)

Colanthura Richardson 1902
C. bruscai Poore 1984a Mexico intertidal-27
C. gauguini Müller 1993b Society Is. 0.5
C. kensleyi Poore 1984a Philippines intertidal-6
C. nigra Nunomura 1975 Japan 45
C. ornata Carvacho 1977 Guadeloupe 2
C. pigmentata Kensley 1980b Madagascar 1.5
C. setouchiensis Nunomura 1993 Japan
C. tenuis Richardson 1902 Bermuda; Society Is. intertidal
C. uncinata Kensley 1978d South Africa 150-775

Cruranthura Thomson 1946
C. caeca (Mezhov 1976) Kurile I., Russia 15-30 (*Colanthura*)
C. furneauxi (Poore 1981b) Tasmania, Australia intertidal (*Colanthura*)
C. peroni (Poore 1981b) New South Wales, Australia intertidal-20 (*Colanthura*)

C. simplicia Thomson 1946 Western Australia intertidal
Cruregens Chilton 1882b
C. fontanus Chilton 1882b New Zealand fresh water

Paranthura Bate & Westwood 1866
P. acacia Poore 1984b Victoria, Australia 5-18
P. alba Nunomura 1977 Japan intertidal-15
P. algicola Nunomura 1979b California, USA intertidal
P. antarctica Kussakin 1967 Antarctic Ocean 3-334
P. antillensis Barnard 1925 Virgin Is. 29
P. argentinae Kussakin 1967 Argentine Basin, South Atlantic Ocean 400-508
P. astrolabium Kensley 1979a Fiji 5-13
P. australis Haswell 1881 New South Wales, Australia
P. barnardi Paul & Menzies 1971 Venezuela 95
P. bellicauda Miller & Menzies 1952 Hawaii, USA shelf

P. boronia Poore 1984b Victoria, Australia intertidal-8
P. brucei Negoescu 1999 Fiji 0-8
P. bunakenensis Negoescu 1997b Celebes Sea, Indonesia 4.5
P. caesia Poore 1984b Victoria, Australia intertidal-6
P. californiae Nunomura 1979b California, USA, 18-27
P. caribbiensis Kensley 1982a Belize shelf
P. ciliata Whitelegge 1901 New South Wales, Australia 98-329
P. costana Bate & Westwood 1866 English Channel, UK 0-355
P. deodata Müller 1990a Réunion I. 0.5-1
P. dryandra Poore 1984b Victoria, Australia intertidal-3
P. elegans Menzies 1951 California, USA 0-55
P. epacris Poore 1984b Victoria, Australia intertidal-80
P. flagellata (Chilton 1882a) New Zealand 0-107 (*Anthura*)
P. floridensis Menzies & Kruczynski 1983 Gulf of Mexico 73
P. gracilipes Nordenstam 1930 Juan Fernandez I. 30-40
P. grevillea Poore 1984b New South Wales, Australia intertidal-27
P. hasticauda Nunomura 1974a Japan shallow
P. infundibulata Richardson 1902 Bermuda intertidal-12
P. involuta Whitelegge 1901 New South Wales, Australia 90-93
P. japonica Richardson 1909 Japan intertidal-15
P. kagawaensis Nunomura 1993 Japan
P. kobensis Nunomura 1975 Japan intertidal
P. kunzea Poore 1984b Tasmania, Australia intertidal-3
P. laticauda Nunomura 1975 Japan 0-15
P. latipes Barnard 1955 Mozambique 0-15
P. lifuensis Stebbing 1900 Loyalty Is. 72-146
P. linearis (Boone 1923) California, USA (*Edanthura*)
P. lineata Nunomura 1977 Japan 30
P. lobelia Poore 1984b New South Wales, Australia intertidal
P. longa Wägele 1985b New Zealand 514
P. longitelson Wägele 1984c Baja California, Mexico intertidal
P. maculosa Nunomura 1974a Japan 4
P. microtis Poore 1984b South Australia intertidal
P. nana Nordenstam 1930 Juan Fernandez Is. 30-40
P. neglecta Beddard 1886 Kerguelen I. 131-283
P. nigrocaudata (Nunomura 1975) Japan intertidal (*Leptanthura*)
P. nigropunctata (Lucas 1846) Algeria 2-50 (*Anthura*)
P. ostergaardi Miller & Menzies 1952 Hawaii, USA shelf
P. plumosa Pillai 1966 India shelf
P. polynesica Kensley 1979a Cook Is. 1
P. porteri (Boone 1920) Chile intertidal (*Calamura*)
P. possessia Kensley 1980b Crozet I. 508-2707
P. punctata (Stimpson 1855) South Africa 0-200 (*Anthura*)
P. senecio Poore 1984b New South Wales, Australia 3-8
P. setigera Negoescu 1997b Bali Sea, Indonesia 3-4
P. skottsbergi Nordenstam 1930 Juan Fernandez I. 30-46
P. societensis Müller 1993b Society Is. 0.5

P. telopea Poore 1984b Victoria, Australia intertidal
P. urochroma Pires 1981 Brazil 6
P. verrillii Richardson 1902 Bermuda intertidal

Pseudanthura Richardson 1911
P. albatrossae Kensley 1978e Flores Sea, Indonesia 1380
P. baeckea Poore & Lew Ton 1986b New South Wales, Australia 79-920
P. lateralis Richardson 1911 West Africa 930-3200
P. recifensis Kensley 1982b Brazil 587
P. tenuis Kensley 1978e South Africa 560-850

REFERENCES

Amar, R. 1953. Isopodes marin du littoral Corse. *Bulletin de la Société Zoologique de France* 77: 345-355.

Amar, R. 1966. Notes sur les anthurides (Crustacés Isopodes) méditerranéens. 1. *Haliophasma alaticauda* n. sp. *Recueil de Travaux de la Station Marine d'Endoume* 41: 193-199.

Andreev, S. 1982a. Une *Cyathura* cavernicole nouvelle de Sarawak - Kalimantan du Nord (Isopoda, Anthuridae). *Bulletin Zoölogisch Museum, Universiteit van Amsterdam* 8: 149-155.

Andreev, S. 1982b. Sur une nouvelle espèce cavernicole du genre *Cyathura* (Isopoda, Anthuridae), (Résultats Zoologiques de la Mission Spéleologique Brittanique en Papouasie-Nouvelle Guinée, 1975, 7). *International Journal of Speleology* 12: 55-62.

Argano, R. 1971. *Cyathura sbordonii*, nuova specie cavernicola del messico sudorientale diagnosi preliminare (Crustacea, Isopoda, Anthuridae). *Fragmenta Entomologica* 7: 303-305.

Bamber, R.N. 1997. Peracarid crustaceans from Cape d'Aguilar and Hong Kong, I. Mysidacea, and Isopoda: Anthuridea. In Morton, B. (ed.), *The marine flora and fauna of Hong Kong and southern China IV. Proceedings of the Eight International Marine Biological Workshop: 2-20 April 1995* : pp. 77-86. Hong Kong : Hong Kong University Press.

Barnard, K.H. 1914. Contributions to the crustacean fauna of South Africa. 3. Additions to the marine Isopoda, with notes on some previously incompletely known species. *Annals of the South African Museum* 10: 325a-358a, 359-442.

Barnard, K.H. 1920. Contributions to the crustacean fauna of South Africa. No. 6. Further additions to the list of marine Isopoda. *Annals of the South African Museum* 17: 319 438.

Barnard, K.H. 1925. A revision of the family Anthuridae (Crustacea, Isopoda), with remarks on certain morphological peculiarities. *Zoological Journal of the Linnean Society* 36: 109-160.

Barnard, K.H. 1935. Report on some Amphipoda, Isopoda, and Tanaidacea in the collections of the Indian Museum. *Records of the Indian Museum* 37: 279-319.

Barnard, K.H. 1940. Contribution to the crustacean fauna of South Africa. XII. Further additions to the Tanaidacea, Isopoda, and Amphipoda, together with keys for the identification of hitherto recorded marine and freshwater species. *Annals of the South African Museum* 32: 381-543.

Barnard, K.H. 1955. Additions to the fauna-list of South African Crustacea and Pycnogonida. *Annals of the South African Museum* 43: 1-107.

Barnard, K.H. 1957. Additions to the fauna-list of South African Crustacea. *Annals and Magazine of Natural History* (12) 10: 1-12.

Bate, C.S. & J.O. Westwood 1866. *A history of the British sessile-eyed Crustacea*. Vol. 2. London.: John Van Voorst. 536 pp.

Beddard, F.E. 1886. Report on the Isopoda collected by H.M.S. Challenger during the years 1873-76. Part II. *Report on the Scientific Results of the Voyage of H.M.S. Challenger during the years 1873-76. Zoology* 17: 1-178, pls I-XXV, 1 chart.

Birstein, J.A. 1963. *Deep water isopods (Crustacea. Isopoda) of the north-western part of the Pacific Ocean.* Moscow: Akademia Nauk, SSSR. 213 pp. English translation by the Indian National Scientific Documentation Centre, New Dehli, 1973.

Bonnier, J. 1896. Résultats Scientifiques de la Campagne du 'Caudan' dans le Golfe de Gascogne - Août-septembre 1895 - Édriophthalmes. *Annales de l'Université de Lyon*: 527-689.

Boone, P.L. 1920. *Calamura porteri*, a new genus and species of isopod from Chile. *Revista Chilena de Historia Natural* 24: 25-31, pl 2.

Boone, P.L. 1923. New marine tanaid and isopod Crustacea from California. *Proceedings of the Biological Society of Washington* 36: 147-156.

Botosaneanu, L. 1983. First record of an anthurid isopod, *Cyathura univam* sp. n., on the South American continent. *Bijdragen tot de Dierkunde* 53: 247-254.

Botosaneanu, L. 1987. A new thalassostygobiont species of *Cyathura* (Isopoda, Anthuridea) from the south-east Pacific. *Stygologia* 3: 296-304.

Botosaneanu, L. & B. Sket 1999. A new freshwater stygobiotic species of Cyathura (Isopoda: Anthuridae) from Bohol Island, the Philipinnes. *Acta Biologica Slovenica* 42: 27-33.

Botosaneanu, L. & J.H. Stock 1982. Amsterdam expeditions to the West Indian Islands, Report 17. Les *Cyathura* stygobies (Isopoda, Anthuridea) et surtout celles des Grandes et des Petites Antilles. *Bijdragen tot de Dierkunde* 52: 13-42.

Botosaneanu, L. & J.H. Stock 1997. Stygofauna of Oman, 1. A new freshwater stygobiont *Cyathura* (Isopoda, Anthuridae), from interstitial of coarse wadi sediments in Oman. *Annales de Limnologie* 33: 79-84.

Bowman, T.E. 1965. *Cyathura specus*, a new cave isopod from Cuba (Anthuroidea: Anthuridae). *Studies on the fauna of Curaçao and other Caribbean Islands* 22: 88-97.

Brandt, A. & I. Negoescu 1997. Redescription of the Arctic-boreal species *Calathura brachiata* (Stimpson, 183) (Isopoda, Anthuridea, Paranthuridae). *Crustaceana* 70: 418-453.

Brusca, R.C. & E.W. Iverson 1985. A guide to the marine isopod Crustacea of Pacific Costa Rica. *Revista de Biologia Tropical (Universidad de Costa Rica)* 33: 1-77.

Brusca, R.C. & H.G. Müller 1991. *Skuphonura kensleyi* (Crustacea: Isopoda), a new anthuridean species from the Caribbean coast of Columbia. *Proceedings of the Biological Society of Washington* 104: 593-602.

Brusca, R.C. & G.D.F. Wilson 1991. A phylogenetic analysis of the Isopoda with some classificatory recommendations. *Memoirs of the Queensland Museum* 31: 143-204.

Carvacho, A. 1977. Isopodes de la mangrove de la Guadeloupe, Antilles Françaises. *Studies on the fauna of Curaçao and other Caribbean Islands* 54: 1-24.

Chappuis, P.-A., C. Delamare Debouteville & R. Paulian 1956. Crustacés des eaux souterraines littorales d'une résurgence d'eau douce à la Réunion. *Mémoires de l'Institut Scientifique de Madagascar* 11: 51-78.

Chilton, C. 1882a. Additions to the New Zealand Crustacea. *Transactions and Proceedings of the New Zealand Institute, Zoology* 14: 171-174.

Chilton, C. 1882b. On some subterranean Crustacea. *Transactions and Proceedings of the New Zealand Institute, Zoology* 14: 174-180.

Chilton, C. 1894. The subterranean Crustacea of New Zealand. *Transactions of the Linnean Society of London* (2) 6: 163-284.

Cleret, J.J. 1960. Étude de *Cyathura carinata* (Kröyer) (Isopode Anthuridae). *Cahiers de Biologie Marine* 1: 433-452.

Dallwitz, M.J., T.A. Paine & E.J. Zurcher 1997. *User's guide to the DELTA system. A general system for processing taxonomic descriptions.* 4.08 edition. Canberra: CSIRO Division of Entomology. 160 pp.

Frankenberg, D. 1965. A new species of *Cyathura* (Isopoda, Anthuridae) from coastal waters off Georgia, U.S.A. *Crustaceana* 8: 206-212.

George, R.Y. & I. Negoescu 1982. *Ocsanthura bacescui* a new anthuridean isopod (Isopoda, Anthuridea) from the outer continental shelf of North Carolina. *Travaux du Muséum d'Histoire Naturelle 'Grigore Antipa'* 24: 97-103.

George, R.Y. & I. Negoescu 1985. Anthuridean isopods (Crustacea, Isopoda, Anthuridea) from the subantarctic islands - South Georgia, Elephant, South Orkney and Falkland. *Travaux du Muséum d'Histoire Naturelle 'Grigore Antipa'* 27: 19-47.

Gomez Simes, E. 1979. Algunos isopodos de la Ria Deseado (Santa Cruz, Argentina). *Contribuciones Científica Centro de Investigación de Biología Marina, Estación Puerto Deseado, Buenos Aires* 166: 5-16.

Hackney, C.T. & R.F. Ganucheau 1989. A new species of *Cyathura* (Isopoda, Anthuridae) from a high intertidal march in North Carolina, U.S.A. *Crustaceana* 57: 304-310.

Hansen, H.J. 1916. Crustacea Malacostraca III: V. The Order Isopoda. *Danish Ingolf-Expedition* 3: 1-262 pls 1-16.

Harger, O. 1878. Descriptions of new genera and species of Isopoda, from New England and adjacent regions. *American Journal of Sciences and Arts* (3) 15: 373-379.

Haswell, W.A. 1881. On some new Australian marine Isopoda - Part I. *Proceedings of the Linnean Society of New South Wales* 5: 470-481, pls 16-19.

Haswell, W.A. 1884. On a new crustacean found inhabiting the tubes of *Vermilia* (Serpulidae). *Proceedings of the Linnean Society of New South Wales* 9: 676-679, pls 36, 37.

Haswell, W.A. 1885. A revision of the Australian Isopoda. *Proceedings of the Linnean Society of New South Wales* 9: 1001-1015, pls 50-53.

Hodgson, T.V. 1910. Crustacea. 9. Isopoda. *National Antarctic Expedition, 1901-1904, Natural History* 5, Zoology and Botany: 1-77.

Holthuis, L.B. 1977. The dates of publication of C. Spence Bate and J.O. Westwood's 'A history of British sessile-eyed Crustacea.'. *Crustaceana* 33: 313-316.

Hooker, A. 1985. New species of Isopoda from the Florida Middlegrounds (Crustacea: Peracarida). *Proceedings of the Biological Society of Washington* 98: 255-280.

Kensley, B. 1975a. Marine Isopoda from the continental shelf of South Africa. *Annals of the South African Museum* 67: 35-89.

Kensley, B. 1975b. A new species of anthurid isopod from the Cape. *Zoologica Africana* 10: 209-213.

Kensley, B. 1976. Isopodan and tanaidacean Crustacea from the St Paul and Amsterdam Islands, southern Indian Ocean. *Annals of the South African Museum* 69: 261-323.

Kensley, B. 1978a. Five new genera of anthurid isopod crustaceans. *Proceedings of the Biological Society of Washington* 91: 775-792.

Kensley, B. 1978b. *Guide to the marine isopods of southern Africa*. Cape Town.: Trustees of the South African Museum. 173 pp.

Kensley, B. 1978c. A new genus and species of anthurid isopod from deep water off the east coast of the United States. *Proceedings of the Biological Society of Washington* 91: 558-562.

Kensley, B. 1978d. The South African Museum's Meiring Naude Cruises. Part 8. Isopoda Anthuridea. *Annals of the South African Museum* 77: 1-25.

Kensley, B. 1978e. Two new species of the genus *Pseudanthura* Richardson (Crustacea: Isopoda: Anthuridea). *Proceedings of the Biological Society of Washington* 91: 222-233.

Kensley, B. 1979a. New species of anthurideans from the Cook and Fiji Islands (Crustacea: Isopoda: Anthuridea). *Proceedings of the Biological Society of Washington* 92: 814-836.

Kensley, B. 1979b. Redescription of *Haliophasma barnardi* (Monod) from West Africa (Crustacea: Isopoda: Anthuridae). *Proceedings of the Biological Society of Washington* 92: 658-664.

Kensley, B. 1980a. Anthuridean isopod crustaceans from the International Indian Ocean Expedition, 1960-1965, in the Smithsonian Collections. *Smithsonian Contributions to Zoology* 304: 1-37.

Kensley, B. 1980b. Marine isopods from Marion, Prince Edward, and Crozet Islands (Crustacea, Isopoda). *Annals of the South African Museum* 82: 155-185.

Kensley, B. 1980c. Records of anthurids from Florida, Central America, and South America (Crustacea: Isopoda: Anthuridae). *Proceedings of the Biological Society of Washington* 93: 725-742.

Kensley, B. 1981. Amsterdam Expeditions to the West Indian Islands Report 10. *Curassanthura halma*, a new genus and species of interstitial isopod from Curaçao, West Indies (Crustacea: Isopoda: Paranthuridae). *Bijdragen tot de Dierkunde* 51: 131-134.

Kensley, B. 1982a. Anthuridea (Crustacea: Isopoda) of Carrie Bow Cay, Belize. *Smithsonian Contributions to the Marine Sciences* 12: 321-353.

Kensley, B. 1982b. Deep-water Atlantic Anthuridea (Crustacea: Isopoda). *Smithsonian Contributions to Zoology* 346: 1-60.

Kensley, B. 1982c. Revision of the southern African Anthuridea (Crustacea, Isopoda). *Annals of the South African Museum* 90: 95-200.

Kensley, B. 1984. The Atlantic barrier reef ecosystem at Carrie Bow Cay, Belize, III: New marine Isopoda. *Smithsonian Contributions to the Marine Sciences* 24: 1-81.

Kensley, B. 1987a. Further records of marine isopod crustaceans from the Caribbean. *Proceedings of the Biological Society of Washington* 100: 559-577.

Kensley, B. 1987b. A re-evaluation of the systematics of K.H. Barnard's review of anthuridean isopods. *Steenstrupia* 13: 101-139.

Kensley, B. 1992. New records of interstitial Anthuridea including *Curassanthura*, from the Western Atlantic (Crustacea: Isopoda). *Stygologia* 7: 187-192.

Kensley, B. 1994. Records of shallow-water marine isopods from Bermuda with descriptions of four new species. *Journal of Crustacean Biology* 14: 319-336.

Kensley, B. 1996. The genus *Ptilanthura* in the western Atlantic: evidence for primary males and description of a new species (Isopoda: Anthuridae). *Journal of Crustacean Biology* 16: 763-781.

Kensley, B. & R.W. Heard 1991. Studies on the Crustacea of the Turks and Caicos Islands, British West Indies I. Four new marine isopod crustaceans from the vicinity of Pine Cay. *Gulf Research Reports* 8: 237-246.

Kensley, B. & M.L. Koening 1979. Two new species of *Quantanthura* from Brazil (Crustacea, Isopoda, Anthuridae). *Proceedings of the Biological Society of Washington* 91: 953-962.

Kensley, B. & G.C.B. Poore 1982. Anthurids from the Houtman Abrolhos Islands, Western Australia (Crustacea: Isopoda: Anthuridae). *Proceedings of the Biological Society of Washington* 95: 625-636.

Kensley, B. & J. Reid 1984. *Arabanthura enigmatica*, a new genus and species of anthurid isopod from the Arabian Gulf. *Proceedings of the Biological Society of Washington* 97: 674-680.

Kensley, B. & M. Schotte 1987. New records of isopod Crustacea from the Caribbean, the Florida Keys, and the Bahamas. *Proceedings of the Biological Society of Washington* 100: 216-247.

Kensley, B. & M. Schotte 1989. *Guide to the marine isopod crustaceans of the Caribbean*. Washington, D.C., and London: Smithsonian Institution Press. 308 pp.

Kensley, B. & P. Snelgrove 1987. Records of marine isopod crustaceans associated with the coral *Madracis mirabilis* from Barbados. *Proceedings of the Biological Society of Washington* 100: 186-197.

Kensley, B., M. Ortiz & M. Schotte 1997. New records of marine Isopoda from Cuba (Crustacea: Peracarida). *Proceedings of the Biological Society of Washington* 110: 74-98.

Koening, M.L. 1980. Ocorrencia do genro *Mesanthura* Barnard, 1914 no Brasil. Descricao de *Mesanthura brasiliensis* n. s. (Isopoda-Paranthuridae). *Trabahlos Oceanografico da Universidade Federale de Pernambuco* 15: 357-363.

Kruczynski, W.L. & G.J. Myers 1976. Occurrence of *Apanthura magnifica* Menzies and Frankenberg, 1966 (Isopoda: Anthuridae) from the west coast of Florida, with a key to the species of *Apanthura* Stebbing, 1900. *Proceedings of the Biological Society of Washington* 89: 353-360.

Krøyer, H. 1847. Karcinologiste Bidrag. *Naturhistorisk Tidsskrift* (2) 2: 366-446.

Kussakin, O.G. 1967. Fauna of Isopoda and Tanaidacea in the coastal zones of the Antarctic and Subantarctic waters. [Translation from Russian by the Israel Program for Scientific Translations, Jerusalem, 1968.]. *Biological Reports of the Soviet Antarctic Expedition (1955-1958)* 3: 220-389.

Kussakin, O.G. 1982. Marine and brackish-water Crustacea (Isopoda) of cold and temperate waters of the Northern Hemisphere. Suborders Anthuridea, Microcereberidea, Valvifera, Tyloidea. *Opredeliteli po Faune SSR, Akademiya Nauk, SSSR* 131: 1-461. In Russian.

Kussakin, O.G. & G.S. Vasina 1982. Addition to the fauna of benthic Isopoda and Gnathiida (Crustacea) of subantarctic waters of the Indian Ocean. 1. Isopoda (Flabellifera and Anthuridea). *Tethys* 10: 261-273.

Leach, W.E. 1814. Crustaceology. *Brewster's Edinburgh Encyclopedia* 7: 383-437, pl. 221.

Lowry, J.K. 1986. The callynophore, a eucaridan/peracaridan sensory organ prevalent among the Amphipoda (Crustacea). *Zoologica Scripta* 15: 333-349.

Lucas, P.H. 1846. *Histoire naturelles des animaux articulés, par H. Lucas. Pt. 1. Crustacés, Arachnides, Myriapodes et Hexapodes. In Exploration scientifique de l'Algerie pendent les années 1840, 1841, 1842.* Paris: Imprimerie Nationale.

Menzies, R.J. 1951. New marine isopods, chiefly from northern California, with notes on related forms. *Proceedings of the United States National Museum* 101: 105-156.

Menzies, R.J. 1956. New bathyal Isopoda from the Caribbean with observations on their nutrition. *Breviora* 63: 1-10.

Menzies, R.J. 1962. The isopods of abyssal depths in the Atlantic Ocean. *Vema Research Series* 1: 79-206.

Menzies, R.J. & J.L. Barnard 1959. Marine Isopoda on coastal shelf bottoms of Southern California: systematics and ecology. *Pacific Naturalist* 1: 3-35.

Menzies, R.J. & D. Frankenberg 1966. *Handbook on the common marine isopod Crustacea of Georgia.* Athens.: University of Georgia Press. vii + 93 pp.

Menzies, R.J. & R.Y. George 1972. Isopod Crustacea of the Peru-Chile Trench. *Anton Bruun Report* 9: 1-124.

Menzies, R.J. & P.W. Glynn 1968. Studies on the fauna of Curaçao and other Caribbean Islands No. 27. The common marine isopod Crustacea of Puerto Rico. A handbook for marine biologists. *Uitgaven van de Natuurwetenschappelijke Kring voor Suriname en de Nederlandse Antillen* 51: 1-133.

Menzies, R.J. & W.L. Kruczynski 1983. Isopod Crustacea (exclusive of Epicaridea). *Memoirs of the Hourglass Cruises* 6: 1-126.

Mezhov, B.V. 1976. New species of Anthuroidea from the upper part of the sublittoral zone of the middle Kurile Islands. *Biologiya Morya, Vladivostok* 5: 19-27.

Miller, M.A. & R.J. Menzies 1952. The isopod Crustacea of the Hawaiian Islands, III. Superfamily Flabellifera, family Anthuridae. *Occasional Papers of the Bernice P. Bishop Museum* 21: 1-15.

Monod, T. 1922. Sur un essai de classification rationelle des isopodes. *Bulletin de la Société Zoologique de France* 47: 134-140.

Monod, T. 1925. Tanaidacés et Isopodes aquatiques de l'Afrique Occidentale et Septentrionale. *Bulletin de la Société des Sciences Naturelles et Physiques du Maroc* 5: 61-77, 84-85, pls 4-16.

Monod, T. 1927. Notes Isopodologiques. II. Sur un anthuride nouveau du Cameroen, *Notanthura barnardi* nov. gen. nov. spec. *Bulletin de la Société Zoologique de France* 52: 200-211.

Montagu, G. 1808. Description of several marine animals found on the south coast of Devonshire. *Transactions of the Linnean Society of London* 9: 81-144.

Moore, W. 1996. Biogeography of anthuridean isopods (Crustacea: Isopoda: Anthuridea) of the central tropical Pacific. MSc Thesis, University of Charleston: Charleston. 111 pp.

Müller, H.-G. 1990a. Anthuridea from coral reefs at Réunion Island, southern Indian Ocean (Crustacea: Isopoda). *Senckenbergiana Biologica* 70: 359-395.

Müller, H.-G. 1990b. Hyssuridae from the Caribbean coast of Colombia, with descriptions of four new species (Isopoda, Anthuridea). *Bijdragen tot de Dierkunde* 60: 65-78.

Müller, H.-G. 1990c. Paranthurid isopods from the Caribbean Sea of Colombia (Crustacea). *Bulletin Zoölogisch Museum, Universiteit van Amsterdam* 12: 181-191.

Müller, H.-G. 1990d. Two new species of *Eisothistos* and *Anthomuda* from coral reefs at Moorea and Bora Bora, Society Islands (Isopoda, Anthuridea: Hyssuridae, Paranthuridae). *Zoologische Abhandlungen, Staatliches Museum für Tierkunde, Dresden* 45: 111-119.

Müller, H.-G. 1991a. Marine Anthuridea from Martinique, French Antilles, with redescriptions of some species (Crustacea: Isopoda). *Revue Suisse de Zoologie* 98: 739-768.

Müller, H.-G. 1991b. New species and records of *Amakusanthura*, *Cyathura* and *Haliophasma* from Sri Lanka (Crustacea, Isopoda, Anthuridae). *Revue Suisse de Zoologie* 98: 589-612.

Müller, H.-G. 1991c. Three new species and a new genus of eyeless isopods from coral reefs at Moorea, Society Islands. *Senckenbergiana Biologica* 71: 289-301.

Müller, H.-G. 1992a. Anthuridae of the genera *Amakusanthura*, *Cortezura* and *Mesanthura* from the Caribbean Sea of Columbia (Crustacea: Isopoda. *Revue Suisse de Zoologie 99: 31-79.*

Müller, H.-G. 1992b. Anthuridae of the genera *Apanthura* and *Cyathura* from Malaysian coral reefs, with descriptions of two new species (Crustacea: Isopoda: Anthuridae). *Zoologischer Anzeiger* 228: 156-166.

Müller, H.-G. 1992c. *Eisothistos besar* n. sp. from a coral reef in Tioman Archipelago, first member of the genus from southeast Asia (Crustacea: Isopoda). *Revue Suisse de Zoologie* 99: 369-376.

Müller, H.-G. 1992d. A review of *Leptanthura* Sars, 1899, with description of new species from the Tioman Archipelago, Malaysia (Crustacea: Isopoda: Paranthuridae). *Cahiers de Biologie Marine* 33: 179-200.

Müller, H.-G. 1993a. The genus *Mesanthura* Barnard, 1914 from coral reefs in Kenya and Malaysia, with descriptions of three new species (Crustacea: Isopoda: Anthuridae). *Mitteilungen aus dem Zoologischen Museum in Berlin* 69: 19-44.

Müller, H.-G. 1993b. Paranthurid isopods from French Polynesian coral reefs, including descriptions of six new species (Crustacea: Peracarida). *Cahiers de Biologie Marine* 24: 289-341.

Negoescu, I. 1979a. *Cyathura cubana* sp. n. (Isopoda, Anthuridea) from the Caribbean Sea (Cuban Waters). *Travaux du Muséum d'Histoire Naturelle 'Grigore Antipa'* 20: 157-164.

Negoescu, I. 1979b. *Xenanthura bacescui* sp. n. (Isopoda, Anthuridea) from the South-western Atlantic Ocean (Brazil). *Travaux du Muséum d'Histoire Naturelle 'Grigore Antipa'* 20: 165-169.

Negoescu, I. 1980a. Contribution to the study of anthurid isopods (Isopoda, Anthuridea) from the Mediterranean (Libya) with the description of two new species. *Travaux du Muséum d'Histoire Naturelle 'Grigore Antipa'* 21: 89-102.

Negoescu, I. 1980b. Littoral anthuridean isopods (Isopoda, Anthuridea) from the North-western Indian Ocean. *Travaux du Muséum d'Histoire Naturelle 'Grigore Antipa'* 22: 401-420.

Negoescu, I. 1981. New data about the fauna of anthuridean isopods isopods (Isopoda, Anthuridae) from the Libyan coast and the Suez Canal. *Travaux du Muséum d'Histoire Naturelle 'Grigore Antipa'* 22: 401-420.

Negoescu, I. 1985. Isopodes anthurides (Crustacea, Isopoda, Anthuridea). In Laubier, L. & C. Monniot (eds), *Peuplements profonds du Golfe de Gascoyne* : pp. 449-463. Paris: IFREMER.

Negoescu, I. 1989. New data on the species *Pilosanthura fresii* (Wägele 1980) (Isopoda, Anthuridea, Anthuridae). *Travaux du Muséum d'Histoire Naturelle 'Grigore Antipa'* 30: 181-186.

Negoescu, I. 1991. *Leptanthura geocostarioi* n. sp. (Isopoda, Anthuridea, Paranthuridae) from the coastal waters off Brazil. *Travaux du Muséum d'Histoire Naturelle 'Grigore Antipa'* 31: 365-374.

Negoescu, I. 1992a. *Leptanthura baliensis* n. sp. (Crustacea, Isopoda, Anthuridea) of SE Bali Island (Indonesia). In *Guide-book of temporary exhibition 'Four Romanian Naturalists in Indonesia.'* Bucharest.

Negoescu, I. 1992b. *Leptanthura chardyi* n. sp. (Isopoda, Anthuridea, Paranthuridae) of the Atlantic Ocean. *Travaux du Muséum d'Histoire Naturelle 'Grigore Antipa'* 32: 263-266.

Negoescu, I. 1994. Isopoda Anthuridea (Crustacea: Peracarida) from New Caledonia and Loyalty Islands (south-western Pacific Ocean). I. *Travaux du Muséum d'Histoire Naturelle 'Grigore Antipa'* 34: 147-225.

Negoescu, I. 1997a. Preliminary zoogeographical considerations of suborder Anthuridea Monod,

1922 (Crustacea: Isopoda). *Travaux du Muséum d'Histoire Naturelle 'Grigore Antipa'* 34: 257-290.

Negoescu, I. 1997b. Results of the Zoological Expedition organised by 'Grigore Antipa' Museum, in the Indonesian Archipelago (1991). I. Peracarida (Crustacea). Isopoda Anthuridea. *Travaux du Muséum d'Histoire Naturelle 'Grigore Antipa'* 38: 177-251.

Negoescu, I. 1999. Isopoda Anthuridea (Crustacea) from Fiji Islands. Three new species. First record of primary and secondary males in Paranthuridae family. *Travaux du Museum d'Histoire Naturelle 'Grigore Antipa'* 41: 199-280.

Negoescu, I. & J. Svavarsson 1997. Anthurideans (Crustacea, Isopoda) from the North Atlantic and the Arctic Ocean. *Sarsia* 82: 159-202.

Negoescu, I. & J.W. Wägele 1984. World list of the anthuridean isopods (Crustacea, Isopoda, Anthuridea). *Travaux du Muséum d'Histoire Naturelle 'Grigore Antipa'* 25: 99-146.

Negoescu Vladescu, I. 1983. A study of genus *Cyathura* from the Cuban freshwaters with the description of a new cave species: *C. orghidani* (Isopoda, Anthuridae). *Résultats des Expéditions Biospéologique Cubano-Roumaines à Cuba* 4: 39-45.

Nierstrasz, H.F. 1941. Die Isopoden der Siboga-Expedition. IV. Isopoda Genuina. III. Gnathiidea, Anthuridea, Valvifera, Asellota, Phreatocoidea. *Siboga Expéditie* 19: 235-308.

Nordenstam, A. 1930. Tanaidacea and marine Isopoda from Juan Fernandez. In Skottsberg, C. (ed.), *The natural history of Juan Fernandez and Easter Island* Vol. 3: pp. 525-552, pl. 20. Uppsala: Almqvist & Wiksells.

Norman, A.M. & T.R.R. Stebbing 1886. On the Crustacea of the 'Lighting', 'Porcupine' and 'Valorous' Expeditions. *Transactions of the Zoological Society of London* 12: 77-141.

Nunomura, N. 1974a. Marine Isopoda from the coast of Hikigawa Town, Kii Peninsula, Middle Japan (1). *Bulletin of the Osaka Museum of Natural History* 28: 1-12.

Nunomura, N. 1974b. A new anthurid isopod from the estuary of the Muromi river, Northern Kyushu, Japan. *Bulletin of the Osaka Museum of Natural History* 28: 13-16.

Nunomura, N. 1975. Marine Isopoda from the rocky shore of Osaka Bay, Middle Japan (1). *Bulletin of the Osaka Museum of Natural History* 29: 15-35.

Nunomura, N. 1976a. *Anathura rigida*, n. sp., a new deep-sea isopod from Suruga Bay, Japan. *Bulletin of the Osaka Museum of Natural History* 30: 93-96.

Nunomura, N. 1976b. A new isopod of the genus *Accalathura* (Crustacea) from the Okhotsk Sea. *Bulletin of the Osaka Museum of Natural History* 30: 27-30.

Nunomura, N. 1977. Marine Isopoda from Amakusa, Kyushu (I). *Publications of the Amakusa Marine Biological Laboratory* 4: 71-90.

Nunomura, N. 1979a. *Mesanthura miyakoensis*, a new anthurid isopod from Miyakojima Island, Ryukyu Islands, Japan. *Bulletin of the Toyama Science Museum* 1: 31-35.

Nunomura, N. 1979b. Tanaidaceans and anthuridean isopods collected on the Presidential Cruise of 1938. *Proceedings of the Biological Society of Washington* 91: 936-952.

Nunomura, N. 1984. A new species of the anthurid isopod genus *Eisothistos* from Osaka Bay, Middle Japan. *Bulletin of the Toyama Science Museum* 6: 51-55.

Nunomura, N. 1985. Marine isopod crustaceans in the coast of Toyama Bay. *Memoires of the National Science Museum, Tokyo* 18: 121-139.

Nunomura, N. 1992a. Anthuridea (Crustacea: Isopoda) from the Ryukyu Archipelago. *Bulletin of the Toyama Science Museum* 15: 47-56.

Nunomura, N. 1992b. A new species of the genus *Amakusanthura* (Crustacea, Isopoda) from Shinminato, Toyama Prefecture, Middle Japan. *Bulletin of the Toyama Science Museum* 15: 25-29.

Nunomura, N. 1993. Marine isopod crustaceans of Seto Inland Sea deposited at the Toyama Science Museum 1. Suborder Anthuridea - 1. *Bulletin of the Toyama Science Museum* 16: 15-30.

Nunomura, N. & M. Hagino 2000. A new species of the genus Cyathura from Lake Kasumigaura, Ibaraki Prefecture, Japan. *Bulletin of the Toyama Science Museum* 23: 5-9.

Pasternak, F.A. 1982. Composition, origin and peculiarities of distribution of the Mediterranean deep-sea isopod fauna. *Trudy Instituta Okeaonogiya, Akademiya Nauk SSSR, Moscow* 117: 163-177.

Paul, A.Z. & R.J. Menzies 1971. Sub-tidal isopods of the Fosa de Cariaco, Venezuela, with descriptions of two new genera and twelve new species. *Boletín del Instituto Oceanografico, Universidad de Oriente* 10: 29-48.

Pillai, N.K. 1954. A preliminary note on the Tanaidacea and Isopoda of Travancore. *Bulletin of the Central Research Institute, University of Tranvancore (series C)* 3: 1-21.

Pillai, N.K. 1966. Littoral and parasitic isopods from Kerala: family Anthuridae. - 1. *Journal of the Bombay Natural History Society* 63: 152-161.

Pillai, N.K. & K.C. Eapen 1966. On a new genus of anthurid isopod *Indanthura* from south India. *Bulletin of the Department of Marine Biology and Oceanography, University of Kerala* 2: 13-18.

Pires, A.M.S. 1981. Anthurids from intertidal and shallow infralittoral waters from southeastern Brazil (Isopoda: Anthridea). *Journal of Crustacean Biology* 1: 211-226.

Poore, G.C.B. 1975. Australian species of *Haliophasma* (Crustacea: Isopoda: Anthuridae). *Records of the Australian Museum* 29: 503-533.

Poore, G.C.B. 1978. *Leptanthura* and new related genera (Crustacea, Isopoda, Anthuridea) from eastern Australia. *Memoirs of the National Museum of Victoria* 39: 135-169.

Poore, G.C.B. 1980. A revision of the Paranthuridae (Crustacea: Isopoda: Anthuridea) with a catalogue of species. *Zoological Journal of the Linnean Society* 68: 53-67.

Poore, G.C.B. 1981a. Marine Isopoda of the Snares Islands, New Zealand - 1. Gnathiidea, Valvifera, Anthuridea, and Flabellifera. *New Zealand Journal of Zoology* 8: 331-348.

Poore, G.C.B. 1981b. Paranthurid isopods (Crustacea, Isopoda, Anthuridea) from southeastern Australia. *Memoirs of the National Museum of Victoria* 42: 57-88.

Poore, G.C.B. 1984a. *Colanthura, Califanthura, Cruranthura* and *Cruregens*, related genera of the Paranthuridae. *Journal of Natural History* 18: 697-715.

Poore, G.C.B. 1984b. *Paranthura* (Crustacea, Isopoda, Paranthuridae) from south-eastern Australia. *Memoirs of the National Museum of Victoria* 45: 33-69.

Poore, G.C.B. & T.M. Bardsley 1990. *Chelanthura* (Crustacea: Isopoda: Anthuridae), a new genus from southern Australia. *Memoirs of the Museum of Victoria* 51: 109-119.

Poore, G.C.B. & B. Kensley 1981. *Coralanthura* and *Sauranthura*, two new genera of anthurideans from north-eastern Australia (Crustacea: Isopoda: Anthuridae). *Proceedings of the Biological Society of Washington* 94: 503-513.

Poore, G.C.B. & H.M. Lew Ton 1985a. *Apanthura, Apanthuretta* and *Apanthuropsis* gen. nov. (Crustacea: Isopoda: Anthuridae) from south-eastern Australia. *Memoirs of the Museum of Victoria* 46: 103-151.

Poore, G.C.B. & H.M. Lew Ton 1985b. New species of *Cyathura* (Crustacea: Isopoda: Anthuridae) from estuaries of eastern Australia. *Memoirs of the Museum of Victoria* 46: 89-101.

Poore, G.C.B. & H.M. Lew Ton 1986a. *Mesanthura* (Crustacea: Isopoda: Anthuridae) from south-eastern Australia. *Memoirs of the Museum of Victoria* 47: 87-104.

Poore, G.C.B. & H.M. Lew Ton 1986b. New species of *Aenigmathura* and *Pseudanthura* (Crustacea: Isopoda: Paranthuridae) from eastern Australia. *Memoirs of the Museum of Victoria* 47: 59-73.

Poore, G.C.B. & H.M. Lew Ton 1986c. *Quantanthura* (Crustacea: Isopoda: Anthuridae) from south-eastern Australia and New Zealand. *Memoirs of the Museum of Victoria* 47: 75-85.

Poore, G.C.B. & H.M. Lew Ton 1988a. *Amakusanthura* and *Apanthura* (Crustacea: Isopoda: Anthuridea) with new species from tropical Australia. *Memoirs of the Museum of Victoria* 49: 107-147.

Poore, G.C.B. & H.M. Lew Ton 1988b. Antheluridae, a new family of Crustacea (Isopoda: Anthuridea) with new species from Australia. *Journal of Natural History* 22: 489-506.

Poore, G.C.B. & H.M. Lew Ton 1988c. A generic review of the Hyssuridae (Crustacea: Isopoda) with a new genus and new species from Australia. *Memoirs of the Museum of Victoria* 49: 169-193.

Poore, G.C.B. & H.M. Lew Ton 1988d. More Australian species of *Haliophasma* (Crustacea: Isopoda: Anthuridae). *Memoirs of the Museum of Victoria* 49: 85-106.

Poore, G.C.B. & H.M. Lew Ton 1990. *Accalathura* (Crustacea: Isopoda: Paranthuridae) from northern Australia and adjacent seas. *Memoirs of the Museum of Victoria* 50: 379-402.

Reboreda, P. & J.W. Wägele 1992. *Amakusanthura iberica* n. sp., first record of the genus *Amakusanthura* in Europe (Crustacea, Isopoda, Anthuridea). *Vie et Milieu* 42: 289-293.

Richardson, H. 1901. Key to the isopods of the Atlantic coast of North America with descriptions of new and little known species. *Proceedings of the United States National Museum* 23: 493-579.

Richardson, H. 1902. The marine and terrestrial isopds of the Bermudas, with descriptions of new genera and species. *Transactions of the Connecticut Academy of Arts and Sciences* 11: 277-310.

Richardson, H. 1905. A monograph on the isopods of North America. *United States National Museum Bulletin* 54: VII-LIII, 1-727.

Richardson, H. 1909. Isopods collected in the northwest Pacific by the U.S. Bureau of Fisheries Steamer 'Albatross' in 1906. *Proceedings of the United States National Museum* 37: 75-129.

Richardson, H. 1911. Les crustacés isopodes du Travailleur et du Talisman: formes nouvelles. *Bulletin Mensuel de la Société Linnéenne de Lyon* 7: 518-534.

Sars, G.O. 1873. Bidrag till Kundskaben om Dyrelivet pea vore Havbanken. *Forhandlinger i Videnskaps-Selskapet in Kristiania* : 73-119.

Sars, G.O. 1897. *An account of the Crustacea of Norway with short descriptions and figures of all the species.* Vol. 2 Isopoda. Parts 3, 4. Anthuridae, Gnathiidae, Aegidae, Cirolanidae, Limnoriidae. Pp. 41-80, pls 17-32. Bergen: Bergen Museum. 270 pp.

Schotte, M., R.W. Heard & B. Kensley 1991. Studies on the Crustacea of the Turks and Caicos Islands, British West Indies. III. Records of marine Isopoda from Pine Cay, Fort George Cay, Water Cay, and adjacent waters. *Gulf Research Reports* 8: 251-257.

Schultz, G.A. 1964. Some marine isopod crustaceans from off the southern California coast. *Pacific Science* 18: 307-314.

Schultz, G.A. 1966. Submarine canyons of southern California. Part IV. Systematics: Isopoda. Marine isopods of the submarine canyons of the southern Californian continental shelf. *Allan Hancock Pacific Expeditions* 27: 1-56.

Schultz, G.A. 1977. Anthurids from the west coast of North America, including a new species and three new genera (Crustacea, Isopoda). *Proceedings of the Biological Society of Washington* 90: 839-848.

Schultz, G.A. 1979. A new Asellota (Stenetriidae) and two, one new, Anthuridea (Anthuridae) from Bermuda (Crustacea, Isopoda). *Proceedings of the Biological Society of Washington* 91: 904-911.

Shyamasundari, K., C.J. Kumari, K.H. Rao & A. Mary 1991. A new species of the genus *Heteranthura* Kensley (Crustacea: Isopoda: Anthuridae) from Visakhapatnam coast. *Journal of the Bombay Natural History Society* 88: 262-264.

Sivertsen, E. & L.B. Holthuis 1980. The marine isopod Crustacea of the Tristan da Cunha Archipelago. *Gunneria* 35: 1-128.

Stebbing, T.R.R. 1900. On Crustacea brought by Dr Willey from the South Seas. In Willey, A. (ed.), *Zoological Results based on material from New Guinea, New Britain, Loyalty Islands and elsewhere ... 1895, 1896 and 1897* Vol. 5: pp. 609-690. Cambridge..

Stebbing, T.R.R. 1904. Gregarious Crustacea from Ceylon. *Spolia Zeylanica, Bulletin of the National Museum, Ceylon* 2: 1-29.

Stebbing, T.R.R. 1910. Isopoda from the Indian Ocean and British East Africa. *Transactions of the Linnean Society of London* (2) 14: 83-122.

Stimpson, W.M. 1853. Synopsis of the marine Invertebrata of Grand Manan: or the region about the mouth of the Bay of Fundy, New Brunswick. *Smithsonian Contributions to Knowledge* 6: 1-66.

Stimpson, W. 1855. Descriptions of some new marine invertebrates. *Proceedings of the Academy of Natural Science of Philadelphia* 7: 384-397.

Stork, A.H. 1940. A new fresh-water isopod from Curacao. *Studies on the fauna of Curaçao and other Caribbean Islands* 2: 147-150.

Thomson, J.M. 1946. New crustacea from the Swan River estuary. *Journal and Proceedings of the Royal Society of Western Australia* 30: 35-53.

Thomson, J.M. 1951. The fauna of Rottnest Island. X. Anthuridae. *Journal of the Royal Society of Western Australia* 35: 1-8.

Vanhöffen, E. 1914. Die Isopoden der Deutschen Südpolar-Expedition 1901-1903. *Deutsche Südpolar-Expedition, 1901-03* 15: 449-598.

Wägele, J.W. 1979a. Der Fortpflanzungszyklus von *Cyathura carinata* (Isopoda, Anthuridea) im Nord-Ostee-Kanal. *Helgoländer Meeresuntersuchungen* 32: 295-304.

Wägele, J.W. 1979b. Die Homologie der Mundwerkzeuge von *Cyathura carinata* (Kroyer 1847) (Crustacea, Isopoda, Anthuridea). *Zoologischer Anzeiger* 203: 334-341.

Wägele, J.W. 1979c. Morphologische Studien an *Eisothistos* mit Beschreibung von drie neuen Arten (Crustacea, Isopoda, Anthuridea). *Mitteilungen aus dem Zoologischen Museum der Universitat zu Kiel* 1: 1-19.

Wägele, J.W. 1979d. *Stellanthura* nov. gen. *cryptobia* nov. spec.: ein neuer Isopode aus dem Tyrrhenischen Meer (Crustacea, Isopoda, Anthuridea). *Mitteilungen aus dem Zoologischen Museum der Universität zu Kiel* 1: 21-27.

Wägele, J.W. 1980a. Anthuridea (Crustacea, Isopoda) aus dem Tyrrenischen Meer. *Zoologica Scripta* 9: 53-66.

Wägele, J.W. 1980b. Zur phylogenie der Anthuridea (Crustacea, Isopoda) mit Beitragen zur Lebensweise, Morphologie, Anatomie und Taxonomie. Dissertation zur Erlangung des Doktorgrades Thesis, Christian-Albrechts-Universität zu Kiel : Kiel.\ 192 pp.

Wägele, J.W. 1981a. Study of the Anthuridae (Crustacea: Isopoda: Anthuridea) from the Mediterranean and Red Sea. *Israel Journal of Zoology* 30: 113-159.

Wägele, J.W. 1981b. Study of the Hyssuridae (Crustacea: Isopoda: Anthuridea) from the Mediterranean and the Red Sea. *Israel Journal of Zoology* 30: 47-87.

Wägele, J.W. 1981c. Study of the Paranthuridae (Crustacea: Isopoda: Anthuridea) from the Mediterranean. *Israel Journal of Zoology* 30: 211-229.

Wägele, J.W. 1981d. Zur phylogenie der Anthuridea (Crustacea, Isopoda) mit Beitragen zur Lebensweise, Morphologie, Anatomie und Taxonomie. *Zoologica (Stuttgart)* 132: 1-127.

Wägele, J.W. 1982a. The hypogean Paranthuridae *Cruregens* Chilton and *Curassanthura* Kensley (Crustacea, Isopoda), with remarks on their morphology and adaptations. *Bijdragen tot de Dierkunde* 52: 49-59.

Wägele, J.W. 1982b. Neubeschreibung und Vergleich der mediterranen *Paranthura*-Arten (Crustacea, Isopoda, Anthuridea). *Marine Ecology* 3: 109-132.

Wägele, J.W. 1982c. A new hypogean *Cyathura* from New Caledonia (Crustacea, Isopoda, Anthuridea). *Bulletin Zoölogisch Museum, Universiteit van Amsterdam* 8: 189-197.

Wägele, J.W. 1982d. On *Apanthuretta lathridia* n. sp. (Crustacea, Isopoda, Anthuridea) from Cuba. *Bijdragen tot de Dierkunde* 52: 43-48.

Wägele, J.W. 1984a. On a small collection of littoral Crustacea Isopoda Anthuridea (family Anthuridae) from the Far East. *Journal of Natural History* 18: 739-757.

Wägele, J.W. 1984b. Redescription of K.H. Barnard's three species of *Mesanthura* (Crustacea: Isopoda: Anthuridea). *Journal of Natural History* 18: 389-403.

Wägele, J.W. 1984c. Two new littoral Anthuridea from Baja California and redescription of *Mesanthura occidentalis* (Crustacea, Isopoda). *Zoologica Scripta* 13: 45-57.

Wägele, J.W. 1985a. On the tethyan origin of the stygobiont Anthuridea *Curassanthura* and *Cyathura* (*Stygocyathura*), with description of *Curassanthura canariensis* n. sp. from Lanzarote (Crustacea, Isopoda). *Stygologia* 1: 258-269.

Wägele, J.W. 1985b. Two new genera and twelve new species of Anthuridea (Crustacea: Isopoda) from off the west coast of New Zealand. *New Zealand Journal of Zoology* 12: 363-423.

Wägele, J.W. 1987. *Neohyssura atlantica* n. sp. from the Cape Verde Islands (Crustacea: Isopoda: Anthuridea). *Bulletin Zoölogisch Museum, Universiteit van Amsterdam* 11: 13-19.

Wägele, J.-W. 1989. Evolution und phylogenetisches System der Isopoda. Stand der Forschung und neue Erkenntnisse. *Zoologica (Stuttgart)* 140: 1-262.

Wägele, J.W. & A. Brandt 1985. New west Atlantic localities for the stygobiont paranthurid *Curassanthura* (Crustacea, Isopoda, Anthuridea) with description of *C. bermudensis* n. sp. *Bijdragen tot de Dierkunde* 55: 324-330.

Wägele, J.W. & D. Platvoet 1982. Desciption of *Monodanthura maroccana* nov. gen., nov. spec. (Crustacea, Isopoda, Anthuridea). *Bulletin Zoölogisch Museum, Universiteit van Amsterdam* 8: 213-220.

Wägele, J.-W., O. Coleman & U. Hosse 1987. Two new hypogean species of *Cyathura* from Melanesia (Crustacea, Isopoda, Anthuridea: further Tethyan relicts? *Stygologia* 3: 89-106.

Wagner, H.P. 1990. The stygobiont isopods of the genus *Cyathura* in the Dominican Republic (Crustacea; Isopoda; Anthuridae). *Bulletin Zoölogisch Museum, Universiteit van Amsterdam* 12: 145-159.

Walker, A.O. 1901. Contributions to the malacostracan fauna of the Mediterranean. *Journal of the Linnean Society of London (Zoology)* 28: 290-307.

Whitelegge, T. 1901. Scientific results of the trawling expedition of H.M.C.S. 'Thetis' off the coast of New South Wales... 1898. Crustacea. Part II. Isopoda. Part I. *Memoirs of the Australian Museum* 4: 203-246.

Wolff, T. 1956. Isopoda from depths exceeding 6000 m. *Galathea Reports* 2: 85-157.

Systematics of the Phreatoicidea

GEORGE D.F. WILSON & STEPHEN J. KEABLE
Center for Evolutionary Research, Australian Museum, Sydney, New South Wales, Australia

ABSTRACT

We provide a new cladistic analysis of the Phreatoicidea, revising previously employed characters and including additional generic exemplars. This analysis reinforces the conclusion that many currently recognized families and subfamilies are not monophyletic. The 'primitive' family Amphisopodidae is paraphyletic and the Nichollsiidae (with species from India) appears to be a clade within the Australian subfamily Hypsimetopodinae. The results also suggest changes to the composition of some subfamilies. By removal of the southeastern Australian genus *Crenoicus*, the Phreatoicinae is restricted to New Zealand taxa. Approximate ages of tectonic rifting of the super-continent Gondwana were used to biogeographically constrain the timing of events in phreatoicidean phylogeny. Because the Indian genus *Nichollsia* is nested within the Australian Hypsimetopodinae, this clade has a minimum age of 130 million years. Phreatoicidean isopods were classified into three ecotypes: surface-cryptic, surface-burrower, and groundwater dwelling forms. The phylogenetic distribution of these ecotypes provides evidence for at least two separate colonizations of groundwater habitats. One colonization may be older than the rifting of Indian subcontinent and the other is approximately dated by the separation of New Zealand from East Gondwana. The phylogenetic age, endemism and localized diversity of phreatoicidean isopods confer a high conservation value upon this group. This phylogenetic and biogeographic knowledge of phreatoicidean isopod crustaceans may assist informed decisions on their conservation.

1 INTRODUCTION

Phreatoicidean isopod crustaceans are an ancient group, as evident from their fossil record (Schram 1974) and phylogenetic origin as the sister group to all other isopods (Brusca & Wilson 1991). Marine fossil phreatoicidean isopods are known from the late Carboniferous (Schram 1974), and fresh water fossils of Triassic age in Australia are similar to living taxa (Chilton 1918, Nicholls 1943), indicating that the group is morphologically conservative. Extant taxa have a classic Gondwanan distribution (Banarescu 1990). Recent evaluations of their phylogenetic biogeography (Wilson & Johnson 1999) indicate that ancient relationships emerge at the generic level. For example, *Crenisopus* Wilson & Keable, 1999 appears to be basally derived, and *Pilbarophreatoicus* Knott & Halse, 1999 ('New

Genus X5' in Wilson & Johnson 1999) is related to *Nichollsia* Chopra & Tiwari, 1950 from India.

The classification of the Phreatoicidea needs revision because the suborder is much more diverse than currently recognized. Currently, fewer than 50 Australian species are described. The monotypic genus *Eophreatoicus* Nicholls, 1943, found in Kakadu National Park and Arnhemland (Northern Territory), contains minimally ten additional undescribed species (unpublished data). Similarly, *Crenoicus* Nicholls, 1943, has two or three undescribed species in each large watershed region in New South Wales above 1000 m (e.g., Kosciusko Plateau, Boyd Plateau, Barrington Tops) (Wilson & Ho 1996, Adlem 1996), possibly reaching a total of 14-16 species in this state alone. A linear extrapolation of these figures for *Crenoicus* yields approximately 200 undescribed species of phreatoicideans in Australia. The recognition of these additional species also suggests that many phreatoicideans are short-range endemics with highly relictual distributions. Where Phreatoicidea are mentioned in faunal surveys, however, the specimens are rarely identified to species categories (e.g., recent cave surveys of Eberhard et al. 1991 and Eberhard & Spate 1995). Australia has nineteen nominal genera of phreatoicidean isopods (Table 1) and at least three new undescribed genera. Many existing genera are poorly defined and difficult to identify based on the existing monographic works. Because the existing classification of phreatoicideans is not based on phylogenetic relationships (Wilson & Keable 1999, Wilson & Johnson 1999), these problems cannot be adequately treated until a well corroborated phylogeny is estimated. Clearly, new tools to classify and identify phreatoicidean species are needed.

We report a new evaluation of the phylogeny and classification of the phreatoicideans, with evidence for diversification of endemic taxa during the Mesozoic Era in Gondwana. This paper extends the study of Wilson & Johnson (1999), providing a more detailed analysis of phreatoicidean relationships by including additional taxa from Australia and New Zealand, new out-groups and new data on specimens of *Nichollsia* from India. We also use this inferred phylogeny to evaluate the evolution of generalized ecotypes within the Phreatoicidea.

2 MATERIALS AND METHODS

The matrix for phylogenetic analysis was generated from a DELTA database as described by Wilson & Keable (1999) and Wilson & Johnson (1999). The character list (Appendix 1) excludes several multi-state characters in the DELTA database, particularly those involving eyes, owing to the variability in these features. Setal counts and ratios (describing the relative sizes of body parts) are also excluded from the character set. The trimmed character set consists of 79 characters, 77 of which are variable and parsimony informative (Appendix 1). Only taxa for which we have completed data entry have been used in the analysis (Table 1, which should be consulted for authors of phreatoicidean genera; Appendix 2). Some congeneric taxa were omitted (e.g., some *Crenoicus*, species) because preliminary analyses indicate that they cluster together. Several potentially congeneric taxa in *Metaphreatoicus* and *Colubotelson*, however, were included in the analysis to constrain the position of these taxa. Unlike previous analyses of the phreatoicideans, we have included two isopod outgroups, *Tainisopus* Wilson & Ponder, 1992 and *Stenasellus* (Dollfus 1897). All possible living isopod out-groups are 'upstream' taxa – derived later than

Table 1. Genera of the Phreatoicidea. Key: * = not Australian; Xn = new genera observed by the authors; # = not in the DELTA database (July 1998). Australian State Abbreviations: NSW, New South Wales; WA, Western Australia; SA, South Australia; NT, Northern Territory; VIC, Victoria; TAS, Tasmania. [1]Based on our evaluation of specimens in hand.

Genus	Described Species (est. new species[1])	Distribution
**Mesamphisopus* Nicholls, 1943	4	South Africa
**Neophreatoicus* Nicholls, 1944	1	New Zealand
**Nichollsia* Chopra & Tiwari, 1950	2	India
#**Notamphisopus* Nicholls, 1944	6	New Zealand
**Phreatoicus* Chilton, 1882	2	New Zealand
Amphisopus Nicholls, 1926	2	WA
# *Colacanthotelson* Nicholls, 1944	1 (1?)	NSW, (TAS?)
Colubotelson Nicholls, 1944	15 (1?)	VIC, NSW, TAS
Crenoicus Nicholls, 1944	4 (>20)	VIC, NSW
Crenisopus Wilson & Keable, 1999	1	NE Kimberleys, WA
Eophreatoicus Nicholls, 1926	1 (>10)	NT
Hyperoedesipus Nicholls & Milner, 1923	1	WA
Hypsimetopus Sayce, 1902	1	TAS
# *Mesacanthotelson* Nicholls, 1944	4	TAS
Metaphreatoicus Nicholls, 1944	4	NSW, TAS
# *Onchotelson* Nicholls, 1944	2	TAS
# *Paraphreatoicus* Nicholls, 1944	1	TAS
Paramphisopus Nicholls, 1943	3	WA
Phreatoicoides Sayce, 1900	3	TAS, VIC
Phreatoicopsis Spencer & Hall, 1896	1 (1?)	VIC
Pilbarophreatoicus Knott & Halse, 1999	1 (1-2)	Pilbara, WA
Phreatomerus Sheppard, 1927	1 (1?)	SA
# *Protamphisopus* Nicholls, 1943	1	NSW (fossil)
Synamphisopus Nicholls, 1943	1	VIC
# *Uramphisopus* Nicholls, 1943	1	TAS
# (New Genus X2)	(1)	SE WA
# (New genus X3)	(1)	Kimberleys, WA
# (New genus X4)	(1)	SW WA

the ancestor of the phreatoicideans (cf. Brusca & Wilson 1991). The chosen outgroup taxa have some possibly plesiomorphic traits, such as a basal article of the antennal protopod and a rudimentary second flagellum on the antennula. Many character states of *Tainisopus* and *Stenasellus* were necessarily inapplicable and were scored as unknowns. We make no claims about the phylogenetic position of these two taxa relative to the phreatoicideans.

Phylogenetic analysis was performed with 'Portable PAUP*' (beta version 4.0.0d64 for DOS; Swofford, 1999), applying 10 random addition sequence iterations using MulPars with the characters unordered and equally weighted. Taxa scored as multi-state for particular characters were interpreted as displaying polymorphism. Bremer support (decay values) was obtained using constraint trees generated by AutoDecay (version 4.0; Eriksson 1998) and analysed using PAUP*. Character distributions were evaluated using MacClade (ver. 3.07; Maddison & Maddison 1992).

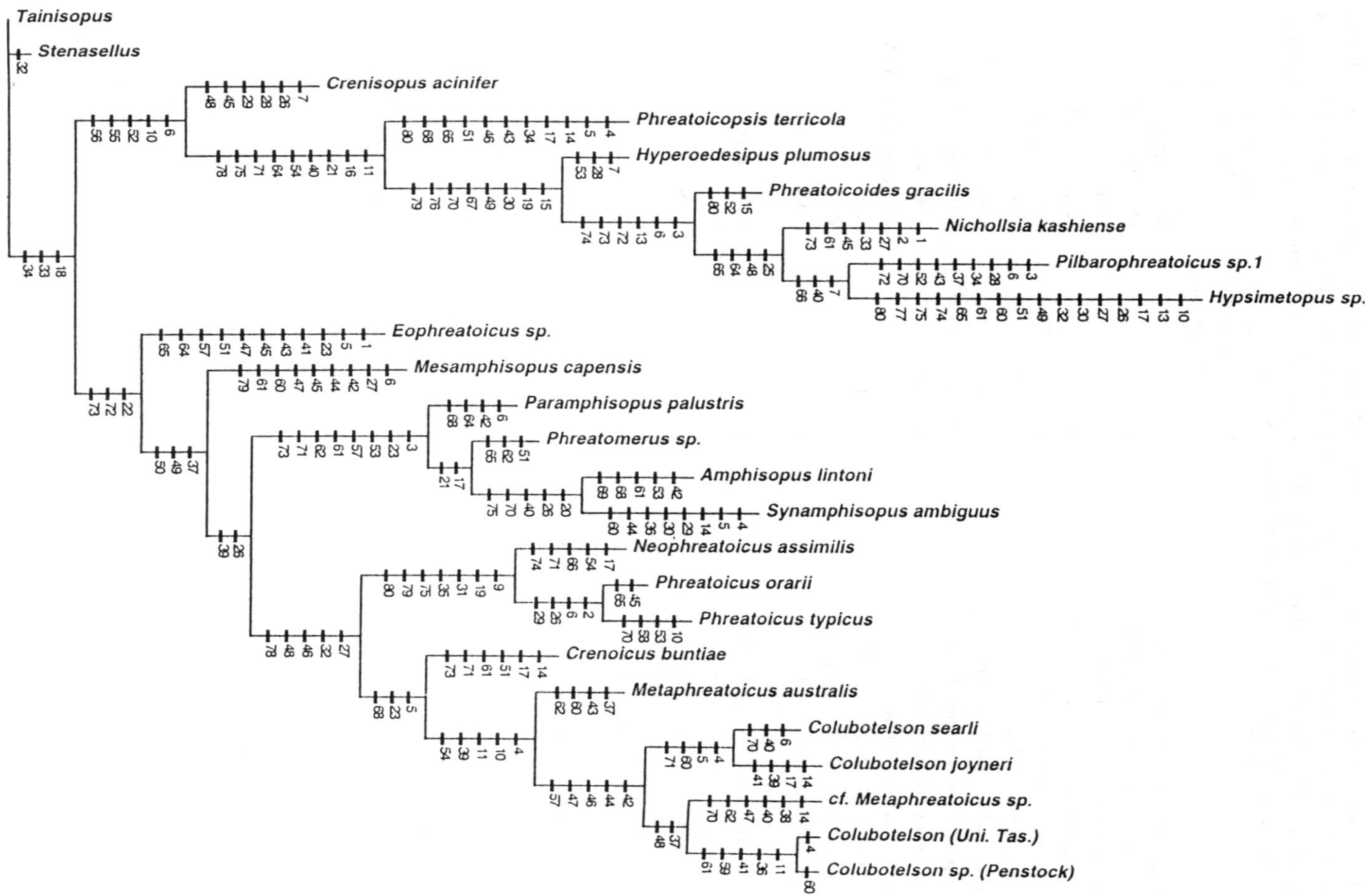

Figure 1. The single tree resulting from a PAUP* analysis of the data in Table 2: tree length 335, consistency index 0.394, retention index 0.561, rescaled consistency index 0.221. Character 80 is a non phylogenetic ecotype character: see Figure 3.

3 RESULTS AND DISCUSSION

3.1 *Phylogenetic Analysis*

A single tree results from the analysis of the data in Table 2, (length 335, consistency index 0.394, rescaled consistency index 0.221; Fig. 1). Four clades relating to the status of the current classification are well supported, based on their branch or Bremer support (Bremer 1994) values (Fig. 2 with node labels and branch support). An additional clade (node 4) is marked; although it was supported by only three steps because of its correspondence to the family Phreatoicidae and because its position is constrained owing to well-supported clades at nodes 3 and 5. The character changes defining these five nodes are summarized in Table 3. The well supported clade '*Crenisopus*+node 1' is not labeled because it has no implications for the existing classification. After this paper was completed, the coding of uropodal protopodal medial ridge characters (75-76) was revised owing to new observations on Tasmanian taxa not included in this analysis. These changes did not result in new topologies for the well-supported branches, although some Phreatoicidae in-group relationships were changed, owing to low levels of character support in those branches.

3.2 *Apomorphies*

Consistent characters (character consistency index = 1.0) define well-supported nodes in the single cladogram (Figs 1-2; Table 3). The in-group node for the Phreatoicidea is identified by three apomorphies. A vaulted pleotelson (higher than broad; char. 18) is a synaphomorphy of all species of phreatoicideans. Many species have a dorsally reflexed distal tip of the pleotelson (char. 22), but this feature is not universal for the suborder, and is absent in the taxa of nodes 1 and 2 (Fig. 2). The bifurcate spines in the mandibular spine row (char. 33) are found in nearly all phreatoicideans. *Nichollsia* has undivided mandibular spines, an apparent reversion. Phreatoicideans are also basally defined by the projecting ridge of the spine row (char. 34). Most species, however, have a pedunculate spine row.

The subtree including *Crenisopus*, *Phreatoicopsis* and the Hypsimetopodinae Nicholls, 1943 (name corrected from Nicholls' original spelling, Hypsimetopinae) is defined by the absence of a propodal articular plate on most but not all pereiopods (chars. 52, 55 and 56). Node 1 (hypsimetopodines + *Phreatoicopsis*) is defined by the consistent characters of an indented distal tip of the pleotelson (char. 21) and a pleonite 1 pleural depth that is similar to pleurae on the more posterior pleonites. Node 2 clade members (hypsimetopodines) have pleopodal exopods with elongate lateral lobes on the proximal article (char. 63) and lack protopodal epipods (char. 67). The partial amphisopodine clade (node 3) and the phreatoicid clade (node 4) have no consistent characters defining them. Node 3, however, is strongly supported by at least 6 character changes. An unusually thin mandibular incisor process (char. 31) and a broad maxillular medial lobe (char. 35) define the New Zealand phreatoicine clade (node 5).

3.3 *Classification*

Nicholls (1943) placed undue influence on the presence of a lacinia mobilis on the right mandible as the defining feature of the Amphisopodidae, even though it is absent in *Hyp-*

Table 2. Data used for parsimony analysis. '?' = unknown or inapplicable.

Character number (10)										1										2										3										4
Character number	1	2	3	4	5	6	7	8	9	0	1	2	3	4	5	6	7	8	9	0	1	2	3	4	5	6	7	8	9	0	1	2	3	4	5	6	7	8	9	0
Tainisopus	1	?	?	3	2	1	1	1	2	1	1	?	?	3	2	?	1	1	?	?	1	1	1	?	?	?	1	2	1	2	1	1	1	1	1	?	?	?	?	?
Stenasellus	1	?	?	?	?	?	?	1	2	1	1	?	?	3	?	?	?	1	?	?	1	1	1	?	?	?	1	2	1	2	1	2	1	1	1	?	?	?	?	?
Mesamphisopus capensis	1	1	1	3	2	2	1	1	2	1	1	1	2	3	1	1	2	2	1	1	1	2	1	?	2	2	3	2	1	2	1	1	2	3	1	2	2	2	2	2
Metaphreatoicus australis	2	1	1	2	1	1	1	1	2	2	2	1	2	3	1	1	2	2	1	1	1	2	3	?	2	1	2	2	1	3	1	2	2	3	1	2	3	2	2	2
cf. *Metaphreatoicus* sp.	1	1	1	2	1	1	1	2	2	2	2	1	2	2	1	1	2	2	1	1	1	2	3	?	2	1	2	2	1	2	1	2	2	3	1	2	1	1	(12)	3
Crenoicus buntiae	2	1	1	3	1	1	1	1	2	1	1	1	2	2	1	1	1	2	1	1	1	2	3	?	2	1	2	2	1	3	1	2	2	3	1	2	2	2	1	2
Eophreatoicus sp.	3	1	1	3	1	1	1	1	2	1	1	1	2	3	1	1	2	2	1	1	1	2	2	?	2	2	1	2	1	2	1	1	2	2	1	2	3	3	2	2
Phreatomerus sp.	1	1	2	3	2	1	2	2	2	3	3	2	2	3	1	1	1	2	1	1	2	2	2	?	2	1	1	2	1	3	1	1	2	2	1	2	2	2	(12)	2
Hypsimetopus sp.	3	1	1	3	2	1	2	1	2	2	2	1	2	3	3	2	1	2	2	1	3	1	1	1	1	1	2	2	1	2	1	2	2	2	1	2	1	2	2	1
Crenisopus acinifer	1	1	2	3	2	2	2	1	2	3	1	1	2	3	1	1	2	2	1	1	1	1	1	?	2	1	3	1	2	2	1	1	2	2	1	2	3	2	2	2
Neophreatoicus assimilis	3	1	1	?	2	1	1	1	3	1	1	1	?	?	1	1	1	2	2	1	1	2	1	?	2	1	2	2	1	?	2	2	2	3	2	2	2	?	?	?
Paramphisopus palustris	1	1	2	3	2	2	2	1	2	2	2	1	2	3	1	1	2	2	1	1	1	2	2	?	2	1	1	2	1	3	1	1	2	2	1	2	2	2	1	2
Phreatoicus typicus	3	4	1	3	2	2	1	1	3	3	1	1	2	4	1	1	2	2	2	1	1	2	1	?	2	2	2	2	3	2	2	2	2	3	2	2	2	2	1	2
Nichollsia kashiense	2	4	1	3	2	1	1	1	1	3	3	1	1	3	3	2	2	2	2	1	3	1	1	1	1	2	1	2	1	3	1	1	1	2	1	2	1	1	2	3
Phreatoicopsis terricola	1	1	2	1	1	2	1	1	2	3	3	1	2	1	1	2	1	2	1	1	3	1	1	1	2	2	1	2	1	2	1	1	2	4	1	2	3	3	2	3
Synamphisopus ambiguus	(23)	1	2	1	1	1	1	1	2	3	3	1	2	1	1	1	1	2	1	2	2	2	2	?	2	2	1	2	2	2	1	1	2	4	1	1	3	3	1	3
Pilbarophreatoicus sp.1	3	2	2	3	2	2	2	1	4	3	1	1	1	3	3	2	2	2	2	1	3	1	1	1	1	2	3	1	1	3	1	1	2	1	1	2	2	2	2	1
Hyperoedesipus plumosus	(123)	1	2	3	2	2	2	1	2	3	3	1	2	3	3	2	(12)	2	2	1	3	1	1	1	2	2	3	1	1	3	1	1	2	2	1	2	1	3	2	3
Phreatoicoides gracilis	3	2	1	3	2	1	1	1	1	3	3	1	1	3	2	2	2	2	2	1	3	1	1	1	2	2	3	2	1	3	1	1	2	2	1	2	3	3	2	3
Amphisopus lintoni	1	1	2	3	2	1	1	2	2	3	3	1	2	3	1	1	1	2	1	2	2	2	2	?	2	2	1	2	1	3	1	1	2	1	1	2	1	1	1	3
Colubotelson joyneri	1	1	1	1	2	1	1	1	2	2	2	1	2	4	1	1	1	2	1	1	1	2	3	?	2	1	2	(12)	1	2	1	2	2	3	1	2	2	2	1	2
Colubotelson searli	1	1	1	1	2	2	1	1	2	2	2	1	2	3	1	1	2	2	1	1	1	2	3	?	2	1	2	(12)	1	3	1	2	2	3	1	2	2	2	2	3
Colubotelson sp.(Uni. Tas.)	1	1	1	1	1	1	1	1	2	2	1	1	2	3	1	1	2	2	1	1	1	2	3	?	2	1	2	2	1	2	1	2	2	3	1	1	1	2	2	2
Colubotelson sp. (Penstock)	1	1	1	2	1	1	1	2	2	2	1	1	2	3	1	1	(12)	2	1	1	1	2	3	?	2	1	2	2	1	2	1	2	2	3	1	1	1	2	2	2
Phreatoicus orarii	2	4	1	3	2	2	1	1	3	1	1	1	2	?	1	1	2	2	2	1	1	2	1	?	2	2	2	2	3	?	2	2	2	3	2	2	2	2	?	2

Table 2. Continued.

Character number (10)	4									5										6										7									
Character number	1	2	3	4	5	6	7	8	9	0	1	2	3	4	5	6	7	8	9	0	1	2	3	4	5	6	7	8	9	0	1	2	3	4	5	6	7	8	9
Tainisopus	?	?	?	?	?	?	?	?	?	?	?	1	?	?	1	1	?	?	?	?	?	?	?	?	?	1	1	?	?	?	1	2	2	?	?	?	?	2	1
Stenasellus	?	?	?	?	?	?	?	?	?	?	?	1	?	?	1	1	?	?	?	?	?	?	?	?	?	?	?	?	?	?	?	?	?	?	?	?	?	2	3
Mesamphisopus capensis	2	3	3	2	2	2	2	4	1	1	2	1	1	2	1	1	2	2	1	3	2	1	1	1	1	1	1	2	1	1	1	1	1	1	1	?	2	2	3
Metaphreatoicus australis	1	2	2	3	4	1	3	1	1	1	2	1	1	1	1	1	2	2	1	4	3	3	1	1	1	1	1	1	2	1	1	1	1	1	1	?	1	1	2
cf. *Metaphreatoicus* sp.	1	3	3	1	4	2	2	4	1	1	2	1	1	1	1	1	4	2	1	1	3	4	1	1	1	1	1	1	1	2	1	1	1	1	2	2	1	1	3
Crenoicus buntiae	1	2	3	3	4	1	3	1	1	1	1	1	1	2	1	1	2	2	2	1	1	1	1	1	1	1	1	1	2	1	2	1	4	1	1	?	1	1	2
Eophreatoicus sp.	1	2	2	3	1	2	1	4	2	2	1	1	1	2	1	1	4	2	1	1	3	1	1	2	2	1	1	2	1	1	1	1	1	1	1	?	1	2	2
Phreatomerus sp.	2	2	3	3	4	2	1	4	1	1	1	1	2	2	1	1	4	2	2	1	1	2	1	1	2	1	1	2	1	1	3	1	3	1	1	?	1	2	2
Hypsimetopus sp.	2	2	3	3	4	2	3	1	2	2	1	2	2	1	1	2	2	2	1	3	2	4	(12)	1	2	2	2	2	1	2	1	1	3	1	1	?	1	1	1
Crenisopus acinifer	2	2	3	3	3	2	3	3	2	2	2	2	2	2	2	2	2	1	1	1	3	1	1	1	1	1	1	2	1	1	1	2	2	1	1	?	1	2	2
Neophreatoicus assimilis	?	2	?	3	4	1	?	?	1	?	?	1	1	1	1	?	2	2	2	?	3	1	1	1	?	2	1	2	1	?	3	?	1	2	2	1	2	1	1
Paramphisopus palustris	2	3	3	3	4	2	3	4	1	1	2	1	2	2	1	1	4	2	1	1	1	3	1	2	1	1	1	1	1	1	3	1	3	1	1	?	2	2	2
Phreatoicus typicus	1	2	3	3	4	1	3	1	1	1	2	1	2	2	1	1	2	1	2	1	3	1	1	1	1	1	1	2	1	2	1	1	1	1	2	1	2	1	1
Nichollsia kashiense	2	3	3	3	1	2	3	1	1	2	2	2	2	1	2	2	2	1	1	1	1	4	2	1	3	1	2	2	1	2	1	1	4	2	2	3	2	(12)	1
Phreatoicopsis terricola	2	2	2	3	4	1	3	4	2	2	1	2	2	1	2	2	2	2	(12)	1	3	(14)	1	4	4	1	1	1	1	1	3	2	2	1	2	2	2	1	2
Synamphisopus ambiguus	?	2	3	1	4	2	1	4	1	(12)	2	1	2	1	1	1	4	2	2	4	1	3	1	1	1	1	1	2	1	2	3	1	3	1	2	2	1	2	2
Pilbarophreatoicus sp.1	2	2	2	3	4	2	3	1	1	2	2	1	2	1	1	2	2	1	1	1	3	4	2	1	3	2	2	2	1	1	3	2	3	2	2	1	2	1	1
Hyperoedesipus plumosus	2	1	3	3	4	2	3	4	1	1	2	2	1	1	2	2	2	1	1	1	3	4	(12)	4	1	1	2	2	1	2	3	2	2	1	2	3	2	1	1
Phreatoicoides gracilis	2	3	3	3	4	2	3	4	1	1	2	1	2	1	1	2	2	2	1	1	3	4	(12)	4	1	1	2	2	1	2	3	1	3	2	2	3	2	1	1
Amphisopus lintoni	2	3	3	3	4	2	3	4	1	1	2	1	1	(12)	1	1	4	2	2	1	3	3	1	1	1	1	1	1	2	2	3	1	3	1	2	2	1	2	2
Colubotelson joyneri	2	3	3	1	1	2	1	1	1	1	2	1	1	1	1	1	4	2	1	3	3	1	1	1	1	1	1	1	1	1	3	1	1	1	1	?	1	1	1
Colubotelson searli	1	3	3	1	1	2	1	1	1	1	2	1	1	1	1	1	4	2	1	3	3	1	1	1	1	1	1	1	1	2	3	1	1	1	1	?	1	1	1
Colubotelson sp.(Univ. Tas.)	2	3	3	1	1	2	1	4	1	1	2	1	1	1	1	1	4	2	2	1	1	1	1	1	1	1	1	1	1	1	1	1	1	1	1	?	1	1	3
Colubotelson sp.(Penstock)	2	3	3	1	1	2	1	4	1	1	2	1	1	1	1	1	4	2	2	3	1	1	1	1	1	1	1	1	1	1	1	1	1	1	2	1	1	1	3
Phreatoicus orarii	2	2	3	3	1	1	3	1	1	1	?	?	1	2	?	?	2	2	2	1	3	1	1	1	2	1	1	2	1	1	1	?	1	1	2	1	2	1	1

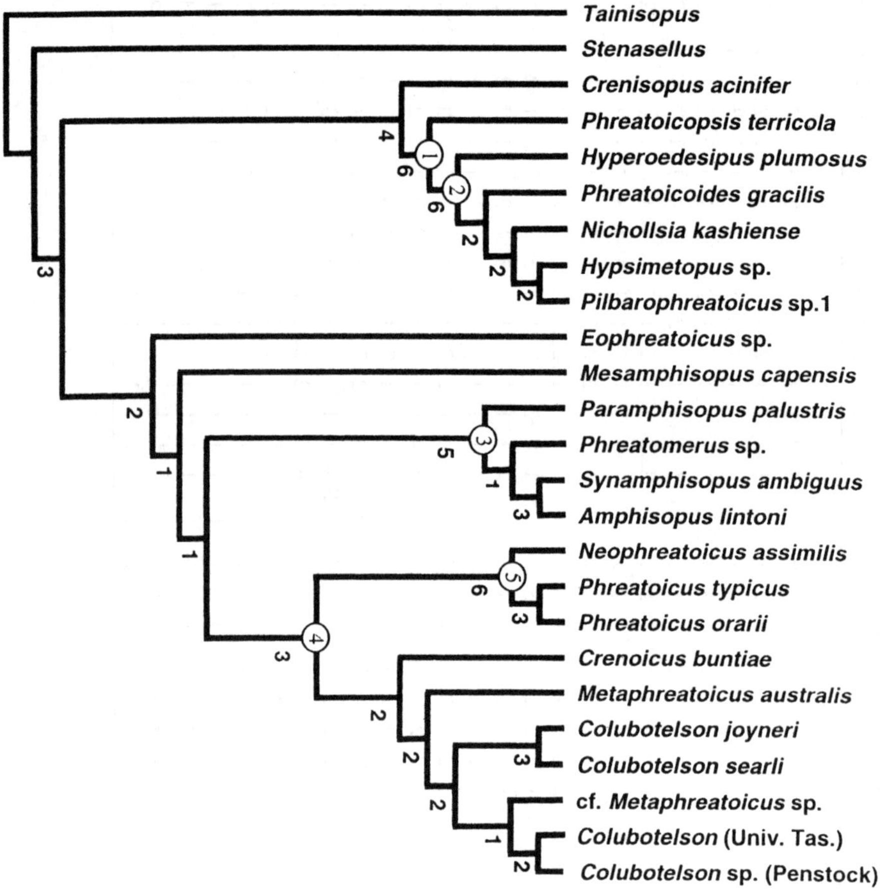

Figure 2. Branch support for cladogram in Figure 1 (Bremer support numbers to the left of the branches), with nodes of well supported clades (numbers in circles). Apomorphies for marked nodes are shown in Table 3.

simetopus. The Amphisopodidae, although almost all members have the right lacinia, consists of at least two distinct clades in our analysis and is paraphyletic owing to the nested position of the Phreatoicidae. Moreover, the right lacinia mobilis may be a plesiomorphy, because it occurs in the peracaridan outgroup Spelaeogriphacea as well as in other isopods, such as *Tainisopus*. Such a distribution renders the right lacinia mobilis uninformative for defining clades of Phreatoicidea. The remainder of Nicholls' (1943:25) definition of the Amphisopodidae consists of variable features.

The cladogram resulting from the analysis (Figs 1-2) shows that the current family-level classification is not consistent with the phylogenetic relationships of the Phreatoicidea (Table 4). These results, however, must be considered preliminary because some genera classified by Nicholls (1943, 1944) have not been evaluated. Therefore, a new arrangement cannot be proposed at this time. Several changes, however, will be necessary because the positions of some taxa are likely to be stable under larger taxon samples. The Indian genus *Nichollsia* is nested in a clade containing the Australian genera *Hypsimeto-*

Table 3. Apomorphies of well-supported branches in Figures 1-2. See appendix 1 for character and character state definitions. Single-lined arrows indicate transitions assigned to a branch in some, but not all, optimisations. Double-lined arrows indicate unambiguous transitions. Consistency indices for each character shown in parantheses.

Branch	Character, state change & (consistency index)
Ingroup node to (*Crenisopus*+node 1)	03, 1->2 (0.250); 06, 1=>2 (0.143); 10, 1=>3 (0.333); 27, 1->3 (0.333); 52, 1=>2 (0.333); 53, 1->2 (0.200); 55, 1=>2 (0.333); 56, 1=>2 (1.000).
Branch to node 1	11, 1=>3 (0.286); 16, 1=>2 (1.000); 21, 1=>3 (1.000); 38, 2->3 (0.286); 40, 2=>3 (0.400); 54, 2=>1 (0.400); 62, 1->4 (0.667); 64, 1=>4 (0.500); 71, 1=>3 (0.286); 75, 1=>2 (0.167); 77, 1->2 (0.200); 78, 2=>1 (0.667).
Branch to node 2	01, 1->3 (0.455); 15, 1=>3 (0.667); 19, 1=>2 (0.500); 30, 2=>3 (0.143); 37, 3->1 (0.250); 42, 2->1 (0.286); 49, 2=>1 (0.333); 50, 2->1 (0.500); 63, 1->2 (1.000); 67, 1=>2 (1.000); 70, 1=>2 (0.167); 76, 2=>3 (0.500); 79, 2=>1 (0.286).
Branch to node 3	03, 1=>2 (0.250); 07, 1->2 (0.200); 10, 1->2 (0.333); 11, 1->2 (0.286); 23, 1=>2 (0.667); 34, 3->2 (0.429); 53, 1=>2 (0.200); 57, 2=>4 (0.333); 61, 3=>1 (0.286); 62, 1=>3 (0.667); 71, 1=>3 (0.286); 73, 1=>3 (0.600).
Branch to node 4	01, 1->2 (0.455); 27, 1=>2 (0.333); 32, 1=>2 (0.333); 41, 2->1 (0.200); 46, 2=>1 (0.333); 48, 4=>1 (0.500); 76, 2->1 (0.500); 78, 2=>1 (0.667).
Branch to node 5	01, 2->3 (0.455); 09, 2=>3 (0.750); 14, 3->4 (0.500); 19, 1=>2 (0.500); 30, 3->2 (0.143); 31, 1=>2 (1.000); 35, 1=>2 (1.000); 75, 1=>2 (0.167); 79, 2=>1 (0.286).

pus, *Phreatoicoides*, *Hyperoedesipus* and *Pilbarophreatoicus* (node 2 in Fig. 2). Because this clade is strongly supported (Bremer support of six steps), the eventual classification will recognize the junior synonymy of Nichollsiidae. The name of this clade could be based on Hypsimetopodinae although the placement of the more basal genera *Phreatoicopsis* and *Mesamphisopus* is relevant, because Nicholls (1943:28) based a subfamily name on each of these genera (Table 4). These names would have precedence pending their inclusion into the hypsimetopodine clade. The sister group relationship between *Phreatoicopsis* and the hypsimetopodines is well supported (six steps for node 1, Fig. 2), so a revised classification may recognize a monotypic phreatoicopsidine clade and a hypsimetopodine clade.

The composition of the Amphisopodidae (Table 4) may be reduced considerably with only *Amphisopus*, *Synamphisopus*, *Paramphisopus* and *Phreatomerus* included in a well-supported clade (node 3, Fig. 2). The Phreatoicinae (node 5) appears to be confined to those taxa that occur in New Zealand with *Crenoicus* falling outside this clade. Most of the southeastern Australian phreatoicidean genera may be placed in either the Mesacanthotelsoninae or the Paraphreatoicinae, pending analysis of the mesacanthotelsonine genera. Other available family names, such as the Mesamphisopodinae, have an uncertain status because the basal parts of the cladogram are weakly supported and unstable compared to previous analyses (Wilson & Keable 1999, Wilson & Johnson 1999).

3.4 *Biogeography and Cladogenic events*

The addition of more taxa and characters produced a result that supported the conclusions of Wilson & Johnson (1999). The clade *Nichollsia+Hypsimetopus+Pilbarophreatoicus* is

Table 4. Family-level phreatoicidean names from Nicholls (1943, 1944) compared to well-supported clades in phylogenetic analysis.

Species	Family name & subfamily	Node Number in Figure 2
	AMPHISOPODIDAE	
Amphisopus lintoni	Amphisopodinae	3
Paramphisopus palustris	Amphisopodinae	3
Hypsimetopus sp.	Hypsimetopodinae	2
Phreatoicoides gracilis	Hypsimetopodinae	2
Hyperoedesipus plumosus	Mesamphisopodinae	2
Mesamphisopus capensis	Mesamphisopodinae	–
Eophreatoicus sp.	Phreatoicopsinae	–
Phreatoicopsis terricola	Phreatoicopsinae	1
Synamphisopus ambiguus	Phreatoicopsinae	3
Phreatomerus latipes	Phreatomerinae	3
Crenisopus acinifer	–	–
Pilbarophreatoicus sp. 1	–	2
Nichollsia kashiense	NICHOLLSIIDAE	2
	PHREATOICIDAE	
Colubotelson joyneri	Paraphreatoicinae	4
Colubotelson searli	Paraphreatoicinae	4
Colubotelson sp. (Penstock Lagoon)	Paraphreatoicinae	4
Colubotelson sp. (Univ. Tasmania)	Paraphreatoicinae	4
Metaphreatoicus australis	Paraphreatoicinae	4
cf. *Metaphreatoicus* sp.	Paraphreatoicinae	4
Crenoicus buntiae	Phreatoicinae	4
Neophreatoicus assimilis	Phreatoicinae	5
Phreatoicus orarii	Phreatoicinae	5
Phreatoicus typicus	Phreatoicinae	5

deeply nested within the resulting cladogram (Figs 1-2) providing a minimum age constraint of ~120mybp (million years before present) on the events depicted by our cladogram (Indian subcontinental rifting; Storey 1995). The South African genus *Mesamphisopus* is no longer derived basally as in previous cladograms (Wilson & Keable 1999, Wilson & Johnson 1999), although the position of the root in the phreatoicidean cladogram is not strongly supported. If this basal topology proves to be better supported with more data, the minimum age of phreatoicidean diversification must be prior to the initial rifting between West and East Gondwana (~180mybp; Storey 1995). This latter age is consistent with ~237mybp age for the fossil phreatoicidean *Protamphisopus wianamattensis* (Chilton, 1918), known from freshwater lacustrine habitats preserved in the Middle Triassic Wianamatta shales of Sydney (Herbert 1997). The freshwater diversification of the phreatoicideans, therefore, may have taken place in Gondwana during the Triassic. Later biogeographic patterns may have been further influenced by later Mesozoic events, such as the inundation of East Gondwana during mid Cretaceous times (Wilson & Johnson 1999).

A more detailed biogeographic picture is somewhat less obvious. Biogeographic paralogy (Nelson & Ladiges 1996), wherein several independently evolving subclades replicate the same biogeographic patterns, may complicate phreatoicidean distributions because their relationships may have been established prior to Gondwanan rifting. The hypsimetopodine clade (node 2, Fig. 2) has transoceanic and trans-Australian relationships (Table 1), although the precise sister group relationships within this clade are weak. The amphisopodine clade (node 3, Fig. 2) shows several trans-Australian patterns and the large phreatoicid clade (node 4, Fig. 2) has both trans-Tasman and trans-Bassian relationships. The distribution of the Phreatoicidae may be the only restricted biogeographic pattern, with this family being confined to southeastern Australia, Tasmania and New Zealand. Confirmation of this pattern requires further phylogenetic analysis of the remaining genera in this family. The distributions of taxa in nodes 1-3 overlaps those in node 4, so a better understanding of these patterns may be improved by resolution of the basal phylogenetic relationships.

3.5 *Evolution of ecotypes*

With few exceptions, extant phreatoicideans are freshwater animals with limited swimming abilities. The earliest known ecotype for these animals is exemplified by *Protamphisopus wianamattensis* (Chilton, 1918), found in shales containing 'fishes, labrinthodonts, insects, coprolitic fragments and plants' (Tillyard 1916, quoted in Chilton 1918). This animal probably lived much as *Paramphisopus palustris* or the Tasmanian species of *Uramphisopus*, *Onchotelson*, or *Mesacanthotelson* do today, among vegetation in freshwater lakes and ponds. In more arid environments where permanent lakes and streams are absent, phreatoicidean species are associated at the surface with spring-fed brooks or seeps. In all cases, they prefer to live among fallen vegetation or aquatic roots and, if the substrate consists of coarse gravel, will live in-faunally as well. Most, but not all, such species have eyes and are pigmented. Species of *Crenoicus*, common denizens of highland *Sphagnum* swamps in New South Wales and Victoria, are exceptions: they are blind but have some color. We designate this ecotype as 'surface-cryptic,' owing to their surface-living habits but with an inclination to hide.

A second ecotype, which we designate as 'surface-burrower', includes species that burrow in fine sediments and are found alongside submerged roots of living vegetation such as button grass or sedges. Species of this ecotype also may be found in the burrows of crayfish (pholeteros: Richardson & Swain 1978, Horwitz 1989). Species of *Phreatoicoides* and *Hypsimetopus* are included in this category. Another 'surface burrower,' *Phreatoicopsis terricola*, creates small oval burrows in the moist soil among tree fern roots but without overlying free-running water. Many 'surface-burrowers' are white and pigmentless, as well as blind, although *Phreatoicopsis* has tiny eyes and some color in the head region.

The third ecotype, designated 'ground water,' includes truly hypogean phreatoicideans that are found only in permanent ground water, such as wells or springs, and occur in surface pools and spring outflows only during the wet season. All such taxa are translucent, completely without color and blind. The New Zealand phreatoicideans, *Nichollsia* from India, and *Hyperoedesipus*, *Crenisopus* and *Pilbarophreatoicus* from Western Australia are examples of this ecotype. The richness of Western Australian genera in this ecotype may reflect the relative aridity of this region. Appendix 2 includes observations on living conditions of most species included in this analysis.

The ecotype classification was plotted on the phylogenetic tree (Fig. 3). The majority

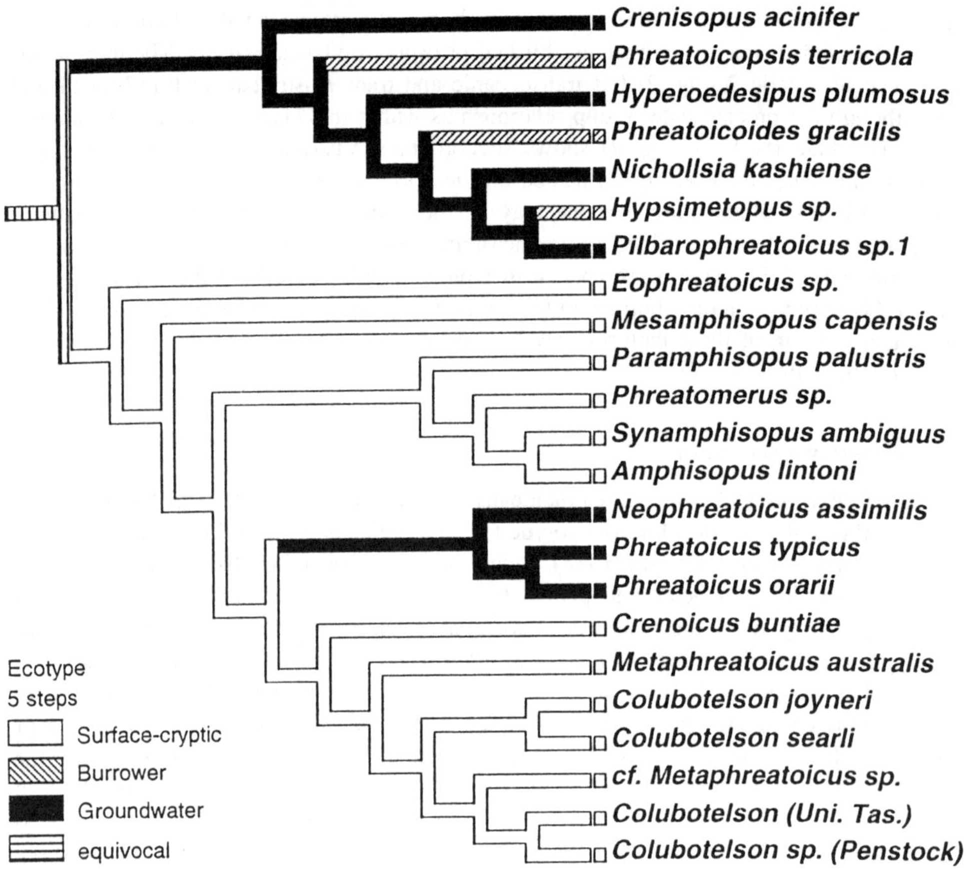

Figure 3. Phylogenetic distribution of ecotypes for in-group pheatoicideans, using the topology shown in Figures 1-2.

of the species are surface-cryptic, which may be the plesiomorphic habit for this suborder. The groundwater ecotype arose at least twice in the evolution of the phreatoicideans, because the clade containing New Zealand representatives of this type (node 5) is well separated from nodes 1 and 2, with surface-cryptic taxa placed basally. The New Zealand genus *Notamphisopus* has not been included in this analysis, although information in Nicholls (1944) indicates that its eyeless species are surface-cryptic. Species of *Crenoicus* occurring in Australia are also surface-cryptic and blind but several undescribed species also occur in the groundwater of caves (Eberhard & Spate 1995, Wilson & Ho 1996, unpublished data). Therefore, both *Crenoicus* and *Notamphisopus* species may represent instances of re-invasion of surface habitats from the groundwater. Emergent adaptation to surface habitats is also supported in the clade *Crenisopus*+node 1, with burrowing ecotype taxa apparently showing multiple derivations from groundwater ancestors. Of the taxa in this clade, only *Phreatoicopsis terricola* has eyes, although they are tiny compared to the large eyes of the basally derived genera *Eophreatoicus* or *Mesamphisopus*.

The timing of phreatoicidean adaptations to groundwater habitats may be addressed using biogeographic constraints. Because *Nichollsia* is deeply nested into the groundwater/burrowing clade (nodes 1 & 2 in Figs 2 and 3), these taxa adapted to this habitat minimally by ~120mybp, before the rifting of the Indian subcontinent from West Gondwana. The New Zealand phreatoicine clade (node 5, Fig. 2) may have adapted to hypogean habitats after this subcontinent separated from Australia and Antarctica (~90mybp). The adjacent clade position of *Crenoicus*, which includes some cavernicolous species, suggests that the hypogean adaptations of the New Zealand taxa could be older.

4 CONCLUDING REMARKS

Improving our knowledge of the Phreatoicidea requires urgent action owing to continuing degradation of their habitats, especially in areas where ground water is used for mining, agriculture and industry. The existing phreatoicidean classification is inadequate, despite their potential value in biogeography and conservation. As an interim source of information, this paper provides a well-resolved cladistic estimate with immediate applications. The phylogenetic relationships of phreatoicidean ecotypes suggest that these adaptations to ground water habitats are ancient. The phreatoicidean dependence on groundwater may greatly constrain their ability to survive rapid human-induced changes to the ground water environments. Admittedly, this observation is general for hypogean faunas on any continent but in the Australian context, where water is a limited and unpredictable resource, it takes on much more importance. In pre-European times, extensive wetlands associated with forests on the Great Dividing Range in southeastern Australia may have provided many habitats for phreatoicideans and other invertebrates (cf. Mollusca: Ponder 1994, 1997). Over the last 200 years, upland groundwater-fed springs and swamps associated with woodlands has been extensively modified or removed by the activities of graziers and foresters. Cattle, sheep, and feral domestic animals, such as pigs, goats and horses, further degrade remaining freshwater habitats by feeding and trampling on associated wetland vegetation. Many regions of the Great Dividing Range become arid during part of the year owing to vegetation removal, and no longer support surface exposures of ground water. Because *Crenoicus* species appear to have small geographic distributions (Wilson & Ho 1996), many species may have been lost along with these decimated woodlands. Extinct *Crenoicus* species in New South Wales alone may exceed the estimated number of extant species (approximately 20 spp.). The threat of more extinctions continues today owing to extensive logging in some state forests, adding to the exaggerated wet-dry cycles caused by vegetation removal. This example has parallels in other regions, such as in Western Australia where direct use of ancient aquifers for mining and agriculture threatens the groundwater faunas (Humphreys 1994).

The first step in preventing extinctions is informing those in charge of the management of these habitats on the diversity and relevance of their native fauna and flora. Our phreatoicidean research will attempt this first step because conservation choices and priorities could favor those taxa that provide the highest phylogenetic diversity for a region (e.g. Faith 1992). As demonstrated here, phreatoicidean isopods, with their ancient relationships and adaptations to sensitive habitats, provide great potential for defining the phylogenetic diversity of the regions in which they occur.

ACKNOWLEDGEMENTS

Greg Edgecombe and Regina Wetzer commented on an earlier version of this paper. Further discussions with Winston Ponder and Richard Major helped to clarify conservation issues. Rick Johnson assisted with taxonomic data collection. Bill Humphreys sent specimens of the *Pilbarophreatoicus*. This research was supported by a grant from the Australian Biological Resources Survey and by the Trust of the Australian Museum. We thank these people and institutions for their contributions to this research.

APPENDIX 1 – LIST OF CHARACTERS AND CHARACTER STATES

1. Head length: (1) length shorter than width in dorsal view; (2) length subequal to width in dorsal view; (3) length greater than width in dorsal view.
2. Head lateral profile of dorsal surface: (1) smoothly curved; (2) angularly curved; (3) indented; (4) flattened curve.
3. Cervical groove: (1) straight, or undulating, or smoothly curved; (2) absent.
4. Mandibular genal groove: (1) present, smoothly indented; (2) present, with acute indentation; (3) absent.
5. Mandibular notch: (1) present; (2) absent.
6. Clypeal notch: (1) present; (2) absent.
7. Antennal notch: (1) present; (2) absent.
8. Pereion relative width in dorsal view: (1) narrow, width near head width; (2) broad, width exceeding head width.
9. Pereionites 2-7 in dorsal view: (1) longer than wide; (2) wider than long; (3) anteriorly longer than wide decreasing posteriorly to wider than long; (4) anteriorly and posteriorly wider than long, but longer than wide in the middle.
10. Coxal articulation of pereionites 2-4: (1) free, with a pronounced v-shaped suture; (2) nearly fused, with a u-shaped suture; (3) fused, with no suture.
11. Coxal articulation of pereionites 5-7: (1) free; (2) nearly fused; (3) fused.
12. Lateral tergal plates of pereionites 2-4: (1) absent (coxae visible in lateral view); (2) extending ventrally (coxae not visible in lateral view).
13. Sternal processes: (1) present; (2) absent.
14. Typhlosole form in cross-section: (1) well developed, ventral invagination forming double spiral; (2) minimal, ventral invagination inverted, 'U' shaped; (3) absent, gut round; (4) intermediate, ventral invagination 'Y' shaped
15. Pleonites in lateral view (depth of pleurae/epimera): (1) much deeper than pereionites, with large ventrolateral plates, basal region of pleopods not visible; (2) having depth approximating depth of pereionites, with small ventro-lateral plates, basal region of pleopods visible; (3) having depth equal to depth of pereionites, without ventrolateral plates.
16. Pleonite 1 pleura size: (1) distinctly shallower than pleurae of pleonites 2-5; (2) near depth of pleurae of pleonites 2-5.
17. Pleonites 1-4 relative lengths: (1) subequal; (2) unequal, pleonite 4 length greater than pleonites 1-3.
18. Pleotelson form: (1) broader than high; (2) vaulted, higher than broad.
19. Pleotelson lateral length: (1) less than depth, or subequal to depth; (2) greater than depth.

20. Pleotelson lateral dorsal ridges: (1) absent; (2) continuous with telsonic region.
21. Telsonic region or tail piece median lobe in dorsal view: (1) entire; (2) cleft (i.e. two horizontal projections at posterolateral margins, narrowly separated); (3) broadly indented (i.e. no real horizontal projections); (4) broadly incised (i.e. two horizontal projections at posterolateral margins but more widely separated).
22. Telsonic region median lobe: (1) not reflexed; (2) reflexed dorsally.
23. Telsonic region lateral lobes: (1) absent; (2) forming a vertical plate; (3) forming a rounded setose lobe in dorsal view.
24. Telsonic region (if medial lobe indented) sensillate setae: (1) absent, or fine; (2) robust setae distinctly heavier than setae on dorsal surface.
25. Telsonic region posterior margin of pleotelson irregular denticulations: (1) present; (2) absent.
26. Dorsal uropodal ridge of pleotelson: (1) present; (2) absent.
27. Antennula terminal article shape: (1) tubular (sides approximately parallel, length greater than width), or conical; (2) inflated and bulbous (sides curved, width distinctly greater than previous article), or globular (sides curved, length approximates width, subequal or narrower than previous article); (3) tiny, vestigial.
28. Antennula penultimate article length: (1) distinctly longer than any other article; (2) subequal or shorter than other articles.
29. Antennula penultimate article width: (1) width subequal or narrower than proximal article; (2) inflated, width much greater than proximal article; (3) broad and flattened.
30. Antenna propodal article 1: (1) present, large; (2) present, forming thin ring; (3) absent.
31. Mandible incisor processes shape: (1) broad, width greater than thickness; (2) thin, resembling denticulate spines, width near thickness.
32. Mandible right lacinia mobilis: (1) present; (2) absent.
33. Mandible spine rows bifurcate spines: (1) absent; (2) present.
34. Mandible spine rows shape: (1) in smooth low arc between incisor and molar process; (2) on projecting ridge between incisor and molar processes; (3) on linear pedunculate projection between incisor and molar processes; (4) on round pedunculate projection between incisor and molar processes.
35. Maxillula medial lobe width: (1) less than lateral lobe; (2) subequal to lateral lobe, or greater than lateral lobe.
36. Pereiopods ischium dorsal margin with single elongate robust simple seta: (1) present; (2) absent.
37. Pereiopod I dactylus length measured along inner margin of dactylus: (1) projecting beyond palm in male; (2) subequal to palm in male; (3) shorter than palm in male.
38. Pereiopod I dactylus length: (1) projecting beyond palm in female; (2) subequal to palm in female; (3) shorter than palm in female.
39. Pereiopod I dactylus ventral margin distal cuticular fringe: (1) present; (2) absent.
40. Pereiopod I dactylus accessory claws: (1) with several distal accessory claws; (2) with 1 distal accessory claw; (3) without distal accessory claws.
41. Pereiopod I dactylus distal accessory spines: (1) present; (2) absent.
42. Pereiopod I propodus dorsal margin proximal region in male: (1) protruding to or beyond distodorsal margin of ischium; (2) protruding beyond distodorsal margin of carpus; (3) not protruding.
43. Pereiopod I propodus dorsal margin proximal region in female: (1) protruding to or

beyond distodorsal margin of ischium; (2) protuding beyond distodorsal margin of carpus; (3) not protuding.

44. Pereiopod I propodal palm in male stout denticulate setae: (1) present, serrate (multiple denticulations); (2) present, bifid (two spinules only); (3) absent.
45. Pereiopod I propodal palm in male stout robust simple setae not on ridge: (1) present, conical; (2) present, basally inflated; (3) present, tooth-like; (4) absent.
46. Pereiopod I propodal palm in male low conical setae on ridge: (1) present; (2) absent.
47. Pereiopod I propodal palm in female stout denticulate setae: (1) present, serrate (multiple denticulations); (2) present, bifid (i.e. two spinules only); (3) absent.
48. Pereiopod I propodal palm in female stout robust simple setae not on ridge: (1) present, conical; (2) present, basally inflated; (3) present, tooth-like; (4) absent.
49. Pereiopod I merus dorsal margin projection in anterior-dorsal view: (1) shelf-like and U-shaped in male; (2) spine-like and pointed in male.
50. Pereiopod I merus dorsal margin projection: (1) with numerous elongate fine simple setae; (2) with 1 or 2 robust simple setae.
51. Pereiopods II-III dactylus spines on ventral margin: (1) present; (2) absent.
52. Pereiopods II-III propodus articular plate: (1) present; (2) absent.
53. Pereiopod IV: (1) sexually dimorphic; (2) not sexually dimorphic.
54. Pereiopod IV dactylus: (1) shorter than propodal palm; (2) subequal to propodal palm; (3) longer than propodal palm.
55. Pereiopod IV propodus articular plate on posterior side of limb: (1) present; (2) absent.
56. Pereiopods V-VII propodus articular plate on posterior side of limb: (1) present; (2) absent.
57. Pereiopods V-VII basis dorsal ridge in cross section: (1) rounded; (2) angular; (3) produced but not forming distinct plate; (4) produced and forming distinct plate.
58. Penes form: (1) straight; (2) curved posteriorly.
59. Penes length: (1) extending near midline; (2) extending past midline and onto pleonite 1.
60. Penes armature: (1) smooth, lacking setae; (2) shaft denticulate; (3) with setae on shaft; (4) with setae on tip.
61. Penes shaft distal shape: (1) tapering; (2) broadening; (3) tubular.
62. Penes distal tip shape: (1) rounded; (2) pointed; (3) flattened; (4) truncate.
63. Pleopod exopods II-V proximal article distolateral lobes: (1) shorter than distal article; (2) subequal or longer than distal article.
64. Pleopod exopods I-V lateral proximal lobes: (1) present on exopods II-V; (2) present on exopods I and III-V; (3) absent; (4) present on exopods I-V.
65. Pleopod exopods I-V medial proximal lobes: (1) present on exopods II-V; (2) present on exopods III-V; (3) absent; (4) present on exopods I-V.
66. Pleopod protopod medial epipods: (1) with medial epipods or with small medial projections; (2) without medial epipods.
67. Pleopod protopod lateral epipods: (1) protopods III-V with lateral epipods or protopods I-V with lateral epipods; (2) protopods without lateral epipods.
68. Male pleopod I exopod distal margin: (1) pointed; (2) rounded.
69. Male pleopod I exopod lateral margin: (1) rounded; (2) broadly angular.
70. Male pleopod I exopod dorsal surface setae: (1) with setae; (2) lacking setae.
71. Male pleopod II endopod appendix masculina ventral shape of cross section of proximal half of shaft: (1) concave (forming an elongate trough); (2) convex (rod-like);

(3) convex (flattened in cross section).

72. Male pleopod II endopod appendix masculina basal musculature: (1) pronounced; (2) not pronounced.
73. Male pleopod II endopod appendix masculina distal tip: (1) truncate, or broadly rounded; (2) acutely rounded; (3) sharply pointed and spine-like; (4) pointed and distally spatulate.
74. Male pleopod II endopod appendix masculina length: (1) distal tip extending near to distal margin of endopod (either laterally or medially extending beyond half length of exopod); (2) distal tip not extending beyond half length of exopod.
75. Uropod protopod dorsomedial ridge: (1) not produced; (2) produced.
76. Uropod protopod dorsomedial ridge bump with robust setae: (1) bump with robust setae; (2) plate-like; (3) spur or spine-like.
77. Uropod protopod robust spinose setae on distoventral margin: (1) present; (2) absent.
78. Uropod rami distal tips: (1) pointed; (2) rounded.
79. Uropod endopod: (1) shorter than protopod; (2) subequal to protopod length; (3) longer than protopod.

APPENDIX 2 – MATERIAL EXAMINED AND HABITAT

(Abbreviations: AM, Australian Museum; WAM, Western Australian Museum; TMAG, Tasmanian Museum & Art Gallery; ZSIC, Zoological Survey of India Calcutta; bl, body length)

Amphisopus lintoni (Nicholls, 1926): AM P8795, 1 primary male (bl 8.7 mm) and 1 preparatory II female (bl 15.3 mm), from 6 males, 15 females, Albany, W.A. 35°00'S, 117°52'E (estimated from map), 'on submerged stakes or snags or under vegetation growing in water' (Nicholls 1943:81), G.E. Nicholls, collected prior to March 1927.

Colubotelson sp. (Penstock Lagoon, Tasmania): AM P54096, several males and females from approximately 172 specimens taken from trout gut, Penstock Lagoon, central Tasmania, Australia, 42°06'S 146°46'E, T. Hume, 25 April, 1977.

Colubotelson sp. (University of Tasmania, Hobart): AM P54097, male (bl 14.1 mm), female (bl 10.5 mm). Among fallen leaves in stream on University of Tasmania campus, Hobart, Tasmania, Australia, 42°53'S 147°19'E, R. Swain, 14 July 1990.

Colubotelson joyneri (Nicholls, 1926): AM P8796, male (bl 12.8 mm), preparatory II female (bl 10.5 mm), from 14 males and 5 females, Mt. Buffalo, Victoria 36°45'S, 146°35'E, 'in black ooze in shallow ditches, and puddles, beneath sphagnum in several boggy areas, and ... in fairly fast-flowing creeks. ... It would seem that it requires a sheltered situation...' (Nicholls 1944:110), G.E. Nicholls, Feb 1925.

Colubotelson searli Nicholls, 1944 (originally described as subspecies *C. joyneri searli*): AM P54098, male (bl 14 mm) and preparatory I female (bl 9.6 mm), stn. SC 418, Mt Baw Baw, Victoria 37°50.18'S, 146°15.39'E (GPS), in *Sphagnum* moss, altitude 1460 m, S. Clark, 4 Jan 1997.

Crenisopus acinifer Wilson & Keable, 1999: WAM C23229, holotype male (bl 4.39 mm), WAM C23230, paratype male (bl 5.25 mm dissected), WAM C23231 paratype male (bl 4.20 mm), WAM C23232 paratype preparatory female (bl 4.20 mm, dissected), WAM C23233 paratype preparatory female, (bl 4.44 mm), Zebedee Springs, El Questro Station, north Kimberley, W.A., Australia, 16°00'S 128°01'E, collected in net over spring outlet, 100 cubic meters of water strained, water temperature 35°C, W.F. Humphreys & B. Vine, 13 June 1994.

Crenoicus buntiae (Wilson & Ho, 1996): Data from Wilson & Ho (1996). Female characters not included in Wilson & Ho (1996) were scored from AM P44463, preparatory II female (bl 11.7 mm) from Mumbedah Swamp at head of Mumbedah Creek, Kanangra-Boyd National Park, NSW, Australia, at and near 33°53.73'S, 150°4.05'E, altitude 1200 m, in submerged *Sphagnum* moss, G.Wilson & party, 2 April 1992 & 18 Nov 1992.

Eophreatoicus 'sp.6': AM P54099, male (bl 21.3 mm) and 1 preparatory II female (bl 11.8 mm), ~1 km upstream of Gubara Pools, 12°49.65'S 132°52.96'E (GPS), Kakadu National Park NT, under leaves and among roots in gravel, G. Wilson, W. Ponder & V. Kessner, 15 August 1994.

Hyperoedesipus plumosus Nicholls & Milner, 1923: Combination of literature (Nicholls & Milner 1923, Nicholls 1943), 'cotype' primary male (bl 6.7 mm) and and preparatory II female (bl 5.4 mm), WAM 10665/6 and AM P8799 primary male (bl 4.7 mm), spring-fed pool at foot of Lesmurdie Falls, Darling Range, W.A, Nicholls, collection date unknown.

Hypsimetopus sp.: AM P54100, 1 primary male (bl 9.6 mm) and 1 preparatory II female (bl 10.2 mm), specimens scored from 8 individuals from near Zeehan, Tasmania, Australia; 41°53'S, 145°21'E (estimated from map), from roots of submerged sedges and button grass, A.M.M. Richardson & G.D.F. Wilson, 12 July 1990.

Mesamphisopus capensis (Barnard, 1914): TMAG G678, 1 primary male (bl 10.33 mm) and 1 preparatory female (bl 11.78 mm), near reservoir on Table Mountain, Cape Town, South Africa, altitude approx. 3000 ft., from a swiftly running stream under the moss growing on stones in the river bed, K.H. Barnard, circa 1936.

Metaphreatoicus australis (Chilton, 1891): AM P3347 Syntypes: 1 primary male (bl 17.4 mm) and 1 preparatory II female (bl 9.4 mm) specimens scored from 110 individuals from Upper Pipers Creek, Mt Kosciusko National Park upper Pipers Creek, NSW, 36°25'S,148°25.5'E (estimated from collector's account), in pools under rocks, R. Helms, March 1889.

cf. *Metaphreatoicus* sp. (Thredbo River, NSW): AM P52667, 1 primary male (bl 15.8 mm) and prep II female (bl 14.6 mm) from 5 males, 3 females and 3 juveniles. Stn. NSW 497, Spring-fed tributary of Thredbo River on Cascade Trail, Kosciusko National Park, NSW 36°31.81'S 148°16.07'E, altitude 1620 m and temp 10.1°C, under moss in roots with gravel, G. Wilson & W. Vader, 20 Feb 1993.

Neophreatoicus assimilis (Chilton, 1894): Data from literature, Chilton (1894) and Nicholls (1944), Winchester, South Canterbury, New Zealand, in wells and from a spring discharging into Trout Hatchery at Temuka NZ (Nicholls 1944).

Nichollsia kashiense Chopra & Tiwari, 1950: ZSIC C4516/1, male (bl 27.7 mm), and supplemented by female (bl ~16.96 mm), male (bl 21.8 mm), female (bl 15.8 mm), male (bl 20.8 mm), ZSIC C4517/1, nontype specimens from type locality, well, Benaras Cant (UP).

Paramphisopus palustris (Glauert, 1924): AM P44487, 1 primary male (bl 7.8 mm) and 1 preparatory II female (bl 7.4 mm), from 112 individuals from Lake Monger, Perth, W.A. 31°55'S, 115°50'E (est from map), under fallen reeds and vegetation along shoreline, D. Jones & G.D.F. Wilson, 22 March 1995.

Phreatoicoides gracilis Sayce, 1900: AM P3348, 'normal' male (bl 17.7 mm) and brooding female (bl 12.9 mm), Narracan River, 30 km SE of Warragul, Victoria, Gippsland, Victoria, O.A. Sayce, collected before Sept 1912.

Phreatoicopsis terricola Spencer & Hall, 1896: AM P54101, 1 primary male (bl 41.2 mm) and hermaphrodite (bl 36 mm), from 16 individuals [largest male bl 53.8 mm], Stn V19, 900 m south west of Cowley Track, Otway Ranges, Vic, 38°33.36'S, 143°50.48'E (map), dug from oval burrows in damp sediment under tree ferns roots – no free water observed,W.F. Ponder & G.D.F. Wilson, 15 July 1991.

Phreatoicus orarii Nicholls, 1944: Data from Nicholls, 1944, Figures 35-36; in wells of the Orari River valley.

Phreatoicus typicus Chilton 1883: Lectotype female, Canterbury Museum catalogue number IZ 3550 (bl approximately 18.5 mm), type locality, Templeton, Canterbury Plains, New Zealand. AM P52733, male (bl 12.8 mm). AM P52734, preparatory female (bl 13.8 mm); male (bl 11.5 mm); male (bl 12.8 mm): Templeton, 43°33.11'S, 172°26.38'E (map), from sediments at bottom of well 2, depth approximately 30 m, G.D. Fenwick, 11 June 1986 (detailed data in Wilson & Fenwick 1999).

Phreatomerus sp. (cf. *latipes*): AM P54102, 1 primary male (bl 14.6 mm) and 1 preparatory II female (bl 11.4 mm) of possibly undescribed species from Dead Woman Springs, South Australia. 29°35.37'S, 137°27.28'E (map), W.F. Ponder & B. Jenkins, 30 Aug 1983.

Synamphisopus ambiguus (Sheard, 1936): AM P54103, male (bl 26.2 mm) and preparatory II female (bl 24.9 mm), site GR19, in Stony Creek, off Silverbrand Rd, the Grampians, Vic, 37°09.20'S 142°29.67'E (map), T. Doeg & J. Read, 15 Nov 1994.

Pilbarophreatoicus n.sp.1: AM P54104, male (bl 9.2 mm) and prepartory II female (bl 8.5 mm), Stn BES 4873, Millstream aquifer, Pilbara, Western Australia, 21°35'S, 116°58'E, W.F. Humphreys & S.M. Eberhard, 30 July 1997

REFERENCES

Adlem, L.T. 1996. Biogeography of the freshwater Peracarida (Crustacea) from Barrington Tops. Unpublished Honors Thesis, N.S.W. Bachelor of Science. Newcastle: Department of Geography, University of Newcastle.

Banarescu, P. 1990. *Zoogeography of fresh waters. General distribution and dispersal of freshwater animals.* Vol.1 (with 208 maps). Wiesbaden: Aula-Verlag.

Bremer, K. 1994. Branch support and tree stability. *Cladistics* 10: 295-304.

Brusca, R.C. & G. D. F. Wilson 1991. A phylogenetic analysis of the Isopoda with some classificatory recommendations. *Mem. Queensland Mus.* 31: 143-204.

Chilton, C. 1918. A fossil isopod belonging to the fresh-water genus *Phreatoicus*. *J. Proc. Roy. Soc. New South Wales* 51: 365-388.

Eberhard, S. & A. Spate 1995. *Cave Invertebrate Survey: Towards an Atlas of NSW Cave Fauna.* A report prepared under NSW Heritage Assistance Program NEP 94 765. Queanbeyan: National Parks & Wildlife Service, NSW.

Eberhard, S.M., A.M.M. Richardson & R. Swain 1991. *The invertebrate cave fauna of Tasmania.* Hobart: Zoology Department, University of Tasmania.

Eriksson, T. 1998. *AutoDecay ver. 4.0.* Program distributed by the author. Stockholm: Department of Botany, Stockholm University.

Faith, D.P. 1992. Systematics and conservation on predicting the feature diversity of subsets of taxa. *Cladistics* 8: 361-373.

Herbert, C. 1997. Sequence stratigraphic analysis of Early and Middle Triassic alluvial and estuarine facies in the Sydney Basin, Australia. *Austr. J. Earth Sci.* 44: 125-143.

Horwitz, P.H.J. 1989. The faunal assemblage (or pholeteros) of some freshwater crayfish burrows in southwestern Tasmania. *Bul. Austr. Soc. Limnology* 12: 29-36.

Humphreys, W.F. 1994. *The subterranean fauna of the Cape Range coastal plain, northwestern Australia.* Report to the Australian Heritage Commission and the Western Australia Heritage Committee. Perth: Western Australian Museum.

Knott, B. & S.A. Halse 1999. *Pilbarophreatoicus platyarthricus* n.gen., n.sp. (Isopoda: Phreatoicidea: Amphisopodidae) from the Pilbara region of Western Australia. *Records of the Australian Museum* 51(1): 33-42.

Maddison, W.P. & D.R. Maddison 1992. *MacClade, analysis of phylogeny and character evolution, version 3.* Sunderland, Massachusetts: Sinauer.

Nelson, G. & P.Y. Ladiges 1996. Paralogy in cladistic biogeography and analysis of paralogy-free subtrees. *Amer. Mus. Novitates* 3167: 1-58.

Nicholls, G.E. 1943. The Phreatoicoidea. Part I. The Amphisopidae. *Pap. Proc. Roy. Soc. Tasmania* 1942: 1-145.

Nicholls, G.E. 1944. The Phreatoicoidea. Part II. The Phreatoicoidae. *Pap. Proc. Roy. Soc. Tasmania* 1943: 1-156.

Ponder, W.F. 1994. Australian freshwater Mollusca: Conservation priorities and indicator species. *Rec. Queensland Mus.* 36: 191-196.

Ponder, W.F. 1997. Conservation status, threats and habitat requirements of Australian terrestrial and freshwater Mollusca. *Mem. Mus. Victoria,* 56: 421-430.

Richardson, A.M.M. & R. Swain 1978. The freshwater invertebrates. Lower Gordon River scientific survey, report to the Hydro-Electricity Commission, Tasmania. Hobart: University of Tasmania.

Schram, F.R. 1974. Paleozoic Peracarida of North America. *Fieldiana, Geology* 33: 95-124.

Storey, B.C. 1995. The role of mantle plumes in the continental breakup: case histories from Gondwanaland. *Nature* 377: 301-308.

Swofford, D.L. 1999. *PAUP* 4.0, Phylogenetic Analysis Using Parsimony (and Other Methods).* Sunderland, Massachusetts: Sinauer.

Tillyard, R.J. 1916. Mesozoic and Tertiary Insects of Queensland and New South Wales: Descriptions of the Fossil Insects. *Queensland Geol. Surv. Publ.* 253: 11-60.

Wilson, G.D.F. & G.D. Fenwick 1999. Taxonomy and Ecology of *Phreatoicus typicus* Chilton, 1883 (Crustacea, Isopoda, Phreatoicidae). *J. Roy. Soc. New Zealand* 29(1): 41-64.

Wilson, G.D.F. & E.L. Ho 1996. *Crenoicus* Nicholls, 1944, (Crustacea, Isopoda, Phreatoicidea): Systematics and biology of a new species from New South Wales. *Rec. Austr. Mus.* 48: 7-32.

Wilson, G.D.F. & R.T. Johnson 1999. Ancient endemism among freshwater isopods (Crustacea, Phreatoicidea). In: The Other 99%. The conservation and Biodiversity of Invertebrates, W.F. Ponder & D. Lunney (eds) 264-268. Mosma: *Trans. Roy. Zool. Soc. New South Wales.*

Wilson, G.D.F. & S.J. Keable 1999. A new genus of phreatoicidean isopod (Crustacea) from the North Kimberley Region, Western Australia. *Zool. J. Linn. Soc.* 126: 51-79.

A review of the bopyrid isopods parasitic on thalassinidean decapods

JOHN C. MARKHAM
Arch Cape Marine Laboratory, Arch Cape, Oregon, USA

ABSTRACT

The epicaridean isopods currently contains just over 500 described species, of which 52 species infest at least 48 species of hosts in the Thalassinidea. This proportion of 10% of the species in the Bopyridae is much greater than expected from the number of known species of thalassinideans. Aside from the peculiar *Phyllodurus abdominalis*, all of the parasites of thalassinideans belong to the bopyrid subfamilies Pseudioninae and Ioninae and infest their hosts branchially. This report presents tabulations of all the known bopyrids infesting thalassinideans with their hosts and localities and of the host species and their respective parasites and localities. Several hosts bear numerous species of parasites, while many parasites infest multiple species of hosts in patterns unusual for the Bopyridae. A discussion of the tabulations presents possible interpretations of the known occurrence of these species.

1 INTRODUCTION

The epicaridean family Bopyridae currently contains just over 500 recognized described species, all of them, as adults, parasitic on decapod Crustacea (Markham 1994). Of these, 52 species (plus two reported by genus only and five more mentioned in the literature without any indication of identification), or about one tenth of the total, are parasites of thalassinidean decapods. The most recently published estimate of the number of described decapod species is 10,000 (Bowman & Abele 1982), of which thalassinideans, in the most recent count (Nguyen Ngoc-Ho pers. comm.), are 516 described species or 5%. Thus the number of known bopyrid species as infesting thalassinideans is about twice that expected from host-species numbers alone.

The family Bopyridae contains 10 subfamilies, most of whose species infest their hosts branchially, in what is evidently the ancestral condition. All but a tiny number of the others are abdominal parasites of their hosts, either dorso-abdominally on paguroids, or ventro-abdominally (with a few exceptions) on caridean shrimps; these seem to be separately derived from branchially-infesting ancestors (Markham 1986). Each subfamily is largely or entirely limited to hosts in a single decapodan infraorder or lower taxon. The largest subfamilies, all containing branchial parasites, are of world wide distribution. The most speciose, the Pseudioninae, is characteristically found on anomurans, the second, the

Bopyrinae, contains parasites of carideans, and the third, the Ioninae, is known mostly from brachyuran hosts. On the bases of generalized morphology, large number of genera infesting genera of hosts, wide geographical distribution and fossil evidence (Markham 1986), there is reason to believe that the first bopyrids were members of the Pseudioninae infesting anomurans of the family Galatheidae.

2 PARASITES OF THALASSINIDEA

2.1 *General remarks*

Bopyrids infesting thalassinideans do not follow the generalizations above. Aside from the peculiar *Phyllodurus abdominalis*, the sole member of its subfamily, with a structure similar to that of branchial bopyrids but infesting its hosts abdominally, all are branchial parasites belonging to the subfamilies Pseudioninae and Ioninae. The genera represented I have designated the 'Thalassinidean Transition' (Markham 1986) because they appear to represent an evolutionary link between these two subfamilies and may possibly also represent a connection to the Bopyrinae as well as transfer of parasitism from the Anomura to the Brachyura and Caridea. The species in this group are readily assigned to genera, but those genera are often difficult to assign with certainty to subfamilies. This is evidently because they are more generalized and probably closer to the ancestral stock of the Bopyridae than are the more distinctive and characteristic genera in their respective subfamilies infesting hosts of the 'proper' infraorders. This problem is compounded by the systematic histories of the two most prominent genera involved. These are the type-genera of the two subfamilies, *Pseudione* (Kossman 1881), and *Ione* (Latreille 1817). The type-localities of the type-species of both genera, *Pseudione callianassae* and *Ione thoracica*, are in the Mediterranean, and both of these species infest *Callianassa subterranea* there, among other hosts. Regrettably, *P. callianassae* is regarded as a nomen nudum (Bourdon 1981b), thus making this largest of all bopyrid genera difficult to define. Of the 18 genera currently assigned to the Thalassinidean Transition, all but two contain only species known to infest thalassinideans. One genus with other hosts is *Pseudione*, most of whose 50+ described species infest a great range of anomuran hosts, while a few species infest large deep-water carideans (Markham 1994). The other such genus is *Gigantione*, of which three species infest thalassinideans, while the other nine described species are parasites of a broad range of brachyuran hosts (Markham 1994).

Host-species specificity within the Bopyridae exhibits so many different patterns that generalizations are difficult to make. There seem to be variations in the patterns, which may be correlated with parasite subfamily (branchial parasites tend to be more restricted than abdominal ones), position in parasite subfamily (more primitive genera usually share their host genera and even species with more members of other genera), host group (parasites of brachyurans cover wider ranges of genera and families of hosts), and latitude (tropical species occur on fewer species of hosts than temperate or polar species). Parasites of thalassinideans are difficult to characterize in terms of these patterns.

2.2 *Detailed remarks*

Aside from those specimens not identified to species, the bopyrid parasites of thalassinideans (Table 1) belong to 36 species in 11 genera of Pseudioninae, 15 species in 6 genera of Ioninae, and the single monotypic genus and species of the Phyllodurinae, *Phyllodurus abdominalis*, the last known to infest only two species of *Upogebia* along the west coast of North America. Most parasite species, except for those in *Pseudione* and *Ione*, infest hosts predominantly in a single family. The three species of *Gigantione* found on thalassinidean hosts infest only members of the family Axiidae. The three species of *Castrione* infest hosts only in the Micheleidae, which bear no other species of parasites. At the same time, the genera *Pseudione* and *Ione* both contain species found on many host species, including members of several thalassinidean families. There is no evident correlation between the subfamilies Pseudioninae and Ioninae and the families of their respective hosts.

Most of the bopyrid species considered here are known from only rather limited geographic ranges, and many of those that infest more than single host species do so in single localities. There are, however, a few marked exceptions, all of them species infesting different host species in far ranges of their distributions. *Ione cornuta*, the most widespread species, infests different callianassid hosts in the western and eastern Pacific Ocean, as does *Pseudione borealis* between the North Sea and Africa. Similarly, *Progebiophilus upogebiae* infests different species of *Upogebia* in ranging from North Carolina to Brazil, and *Phyllodurus abdominalis* does the same over its range from Canada to Mexico. *Ione thoracia*, while infesting many different hosts, follows two of them from Britain through the Mediterranean, though it is known to infest *Callianassa balssi* only in Nigeria. A similar case is *Gyge branchialis*, which infests four host species, but follows two of them from the North Sea to the Black Sea.

Of the thalassinidean families (Table 2) with species known to bear bopyrid parasites, best represented are the two largest families Callianassidae and Upogebiidae, the latter especially by its largest genus *Upogebia*. In a pattern otherwise unusual within the Bopyridae, many of the host species in the Callianassidae and the Upogebiidae bear parasites belonging to more than one species, genus and even subfamily in a single geographical locality. The extreme is the Mediterranean species *Upogebia pusilla*, which hosts five species of parasites in as many genera in two subfamilies. *U. issaeffi, U. major, Wolffogebia phuketensis* and the callianassids *Callianassa truncata* and *Neotrypaea uncinata* each bear three noncongeneric bopyrid species of both the Pseudioninae and Ioninae. In many of the cases of multiple parasite species, the parasites include species of both *Pseudione* and *Ione*.

The genus *Upogebia*, with 29 species (counting the three identified only to genus as separate species) reported as bopyrid hosts, is exceeded only by the anomuran genera *Pagurus* (at least 32 species, bearing both branchial and abdominal parasites) and *Munida* (also at least 32 species, nearly all branchially infested but four also bearing the monotypic but very widespread internal bopyrid, *Entophilus omnitectus* Richardson). In both of those genera, infestation of a single host species by more than one parasite species is fairly frequent. These large numbers of parasitized species in a host genus can be interpreted as an indication that bopyrid infestation is ancient in these genera. Another indication of that is the occurrence of a large number of parasite genera in a single host genus. Species in the genus *Upogebia* bear a total of 15 bopyrid genera, a number comparable to those recorded from *Pagurus* and *Munida*. Although diagnostic branchial swellings are fairly well known from species of *Munida* back to the Upper Jurassic and Cretaceous (Hessler 1969),

Table 1. Bopyridae infesting Thalassinidea.

Parasite species	Host species	Locality (ies)	References
SUBFAMILY PSEUDIONINAE			
Aporobopyroides upogebiae	*Upogebia savignyi*	Red Sea	Nobili 1906
Gigantione elconaxii	*Eiconaxius* sp.	New Caledonia	Markham 1994
G. pikei	*Axiopsis cf. australiensis*	New Zealand	Page 1985
G. uberlackerae	*Axiopsis n.* sp.	Gulf of Mexico	Adkison 1984
Gyge angularis	*U. hirtifrons*	New Zealand	Page 1985
G. branchialis	*Callichirus tyrrhea*	Channel Is.	LeSueur 1954
	U. deltaura	England to Black Sea	Bourdon 1968
	U. pusilla	Medit.; Black Sea	Bourdon 1968
	U. stellata	Denmark to Bl. Sea	Bourdon 1968
Ionella agassizi	*Neotrypaea uncinata*	S. Chile	Muñoz 1997
I. maculata	*Callianassa* sp.	New Caledonia	Markham 1994
I. murchisoni	*Callianassa* sp.	Hawai'i	Danforth 1970
Metabopyrus irregularis	*Wolffogebia phuketensis*	Phuket, Thailand	Markham 1985
M. ovalis	*U. issaeffi*	Kyûsyû, Japan	Shiino 1939a
	U.major	SW Korea	Kim & Kwon 1988
Orthione furcata	*U. affinis*	E coast USA	Markham 1988
Pauperella rotunda	*U. hexaceras*	Kei Is., Indonesia	Nierstrasz & Brender à Brandis 1929
Progebiophilus brevis	*U. contigua*	Sierra Leone	Bourdon 1981b
P. bruscai	*U. dawsoni*	Gulf of California	Salazar-Vallejo & Leija-Tristán 1989
P. chapini	*U. furcata*	Congo	Van Name 1920
P. euxinicus	*U. deltaura*	France, Medit.	Bourdon 1968
	U. pusilla	Black Sea; France	Popov 1927; Bourdon 1968
P. filicaudatus	*U. issaeffi*	Mie, Japan	Shiino 1958
P. sinicus	*U. major*	Hong Kong	Markham 1982
P. upogebiae	*U. affinis*	North Carolina, USA	Hay 1917
	U. felderi	Tamaulipas, Mexico	Williams 1993
	U. omissa	Ceará, Brazil	Lemos de Castro 1970
	Upogebia sp.	Ceará, Brazil	Lemos de Castro 1970
P. villosus	*U. pugnax*	Amami, Japan	Shiino 1964
Progebiophilus assisi	*Upogebia assisi*	Karachi, Pakistan	Kazmi & Bourdon 1997
Pseudione bakeri	*U. bowerbankii*	S. Australia	Hale 1929
P. borealis	*Callianassa subterranea*	Helgoland, Germany	Caspers 1939
		Bay of Biscay	Bourdon 1981a
	C. marchali	Ivory Coast	Bourdon 1981b
P. brattstroemi	*Neotrypaea uncinata*	Central Chile	Stuardo et al. 1986
P. callianassae	*Callianassa subterranea*	Britain; Mediterr.	Isaac et al. 1990
P. compressa	*C. jousseamei*	Amami, Japan	Shiino 1964
P. dohrni	*Necallianassa truncata*	Italy	Bourdon 1968
P. filicaudata	*Upogebia issaeffi*	Mie, Japan	Shiino 1958
P. hanseni	*Axiopsis brocki*	Karakelong, Indonesia	Nierstrasz & Brender à Brandis 1923
P. longicauda	*Nihonotrypaea japonica*	Seto & Mie, Japan	Shiino 1972
	N. petalura	Hong Kong	Markham 1992

Table 1. Continued.

Parasite species	Host species	Locality (ies)	References
P. murawaiensis	*C. filholi*	Aukland, N. Z.	Page 1985
Pseudione sp.	*Upogebia africana*	South Africa	Barnard 1940
	U. pusilla	Italy	Bonnier 1900
Upogebione ovalis	*Upogebia* sp.	Gulf of Thailand	Nierstrasz &Brender à Brandis 1931
U. phuketensis	*Wolffogebia phuketensis*	Phuket, Thailand	Markham 1985
U. tropica	*Upogebia omissa*	Paraíba, Brazil	Lemos de Castro & Brasil Lima 1975
Upogebiophilus rhadames	*U. rhadames*	Djibouti, Red Sea	Nobili 1906
Upogebiophilus sp.	*Upogebia* sp.	Karachi, Pakistan	Kazmi & Tirmizi 1996
SUBFAMILY IONINAE			
Acrobelione anisopoda	*Upogebia talismani*	Bight of Biafra	Bourdon 1981a
A. langi	*U. furcata*	Congo	Van Name 1920
A. reverberii	*U. pusilla*	Italy	Restivo 1970
	Necallianassa truncata	Italy	Restivo 1970
Castrione digitcaudata	*Marcusiaxus wamsoi*	Irian Jaya	Markham 1995
C. longicaudata	*M. lemoscastroi*	Amapá, Brazil	Brasil Lima 1980
	Meticonaxius sp.	Pará, Brazil	Brasil Lima 1980
Coxalione inaequalis	*Laomedia astacina*	Koh Boi Tai,Thailand	Bourdon 1977
Hemicepon muelleri	*Upogebia. brasiliensis*	Rio de Janiero, Brazil	Lemos de Castro & Brasil Lima 1980
Ione cornuta	*Nihonotrypaea japonica*	Seto, Japan; Korea	Shiino 1939b; Kim & Kwon 1988
	Neotrypaea californiensis	W Canada & USA	Kozloff 1983
	N. gigas	W Canada & USA	Hatch 1947
I. ovata	*N. uncinata*	Mid to S Chile	Muñoz 1997
I. taiwanensis	*Callianidea typa*	Taiwan	Markham 1995
I. thompsoni	*Gilvossius setimanus*	New England, USA	Markham 1995
I. thoracica	*Callianassa balssi*	Nigeria	Olaosebikan 1986
	C. subterranea	Britain thru Medit.	Bourdon 1968
	C. tyrrhena	Britain thru Medit.	Bourdon 1968
	Upogebia pusilla	Italy	Giard & Bonnier 1890
	U. stellata	Mediterranean	Bonnier 1900
I. vicina	*Necallianassa truncata*	Italy	Bourdon 1968
Procepon horridulum	*Wolffogebia phuketensis*	Phuket, Thailand	Markham 1995
P. insolitum	*Upogebia major*	Tanabe Bay, Japan	Shiino 1937
	U. issaeffi	Mie, Japan	Shiino 1958
	Upogebia sp.	Phuket, Thailand	
SUBFAMILY PHYLLODURINAE			
Phyllodurus abdominalis	*Upogebia pugettensis*	Br. Columb. to Calif.	Markham 1977
	U. macginitieorum	Baja Calif., Mexico	Campos Gonzalez & Campoy-Favela 1987
UNIDENTIFIED SUBFAMILY			
Unidentified	*Upogebia felderi*	Tamaulipas, Mexico	Williams 1993
Unidentified	*U. major*	Akkeshi, Japan	Oguro 1957
Unidentified	*U. omissa*	Trinidad	Williams 1993
Unidentified	*U. spinigera*	Colombia (Pacific)	Holthuis 1954
Unidentified	*U. vasquezi*	Atlantic Panama; Turks	Williams 1993

Table 2. List of Thalassinidea hosting Bopyridae. Numbers are species infested over number described in each family. References cited in Table 1 only.

Host species	Parasite species	Locality (ies)
SUPERFAMILY CALLIANASSOIDEA		
FAMILY CALLIANASSIDAE (15/166)		
Callianassa balssi	*Ione thoracica*	Nigeria
C. filholi	*Pseudione muraiensis*	New Zealand
C. jousseaumei	*P. compressa*	Amami, Japan
C. marchali	*P. borealis*	Ivory Coast
C. subterranea	*Ione thoracica*	North Sea to Black Sea
	Pseudione borealis	Helgoland; W Iberia
	P. callianassae	Italy
C. tyrrhena	*Ione thoracica*	France; Italy
Callianassa sp.	*Ionella maculata*	New Caledonia
Callianassa sp.	*I. murchisoni*	Hawai'i
Gilvossius setimanus	*I. thompsoni*	Massachusetts, USA
Necallianassa truncata	*Acrobelione reverberi*	Italy
	Ione vicina	Italy
	Pseudione dohrni	Italy
Neotrypaea californiensis	*I. cornuta*	Brit. Col. to California
N. gigas	*I. cornuta*	California to Baja Calif.
N. uncinata	*I. ovata*	S Chile
	Ionella agassizi	S Chile
	Pseudione brattstroemi	S Chile
Nihonotrypaea japonica	*Ione cornuta*	Japan; Korea
	Pseudione longicauda	Seto & Mie, Japan
N. petalura	*P. longicauda*	Hong Kong
FAMILY CALLIANIDEIDAE (1/2)		
Callianidea typa	*Ione taiwanensis*	Taiwan
FAMILY CTENOCHELIDAE (0/19)		
No records of infestation		
FAMILY LAOMEDIDIIDAE (1/19)		
Laomedia astacina	*Coxalione inaequalis*	Koh Boi Tai, Thailand
FAMILY THOMASSINIDAE (0/3)		
No records of infestation		
FAMILY UPOGEBIIDAE (30/141)		
Upogebia affinis	*Orthione furcata*	Virginia to Florida., USA
	Progebiophilus upogebiae	E coast USA
U. africana	*Pseudione* sp.	South Africa
U. assisi	*Progebiophilus assisi*	Karachi, Pakistan
U. bowerbankii	*Pseudione bakeri*	South Australia
U. brasiliensis	*Hemicepon muelleri*	Rio de Janiero, Brazil
U. contigua	*Progebiophilus brevis*	Sierra Leone
U. dawsoni	*P. bruscai*	Gulf of California
U. deltaura	*P. euxinicus*	France (Mediterranean)
	Gyge branchialis	France; England
U. felderi	*Progebiophilus upogebiae*	Tamaulipas, Mexico
	unidentified	Tamaulipas, Mexico
U. furcata	*Progebiophilus chapini*	Congo
	Acrobelione langi	Congo
U. hexaceras	*Pauperella rotunda*	Kei Is., Indonesia

Table 2. Continued.

Host species	Parasite species	Locality (ies)
U. hirtifrons	*Gyge angularis*	New Zealand
U. issaeffi	*Metabopyrus ovalis*	Kyûsyû, Japan
	Procepon insolitum	Mie, Japan
	Pseudione filicaudata	Mie, Japan
U. macginitieorum	*Phyllodurus abdominalis*	California; B. C. Mexico
U. major	*Procepon insolitum*	Tanabe Bay, Japan
	Progebiophilus sinicus	Hong Kong
	Metabopyrus ovalis	Korea
	unidentified	Japan
U. omissa	*Progebiophilus upogebiae*	Ceará, Brazil
	Upogebione tropica	Paraíba, Brazil
	unidentified	Trinidad
U. pugettensis	*Phyllodurus abdominalis*	Brit. Columbia – Calif.
U. pugnax	*Progebiophilus villosus*	Amami, Japan
U. pusilla	*Acrobelione reverberii*	Italy
Gyge branchialis	*Progebiophilus euxinicus*	Mediterranean; Black Sea
	Ione thoracica	Italy
	Pseudione sp.	Italy
U. quddusi	*Upogebiophilus* sp.	Karachi, Pakistan
U. rhadames	*Upogebiophilus rhadames*	Djibouti, Red Sea
U. savignyi	*Aporobopyroides upogebiae*	Massaua, Red Sea
U. spinigera	unidentified	Colombia (Pacific)
U. stellata	*Gyge branchialis*	Denmark to Black Sea
	Ione thoracica	Mediterranean
U. talismani	*Acrobelione anisopoda*	Bight of Biafra
U. vasquezi	unidentified	Atlantic Panama; Turks Is.
Upogebia sp.	*Procepon insolitum*	Phuket, Thailand
Upogebia sp.	*Progebiophilus upogebiae*	Ceará, Brazil
Upogebia sp.	*Upogebione ovalis*	Gulf of Thailand
Wolffogebia phuketensis	*Metabopyrus irregularis*	Phuket, Thailand
	Procepon horridulum	Phuket, Thailand
	Upogebione phuketensis	Phuket, Thailand
SUPERFAMILY AXIOIDEA		
FAMILY AXIIDAE (4/114)		
Axiopsis brocki	*Pseudione hanseni*	Karakelong, Indonesia
A. (Paraxiopsis) n. sp.	*Gigantione uberlackerae*	Gulf of Mexico
A. cf *australiensis*	*G. pikei*	New Zealand
Eiconaxius sp.	*G. elconaxii*	New Caledonia
FAMILY CALOCARIDIDAE (0/25)		
No records of infestation		
FAMILY MICHELEIDAE (3/16)		
Marcusiaxus lemoscastroi	*Castrione longicaudata*	Amapá, Brazil
M. wamsoi	*C. digiticaudata*	Irian Jaya
Meticonaxius sp.	*C. longicaudata*	Pará, Brazil
FAMILY STRAHLAXIIDAE (0/9)		
No records of infestation		

the fossil record of bopyrid infestation in the other two genera is lacking because 'most thalassinids and all pagurids are known from claws only, as their bodies are soft' (Glaessner 1969: R434). In all three of these host genera, the most numerous species of parasites belong to the subfamily Pseudioninae, considered, on the basis of morphology, to be the most primitive of the Bopyridae (Shiino 1965).

ACKNOWLEDGEMENTS

Dr Nguyen Ngoc-Ho (Muséum National d'Histoire Naturelle, Paris) provided numbers of currently described species of the families of the Thalassinidea, used in Table 2. Drs Brian Kensley and Raymond B. Manning (National History Museum, Smithsonian Institution) helped me to cite the thalssinidean host species correctly. And many people worldwide responded to my e-mail request for names of the authors of species of *Upogebia*.

REFERENCES

Adkison, D.L. 1984. Two new species of *Gigantione* Kossmann (Isopoda: Epicaridea: Bopyridae) from the western North Atlantic. *Proc. Biol. Soc. Washington* 97: 761- 772.

Barnard, K.H. 1940. Contributions to the crustacean fauna of South Africa. 12. Further additions to the Tanaidacea, Isopoda and Amphipoda, together with keys for the identification of hitherto recorded marine and fresh-water species. *Ann. South African Mus.* 32: 381-543.

Bonnier, J. 1900. Contribution a l'étude des Épicarides: Les Bopyridae. *Trav. Inst. Zool. Lille et Labor. Zool. Marit. Wimereux* 8: 478.

Bourdon, R. 1968. Les Bopyridae des mers européennes. *Mem. Mus. Natl. Hist. Nat. Paris, Nouv. Ser. (A)* 50(2): 77-424.

Bourdon, R. 1977. Sur un nouveau genre de Bopyridae parasite des Thalassinides (Crustacea, Isopoda, Epicaridea). *Steenstrupia* 4: 167-170.

Bourdon, R. 1981a. Bopyriens nouveaux pour la faune européenne de l'Atlantique (Isopoda Epicaridea). *Bull. Mus. Natl. Hist. Nat. Paris* (4)3 (Sect. A, 2): 615-634.

Bourdon, R. 1981b. Sur cinq Bopyrides parasites de Thalassinides ouest-africaines. *Bull. Inst Franç. Afrique Noire* 43 (ser. A, 1-2): 111-134.

Bowman, T. & L.G. Abele 1982. Chapter 1. Classification of the recent Crustacea. In L.G. Abele (ed.), *The biology of Crustacea. Volume 1. Systematics, the fossil record and biogeography.* New York: Academic Press.

Brasil Lima, I.M. 1980. Crustáceos epicarídeos do Brasil. 11. *Castrione longicaudata* g. n. sp. n. do littoral norte do Brasil. *Rev. Brasil. Biol.* 40: 121-123.

Campos-Gonzalez, E. & J.R. Campoy-Favela 1987. Epicarídeos de Baja California I. Primer registro y notas bioecológicas de dos Bopyridae y un Cryptoniscidae (Crustacea, Isopoda) para México. *Cienc. Mar.* 13(3): 39-48.

Caspers, H. 1939. Zwei neue Epicariden-Arten aus der Nordsee (*Pseudione borealis* n. sp. und *Ps. tuberculata* n. sp.). *Zool. Anzeiger* 125: 236-244.

Danforth, C.G. 1970. Epicaridea (Isopoda) of Hawaii. *Bull. Southern California Acad. Sci.* 69: 27-31.

Giard, A. & J.Bonnier 1890. Prodome d'une monographie des épicarides du golfe de Naples. *Bull. Scien., Paris* 22: 367-391.

Glaessner, M.F. 1969. Decapoda. In R.C. Moore (ed.), *Treatise on invertebrate paleontology. Part R. Arthropoda 4.*Volume 2:R399-R533. Lawrence, Kansas: University of Kansas and Geological Society of America.

Hale, H.M. 1929. *The Crustaceans of South Australia.* Adelaide: Government Printer. 380 pp.

Hatch, M.H. 1947. The Chelifera and Isopoda of Washington and adjacent regions. *Univ. Washington Publ. Biol.* 10(5): 155-274.

Hay, W.P. 1917. A new genus and three new species of parasitic isopod crustaceans. *Proc. United States Natl. Mus.* 51: 569-574.

Hessler, R. 1969. Peracarida. In R.C. Moore (ed.), *Treatise on Invertebrate Paleontology, Part R, Arthropoda 4.* Volume 2: R360-393. Lawrence, Kansas: Geological Society of America and University of Kansas Press.

Holthuis, L.B. 1954. On a collection of decapod Crustacea from the Republic of El Salvador (Central America). *Zool. Verhand.* 23: 1-43.

Isaac, M.J, P. Makings, E. Naylor, G. Smaldon & R.G. Withers 1990. Chapter 9: Crustacea II: Malacostraca Peracarida. Pages 362-488. In Hayward, P.J. & J.S. Ryland (eds), *The marine fauna of the British Isles and north-west Europe. Volume I. Introduction and Protozoans to Arthropods.* Oxford: Clarendon Press.

Kazmi, Q.B. & R. Bourdon 1997. A new bopyrid isopod *Progebiophilus assisi* on the mud shrimp *Upogebia (U.) assisi* Barnard (Thalassinoidea) from Pakistan. *Pakistan J. Mar. Sci.* 6: 59-67.

Kazmi, Q.B. & N.M. Tirmizi 1996. Marine invertebrates: associations and partnerships. *MRC Newsletter (Marine Reference Collection and Resource Centre, University of Karachi, Pakistan)* 5(2): 2-3.

Kim, H.S. & D.H. Kwon 1988. Bopyrid isopods parasitic on decapod crustaceans in Korea. *Korean J. Syst. Zool. Spec. Issue* 2: 199-221.

Kozloff, E.N. 1983. *Seashore life of the northern Pacific coast. An illustrated guide to northern California, Oregon, Washington and British Columbia.* Seattle: Univ. Washington Press. 370 pp.

Lemos de Castro, A. 1970. Crustáceos isópodos epicarídeos do Brasil. V. Duas espécies novas de hiperparasitas pertencentes as gênero *Cabirops* Kossmann (Isopoda, Cabiropsidae). *Bolet. Mus. Nacion., Rio de Janiero (Nova Sér.) Zool.* 277: 1-7.

Lemos de Castro, A. & I.M. Brasil Lima 1975. Crustáceos isópodos epicarídeos do Brasil. VII. Descrição de uma espece nova de *Parione* Richardson. *Atas Soc. Biol. Rio de Janiero* 17(3): 101-103.

Lemos de Castro, A. & I.M. Brasil Lima 1980. Crustáceos isópodos epicarídeos do Brasil. 10. Descrição de *Hemicepon mülleri* g. n., sp. n. e ocorrência de *Grapsicepon edwardsi* Giard & Bonnier, em águas brasilieras. *Rev. Bras. Biol.* 40: 115-119.

LeSueur, R.F. 1954. The Cumacea, Mysidacea, Aphipoda, Isopoda and Tanaidacea of the Channel Islands. *Bull. Soc. Jerseyaise* 16: 207-216.

Markham, J.C. 1977. The status and systematic position of the species of the bopyrid isopod genus *Phyllodurus* Stimpson, 1857. *Proc. Biol. Soc. Washington* 90: 813-818.

Markham, J.C. 1982. Bopyrid isopods parasitic on decapod crustaceans in Hong Kong and southern China. In Morton, B.S. & C.K. Tseng (eds), *Proceedings of the First International Marine Biological Workshop. The Marine Flora and Fauna of Hong Kong and Southern China* 1:325-391. Hong Kong: Hong Kong University Press.

Markham, J.C. 1985. Additions to the bopyrid fauna of Thailand. *Zool. Verh.* 224: 1-63.

Markham, J.C. 1986. Evolution and biogeography of the Isopoda Bopyridae, parasites of Crustacea Decapoda. *Crustacean Issues* 4: 143-164.

Markham, J.C. 1988. Descriptions and revisions of some species of Isopoda Bopyridae of the north western Atlantic Ocean. *Zool. Verh.* 246: 1-63.

Markham, J.C. 1990. Crustacea Isopoda: new records of Bopyridae from New Caledonian waters. *Resultats des Campagnes MUSORSTOM Mem. Mus. Natl. Hist. Paris.* 6(A) (145): 55-69.

Markham, J.C. 1992. Second list of additions to the Isopoda Bopyridae of Hong Kong. Pages 277-302, In Morton, B. (ed.), *The marine flora and fauna of Hong Kong and southern China 3. Volume 1: introduction, taxonomy and ecology. Proceedings of the Fourth International Marine Biological workshop: The Marine Flora and Fauna of Hong Kong and Southern China, 11-29 April 1989.* Hong Kong: Hong Kong University Press.

Markham, J.C. 1994. Crustacea Isopoda: Bopyridae in the MUSORSTOM collections from the tropical Indo-Pacific I. Subfamilies Pseudioninae (in part), Argeiinae, Orbioninae, Athelginae and Entophilinae. In Crosnier, A. (ed.), *Resultats des Campagnes MUSORSTOM, Vol. 10. Mem. Mus. Natl Hist. Nat. Paris* 161: 225-253.

Markham, J.C. 1995. Two new species of Isopoda Bopyridae (Crustacea) infesting thalassinideans in the western Pacific. *Raffles Bull. Zool.* 43: 83-89.

Muñoz, G.I. 1997. Primer registro de isópodos bopyridos (Isopoda: Epicaridea) en el nape *Notiax brachyophthalma* (M. Edwards, 1870) y algunos aspectos de la relación hospedador-parásitico. *Gayana Ocean.* 5: 33-39.

Nierstrasz, H.F. & G.A. Brender à Brandis 1923. Die Isopoden der Siboga-Expedition. II. Isopoda Genuina. I. Epicaridea. *Siboga Expeditie Monogr.* 32b:57-121.

Nierstrasz, H.F. & G.A. Brender à Brandis 1929. Papers from Dr Th. Mortensen's Pacific Expedition 1914-16. 48. Epicaridea 1. *Vidensk. Medd.. Dansk Naturh. Foren. Kjobenhavn* 87: 1-44.

Nierstrasz, H.F. & G.A. Brender à Brandis 1931. Papers from Dr. Th. Mortensen's Pacific Expedition 1914-16. 57. Epicaridea 2. *Vidensk. Medd. Dansk Naturh. Foren. Kjobenhavn* 91: 147-225.

Nobili, G. 1906. Nuovi Bopiridi. *Atti Real Accad. Sci. Torino* 41:1-18.

Oguro, C. 1957. Notes on *Upogebia major* infested with an epicaridean parasite. *Annot. Zool. Japon.* 30: 31-32.

Olaosebikan, B.D. 1986. Studies on the parasitic isopod crustaceans of the families Cymothoidae and Bopyridae in the Bonny River. Bachelor of Science Dissertation, Department of Zoology, University of Port Harcourt, Nigeria. 44 pp.

Page, R.D.M. 1985. Review of the New Zealand Bopyridae (Crustacea: Isopoda: Epicaridea). *New Zeal. J. Zool.* 12: 185-212.

Popov, V.K. 1927. [Rhizocephala and Bopyridae of the Bay of Sevastopol]. *Trudy Sevastopol'skoy Biologicheskoy Stantsii Akademii Nauk SSSR* 1: 1-26. [In Russian]

Restivo, F. 1970. A new species of Pseudione (Pseudione reverberii) a parasite of Callianassa truncata. Pubbl. Staz. Zool. Napoli 38: 305-315.

Salazar-Vallejo, S.I & A. Leija-Tristán 1989. *Progebiophilus bruscai* n. sp. a new bopyrid isopod parasitic on the mud shrimp *Upogebia dawsoni* Williams (Thalassinidea) from the Gulf of California. *Cah. Biol. Mar.* 30: 423- 432.

Shiino, S.M. 1937. Bopyrids from Tanabe Bay, IV. *Mem. Coll. Sci., Kyoto Imp. Univ.* (B) 12: 479-493.

Shiino, S.M. 1939a. Bopyrids from Kyûsyû and Ryûkyû. *Recs. Ocean. Works Japan* 10: 79-99.

Shiino, S.M. 1939b. Bopyrids from Tanabe Bay, V. *Annot. Zool. Japon.* 18: 11-16.

Shiino, S.M. 1958. Note on the bopyrid fauna of Japan. *Rep. Fac. Fisheries, Pref. Univ. Mie* 3: 29-73.

Shiino, S.M. 1964. On two species of bopyrid isopods parasitic on *Callianassa uncinata* Milne-Edwards from Chile. *Rep. Fac. Fisheries, Pref. Univ. Mie* 5: 27- 32.

Shiino, S.M. 1965. Phylogeny of the genera within the family Bopyridae. *Bull. Mus. Natl. Hist. Nat, Paris* (2)37: 462-465.

Shiino, S.M. 1972. [The Epicaridea (list of species) from Japan.] *Kansai Shizenkagaku* 24: 7-10 [In Japanese]

Stuardo, J., R. Vega & I. Cespedes 1986. New bopyrid isopod parasitic on *Callianassa uncinata* H. Milne Edwards with functional and ecological remarks. *Guyana, Zool.* 50(1-4): 17-36.

Van Name, W.G. 1920. Isopods collected by the American Museum Congo Expedition. *Bull. Amer. Mus. Nat. Hist.* 43: 42-108.

Williams. A.B. 1993. Mud shrimps, Upogebiidae, from the western Atlantic (Crustacea: Decapoda: Thalassinidea). *Smiths. Contr. Zool.* 544: 1-77.

Biogeography of the marine Isopoda of the Indian Ocean, with a check-list of species and records

BRIAN KENSLEY
Department of invertebrate Zoology, National Museum of Natural History, Smithsonian Institution, Washington, D.C., USA

ABSTRACT

The marine isopod fauna of the Indian Ocean is reviewed in terms of its species diversity and biogeography. For the purposes of the review, the northeastern boundary of the Indian Ocean is defined at Cape Talpot in northern Western Australia and at Rottnest Island/ Perth in southern Western Australia. The history of isopod systematics in the Indian Ocean is reviewed by region, and the great lack of knowledge for Madagascar, parts of Western Australia, the east coast of Africa, and the continental shelf/slope overall, is noted. Previous research on the biogeographical provinces or regions of the Indian Ocean is briefly reviewed. For some groups of organisms, an impression of overall homogeneity for the tropical-subtropical area may prevail. However, close examination dispels the idea of homogeneity, and a number of sub-provinces or regions can be characterized. For a biogeographic review of the Indian Ocean marine isopods, a species list was compiled, which includes as many records for each species as could be found. This list contains approximately 1000 species in 303 genera. About 84% of the species are endemic to the Indian Ocean overall, but only 18% of the genera, a pattern consistent with the geological history especially of the Tethys Sea and the evolution of the Indian Ocean. This pattern is also consistent for a benthic-dwelling group which lacks a pelagic stage and whose juveniles emerge from the brood-pouch as small versions of the adults. More than half the species are known from only a single record. Seven sub-provinces or regions are characterized in terms of their isopod fauna: 1. Sub-Antarctic Region, 79% of 135 species are endemic; 2. South African Region, 68% of 226 species are endemic; 3. Western Australia Region, 58% of 115 species are endemic; 4. East African Region, 46% of 144 species are endemic; 5. Red Sea Region, 63% of 105 species are endemic; 6. Indian Region, 62% of 268 species are endemic; and 7. Madagascar Region, 63% of 168 species are endemic. These levels of endemism are higher than those seen for most organisms for which data exist.

1 INTRODUCTION

The identification of large collections of isopod material from the granitic Seychelles, Aldabra Atoll, scattered collections from several sites in the Indian Ocean in the collections of the National Museum of Natural History, as well as material from Zanzibar Island

and Phuket, Thailand, prompted the compilation of a list of marine isopod species recorded from the Indian Ocean with an accompanying bibliography. Once the list was compiled, it became obvious that at least a preliminary biogeographic analysis of the marine Isopoda for the entire region was possible and perhaps desirable, given the very few published biogeographical reviews for invertebrate groups from the Indian Ocean. In the interest of brevity, new records, represented by material in the NMNH and ZMUC collections, are marked by an asterisk, but are not further elaborated upon, as no section on Systematics is included here.

2 GEOGRAPHIC LIMITS OF THE INDIAN OCEAN

The geographic limits of the Indian Ocean on its northern, western and southwestern margins are not in dispute, being the Indian subcontinent, the eastern coast of Africa, and the eastern coast of South Africa to Cape Agulhas respectively. With the southern margin being open to the Antarctic, the Indian Ocean embraces the southern ocean from south of Africa eastwards to south of Australia, and includes the sub-Antarctic islands of Marion and Prince Edward, Crozet, Kerguelen, MacDonald, Shag and Heard, and St. Paul and Amsterdam. The eastern boundaries, however, are less well defined. Where on the Australian continent the line of demarcation should be drawn, is debatable. Similarly, where the northeastern boundary should be placed in the maze of Indonesian Islands and the northern Australian continent, in uncertain.

The Oceanographic Atlas of the International Indian Ocean Expedition (IIOE) (Wyrtki 1971) sets the eastern boundary at 147°E, the longitude of the South Cape of Tasmania in the south of Australia, and Cape York in the north. This northeastern boundary would encompass the Timor and Arafura seas, all of the coast of the Northern Territory, and the Gulf of Carpentaria. While for oceanographic purposes, these boundaries may be logical, for biologists a more restricted eastern boundary has been used, given the geographic and zoogeographic complexity of the Indonesian region.

Cohen (1973), in a comprehensive discussion of the zoogeography of the fishes of the Indian Ocean, stated in the caption to his Figure 1: 'Cape Talpot about 127°E on the coast of Western Australia is taken as the demarcation between the eastern Indian ocean and the western Pacific.' Klausewitz (1978:25), also discussing fish biogeography (families Chaetodontidae and Pomacanthidae) in the Indian Ocean, stated: 'The border runs west of the Malayan Peninsula and Sumatra and even west of the Andamans, Nicobars and the series of islands – Simueloe-Nias-Mentawei-Nassau.' In this study, Cape Talpot is taken as the limit of the Indian Ocean in northwestern Australia.

The eastern boundary in southern Australia is also problematic. The IIOE limits would include much of South Australia, Victoria, and southern Western Australia, all of which form part of a temperate biogeographic region characterized by high endemism, and having affinities with the rest of Australia, rather than with the subtropical-tropical Indian Ocean. This high endemism is well illustrated in several papers on the isopods of southern Australia (Cohen & Poore 1994, Poore 1987, 1994, Poore & Bardsley 1992, Poore & Lew Ton 1988a, 1988b). Wells (1993), basing his conclusions on a broad floral and faunal survey, noted that the area of Rottnest Island and Perth, Western Australia, lies in the overlap region between the southern temperate and northern tropical regions, and that the maritime fauna is dominated by warm temperate southern Australian species. The Leeuwin

Current, flowing south off Western Australia, carries the larvae of tropical species, some of which eventually settle on the outer continental shelf, but not in-shore where the warm-temperate South Australian species dominate. For this study, the limit of the Indian Ocean is regarded as broadly in the region of Rottnest Island and Perth, Western Australia.

3 HISTORY OF ISOPOD COLLECTION AND RESEARCH IN THE INDIAN OCEAN

Information on the marine isopod fauna of the Indian Ocean is scattered through a host of publications, but there exists no single work that brings together all that is known for the region. There are several isopod works dating from the 19th Century, e.g., Krauss (1843), on southern African isopods, and Beddard (1886), dealing with Challenger Expedition material. The first really substantive works were those of Stebbing (1904a), who contributed a paper on the isopods of the Maldive and Laccadive archipelagoes, in 1905 reported on the isopods of the Gulf of Manaar (India-Sri Lanka), and in 1910a described the isopods collected by the Percy Sladen Trust Expedition to the Indian Ocean, carried out in 1905.

Since Stebbing's work, isopods from various parts of the Indian Ocean have received attention. Although Krauss (1843) was the first to report on the isopods of the southern African region, T.R.R. Stebbing and K.H. Barnard described the bulk of the shallow-water fauna, while Kensley (1975a, 1978a, 1984b) documented portions of the deeper shelf fauna on the southeastern coast, an area dominated by the southward-flowing Agulhas Current. Barnard also contributed to the knowledge of the fauna of Madagascar (1960), Mozambique (1926, 1958), and India (1935, 1936).

The East African coast from Mozambique to Somalia has been relatively poorly sampled. Holdich & Jones (1973), N.L. Bruce (1981), Jones & Icely (1981), and H.-G. Müller (1989a) have described isopods from this region, while N.L. Bruce & Jones (1978) and Wägele (1981a, 1981b) contributed short papers on the Red Sea fauna. Madagascar with its extensive coastline is still poorly documented; Roman (1970, 1974, 1979) produced three reports, but much of the coastline remains to be sampled. Bruce (1997b) lists and illustrates 27 species from East Africa and the western Indian Ocean islands.

The northern areas of the Indian Ocean have received some attention: Javed and Ahmed (1987, 1988a, 1988b) have reported on material from Pakistan, Pillai (1954-1967) on India, Müller (1991a, 1991d, 1991e, 1991f, 1991g, 1991h) on Sri Lanka. Müller (1990a, 1990b, 1991b, 1991e) also documented isopods from Réunion Island, but for neighboring Mauritius there are still only records of 19 species.

Western Australia too, is still poorly collected. Baker (1908, 1910), Thomson (1946, 1951), Poore (1975, 1980, 1987), Kensley & Poore (1982), and N.L. Bruce (1997) have all added to the knowledge of the area, but the northwestern region especially, is largely unexplored below the intertidal. Jones & Morgan (1993) in a checklist of the fauna and flora of Rottnest Island list 18 species of marine isopods.

The sub-Antarctic islands of the Indian Ocean, often explored as an extension to research carried out in the Antarctic, have consequently received some attention: St. Paul and Amsterdam Islands (Brocchi 1875, Kensley 1976, 1989); Crozet Islands (Kensley 1980b); Kerguelen Island (Vanhöffen 1914, Carvacho 1977, Cleret 1973, Kussakin &

Vasina 1980, 1982a, 1982b); Marion and Prince Edward Islands (Cleret 1971, Branch et al. 1991).

Outside of the Antarctic/sub-Antarctic region, the continental shelf has been sampled in very few areas: on the Indian Ocean coast of South Africa, a series of cruises revealed a largely endemic fauna with occasional links to the sub-Antarctic (Kensley 1975a, 1978a, 1978b, 1978f, 1984). Aside from South Africa and a few early stations of the Challenger Expedition only five reports of deep-sea isopods could be found: Robinson & Menzies (1961), Gulf of Aden; Müller (1989), Gulf of Aden; Mezhov (1988, 1989), Arabian Sea; Hobbins & Jones (1993), Red Sea. Thus continental shelf records of isopods have very little bearing on our understanding of the biogeography. Neyman et al. (1973) suggest in some general remarks on the biogeography of benthic invertebrates of the Indian Ocean that up to 50% of the deep sea invertebrates of the northern Indian Ocean are common to the Pacific Ocean, while in the southern Indian Ocean, almost 80% of the benthos are Antarctic forms. Unfortunately, data on specific groups of organisms are not offered to support these views. The present analysis is based almost entirely on shallow-water records.

4 BIOGEOGRAPHIC PROVINCES OF THE INDIAN OCEAN

Ekman (1967) dealt with the Indian Ocean in the broadest terms, speaking of the major oceanographic expeditions such as the John Murray Expedition of 1933-1934, the Danish expeditions of 1937 and 1938 to the Persian Gulf, and referred fleetingly to the endemic crinoid and decapod crustacean fauna. Briggs (1974:14) addressed the diversity of the tropical Indo-West Pacific, which he considered to be the most important and diverse marine center worldwide, and noted that while there is a basic homogeneity caused by many wide-ranging species, there are great differences in diversity among the various parts of the region. This basic homogeneity, however, is only apparent in some organisms, such as decapod crustaceans. For groups such as the isopods, a pattern of localized endemism, with very few wide-ranging species, is the norm.

Several individual groups of animals have received some attention from a zoogeographic point of view. Klausewitz (1972, 1976) recognized eastern and western subprovinces of the Indo-West Pacific region, with the border between them being a line through the Maldives-Laccadives-Chagos islands, and running south almost to Mauritius. In 1978, he noted several important points regarding the zoogeography of littoral fishes in the Indian Ocean, including a decrease in the number of species form east to west, and that the turbidity of the water from rivers along the coasts of India acted as a barrier and led to an absence of reefs. He also noted an increase in the number of endemic species in isolated or peripheral areas. He concluded that: 1. The Indian Ocean was not a homogeneous area, 2. The border between the Malaysian and Indian Provinces was along the middle of the main islands of the Indo-Australian Archipelago, and 3. The Madagascar/Mascarene complex formed a separate sub-province, as did the Red Sea, Gulf of Aden and Persian Gulf.

5 MARINE ISOPOD DISTRIBUTION IN RELATION TO THE BIOGEOGRAPHIC PROVINCES OF THE INDIAN OCEAN: DIVERSITY, ENDEMISM, AFFINITIES

A breakdown of the approximately 1000 species of marine isopods in the attached list is given in Table 1, while a breakdown by biogeographic region is summarized in the Table 2.

From a global point of view, the Indian Ocean may be a single basin, but its different parts have different histories, and this is reflected in the differences in their faunal composition.

The present distribution of groups such as the isopods (and probably all of the Peracarida) is probably the result of the formation and breakup of the Tethys Sea. The dismembering of Pangea in the Mesozoic into Laurasia and Gondwana formed the broad band of the Tethys Sea, which included the current Indo-West Pacific oceanic region. The Tethys was circum-global from the Jurassic. In the Triassic (200-180 m.y.a.) the breakup of Gondwana began, with Africa, India and Antarctica separating; with the Jurassic, the Indian Ocean opened up even further. By the end of the Cretaceous, Madagascar had separated, and the Indian plate was moving still further north (see Hocutt 1987). With major cooling and reduction of the area of shallow seas during the Cretaceous and Cenozoic, the wide distribution of reefs was reduced by subduction along plate edges. The earlier spread of species over wide areas was fractured by these geological events, leaving isolated allopatric populations to speciate yet further. During the early Cenozoic, the marine invertebrate fauna was probably fairly uniform throughout the Tethys Sea (Hocutt 1987), but the northward drift of the Indian plate and its collision with Asia closed the Indo-Mediterranea-Atlantic seaway by the early Miocene. This evolution of the Indian Ocean and the drift of the Indian plate would explain the patterns of isopod endemism of the region today, as well as the wide distribution of many isopod families and genera.

It is no surprise that the southern cold-water province is very different from the rest of the Indian Ocean, nor that both the South African and Australian regions, with their tem-

Table 1. Composition of the Indian Ocean marine Isopoda.

Number of isopod families in the Indian Ocean (none endemic)	44	
Number of genera	303	
Number of endemic genera	53	17.5%
Number of species	998	
Anthuridea	135	
Asellota	163	
Epicaridea	138	
Flabellifera	442	
Gnathiidea	29	
Microcerberidea	7	
Valvifera	84	
Number of endemic species (i.e., confined to Indian Ocean)	835	83.6%
Number of species from > 200 m	93	
Number of single records > 200 m	59	63.4%
Number of species < 200 m	905	
Number of single records < 200 m	508	56.1%

perate components, should also display major differences. It is in the subtropical-tropical region, which is the major part of the Indian Ocean, with its coral reefs distributed roughly within the 20°C surface water isotherm, that we might expect greater uniformity, or at least some overarching similarities. In fact, we find that the tropical-subtropical Indian Ocean can be divided into several more or less distinct areas or sub-provinces (as was shown by Klausewitz 1972), as reflected in Table 2, and especially in the high numbers of endemic species in the various regions.

5.1 *Patterns of endemism*

Some scattered information is available on endemism for several groups of animals in the Indian Ocean, which can be used for comparison with the Isopoda.

Coelenterates: Most of the Indian Ocean corals are of Mesozoic Tethys origin (Hocutt 1987), with most families and genera being widespread by the mid-Tertiary, but the western Indian Ocean may constitute a more recent and separate sub-province (Rosen 1971). Briggs (1974) suggested that 15% of the Red Sea corals are endemic.

Molluscs: Hocutt (1987) suggested that molluscs were distributed pan-tropically throughout the Tethys Sea by the early Tertiary, and that later temperature fluctuations re-

Table 2. Distribution of Indian isopods by zoogeographic area. Sub-Antarctic/Antarctic region: includes Marion and Prince Edward, Crozet, Kerguelen, St. Paul, Amsterdam, Heard, MacDonald, and Shag islands. South African region: from Cape Agulhas to northern Natal. Western Australia: includes coast of Western Australia north of Perth/Rottnest Island, and northwestern shelf to Cape Talpot. East African region: includes coastline from southern Mozambique to Gulf of Aden, and Zanzibar. Red Sea region: includes Red Sea, Gulf of Aden, Persian Gulf, and Gulf of Oman. Indian Region: includes Pakistan, India, Myanmar, Arabian Sea, Bay of Bengal, Sri Lanka, Maldive Islands, Lakshadweep Islands (Laccadives), Gulf of Manaar, and Andaman Sea. Madagascan Region: includes Madagascar, Mauritius, Réunion, Comoros, Seychelles, Aldabra, and Cargados Carajos.

Region	Number of species	Number of endemic species	% endemic species	Number of species > 200 m.	Species in common
sub-Antarctic	135	106	79%	32	18 spp. with S. Africa 10 spp. with Australia
South Africa	226	153	68%	32	25 spp. with East Africa 8 spp. with Australia
Western Australia.	115	67	58%	3	39 spp. with rest of Australia
East Africa	144	66	46%	6	21 spp. with Madagascar 26 sp. with Indian region 12 spp. with Red Sea
Red Sea	105	67	63%	14	9 spp. with Madagascar 16 spp. with Indian region
Indian Region	268	166	62%	4	27 spp. with Madagascar 11 spp. with Red Sea 37 spp. with eastern Indo-Pacific
Madagascan Region	168	106	63%	0	

sulted in a decline in diversity. Briggs (1974) referred to a high degree of endemism for some mollusc groups along the east African coast, the Arabian Peninsula, and the Red Sea.

Crustaceans: Abele (1982) recognized a 21% endemism among the Indo-West Pacific decapods, and 96% for the Hoplocarida, but this very broad geographic area represents an overlap of several of the biogeographic regions of the Indian Ocean. Kurian (1973) provided no biogeographic information on the relatively few samples of Cumacea available from the Indian Ocean.

Bryozoans: Hayward (1983) found a high degree of endemism (65%) off South Africa in the genus *Adeonella*, with 25% restricted to the Indo-West Pacific.

Echinoderms: Briggs (1974) estimated a 24% endemism for the western Indian Ocean, 15% for the Red Sea, and 49% for northwestern Australia.

Chaetognaths: Nair & Rao (1973) suggested that 47% of the epiplanktonic species occurring in the Arabian Sea were endemic.

Fish: Briggs (1974) estimated a 10-15% endemism for marine fishes in the Red Sea, 36-42% for the Western Indian Ocean, while Hocutt (1987) recognized a 14.3% endemism for fishes on the southern African coast.

When these numbers are compared with those for the marine isopods (see Tables 1 and 2), we note for the latter relatively higher levels of endemism, ranging from 46-79%, for the various biogeographic regions of the Indian Ocean. For example, the Red Sea region shows a 15% endemism for corals, fishes, and echinoderms, while isopod endemism is a much higher 63%. These high levels of species endemism for the Indian Ocean isopods are in line with a pattern seen elsewhere, especially with regard to coral reef fauna (Kensley 1998), namely, widespread families and genera (only 17.5% endemic genera in the Indian Ocean), but high local endemism at the species level (83.6% for the overall Indian Ocean, 46-79% for individual biogeographic regions).

5.2 *Widespread species*

From the Species List, it can be seen that there are a number of widespread species, whose distribution may, to some degree, be explained by reference to their lifestyle or habitat preference. Thus some species of Limnoriidae and *Sphaeroma* are wood borers. Their distribution could be complicated by, or in part explained, by their presence in drifting and floating wood. Some species are either fish parasites (all of the Cymothoidae; see Brusca 1981) or micro-predators of fish (several of the Aegidae; see Brusca 1983). Their distribution is closely linked to the distribution of their fish hosts. In some cases, these fish may have very wide distributions throughout the Indo-West Pacific region. Some isopod species (Epicaridea) are parasites of other crustaceans, especially of the Decapoda. Their distribution may reflect the distribution of their crustacean hosts, especially if these latter have some capability of dispersal as adults, i.e., by transporting ovigerous females of the parasite. Some species may have been introduced, e.g., *Paradella dianae* (see Javed & Ahmed 1987) and *Paracerceis sculpta* (see Bruce 1986c) by being transported in the fouling organisms of ships, or, as with *Carpias algicola* (Morris & Mogelberg 1973, Kensley, in press), transported by rafting in floating algae.

6 SPECIES LIST OF MARINE ISOPODS OF THE INDIAN OCEAN

As a basis for a biogeographical discussion, and as a resource for future research on Indian Ocean isopods, it was felt that a list of all isopod records for the region would be a useful tool, and has thus been attempted.

The list has been constructed in the following way: suborders, families within suborders, genera within families, and species within genera are listed alphabetically. For each species, author, date, and type locality are provided. The first geographic record following the author and date, is the type locality. In a few cases, the type locality is unknown. Additional records, including non-Indian Ocean records, with depths where available, are listed after the initial entry, along with author and date. New records are indicated with an asterisk (*). Where known, the host for epicaridean parasites is provided. In several cases, Indian Ocean records have been given specific names of European or Mediterranean species. These cases are indicated by the use of 'non' followed by the original author, to indicate a probable misidentification. While the list is primarily of marine species, a few freshwater records of species with obvious marine affinities (e.g., species of *Annina*) have been included.

ORDER ISOPODA
SUBORDER ANTHURIDEA
Family Antheluridae

Ananthura elegans (Kussakin, 1967). Kerguelen Is.; McDonald Is.

Anthomuda cracens (Kensley, 1980a). Mozambique; Somalia, 42-61 m.

Anthomuda hapla (Kensley, 1980a). Mauritius; Mozambique, 110 m. * Mahé Is., Seychelles, 1 m.

Anthomuda hovea Poore & Lew Ton, 1988a. Western Australia; Great Barrier Reef, intertidal. 18 m. Réunion Is., 0.5-1 m (Müller 1990a).

Anthomuda poorei Müller, 1990a. Réunion Is., 0.5-1 m.

* *Anthomuda* sp. nov. Aldabra Atoll, intertidal 25 m.

Family Anthuridae

Alloanthura sculpta Kensley, 1980a. Mozambique, 135 m.

Amakusanthura africana (Barnard, 1914b). Saldanha Bay, South Africa, 146 m. Namibia to Jeffreys Bay, South Africa, 11-200 m (Kensley, 1982).

Amakusanthura agonis Poore & Lew Ton, 1988c. Northwestern shelf; Queensland, 5-42 m.

Amakusanthura copperingi (Barnard, 1925a). Northern Territory, Australia, 31 m. NW Shelf, Australia, 40 m (Poore & Lew Ton, 1988c).

* *Amakusanthura cosmoledo* sp. nov. Cosmoledo Is., Aldabra Atoll, 1-22 m.

Amakusanthura dubia Barnard, 1914b. False Bay, South Africa, intertidal. Saldanha Bay to Transkei, South Africa, intertidal-200 m (Kensley 1982).

Amakusanthura hibbertia Poore & Lew Ton, 1988c. Great Barrier Reef, 4 m; NW Shelf, Australia, 52 m.

Amakusanthura moragallae Müller, 1991a. Sri Lanka, intertidal.

Amakusanthura pori (Wägele, 1981b). Red Sea.

Amakusanthura stocki Müller, 1991a. Sri Lanka.

Apanthura addui Wägele, 1981b. Maldive Is.

Apanthura insignifica Kensley, 1978c. East London to Natal, South Africa, 90-850 m.
Apanthura microps Kensley, 1980a. E Arabian Sea; off Somalia & Mozambique, 42-95 m.
Apanthura motasi Negoescu, 1980. Gulf of Aden.
Apanthura pultenaea Poore & Lew Ton, 1988c. NW Shelf, Australia, 39-82 m; Queensland, 11-34 m.
Apanthura sandalensis Stebbing, 1900. Loyalty Is. Travancore, India (Barnard 1935). Chilka Lake, India (Chilton 1924). Off Somalia; Mozambique; Mauritius; Sumatra, intertidal 62 m (Kensley 1980a). Madagascar (Roman 1970).
Apanthura stanjeki Wägele, 1981b. Red Sea, Gulf of Suez.
Apanthura wudu Müller, 1990a. Réunion Is., 0.5-1.5 m.
Apanthura xenocheir Stebbing, 1910a. Egmont Reef, Chagos Archipelago.
Apanthura zeewykae Kensley & Poore, 1982. Houtman Abrolhos Is., Western Australia, 1.3 m.
* *Apanthuroides* sp. nov. A. Aldabra Atoll, intertidal 10 m.
* *Apanthuroides* sp. nov. B. Aldabra Atoll, intertidal 6 m; Mahé Is., Seychelles, 8 m.
Apanthuroides foveolata (Kensley 1978c). Off Natal, South Africa, 550-850 m.
Arabanthura enigmatica Kensley & Reid, 1984. Persian Gulf, 1.3-3 m.
Caenanthura indica Negoescu, 1980. Gulf of Oman; Djibouti.
Centranthura caeca (Kensley 1975c). Lambert's Bay to Mossel Bay, South Africa, 5-68 m.
Chelanthura ajuga Poore & Bardsley, 1990. Western Australia; Queensland, Australia, 1.3 m.
Cyathura bentotae Müller, 1991f. Sri Lanka, intertidal 0.5 m.
Cyathura estuaria Barnard, 1914b. Algoa Bay, East London, South Africa, estuarine-intertidal. Langebaan to Natal, South Africa (Kensley 1982).
Cyathura francispori Negoescu, 1981. Suez Canal.
Cyathura indica Barnard, 1925a. Sri Lanka; Singapore; Thailand, 1-26 m. Quilon, Travancore, India (Barnard 1935).
Cyathura milloti Chappuis, Delamare Deboutteville & Paulian, 1956. Réunion Is., interstitial.
Cyathura pusilla Stebbing, 1904b. Sri Lanka; Tanzania. Travancore, India (Pillai 1954).
Cyathura rudloei Kensley, 1980a. Mozambique; Madagascar, intertidal-73 m.
Cyathura terryae Müller, 1990a. Réunion Is., 0.5-1 m.
Exallanthura sexpes Kensley, 1980a. Off Somalia, 49 m; off Singapore, intertidal.
Expanathura ardea (Poore & Kensley, 1981). Great Barrier Reef, intertidal. Houtman Abrolhos Is., Western Australia, 3-5 m (Kensley & Poore 1982).
Expanathura haddae (Kensley & Poore, 1982). Houtman Abrolhas Is., Western Australia, 1-3 m.
Expanathura macronesia (Kensley, 1980a). Madagascar; off Mauritius, intertidal-1.5 m. * Aldabra Atoll, intertidal-16 m. * Mahé Is., Seychelles, intertidal-8 m. * Comoro Is., 3-5 m.
Haliophasma adinae (Negoescu, 1980). Gulf of Oman; Gulf of Aden, 24-110 m.
Haliophasma austroafricana Kensley, 1982. Saldanha Bay to Natal, South Africa, 19-620 m.
Haliophasma beaufortia Poore & Lew Ton, 1988d. Western Australia, NW Shelf, 31-134 m.

Haliophasma blandfordia Poore & Lew Ton, 1988d. Western Australia, NW Shelf, 31-122 m; Queensland, 6-45 m; Bass Strait, 51-329 m.

Haliophasma darwinia Poore & Lew Ton, 1988d. Queensland; NW Shelf, Australia, 24-37 m.

Haliophasma dillwynia Poore & Lew Ton, 1988d. Western Australia; NW Shelf 122-134 m.

Haliophasma palmatum Wägele, 1981b. Red Sea, 1 m.

Haliophasma poorei Kensley, 1980a. Off Bombay, India, 55 m; off Somalia, 95 m. Sri Lanka, intertidal-0.5 m (Müller 1991a).

Haliophasma tricarinata Barnard, 1925a. Agulhas Bank, South Africa, 80 m. Saldanha Bay to East London, South Africa, 48-183 m (Kensley 1982).

* *Heptanthura* sp. nov. Phuket Is., Thailand, intertidal.

Indanthura carinata Pillai & Eapen, 1966. West coast of India, 36 m.

* *Malacanthura* sp. nov. Persian Gulf, 1.5-7 m.

Malacanthura coronicauda (Barnard, 1925a). Saldanha Bay, South Africa, 174 m. Saldanha Bay to Agulhas Bank, South Africa, 26-174 m (Kensley 1982).

Malacanthura foveolata (Barnard, 1940). Algoa Bay, South Africa, intertidal. Saldanha Bay to Algoa Bay, South Africa, intertidal to 148 m (Kensley 1982).

Malacanthura linguicauda (Barnard, 1920). Off Natal, South Africa, 190 m.

Malacanthura mombasa Kensley, 1980a. Mombasa, Kenya 3.5 m.

Malacanthura pseudocarinata (Barnard, 1940). Algoa Bay, South Africa, intertidal. Saldanha Bay to Port Elizabeth, intertidal to 4 m (Kensley 1982).

Malacanthura schotteae Kensley, 1982. Still Bay to Transkei, South Africa, 80-775 m.

Malacanthura serenasinus (Kensley, 1975a). Saldanha Bay to Still Bay, 5-125 m.

Malacanthura transkei Kensley, 1982. Transkei, South Africa, 710-775 m.

Mesanthura albinotata Thomson, 1951. Western Australia, sublittoral.

Mesanthura bipunctata Thomson, 1951. Western Australia, shallow.

Mesanthura catenula (Stimpson, 1855). False Bay, South Africa. Table Bay to East London, South Africa, intertidal to 4 m (Kensley 1982).

Mesanthura dimorpha Kensley, 1982. Natal to Algoa Bay, South Africa, 84-90 m.

Mesanthura gerlachi Wägele, 1981b. Maldive Is., 0-4 m.

Mesanthura maculata (non Haswell, 1881). Travancore, India (Pillai 1954). Sri Lanka; New Zealand (Pillai 1966). Madagascar (Roman 1970).

Mesanthura nigra Müller, 1993. Kenya, 0.5 m.

Mesanthura protei Kensley, 1980a. Madagascar; Mozambique, intertidal-73 m. Houtman Abrolhos Is., Western Australia, 1.3 m (Kensley & Poore 1982). Malaysia; Kenya, intertidal-1 m (Müller 1993). * Phuket, Thailand, 1 m.

Mesanthura pulchra non Barnard, 1925a. Madagascar (Roman 1970).

* *Mesanthura* sp. nov. Mahé Is., Seychelles, intertidal-1 m.

Minyanthura firingae Müller, 1990a. Réunion Is., 0.5-1 m.

Panathura amstelodami Kensley, 1976. St. Paul & Amsterdam Is., upper infralittoral to 80 m. Madagascar; Mozambique, 38-62 m (Kensley 1980a). Natal, South Africa to Mozambique, 710-775 m (Kensley 1982).

Panathura serricauda (Barnard, 1920). Table Bay, False Bay, South Africa, intertidal. Vatta Kayal, Alleppey, India (Barnard 1935). St. Paul & Amsterdam Is., upper infratidal (Kensley 1976). Lüderitz, Namibia to East London, South Africa, intertidal – 90 m (Kensley 1982).

* *Pendanthura* sp. nov. A. Aldabra Atoll, 0.5-18 m.
* *Pendanthura* sp. nov. B. Aldabra Atoll, intertidal-5 m.
* *Pendanthura* sp. nov. C. Thailand, 13-18 m.
* *Quantanthura* sp. nov. Phuket Is., Thailand.
Quantanthura remipes (Barnard, 1914b). Off Cape Peninsula, South Africa, 312 m. Lambert's Bay to Agulhas Bank, South Africa, 78-312 m (Kensley 1982).

Family Hyssuridae
* *Eisothistos* sp. nov. A. Thailand.
* *Eisothistos* sp. nov. B. Aldabra Atoll, intertidal.
Eisothistos anomala (Kensley ,1980a). Madagascar, 1-1.5 m.
Eisothistos antarcticus Vanhöffen, 1914. Antarctic.
Eisothistos bataviae Kensley & Poore, 1982. Houtman Abrolhos Is., Western Australia, 1.3 m.
Eisothistos crateris Kensley, 1976. St. Paul Is., upper infratidal.
Eisothistos maledivensis Wägele, 1979. Maldive Is., 35 m.
Eisothistos neoanomalus Shyamasundari, Kumari, Rao & Mary, 1991. Visakhapatnam, India, intertidal.
Eisothistos vermiformis non Haswell, 1884. Madagascar (Roman 1970).
* *Kupellonura* sp. nov. Aldabra Atoll, intertidal-26 m.
* *Neohyssura* sp. nov. Aldabra Atoll, Madagascar, 1-25 m.
Neohyssura skolops Kensley, 1978c. East London to Natal, South Africa, 90-850 m.
Xenanthura linearis Pillai, 1954. Kerala, India.
Xenanthura orientalis Barnard, 1935. Southwest coast of India.
Xenanthura sinaica Wägele, 1981a. Gulf of Aqaba, Red Sea.
Xenanthura ulawa Poore & Lew Ton, 1988b. Northern and western Australia; Queensland, 1.5-134 m.

Family Paranthuridae
* *Accalathura* sp. nov. A. Aldabra Atoll, intertidal.
* *Accalathura* sp. nov. B. Phuket Is., Thailand, 2-7 m.
* *Accalathura* sp. nov. C. Aldabra Atoll, intertidal-1 m.
Accalathura borradailei (Stebbing, 1910a). Maldive Is., 42 m. Chilka Lake, India (Chilton 1924). Travancore, India (Pillai 1954). Thailand (Pillai 1966). Zanzibar (Monod 1968).
Accalathura indica (Nierstrasz, 1941). Java Sea, 73 m. Natal, South Africa to Mozambique, 17-100 m (Kensley 1982).
Accalathura laevitelson (Kensley, 1975a). Still Bay, South Africa, 30 m.
Accalathura sladeni (Stebbing, 1910a). Cargados Carajos & Saya de Malha, 48-51 m. Andaman Is.; Cargados Carajos; Mozambique; Somalia; Madagascar; Thailand; Sumatra, intertidal-90 m (Kensley 1980a).
Aenigmathura lactanea Thomson, 1951. Western Australia and Victoria, intertidal-24 m.
Bourbonanthura littoralis Müller, 1990a. Réunion Is. 0.5 m.
Califanthura pingouin (Kensley, 1980a). Crozet Is., 155-187 m. Marion & Prince Edward Is. (Branch et al. 1991).
Colanthura anophthalma Kussakin & Vasina, 1982a. Kerguelen Is., 3-15 m.
Colanthura latimana Kussakin & Vasina, 1982a. Kerguelen Is., 268 m.

Colanthura pigmentata Kensley, 1980a. Madagascar, intertidal-1.5 m. * Aldabra Atoll, 6-22 m.
Colanthura uncinata Kensley, 1978c. Natal to Transkei, South Africa, 150-775 m.
Cruranthura simplicia (Thomson, 1946). Western Australia.
* *Leptanthura* sp. nov. A. Aldabra Atoll, intertidal-15 m.
* *Leptanthura* sp. nov. B. Mahé Is., Seychelles, 0.5-12 m.
Leptanthura agulhasensis Kensley, 1975a. Saldanha Bay to East London, South Africa, 26-320 m.
Leptanthura laevigata (Stimpson, 1855). False Bay, South Africa. Orange River mouth, South Africa to Mozambique Channel, 42-1360 m (Kensley 1982).
Leptanthura minuta Kensley, 1978c. Off Natal, South Africa, 550-850 m.
Leptanthura natalensis Kensley, 1978c. Off Natal, South Africa, 90-850 m.
Leptanthura thalassae Negoescu, 1980. Gulf of Aden, 24-110 m.
Leptanthura urospinosa Kensley, 1975a. False Bay to Still Bay, South Africa, 5-200 m.
* *Paranthura* sp. nov. A. Aldabra Atoll, intertidal-8 m.
* *Paranthura* sp. nov. B. Mahé Is., Seychelles, intertidal-6 m.
* *Paranthura* sp. nov. C. Mahé Is., Seychelles, 0.5-5 m.
Paranthura antarctica Kussakin, 1967. Crozet Is., Kerguelen Is., Antarctica.
Paranthura deodata Müller, 1990a. Réunion Is., 0.5-1 m.
Paranthura elegans non Menzies, 1951b. Madagascar (Roman 1970).
Paranthura latipes Barnard, 1955. Mozambique, intertidal. Pakistan (Javed & Yasmeen 1992).
Paranthura neglecta Beddard, 1886. Kerguelen Is., 232 m.
Paranthura plumosa Pillai, 1966. Quilon, India.
Paranthura possessia Kensley, 1980b. Crozet Is., 142-490 m. Marion & Prince Edward Is. (Branch et al. 1991).
Paranthura punctata (Stimpson, 1855). False Bay, South Africa. Orange River mouth to Natal, South Africa, 7-200 m (Kensley 1982). New Zealand; New South Wales; Western Australia; Tasmania; South Australia (Jones & Morgan 1993).
Pseudanthura tenuis Kensley, 1978e. Natal to Transkei, South Africa, 560-850 m.

SUBORDER ASELLOTA

Family Acanthaspidiidae

Acanthaspidia acanthonotus (Beddard, 1886). Kerguelen Is., 3063-3398 m.
Acanthaspidia natalensis (Kensley, 1977). Southern Mozambique, 1360 m.
Ianthopsis beddardi Kussakin & Vasina, 1982b. Kerguelen Is., 200 m.
Ianthopsis certus Kussakin & Vasina, 1982b. Kerguelen Is., 130-305 m.
Ianthopsis studeri Kussakin & Vasina, 1982b. Kerguelen Is., 460 m.

Family Dendrotiidae

Acanthomunna spinipes (Vanhöffen, 1914). Antarctic, 385 m. Natal, South Africa (Kensley 1978a). Crozet Is. (Kensley 1980b).

Family Desmosomatidae

Desmosoma auritum Carvacho, 1977. Kerguelen Is., 20-102 m.

Disparella longimana (Vanhöffen, 1914). Antarctic, 2735 m. St. Paul & Amsterdam Is., 2200 m (Kensley 1989).

Family Eurycopidae

Eurycope glabra Kensley, 1978a. Natal, South Africa, 680-810 m.

Eurycope sarsii Beddard, 1886. Marion Is., 2514-2926 m.

Munneurycope murrayi (Walker, 1903). North Atlantic, 530-780 m. Bay of Bengal (Barnard 1936). Marion Is. (Nierstrasz 1941).

Storthyngura fragilis (Beddard, 1886). Marion Is., 2880 m, southern Indian Ocean; off Japan, 3431 m.

Storthyngura spinosa (Beddard, 1886). Antarctic, 4932 m.

Family Gnathostenetroididae

* *Maresiella* sp. nov. Aldabra Atoll, 15-25 m.

Maresiella indica Müller, 1992. Réunion Is., intertidal.

Family Haploniscidae

Antennuloniscus quadratus Menzies & Schultz, 1968. SW Madagascar, 2270 m.

Haploniscus gernekei Kensley, 1978a. Natal, South Africa, 550-680 m.

Family Ilyarachnidae

Echinozone bispinosa Kussakin & Vasina, 1982b. Kerguelen Is., 460 m.

Echinozone quadrispinosa (Beddard, 1886). Kerguelen Is., 229 m. Antarctica; South Georgia, 200-1500 m (Nierstrasz 1941).

Echinozone cf. *spicata* (Hodgson, 1910). Antarctic. Crozet Is., 36-560 m (Kensley 1980b).

Ilyarachna antarctica Vanhöffen, 1914. Antarctic, 3397-3423 m. Crozet Is., 196-217 m; Kerguelen Is., 305 m (Kussakin & Vasina 1982b).

Ilyarachna crozetensis Kensley, 1980b. Crozet Is., 195-930 m.

Ilyarachna wolffi Kensley, 1977. Off Natal, South Africa, 550 m.

Family Ischnomesidae

Haplomesus zuluensis Kensley, 1984b. Zululand, South Africa, 800-810 m.

Ischnomesus glabra Kensley, 1984b. Transkei, South Africa, 1800-1950 m.

Stylomesus natalensis Kensley, 1984b. Natal, South Africa, 690-850 m.

Family Janiridae

Austrofilius arnaudi Kussakin & Vasina, 1980a. Kerguelen Is., 24 m.

Austrofilius furcatus Hodgson, 1910. Antarctic, < 40 m. Kerguelen Is., 10 m; South Georgia; Falkland Is., Tierra del Fuego (Nierstrasz 1941).

Caecianiropsis ectiformis (Vanhöffen, 1914). Kerguelen Is. St. Paul & Amsterdam Is., sublittoral (Kensley 1976).

* *Caecijaera horvathi* Menzies, 1951a. California. Hawaii (Cooke 1977). Cuba, 0.5 m (Kensley et al. 1997). * Phuket Is., Thailand, 10 m.

* *Carpias* sp. nov. Mozambique, 32 m.

* *Carpias* cf. *algicola* (Miller, 1941). Hawaii. Caribbean; Gulf of Mexico; India (Pires 1982). * Aldabra Atoll, intertidal-22 m. * Comoro Is., 3-5 m. * Red Sea, 0-8 m. * Mauritius.

Carpias brucei (Monod, 1974). Seychelles, 3.2 m.
Carpias crosslandi (Stebbing, 1910b). Red Sea. Egmont Is., Chagos Archipelago (Stebbing 1910a).
Carpias longimanus (Pillai, 1954). Travancore, India. Quilon, India (Pillai 1963a).
Carpias nanus (Stebbing, 1905). Sri Lanka.
Carpias stylodactylus (Nobili, 1906a). Tuamotu Is. Gulf of Suez, Red Sea (Monod 1933). South Pacific (Nordenstam 1946). Caribbean (Menzies & Glynn 1968).
* *Carpias* cf. *triton* (Pires, 1982). Belize, intertidal. * Zanzibar, intertidal.
Ectias angusta (Barnard, 1920). False Bay, South Africa, intertidal. Amsterdam Is., 25 m (Kensley 1976).
Iais pubescens (Dana, 1853). Tierra del Fuego. Sri Lanka (Stebbing 1904b). Travancore, India (Pillai 1954). Kerguelen Is.; Marion Is.; Gough Is.; Tasmania; New Zealand; Tristan da Cunha; Falkland Is.; Antarctic; Namibia to Natal, South Africa (Barnard 1914b, Kensley 1978c).
Ianiropsis palpalis Barnard, 1914a. Table Bay, False Bay, South Africa, Intertidal. St. Paul & Amsterdam Is. (Kensley 1976). Swakopmund, Namibia to East London, South Africa, intertidal-40 m (Kensley 1978d).
Ianisera trepidus Kensley, 1976. St. Paul & Amsterdam Is., 30-100 m.
Iathrippa capensis (Barnard, 1914a). Table Bay, False Bay, South Africa, intertidal. St. Paul & Amsterdam Is., upper infralittoral to sublittoral (Kensley 1976). Lüderitz, Namibia to Algoa Bay, South Africa, 13-84 m (Kensley 1978d).
Iathrippa chilensis (Menzies, 1962a). Southern Chile. Kerguelen Is. (Kussakin & Vasina 1980a).
Iathrippa longicauda (Chilton, 1884). New Zealand. Tierra del Fuego; Falkland Is.; South Georgia Is.; Kerguelen Is.; South Africa (Kussakin & Vasina 1980a).
Iathrippa sarsi (Pfeffer, 1887). South Georgia, 0-700 m. Subantarctic islands; Antarctica (Nordenstam 1933). Kerguelen Is.; South Georgia Is. (Nierstrasz 1941). Crozet Is., 120-600 m (Kensley 1980b).
Neojaera antarctica (Pfeffer, 1887). South Georgia. Kerguelen Is.; Argentina; Tierra del Fuego; Falkland Is. (Nierstrasz 1941).
Neojaera expansa (Kensley, 1984b). Off Natal, South Africa, 90-850 m.
Neojaera serrata (Barnard, 1914b). Table Bay, South Africa, intertidal. Orange River mouth to Port Elizabeth, South Africa, intertidal-320 m (Kensley 1978d).
Protocharon arenicola Chappuis, Delamare-Deboutteville & Paulian, 1956. Réunion, interstitial.
Tole extans (Barnard, 1914b). Table Bay, False Bay, South Africa, intertidal. Möwe Bay, Namibia to East London, South Africa, intertidal (Kensley 1978d).

Family Joeropsididae
* *Joeropsis* sp. nov. A. Mahé Is., Seychelles, intertidal-3 m.
* *Joeropsis* sp. nov. B. Mahé Is., Seychelles, 0.5 m.
* *Joeropsis* sp. nov. C. Mahé Is., Seychelles, 1-3 m.
* *Joeropsis* sp. nov. D. Madagascar, intertidal.
* *Joeropsis* sp. nov. E. Mahé Is., Seychelles, 1-5 m.
* *Joeropsis* sp. nov. F. Mahé Is., Seychelles, 1 m.
* *Joeropsis* sp. nov. G. Aldabra Atoll, 1-6 m.
Joeropsis beuroisi Kensley, 1975b. St. Paul & Amsterdam Is., 0.3-80 m.
Joeropsis bourboni Müller, 1991b. Réunion Is., 0.5-1 m.

Joeropsis ceylonensis Müller, 1991d. Sri Lanka.

Joeropsis curvicornis (Nicolet, 1849). Patagonia. Sri Lanka (Stebbing 1905). Marion Is.; Falkland Is.; Macquarie Island; Chile; Magellan Straits, Tierra del Fuego (Kensley 1975b). Kerguelen Is. (Kussakin & Vasina 1982b).

Joeropsis faurei Müller, 1991b. Réunion Is., 0.5-1 m.

* *Joeropsis indicus* Müller, 1991d. Sri Lanka, intertidal. * Mahé Is., Seychelles, intertidal-3 m. * Zanzibar, intertidal.

Joeropsis integer Kensley, 1984b. East London, South Africa, 90 m.

Joeropsis intermedius Nordenstam, 1933. Argentina & Tierra del Fuego, 100 m; Falkland Is., 22-150 m. Kerguelen Is. (Kussakin & Vasina 1980a).

Joeropsis letourneuri Müller, 1991b. Réunion Is., 0.5-1 m.

Joeropsis marionis Beddard, 1886. Marion Is., 252 m. Kerguelen Is. (Carvacho 1977). Off Crozet Is., 180 m (Kensley 1980b).

Joeropsis meteor Müller, 1991c. Off Aden, 76 m.

Joeropsis paulensis Vanhöffen, 1914. St. Paul & Amsterdam Is., sublittoral. Gough Is. (Kensley 1975b).

Joeropsis sanctipauli Kensley, 1989. St. Paul & Amsterdam Is., 160-165 m.

Joeropsis serrulus Kensley, 1984b. East London, South Africa, 90 m.

Joeropsis waltervadi Kensley, 1975b. Walter's Shoal, SW Indian Ocean, 38-46 m.

Joeropsis wolffi Müller, 1991c. Gulf of Aden, 214-237 m.

Family Katianiridae

Natalianira spinosa Kensley, 1984b. Natal, South Africa, 550-850 m.

Family Macrostylidae

Macrostylis amplinexa Mezhov, 1989. Arabian Sea, 2385-4221 m.

Macrostylis capito Mezhov, 1989. Central Indian Ocean, 2218-4737 m.

Macrostylis carinifera carinifera Mezhov, 1988. Arabian Sea, 3074-4458 m.

Macrostylis carinifera dilatata Mezhov, 1988. Arabian Sea, 2540 m.

Macrostylis porrecta Mezhov, 1988. Off Sumatra, 6433 m.

Macrostylis rectangulata Mezhov, 1989. South-central Indian Ocean, 5220 m.

Macrostylis urceolata Mezhov, 1989. Bay of Bengal, 2596 m.

Macrostylis wolffi Mezhov, 1988. Arabian Sea, 2385-3717 m.

Family Microparasellidae

Angliera cosettae Coineau & Rao, 1972. Andaman Is., interstitial.

Angliera gracilis Ganamuthu, 1954. India, interstitial.

Angliera phreaticola phreaticola Chappuis & Delamare Deboutteville, 1952. Mediterranean, interstitial. Madagascar; India, interstitial littoral (Coineau 1971).

Angliera xarifae Siewing, 1959. Gulf of Aden, interstitial.

Family Munnidae

Astrurus crucicauda Beddard, 1886. Kerguelen Is., 240 m.

Echinomunna uroventralis Kensley, 1976. St. Paul & Amsterdam Is., sublittoral.

Munna antarctica (Pfeffer, 1887). Antarctic, 2-310 m. Kerguelen Is.; South Georgia (Nierstrasz 1941).

Munna concavifrons (Barnard, 1920). Table Bay, Durban, South Africa. Saldanha Bay to Durban, South Africa, intertidal (Kensley 1978d).
Munna crozetensis Kussakin & Vasina, 1982b. Crozet Is., 215 m.
Munna dentata Vanhöffen, 1914. Kerguelen Is.
Munna instructa Cleret, 1971. Marion Is.
Munna kerguelensis Kussakin & Vasina, 1982b. Kerguelen Is., 148 m.
Munna maculata Beddard, 1886. Kerguelen Is., 51 m. Falkland Is. (Nierstrasz 1941).
Munna neglecta Monod, 1931a. South Orkney Is. Falkland Is., 22-40 m (Nordenstam 1933). Crozet Is., 142-600 m (Kensley 1980b). Kerguelen Is. (Kussakin & Vasina 1980a).
Munna pallida Beddard, 1886. Kerguelen Is., 55 m. Falkland Is. (Nierstrasz 1941).
Munna spinifera Robinson & Menzies, 1961. Gulf of Aden, 1201 m.
Munna studeri Hilgendorf, 1893. Kerguelen Is., 0-210 m. South Georgia (Nierstrasz 1941).
Uromunna brevicornis (Thomson, 1946). Western Australia, 1 m. South Australia (Poore 1984).
Uromunna nana (Nordenstam, 1933). Falkland Is., 1-16 m. St. Paul & Amsterdam Is., 0.6-120 m (Kensley 1976).
Uromunna schauinslandi (Sars, 1905). New Zealand. Kerguelen Is.; Chatham Is. (Nierstrasz 1941).

Family Munnopsidae
Acanthocope spinicauda Beddard, 1886. SW Indian Ocean, 3240 m.
Acanthocope pentacornis Müller, 1989a. Gulf of Aden, 1830 m.
Munnopsis mandibularis Wolff, 1962. Off Kenya, 1510 m.
Munnopsoides australis (Beddard, 1885). Marion Is., 2880 m. Antarctica, 400-500 m (Monod 1926b).

Family Paramunnidae
Munnogonium adenensis Müller, 1991c. Gulf of Aden, 228-235 m.
Munnogonium subtilis Kensley, 1976. St. Paul & Amsterdam Is., 50-60 m.
Paramunna dilatata Vanhöffen, 1914. Kerguelen Is.
Paramunna foresti Carvacho, 1977. Kerguelen Is. Crozet Is. (Kensley 1980b).
Paramunna kerguelensis Vanhöffen, 1914. Kerguelen Is. Crozet Is. (Kensley 1980b).
Paramunna laevifrons Stebbing, 1910c. Off East London, South Africa, 75 m.
Paramunna rostrata (Hodgson, 1910). Antarctic. Kerguelen Is.; South Georgia, 0-549 m (Nierstrasz 1941).

Family Pleurocopidae
Pleurocope dasyura Walker, 1901. Mediterranean, 20 m. Sri Lanka (Stebbing 1905).
* *Pleurocope* sp. nov. Phuket Is., Thailand, 77 m; Aldabra Atoll, 6-16 m.

Family Pleurogoniidae
Coulmannia unicornis Kensley, 1976. St. Paul & Amsterdam Is., 80-100 m.
Neasellus kerguelensis Beddard, 1885. Kerguelen Is., 240 m.
Notoxenoides acalama Kensley, 1984b. Off Natal, South Africa, 820 m.
Pleurogonium albidum Beddard, 1886. Kerguelen Is., 240 m.

Pleurogonium serratum Beddard, 1886. Kerguelen Is., 240 m.

Family Pseudojaniridae

* *Pseudojanira* sp. nov. Madagascar, shallow; Mozambique, 42-88 m.; Natal, South Africa, 69 m.

Pseudojanira investigatoris Poore & Just, 1990. South Australia; Victoria; Western Australia, 1-60 m.

Pseudojanira stenetrioides Barnard, 1925b. Natal, South Africa. Natal, South Africa to Mozambique, 55-62 m (Kensley 1978d). Madagascar (Roman 1970).

* *Tangaroa* sp. nov. Rodriques Is., 91 m.

Family Santiidae

Halacarsantia justi Wolff, 1989. Andaman Sea, 8 m.

* *Santia* sp. nov. A. Somalia, 47-49 m; Aldabra Atoll, 25 m.

* *Santia* sp. nov. B. Aldabra Atoll, 1-26 m; Mahé Is., Seychelles, 1-6 m; Zanzibar, intertidal.

Santia bicornis (Cleret, 1971). Marion Is.

Santia concavata (Carvacho, 1977). Kerguelen Is.

Santia dimorpha (Menzies, 1962a). Chile. St. Paul & Amsterdam Is., upper infralittoral to 80 m (Kensley 1976).

Santia hispida (Vanhöffen, 1914). St. Paul & Amsterdam Is., upper infratidal to 60 m. Auckland Is.; Falkland Is.; Antarctica (Nierstrasz 1941).

Santia hofsteni (Nordenstam, 1933). South Georgia, intertidal. St. Paul & Amsterdam Is., 45-100 m (Kensley 1976).

Santia marmorata (Vanhöffen, 1914). St. Paul Is. Kerguelen Is.; South Georgia Is. (Nierstrasz 1941).

Family Stenetriidae

* *Hansenium* sp. nov. A. Aldabra Atoll, 21 m.

* *Hansenium* sp. nov. B. Madagascar, 1-1.5 m; Zanzibar, 9 m.

* *Hansenium remocarpus* sp. nov. Mahé Is., Seychelles, 2-5 m.

Hansenium chiltoni (Stebbing, 1905). Sri Lanka. Amirante Is., Seychelles, 61 m; Gulf of Suez (Monod 1933).

Hansenium dodo (Müller, 1991g). Réunion Is., intertidal.

* *Hansenium monodi* (Nordenstam, 1946). Gulf of Suez. * Mahé Is., Seychelles, 0.5-6 m.

Liocoryphe gertrudae (Müller, 1991g). Réunion Is., intertidal.

* *Liocoryphe siamense* (Hansen, 1905). Thailand. * Zanzibar, 10 m; Phuket Is., Thailand, 3-6 m.

* *Mizothenar* sp. nov. Aldabra Atoll, 12-21 m; Zanzibar, 9-10 m.

Protallocoxa abyssale (Wolff, 1962). Tasman Sea, 4510 m; Kermadec Trench, 4540 m. Natal, South Africa (Kensley 1978d).

* *Stenetrium* sp. nov. A. Zanzibar, intertidal-7 m.

* *Stenetrium* sp. nov. B. Comoro Is., 3-5 m; Aldabra Atoll, 0-10 m; Zanzibar, 16-25 m.

* *Stenetrium* sp. nov. C. Zanzibar, 8-25 m.

Stenetrium bartholomei Barnard, 1940. Still Bay, Algoa Bay, South Africa, intertidal. False Bay to Natal, South Africa, intertidal-22 m (Kensley 1978d).

Stenetrium crassimanus Barnard, 1914a. False Bay, South Africa, intertidal. St. Paul &

Amsterdam Is., sublittoral (Kensley, 1976). False Bay to Natal, South Africa, intertidal-120 m (Kensley 1978d).
Stenetrium dagama Barnard, 1920. Off Cape Point, South Africa, 348-420 m. Natal, South Africa, 680 m (Kensley 1978d).
Stenetrium diazi Barnard, 1920. False Bay, South Africa, intertidal. False Bay to Port Elizabeth, South Africa, 27-80 m (Kensley 1978d).
Stenetrium esquartum Schultz, 1982. False Bay to East London, South Africa, 6 m.
Stenetrium glauerti Nicholls, 1929. Western Australia, intertidal.
Stenetrium macrochirium Nicholls, 1929. Western Australia.
Stenetrium magnimanum Schultz, 1982. False Bay, South Africa, 52 m.
Stenetrium perestrelloi Kensley, 1984b. East London, South Africa, 90 m.
Stenetrium saldanha Barnard, 1920. Saldanha Bay to Cape Point, South Africa, 229-247 m. St. Paul & Amsterdam Is., 50-90 m (Kensley 1976). Saldanha Bay to Still Bay, South Africa, 50-250 m (Kensley 1978d).
Stenetrium spinirostrum Nicholls, 1929. Western Australia, intertidal.
Stenetrium truncatum Nicholls, 1929. Western Australia, sublittoral.
* *Stenobermuda* sp. nov. Zanzibar, 25 m.
Tristenium acutirostrum (Müller, 1991c). Gulf of Aden, 228-235 m.
Tristenium bourboni (Müller, 1991g). Réunion Is., 1 m.

SUBORDER EPICARIDEA
Family Bopyridae
Allokepon hendersoni (Giard & Bonnier, 1888). Madras, India, on *Thalamita callianassa*.
Allokepon sinensis (Danforth, 1972). Hong Kong, on *Lissocarcinus orbicularis*. Phuket, Thailand, on *Portunus pelagicus* (Markham 1985).
Anathelges mossambica Barnard, 1958. Delagoa Bay, Madagascar, on hermit crab.
Aporobopyrina lamellata Shiino, 1934. Japan, on *Petrolisthes coccineus*. Madagascar and Pakistan, on *Petrolisthes* spp. (Bourdon 1983). Karachi, Pakistan, on *Petrolisthes rufescens* (Markham, 1980). Phuket, Thailand, on *Petrolisthes lamarckii* (Markham 1985).
Aporobopyroides upogebiae Nobili, 1906c. Red Sea, on *Upogebia* (*Calliadne*) *savignyi*.
Aporobopyrus aduliticus Nobili, 1906c. Red Sea, on *Petrolisthes rufescens*. Gulf of Suez, on *Polyonyx* sp. (Monod 1933).
Aporobopyrus dollfusi Bourdon, 1980a. Red Sea, on *Aliaporcellana suluensis* and *A. quadrilobata*.
Aporobopyrus ryukyuensis Shiino, 1939. Okinawa, Japan, on *Petrolisthes hastatus* and *P. asiaticus*. Karachi, Pakistan, on *P. boscii* (Markham 1980).
Argeia lowisi Chopra, 1923. Andaman Is., on *Alpheus* sp.
Argeiopsis inhacae Kensley, 1974. Mozambique, on *Stenopus hispidus*.
Asymmetrione asymmetrica Shiino, 1933. Japan, on *Clibanarius bimaculatus*. Phuket, Thailand, on *C. merguiensis* (Markham 1985).
Asymmetrione nossibensis Bourdon, 1976b. Madagascar. Host unknown.
Athelges aegypticus Codreanu, Codreanu & Pike, 1965. Red Sea, on *Diogenes senex*.
Athelges caudalis Barnard, 1955. Mozambique, intertidal, on *Diogenes senex*.
Bopyrella deformans (Hay), var. *indica* Chopra, 1923. Pakistan, on *Synalpheus huluensis*. India; Gulf of Suez; Malaysia; South China Sea (Monod 1933).
Bopyrella malensis Bourdon, 1980b. Maldive Is., on *Synalpheus biunguiculatus*.

Bopyrella tanytelson Markham, 1985. Krabi, Thailand, on *Alpheus* sp.
Bopyrina giardi Bonnier, 1900. Andaman Is., on *Hippolyte ventricosus* (Chopra 1923).
Bopyrina sewelli Chopra, 1930. Nicobar Is., on *Chlorotocella gracilis*.
Bopyrinella albida Shiino, 1958. Japan, on *Athanas kaminatoensis*. Phuket, Thailand, on *Athanas dimorphus* (Markham 1985).
Bopyrinella (?) *stricticauda* Monod, 1933. Gulf of Suez, on *Spongia*.
Bopyrione woodmasoni (Chopra, 1923). Andaman Is., on *Synalpheus* sp.
Bopyrissa liberorum Markham, 1985. Phuket, Thailand, on *Clibanarius merguiensis*.
Bopyrus squillarum Latreille var. *bimaculatus* Chopra, 1923. Ganges Delta, India, on *Leander styliferus*. Near Bombay, India (Chopra 1930).
Dactylokepon catoptri Stebbing, 1910a. Seychelles, 54 m, on *Catoptrus nitidus*.
Dactylokepon holthuisi Bourdon, 1967. Gulf of Aqaba, on *Scyllarus lewinsohni*.
Dactylokepon richardsonae Stebbing, 1910a. Seychelles, on *Trapezia cymodoce*. Bangkok, Thailand, on *Portunus tuberculosus* (Markham 1991).
Dactylocepon seychellensis (Danforth, 1972). Seychelles Is., on *Tetralia glaberrima*.
Dicropleon bifidus (Bourdon, 1967). Maldive Is., on *Periclimenes inornatus*.
Dicropleon processae Markham, 1980. Gulf of Manaar, on *Processa aequimana*.
Diplophyrxus alphei Shiino, 1934. Japan, on *Alpheus* sp. Delagoa Bay, Mozambique, on *Alpheus edwardsii* (Barnard 1958).
Diplophryxus jordani Richardson, 1904. Japan, on *Palaemon serrifer*. Una Is., Mergui Archipelago, on *Leander serrifer* (Chopra 1923). Gulf of Manaar, on *Leander tenuicornis* (Chopra 1930).
Diplophryxus kempi Chopra, 1930. Andaman Is., on *Gnathophyllum fasciolatum*.
Diplophryxus richardsoni Chopra, 1930. Andaman Is., on *Pontophilus lowsii*.
Entophilus omnitectus Richardson, 1903. Hawaii, on *Munida normanni*. Madagascar, on *Munida incerta* (Bourdon 1976b).
Eophrixus brevicauda (Chopra, 1923). Andaman Is., on *Synalpheus* sp.
Eophrixus filiformis (Chopra, 1923). Andaman Is., on *Alpheus* sp.
Eophrixus nigrocinctus (Chopra, 1923). Andaman Is., on *Periclimenes elegans*. Nicobar Is., on *Periclimenes amymone & Leander tenuicornis* (Chopra 1930).
Eophrixus shojii Shiino, 1941. Japan, on *Alpheus japonicus*. Phuket, Thailand, on *Alpheus* sp. (Markham 1991).
Epipenaeon elegans Chopra, 1923. Ganges Delta, India, on *Penaeus carinatus*.
Epipenaeon fissurae Kensley, 1974. Natal, South Africa, on *Parapenaeus fissurae*, 70 m.
Epipenaeon georgei Devi, 1982. India, on *Parapenaeopsis sculptilis, P. stylifera, Metapenaeopsis barbata*. Bay of Bengal, on *Parapenaeus longipes* (Bourdon 1979). Madagascar, on *Parapenaeus longipes* (Bourdon 1979). Indonesia, on *Parapenaeus longipes* (Markham 1994).
Epipenaeon ingens Nobili, 1906c. Red Sea, on *Penaeus* sp. Northern Territory, Australia (Monod 1933).
Epipenaeon ovalis Pillai, 1954. Travancore, India, on *Parapenaeopsis stylifera*.
Epipenaeon qadrii Qasi, 1959. Pakistan, on *Parapenaeopsis stylifera*.
Gigantione moebii Kossmann, 1881. Mauritius, on *Rupellia impressa*.
Gigantione rathbunae Stebbing, 1910a. Chagos Archipelago, on *Actaea* sp.
Gigantione sagamiensis Shiino, 1958. Japan, on *Carpilodes lophopus boninensis*. Mozambique, 42 m, *on Paratergatis longimanus* (Kensley 1974).

Grapsicepon messoris (Kossmann, 1880). Red Sea. Red Sea, on *Metopograpsus messor* (Monod 1933).
Hemiphryxus malindiae (A.J. Bruce, 1974). Kenya, on *Coralliocaris superba.*
Hypoperphryxus latilamellaris Nierstrasz & Brender à Brandis, 1932. Red Sea, on *Synalpheus triunguliculatus.*
Hypophryxus filiformis Chopra, 1923. Andaman Is., on *Alpheus paracyclone.*
Hypophryxus leptochelae Pillai, 1966. Kerala, India on *Leptochela aculeocaudata.*
Hypophryxus pikei A.J. Bruce, 1968. Zanzibar Is., on *Ischnopontonia lophos.*
Ione tubulata Bourdon, 1976b. Madagascar.
Kepon halimi Stebbing, 1910a. Cargados Carajos, 54 m, on *Halimus* sp.
Kepon orientalis Markham, 1985. Phuket, Thailand, on *Metaplax distinctus* & *M. elegans.*
Kepon typus Duvernoy, 1841. Mauritius, on *Grapsus strigosus.*
Megacepon choprai George, 1946. India, on *Sesarma tetragonum.*
Mesophryxus ventralis (A.J. Bruce, 1973a). Zanzibar, on *Harpiliopsis beaupresi.*
Metabopyrus irregularis Markham, 1985. Phuket, Thailand, on *Wolffogebia phuketensis.*
Metaphrixus intutus A.J. Bruce, 1965. Zanzibar Is., on *Palaemonella vestigialis.*
Minicopenaeon apertum (Bourdon, 1979). Madagascar, on *Metapenaeopsis andamanensis.*
Minicopenaeon crosnieri (Bourdon, 1979). Madagascar, on *Metapenaeopsis andamanensis.*
Miophrixus latreutidis Barnard, 1955. Mozambique, intertidal, on *Latreutes pygmaeus.*
Nikione natalensis Kensley, 1974. Natal, South Africa, 43 m, on *Nikoides danae.*
Orbione digitata Bourdon, 1982a. India, on *Solenocera choprai.*
Orbione halipori Nierstrasz & Brender à Brandis, 1923. Indonesia, on *Haliporus sibogae.* Japan, on *Haliporus sibogae* (Shiino 1949). Madagascar, on *H. sibogae madagascariensis* and *Hymenopenaeus halli* (Bourdon 1979). Philippines, on *Haliporoides sibogae* (Markham 1994).
Orbione kempi Chopra, 1923. Andaman Is., on *Sicyonia bispinosa.* India, on *Metapenaeus brevicornis* and *M. lysianassa* (Devi 1982). New Caledonia, on *Sicyonia truncata* (Markham 1994).
Orbione thielemanni Nierstrasz & Brender à Brandis, 1931. Gulf of Thailand, on *Metapenaeopsis stridulans.* Madagascar, on *Metapenaeopsis* sp. (Bourdon 1979).
Orophryxus shiinoi A.J. Bruce, 1972. Zanzibar Is., on *Periclimenes grandis.*
Parabopyrella angulosa (Bourdon, 1980b). Djibouti, on *Alpheus* sp.
Parabopyrella delagoae (Bourdon, 1982b). Mozambique, on *Alpheus* (?) *edwardsi.*
Parabopyrella hodgarti (Chopra, 1923). Bay of Bengal, on *Alpheus crassimanus.* Mozambique, on *Alpheus crassimanus* (Barnard 1955). Natal, South Africa, 200 m, on alpheid shrimp (Kensley 1978d).
Parabopyrella incisa (Bourdon, 1982b). South Africa, on *Alpheus* sp.
Parabopyrella indica (Chopra, 1923). Pakistan, India, on *Synalpheus huluensis.*
Parabopyrella nierstraszi (Chopra, 1930). Andaman Is., on *Hippolysmata vittata.*
Parabopyrella saronae (Bourdon & A.J. Bruce, 1979). Kenya, on *Saron marmoratus.*
Parabopyrella tanyensis (Bourdon, 1979). Madagascar, on *Synalpheus neomeris.*
Parabopyrella thomsoni (Bonnier, 1900). Andaman Is., on *Alpheus* sp.
Paracepon nierstrazi Pillai, 1954. Travancore, India.
Paragigantione papillosa Barnard, 1920. Off East London, South Africa, 540 m, on *Munida sanctipauli.*

Parapagurion calcinicola Shiino, 1933. Japan, on *Calcinus elegans*. Phuket, Thailand, on *Paguristes* sp. (Markham 1985).

Parapenaeon coarctatum coarctatum Bourdon, 1981. Madagascar, on *Metapenaeopsis* sp.

Parapenaeon expansus Bourdon, 1979. Madagascar, on *Penaeus teraoi*. New Caledonia, on *Metapenaeopsis gaillardi*; Indonesia, on *M. sinica*; Seychelles, on *M. faouzii*; Madagascar, on *M mogiensis consobrina* (Markham 1994).

Parapenaeon japonicum (Thielemann, 1910). Japan, on *Penaeus* sp. Burma, on *Parapenaeopsis sculptilis*; Madagascar, on *Metapenaeus monoceros* (Bourdon 1979). Indonesia, on *Metapenaeopsis sinica*; Philippines, on *Metapenaeopsis velutinus*; Madagascar, on *Metapenaeopsis hilarula* (Markham 1994).

Parapenaeonella lamellata Bourdon, 1979. Andaman Is., on *Metapenaeus monoceros* & *M. ensis*.

Parapleurocrypta alphei Chopra, 1923. Andaman Is., on *Synalpheus* sp.

Parapleurocrypta digitata Bourdon, 1976b. Madagascar.

Parathelges weberi Nierstrasz & Brender à Brandis, 1923. Andaman Is.

Parionina pacifica Nierstrasz & Brender à Brandis, 1929. Madagascar, on *Allogalathea elegans*.

Pleurocrypta yatsui (Pearse, 1930). Japan. Persian Gulf & Red Sea, on *Pachycheles natalensis* & *Petrolisthes boscii* (Bourdon 1976a).

Pleurocryptella crassandra Bourdon, 1976b. Madagascar, on *Munidopsis cylindrophthalmus*.

Pleurocryptella infecta tuberculata Bourdon, 1976b. Madagascar, on *Munida tricarinata*.

Portunicepon savignyi (Stebbing, 1910a). Cargados Carajos, on *Actaea granulatus*.

Probopyrus abhoyai (Chopra, 1923). Ganges Delta, India, on *Palaemon* sp.

Probopyrus alcocki (Chopra, 1923). India, on *Palaemon* sp.

Probopyrus bengalensis (Chopra, 1923). Ganges Delta, India, on *Palaemon malcomsoni*; Bangkok, on *Palaemon concinnus* (Chopra 1930).

Probopyrus brachysoma (Chopra, 1923). Ganges Delta, India, on *Palaemon mirabilis*.

Probopyrus buitendijki (Horst, 1910). Ganges Delta, India, on *Palaemon carcinus*. Java, on *P. carcinus* (Chopra 1930). Bangladesh & India, on *Macrobrachium rosenbergii*; Bangkok, Thailand, on *M. rosenbergii*; Papua New Guinea, on *M. rosenbergii* (Markham 1985).

Probopyrus demani (Weber, 1892). India. Godaveri Delta, India, on *Palaemon* sp. (Chopra 1923).

Probopyrus gangeticus Chopra, 1923. Ganges Delta, India, on *Palaemon* sp.

Probopyrus godaveriensis (Chopra, 1923). Godaveri Delta, India, on *Palaemon* sp.

Probopyrus pica (Chopra, 1923). India, on *Palaemon* sp. and *Leander potamiscus*.

Probopyrus prashadi (Chopra, 1923). Ganges Delta, India, on *Palaemon* spp.

Procepon horridulum Markham, 1985. Phuket, Thailand, on *Wolffogebia phuketensis*.

Procepon insolitum Shiino, 1937. Japan, on *Upogebia major*. Phuket, Thailand, on *Upogebia lincolni* (Markham 1985).

Pseudione andamanicae Bourdon, 1976b. Madagascar, on *Munida andamanica*.

Pseudione crenulata Sars, 1898. Norway. East London to Natal, South Africa, on *Galathea dispersa* and *Munida sanctipauli* (Kensley 1978d).

Pseudione indica Chopra, 1930. Andaman Is., on *Pontophilus plebs*. Strait of Makassar, on *Plesionika narval* (Markham 1980).

Pseudione itsindrae Bourdon, 1976b. Madagascar, on *Munidopsis trifida*.

Pseudione munidae Barnard, 1920. East London, South Africa, on *Munida sanctipauli*.
Pseudione minimocrenulata Nierstrasz & Brender à Brandis, 1931. Madagascar, on *Munida incerta*.
Pseudostegias dulcilacuum Markham, 1982. Hong Kong, on *Diogenes edwardsi*. Phuket, Thailand, on *Clibanarius merguiensis* (Markham 1985).
Pseudostegias setoensis Shiino, 1933. Japan, on *Clibanarius bimaculatus*. Phuket, Thailand, on *Clibanarius padavensis* (Markham 1985).
Rhopalione incerta (Bonnier, 1900). Madagascar. Host unknown.
Rhopalione uromyzon Perez, 1920. Persian Gulf, on *Ostracotheres spondyli*.
Schizobopyrina andamanica (Chopra, 1923). Andaman Is., on *Periclimenes elegans*.
Schizobopyrina cochinensis (Chopra, 1923). Cochin, India, on *Periclimenes grandis*.
Schizobopyrina gracilis (Chopra, 1923). Andaman Is., on *Urocaridella gracilis*.
Schizobopyrina kossmanni (Chopra, 1923). Andaman Is., on *Periclimenes elegans*.
Shiinoella gracilipes Bourdon, 1972. Mauritius, on *Synalpheus* sp.
Stegoalpheon choprai Pillai, 1954. Travancore, India, on *Alpheus paludicolla*.
Stegoalpheon kempi Chopra, 1923. Andaman Is., on *Alpheus* sp.
Trapezicepon amicorum Bonnier, 1900. Seychelles, 61 m, on *Actumnus* sp.
Trapezicepon domeciae Danforth, 1972. Maldive Is., on *Domecia glabra*.
Trapezicepon thalamitae Markham, 1985. Phuket, Thailand, on *Thalamita crenata*.
Tylokepon bonnieri Stebbing, 1904a. Maldive Is., on *Tylocarcinus styx*.
Upogebione phuketensis Markham, 1985. Phuket, Thailand, on *Wolffogebia phuketensis*.
Upogebiophilus rhadames Nobili, 1906c. Djibouti, on *Upogebia rhadames*.

Family Cryptoniscidae
Cirolanoniscus willeyi Pillai, 1963b. Kerala, India, on *Cirolana willeyi*.
Danalia longicollis (Kossmann, 1880). Red Sea, on *Chlorodius* (*Leptodius*) *exaratus*.

Family Dajidae
Dajus afromysidis Pillai, 1963c. Trivandrum, India, on *Afromysis macropsis*.
Notophryxus lobatus Pillai, 1963c. Trivandrum, India, on *Rhopalophthalmus tattersallae*.
Prodajus gastrosacci Pillai, 1963c. Trivandrum, India, on *Gastrosaccus simulans*.
Prodajus ovatus Pillai, 1963c. Trivandrum, India, on *Gastrosaccus muticus*.
Prophryxus globicaudatus A.J. Bruce, 1973b. Kenya, on *Coralliocaris graminea*.

Family Entoniscidae
Diogenion vermifactus Codreanu, Codreanu & Pike, 1960. Red Sea, on *Diogenes senex*.

SUBORDER FLABELLIFERA
Family Aegidae
Aega antillensis Schioedte & Meinert, 1879. Cuba. West Indies, 298-422 m (Richardson 1905). Natal, South Africa, 200 m (Barnard 1925b). South Australia (Hale 1929). St. Paul & Amsterdam Is. (Kensley 1976).
Aega chelipous Barnard, 1960. Madagascar, 160 m.
Aega crozetensis Kussakin & Vasina, 1982a. Crozet Is., 280 m.
Aega deshaysiana (H. Milne Edwards, 1840). Mediterranean. Seychelles; Japan; Hawaii;

Australia; Northwest Africa; Cape Verde Is.; Azores; Antilles, 40-6462 m (Nierstrasz 1931).

Aega dollfusi Monod, 1933. Gulf of Suez. * Off Somalia, 59-61 m. * Off Madagascar, 30 m.

Aega edwardsi Dollfus, 1891. Tierra del Fuego. Kerguelen Is. (Kussakin & Vasina 1980a).

Aega falklandica Kussakin, 1967. Falkland Is. Marion & Crozet Is., 185-270 m (Kensley 1980b). Kerguelen Is., 280 m (Kussakin & Vasina 1982a).

Aega koltuni Kussakin, 1967. Crozet Is., 370-400 m.

Aega monilis Barnard, 1914b. South Africa, 90-300 m. St. Paul & Amsterdam Is., 60-70 m (Kensley 1976).

Aega monophthalmus Johnston, 1834. England. Off East London, South Africa, 800 m.; North Atlantic (Kensley 1978d).

Aega pushkini Kussakin & Vasina, 1982a. SW Indian Ocean, 410 m.

Aega semicarinata Miers, 1875a. Kerguelen Is., 45-400 m. Marion & Crozet Is. (Kensley 1980b). Cape Point, South Africa; Cape Horn; Falkland Is. (Nierstrasz 1931).

Aega serripes (H. Milne Edwards, 1840). Japan (Schioedte & Meinert 1879). Western Australia (Bruce 1983).

Aega urotoma Barnard, 1914b. South Africa.

Aega ventrosa Sars, 1859. Norway. Arabian Sea; Maldive Is., 697-1734 m (Barnard 1936).

Aega vigilans (Haswell, 1881). Queensland, Australia. Sri Lanka (Stebbing 1905). Mauritius, 180 m (Stebbing 1910a). Philippines, Richardson 1910. Western Australia (Hale 1925). Andaman Sea, 66 m (Barnard 1936). Western Australia, 127-146 m; Queensland, 6 m (Bruce 1983).

Aega webbi (Guérin Méneville, 1836). Portugal. Cape to Natal, South Africa, 100-300 m (Kensley 1978d).

Alitropus dimorphus Pillai, 1954. Travancore, India, shallow.

Alitropus typus H. Milne Edwards, 1840. Bay of Bengal. India; Sri Lanka; Sumatra; Thailand; Borneo (Pillai 1967).

Barybrotes agilis (Schioedte & Meinert, 1879). Indian Seas. Travancore, India (Pillai 1954).

Rocinela granulosa Barnard, 1914b. Natal, South Africa, 80-200 m.

Rocinela latis Southwell, 1915. Calcutta, India, estuarine.

Rocinela major Brocchi, 1875. St. Paul Is.

Rocinela mundana Lanchester, 1902. Thailand, 108 m.

Rocinela orientalis Schioedte & Meinert, 1879. Philippines, 22-500 m. Natal, South Africa, 100 m (Barnard 1914b). Zanzibar; India; Sri Lanka; Suez Canal; northern & eastern Australia (Pillai 1967).

Syscenus infelix Harger, 1883. Massachusetts, USA, 41-1152 m. Natal, South Africa; North Atlantic; Japan; Philippines (Barnard 1940).

Family Bathynataliidae

Bathynatalia gilchristi Barnard, 1957a. Off Natal, South Africa, 880 m.

Naudea louwae Kensley, 1979. Off Natal, South Africa, 850 m.

Family Cirolanidae

Aatolana schioedtei (Miers, 1884). Arafura Sea, 16-173 m. Northern Territory, 2-20 m;

Western Australia; Queensland (Bruce 1986a). Torres Strait (Keable 1997).

Annina fustis Bowman & Iliffe, 1991. Thailand, freshwater streams & pools.

Annina lacustris Budde Lund, 1908. Zanzibar, freshwater. Kenya (Alluaud & Jeannel 1914). Zanzibar (Fage 1924). Kenya (Chappuis 1927). Kenya (Delamare Deboutteville 1960). Zanzibar (Monod 1968). Kenya (Jones & Icely 1981). Kenya (Jones 1983). Grand Comoro Is. (Messana 1984).

Annina mannai Schotte, 1994. West Bengal, India, freshwater.

Annina mesopotamica (Ahmed, 1971). Persian Gulf, intertidal.

Anopsilana luciae (Barnard, 1940). Zululand, South Africa.

Anopsilana poissoni Paulian & Delamare Deboutteville, 1956. Madagascar.

Anopsilana pustulosa (Hale, 1925). Queensland to Torres Strait, Australia. East Africa; Mozambique; Madagascar; India; Papua New Guinea (Bruce 1986a). * Aldabra Atoll, 0.5 m.

Anopsilana willeyi (Stebbing, 1904b). Sri Lanka. Queensland, Australia; East Africa; Thailand (Bruce 1986a).

Atarbolana exoconta Bruce & Javed, 1987. Off Karachi, Pakistan. * Gulf of Kutch, India.

Atarbolana setosa Javed & Yasmeen, 1989. Off Karachi, Pakistan, intertidal.

Bathylana apalpalis Kensley, 1989. St. Paul & Amsterdam Is., 1000-1430 m

Bathynomus pelor Bruce, 1986a. Western Australia, 297-700 m

Booralana wundurra Bruce, 1986a. Western Australia, 22-130 m.

Cartetolana integra (Miers, 1884). Queensland, Australia. Western Australia, 10-20 m (Bruce 1986a). Northern Territory, Australia; Arafura Sea; Queensland; Papua New Guinea (Keable 1997).

* *Cirolana* sp. nov. A. Aldabra Atoll, intertidal-10 m.

* *Cirolana* sp. nov. B. Off Mozambique, 42 m.

* *Cirolana* sp. nov. C. Off Somalia, 47-95 m.

* *Cirolana* sp. nov. D. Mahé Is., Seychelles, 0.5-1.5 m.

* *Cirolana* sp. nov. E. Mozambique, 5-8 m.

* *Cirolana* sp. nov. F. Off Somalia, 59-61 m.

* *Cirolana* sp. nov. G. Off Somalia, 90 m; off Mozambique, 150-300 m.

Cirolana australiense Hale, 1925. South Australia. Western Australia; Victoria; Tasmania; New South Wales, Australia, sublittoral (Bruce 1986a).

Cirolana bisulcata Hobbins & Jones, 1993. Red Sea, 600-2148 m.

Cirolana bougaardti Kensley, 1984b. E coast of South Africa, 710-775 m.

Cirolana bovina Barnard, 1940. East London, South Africa, intertidal. Kerala & Madras, India (Pillai 1954). Kenya (Jones 1976).

Cirolana brucei Javed & Yasmeen, 1995. Pakistan, intertidal.

Cirolana carina Jones, 1976. Watamu, Kenya.

Cirolana cingulata Barnard, 1920. Still Bay to East London, South Africa, 30 to 170 m.

Cirolana corrugis Jones, 1976. Watamu, Kenya. Red Sea, shallow sublittoral (Bruce & Jones 1978).

Cirolana cranchii Leach, 1818. England. S Angola to Port Elizabeth, South Africa, intertidal to 46 m; N Atlantic, Mediterranean (Kensley 1978d).

Cirolana fluviatilis Stebbing, 1902. South Africa. Chilka Lake & Travancore, India (Barnard 1935). Thailand (Barnard 1940). Knysna to Zululand, estuarine (Kensley 1978d).

Cirolana garuwa Bruce, 1986a. Western Australia, 18 m.

Cirolana harfordi (Lockington, 1877). California. Victoria; Western Australia; New

South Wales, Australia; Japan; USSR (Bruce 1986a).

Cirolana hesperia Bruce, 1986a. Western Australia, 2-60 m.

Cirolana imposita Barnard, 1955. South Africa, 15-360 m.

Cirolana incisicauda Barnard, 1940. East coast of South Africa, intertidal. Hermanus to Port Alfred, South Africa, intertidal (Kensley 1978d).

Cirolana latistylis Dana, 1853. N Borneo. Maldive Is., 64-80 m (Stebbing 1904a).

Cirolana littoralis Barnard, 1920. Saldanha Bay. Saldanha Bay to East London, South Africa (Kensley 1978d).

Cirolana manorae Bruce & Javed, 1987. Off Karachi, Pakistan.

Cirolana mascarensis Müller, 1991e. Réunion Is. * Off Mauritius

Cirolana meinerti Barnard, 1920. Off East London, South Africa, 140 m. Madagascar (Roman 1970).

Cirolana mekista Bruce, 1986a. Northern Territory & Western Australia.

Cirolana meseda Hobbins & Jones, 1993. Red Sea, 731-1544 m.

Cirolana palifrons Barnard, 1920. East London, South Africa, 170 m.

Cirolana perlata Barnard, 1936. Townet, Burma, estuarine.

Cirolana porcellana Barnard, 1936. Andaman Sea off Burma, 14 m.

Cirolana rugicauda Heller, 1861. St. Paul & Amsterdam Is., upper infralittoral. Port Nolloth & St. Helena Bay, South Africa (Kensley 1978d).

Cirolana schioedtei Miers, 1884. Arafura Sea. Northern Territory; Western Australia; Queensland (Bruce 1986a).

Cirolana sulcata Hansen, 1890. South Africa, 5-84 m. * Off Mozambique, 19-22 m.

Cirolana sulcaticauda Stebbing, 1904a. Maldive Is. Gulf of Kutch, India (Monod 1971). Sri Lanka; Kenya; Aldabra Atoll (Jones 1976). * Mahé Is., Seychelles, intertidal-6 m.

Cirolana theleceps Barnard, 1940. Natal, South Africa, intertidal. Gulf of Suez (Bruce & Jones 1978).

Cirolana venusticauda Stebbing, 1902. False Bay, South Africa. Lambert's Bay to East London, South Africa, 11-50 m (Kensley 1978d). Madagascar (Roman 1970).

Conilorpheus blandus Barnard, 1955. Algoa Bay, South Africa.

Conilorpheus herdmani Stebbing, 1905. Gulf of Manaar, Sri Lanka, 16-20 m.

Conilorpheus scutifrons Stebbing, 1908. Algoa Bay, South Africa, 18-29 m. False Bay to Natal, South Africa, 11-80 m (Kensley 1978d).

Dolicholana elongata H. Milne Edwards, 1840. Mouth of the Ganges River.

Eurydice agilis Jones, 1971. Watamu, Kenya.

Eurydice arabica Jones, 1974. Red Sea; Saudi Arabia.

Eurydice cavicaudata Jones, 1971. Watamu, Kenya, sublittoral.

Eurydice chelifer Jones, 1971. Watamu, Kenya.

Eurydice humilis Stebbing, 1910a. Chagos Archipelago.

Eurydice indicis Eleftheriou & Jones, 1976. India, intertidal.

Eurydice inermis non Hansen, 1890. Kerala, India (Pillai 1967). Atlantic; Mediterranean; Red Sea (Bruce & Jones 1978).

Eurydice inornata Jones, 1971. Watamu & Kilife, Kenya.

Eurydice longicornis (Studer, 1883). Lüderitz, Namibia to Algoa Bay, South Africa, intertidal to 11 m.

Eurydice longipes Jones, 1971. Watamu, Kenya.

Eurydice orientalis Hansen, 1890. Java Sea. Andaman Sea off Burma, 22 m; Philippines;

Queensland, Australia (Barnard 1936).
* *Eurydice* sp. nov. Persian Gulf, intertidal.
Eurydice peraticis Jones, 1974. Persian Gulf, Saudi Arabia. Goa, India (Jones 1976).
Eurydice pulchra Leach, 1815. England. Coast of India (Pillai 1967).
Excirolana affinis (Jones, 1971). Turtle Bay, Watamu, Kenya.
Excirolana geniculata Jones, 1971. Turtle Bay, Watamu, Kenya.
Excirolana latipes (Barnard, 1914b). False Bay, South Africa, intertidal. Lüderitz, Namibia to Mozambique, intertidal-13 m (Kensley 1978d).
Excirolana natalensis (Vanhöffen, 1914). Natal, South Africa. Madagascar (Roman 1970). Plettenberg Bay to Natal, South Africa (Kensley 1978d).
Excirolana orientalis (Dana, 1853). Sulu Sea. Nicobar Is.; Watamu, Kenya; Madagascar; Persian Gulf; Red Sea; Malaysia; Papua New Guinea; NE Australia (Jones 1974, 1979).
Gnatholana mandibularis Barnard, 1920. Off East London, South Africa, 155 m. Saldanha Bay to East London, South Africa, 155-170 m (Kensley 1978d).
Limicolana dinjerra Bruce, 1986a. Northern Territory & Western Australia, intertidal.
* *Metacirolana* sp. nov. Off Somalia, 59-70 m.
Metacirolana anocula (Kensley, 1984b). Off Transkei & Natal, South Africa, 690-850 m.
Metacirolana arnaudi Kensley, 1989. St. Paul & Amsterdam Is., 75-1680 m.
Metacirolana bicornis (Kensley, 1977). Off Natal, South Africa, 550 m.
Metacirolana convexissima (Kensley, 1984b). Transkei, southern Africa, 150-200 m.
Metacirolana fishelsoni (Bruce & Jones, 1978). Gulf of Aqaba, Red Sea, shallow infratidal.
Metacirolana japonica non Hansen, 1890. Madagascar (Roman 1970).
Metacirolana mbudya Bruce, 1981. Tanzania, shallow infratidal.
Metacirolana monodi (Jones, 1976). Aldabra Atoll.
Metacirolana rotunda (Bruce & Jones, 1978). Red Sea. E Africa to N Red Sea (Bruce 1981).
Metacirolana sphaeromiformis non Hansen, 1890. Sri Lanka (Stebbing 1905). * Mahé Is., Seychelles, 1 m.
Natatolana albicaudata Stebbing, 1900. Papua New Guinea. Bombay, India (Barnard 1936).
Natatolana anopthalma (Kussakin & Vasina, 1982a). Kerguelen Is. St. Paul & Amsterdam Is., 1430-1600 m (Kensley 1989).
Natatolana hirtipes (H. Milne Edwards, 1840). South Africa. India, Burma, SE Africa (Barnard 1936).
Natatolana insignis Hobbins & Jones, 1993. Red Sea, 731-1825 m.
Natatolana lurur Bruce, 1986a. Western Australia, 150 m.
Natatolana natalensis (Barnard, 1940). Natal, South Africa; shallow infratidal. Madagascar (Roman 1970).
Natatolana nitida (Hale, 1952). Kerguelen Is., 75-290 m. Crozet Is. (Kensley 1980b).
Natatolana pallidocula (Kussakin & Vasina, 1982a). Kerguelen Is., 310 m.
Natatolana pilula (Barnard, 1955). South Africa, 18-66 m.
Natatolana vieta (Hale, 1925). South Australia. Victoria; Western Australia; New South Wales, Australia, 16-156 m (Bruce 1986a).
Natatolana virilis (Barnard, 1940). Algoa Bay, South Africa, 66-80 m.
Neocirolana arabica Javed & Yasmeen, 1990. Karachi, Pakistan, intertidal.

Neocirolana hermitensis (Boone, 1918). Monte Bello Is., Western Australia. Gulf of Carpentaria; Lizard Is., 12-41 m (Bruce 1986a).
Parabathynomus natalensis Barnard, 1924. Off Natal, South Africa, 766 m. Off Mozambique, 230-295 m (Kensley 1978d).
Politolana obtusispina (Kensley, 1975a). Still Bay, South Africa, 120 m.
Pontogelos aselgokeros Stebbing, 1910a. Mauritius.
Pseudolana concinna (Hale, 1925). Western Australia, intertidal. Queensland & Tasmania, Australia (Bruce 1986a).
Seychellana expansa Kensley & Schotte, 1994. Seychelles, 30 m.

Family Corallanidae
Alcirona maldivensis Stebbing, 1904a. Maldive Is. Cargados Carajos Shoals, 54 m (Stebbing 1910a).
Alcirona multidigitata (Dana, 1853). Borneo. * Off Somalia, 60-70 m.
Alcirona pearsoni Monod, 1924. Sri Lanka.
Argathona hirsuta Hobbins & Jones, 1993. Red Sea, 731-760 m.
Argathona macronema (Bleeker, 1857). Java Sea. Kenya; Madagascar; Chagos Archipelago; Red Sea, Gulf of Aqaba; Gulf of Suez; Indonesia; Australia; New Guinea; Fiji; Gilbert Is. (Delaney 1989).
Argathona muraeneae Bal & Joshi, 1959. Bombay, India.
Argathona normani Stebbing, 1905. Sri Lanka. Coast of India, 51-57 m (Barnard 1936).
Argathona rhinoceros (Bleeker, 1857). Zanzibar; Aldabra Atoll; Persian Gulf; India; Sri Lanka; Thailand; Java; Vietnam; Enewetak Atoll (Delaney 1989).
Corallana africana Barnard, 1914b. Algoa Bay. Keurbooms River to Zululand, South Africa (Kensley 1978d). Madagascar (Roman 1970).
Corallana basalis (Heller, 1861). Nicobar Is.
Corallana furcilla Barnard, 1955. Mozambique, estuarine.
Corallana hirsuta Schioedte & Meinert, 1879. Philippines. Maldive Is. (Stebbing 1904a). * Aldabra Atoll; Mahé Is, Seychelles, 0.5-1 m.
Corallana kulai Bruce, 1982. Western Australia, 150 m.
Corallana nodosa Schioedte & Meinert, 1879. Philippines. Sri Lanka (Stebbing 1904b). Travancore, India (Pillai 1954).
Corallana sculpta H. Milne Edwards, 1840. Malabar coast of India.
Corilana erythraea Kossmann, 1880. Red Sea.
Lanocira anasicula Jones, 1982. Watamu, Kenya. * Madagascar, 1 m.
Lanocira gardineri Stebbing, 1904a. Maldive Is. Philippines, Richardson 1910. Sri Lanka (Nierstrasz 1931). Travancore, India (Pillai 1954). South Africa, 4-110 m (Kensley 1978d). Kenya (Jones 1982). * Mozambique; Madagascar, 1 m; Mahé Is., Seychelles, 3-5 m, Kenya, 0-0.5 m (Müller 1991i).
Lanocira glabra Jones, 1982. Watamu, Kenya. Kenya, 0-0.5 m (Müller 1991i).
Lanocira latifrons Stebbing, 1910b. Red Sea. Mozambique (Barnard 1955).
Lanocira rotundicauda Stebbing, 1904a. Maldive Is., 37 m. Sri Lanka (Nierstrasz 1931). Travancore, India (Pillai 1954).
Lanocira zeylanica Stebbing, 1905. Sri Lanka. Travancore, India (Pillai 1954). Kenya.
Tachaea spongillicola Stebbing, 1907. India..

Family Cymothoidae

Agarna malayi Tiwari, 1952. India.
Agarna pustulosa (Pillai, 1954). India.
Amblycephalon indicus Pillai, 1954. Travancore, India.
Anilocra acuminata Haller, 1880. Réunion Is.
Anilocra ankistra Bruce, 1987a. Western Australia. * Kenya, 240-450 m.
Anilocra capensis Leach, 1818. South Africa. Walvis Bay to East London, South Africa (Kensley 1978d).
Anilocra coxalis Schioedte & Meinert, 1881. Zanzibar.
Anilocra dimidiata Bleeker, 1857. Java. Travancore, India (Pillai 1954). Sri Lanka; Java Sea; Papua New Guinea; Philippines (Nierstrasz 1931). * Mozambique Channel, 58 m.
Anilocra leptosoma Bleeker, 1857. Java. Indonesia; Philippines (Schioedte & Meinert 1881). Gulf of Suez (Monod 1933). Travancore, India (Pillai 1954). Natal, South Africa & Delagoa Bay, Mozambique (Kensley 1978d).
Anilocra monoma Bowman & Tareen, 1983. Persian Gulf.
Anilocra nemipteri Bruce, 1987a. Western Australia.
Anilocra pomacentri Bruce, 1987a. Western Australia.
Anilocra soelae Bruce, 1987a. Western Australia.
Catoessa gruneri Bowman & Tareen, 1983. Persian Gulf.
Ceratothoa carinata (Bianconi, 1869). Mozambique.
Ceratothoa guttata (Richardson, 1910). Philippines. Red Sea; Madagascar; Gulf of Carpentaria, Australia; Taiwan (Bruce & Bowman 1989).
Ceratothoa hemiramphi (Pillai, 1954). Travancore, India.
Ceratothoa imbricata (Fabricius, 1787). New Zealand. Table Bay, South Africa to Mozambique (Kensley 1978d). Red Sea; Indonesia; Java; China; Australia (Nierstrasz 1931).
Ceratothoa retusa (Schioedte & Meinert, 1883). Mozambique. South Africa (Barnard 1925b). N Australia; W Irian Jaya (Bruce & Bowman 1989). * Persian Gulf.
Ceratothoa trigonocephala Leach, 1818. Locality unknown. Natal, South Africa; South Australia; Java; China; Japan; New Hebrides (Schioedte & Meinert 1883).
Cinusa tetrodontis Schioedte & Meinert, 1884. Indian Ocean; Pacific Ocean.
Creniola laticauda (Schioedte & Meinert, 1881). SE Australia. Western Australia, 33 300 m (Bruce 1987b).
Creniola saurida (Avdeev, 1977). NW Australia. Western Australia; Northern Territory; Queensland, 220 m (Bruce 1987b).
Cymothoa asymmetrica Pillai, 1954. Travancore, India.
Cymothoa borbonica Schioedte & Meinert, 1884. Réunion Is. Maldive Is. (Stebbing 1904a). Madagascar (Barnard, 1960). Durban, South Africa to Mozambique (Kensley 1978d).
Cymothoa bychowskyi Avdeev, 1979. NW & Western Australia.
Cymothoa carangii Avdeev, 1979. NW Australia.
Cymothoa cinerea Bal & Joshi, 1959. Bombay, India.
Cymothoa eremita (Brunnich, 1783). India. Zanzibar (Stebbing 1910a). Sri Lanka (Monod 1924). Red Sea (Monod 1933).
Cymothoa gadorum Brocchi, 1875. St. Paul Is.
Cymothoa indica Schiodete & Meinert, 1884. Thailand. Australia; India (Hale 1926).
Cymothoa leaschenaultii Leach, 1818. Pondicherry, India (Miers 1880).
Cymothoa paradoxa Haller, 1880. Indian Ocean.

Cymothoa parupenei Avdeev, 1979. NW Australia.
Cymothoa pulchrum Lanchester, 1902. Indonesia. Sri Lanka (Monod 1924).
Cymothoa rotunda Avdeev, 1979. Western Australia.
Cymothoa rotundifrons Haller, 1880. Mauritius.
Cymothoa selari Avdeev, 1978a. Red Sea.
Elthusa foveolata (Hansen, 1897). Sri Lanka.
Elthusa nanoides (Stebbing, 1905). Sri Lanka. Gulf of Suez (Monod 1933).
Elthusa propinqua (Richardson, 1904). Japan. Philippines (Richardson 1910). Arabian Sea; Maldives (Barnard 1936).
Elthusa raynaudii (H. Milne Edwards, 1840). South Africa. Chile (Menzies 1962a). St. Paul & Amsterdam Is. (Kensley 1976). Japan; Tristan da Cunha (Sivertsen & Holthuis 1980). Australia (Bruce 1990).
Elthusa turgidula (Hale, 1926). Western Australia.
Glossobius anctus Bruce & Bowman, 1989. Northwest Shelf, Western Australia. Queensland, Australia; Japan; Hawaii (Bruce & Bowman 1989).
Ichthyoxenus asymmetrica Ahmed, 1970a. Persian Gulf.
Ichthyoxenus micronyx (Miers, 1880). Mauritius.
Ichthyoxenus tanganyikae (Fryer, 1965). Tanzania.
Joryma brachysoma (Pillai, 1954). India.
Joryma engraulidis (Barnard, 1936). Off mouth of Devi River, India.
Joryma sawayah Bowman & Tareen, 1983. Persian Gulf.
Joryma tartoor (Pillai, 1954). Travancore, India.
Lobothorax laevis Richardson, 1910. Philippines. Bay of Bengal, Burma (Barnard 1936).
Mothocya arroser Bruce, 1986b. Kenya.
Mothocya colletti Bruce, 1986b. Singapore, Straits of Malacca. Thailand; Philippines; Caroline Is.; Queensland; Tahiti (Bruce 1986).
Mothocya karobran Bruce, 1986b. Queensland. Pakistan; New South Wales (Bruce 1986).
Mothocya melanosticta (Schioedte & Meinert, 1884). Japan. Chinde, Mozambique, Barnard 1914b. South Australia (Hale, 1929). Hawaii; S Africa (Nierstrasz 1931). Red Sea (Avdeev 1978a).
Mothocya plagulaphora (Haller, 1880). Mauritius. Sri Lanka; Kenya; Zanzibar; Madagascar; Comoro Is.; Somalia; India; Indonesia; Queensland (Bruce 1986).
Mothocya renardi (Bleeker, 1857). Indonesia. Australia; India; Philippines (Hale 1926). Mozambique; New South Wales; Western Australia; Gulf of Thailand; Indonesia; Papua New Guinea; Japan, Philippines (Bruce 1986).
Nerocila aculeata H. Milne Edwards, 1840. Locality unknown. Brazil (Dana 1853). Indian Ocean (Schioedte & Meinert 1883). East Indies (Nierstrasz 1931).
Nerocila armata Dana, 1853. Brazil. South Africa; Mauritius; Senegal; Gabon; Congo (Nierstrasz 1931).
Nerocila blainvillei Leach, 1818. Locality unknown. Mozambique (Hilgendorf 1878). South Australia (Schioedte & Meinert 1881).
Nerocila cephalotes Schioedte & Meinert, 1881. Cape Agulhas, South Africa; Senegal; Gabon.
Nerocila exocoeti Pillai, 1954. Travancore, India.
Nerocila heterozota Ahmed, 1970b. Persian Gulf.
Nerocila kisra Bowman & Tareen, 1983. Persian Gulf.
Nerocila livida Budde Lund, 1908. Zanzibar.

Nerocila longispina Miers, 1880. Malabar, India.

Nerocila madrasensis Ramakrishna & Ramaniah, 1978. India.

Nerocila monodi Hale, 1940. Queensland, Australia. Northern Territory, North West Shelf, Australia (Bruce 1987b).

Nerocila orbignyi (Guérin Méneville, 1832). Greece. South Africa; Western Australia; South Australia; New Zealand (Bruce 1987b). * Off Somalia, 27-31 m. * India, 27-55 m.

Nerocila phaiopleura Bleeker, 1857. Java. Bay of Bengal, Barnard 1936. Natal, South Africa (Kensley 1978d). Western Australia (Bruce 1987b).

Nerocila pigmentata Bal & Joshi, 1959. Bombay, India.

Nerocila pulicatensis Jayadev Babu & Sanjeeva Raj, 1980. India.

Nerocila recurvispina Schioedte & Meinert, 1881. Calcutta, India.

Nerocila serra Schioedte & Meinert, 1881. Indonesia. India (Barnard 1936). St. Lucia, South Africa; Delagoa Bay, Mozambique (Kensley 1978d). Australia (Bruce 1987b).

Nerocila sigani Bowman & Tareen, 1983. Persian Gulf. * Gujurat, India, 34-37 m. * Andaman Sea.

Nerocila sundaica Bleeker, 1857. Java. Indian seas to East Indies (Barnard 1936).

Nerocila trichiura (Miers, 1877). Mauritius. Chagos Archipelago; Durban, South Africa (Kensley 1978d). * Comoro Is., 3500 m.

Ourozeuktes owenii (H. Milne Edwards, 1840). Locality unknown. Australia; Kerguelen Is. (Hale 1929).

Pleopodias elongatus Richardson, 1910. Philippines. Andaman Is. (Barnard 1936).

Renocila curtipinnata Bruce, 1991b. NW Australia.

Renocila heterozota Bowman & Mariscal, 1968. Seychelles.

Renocila ovata Miers, 1880. Malaysia. Andaman Is. (Barnard 1936). Western Australia (Bruce 1987c).

Renocila rogans (Stebbing, 1923). South Africa, 72 m.

Rhiothra callipia Schioedte & Meinert, 1884. Sri Lanka.

Ryukyua circularis (Pillai, 1954). India.

Family Limnoriidae

Limnoria agrostisa Cookson, 1991. Western Australia.

Limnoria andamanensis Rao & Ganapati, 1969. Andaman Is.

Limnoria antarctica (Pfeffer, 1887). Antarctic. Kerguelen Is.; South Georgia Is. (Menzies 1957).

Limnoria bituberculata (Pillai, 1957). Quilon, India.

Limnoria bombayensis Pillai, 1961. Bombay, India.

Limnoria clarkae (Kensley & Schotte, 1987). Belize. * Aldabra Atoll.

Limnoria gibbera Cookson, 1991. South Australia; Western Australia.

Limnoria indica Becker & Kampf, 1958. Mandapam & Madras, India. Mexico; Belize; Hong Kong; Philippines; Australia (Kensley & Schotte 1989). * Aldabra Atoll.

Limnoria insulae Menzies, 1957. Fiji. Andaman Is., Ganapati & Rao 1960. Belize; Guam; Palmyra Is.; Caroline Is. (Kensley & Schotte 1989). * Aldabra Atoll.

Limnoria pfefferi Stebbing, 1904a. Minikoi Atoll. India, Andaman Is. (Ganapati & Rao 1960). Gulf of Mexico; Caribbean; New Guinea; Philippines; Aldabra Atoll (Kensley & Schotte 1989).

Limnoria platycauda Menzies, 1957. Curaçao. Andaman Is. (Ganapati & Rao 1960).

Caribbean; Aldabra Atoll (Kensley & Schotte 1989).

Limnoria quadripunctata Holthuis, 1949. Netherlands. St. Paul & Amsterdam Is., 30 m (Kensley 1976). Australia (Cookson 1991).

Limnoria segnoides Menzies, 1957. Japan. Madagascar (Roman 1970).

Limnoria septima Barnard, 1936. Andaman Is., 93-125 m.

Limnoria stephenseni Menzies, 1957. New Zealand. Marion Is. (Wolff 1990).

Limnoria tripunctata Menzies, 1951b. California. India; Australia (Cookson 1991).

Limnoria uncapedis Cookson, 1991. Northern Territory; South Australia; Western Australia. Andaman Is. (Ganapati & Rao 1960).

Limnoria unicornis Menzies, 1957. Caroline Is. Andaman Is. (Ganapati & Rao 1960). Western Australia (Cookson 1991). * Aldabra Atoll, 0.5 m.

Lynseia annae Cookson & Poore, 1994. Western Australia.

Lynseia dianae Cookson & Poore, 1994. Western Australia.

Paralimnoria andrewsi (Calman, 1910). Christmas Is., 0-4 m. Andaman Is. (Ganapati & Rao 1960). Caribbean; Hawaii; Samoa; Japan (Kensley & Schotte 1989). * Aldabra Atoll.

Family Serolidae

Acutiserolis bromleyana (Suhm, 1876). New Zealand. Kerguelen Is., 450 m (Kussakin & Vasina 1982a).

Caecoserolis apheles (Schotte, 1992). Madagascar Plateau, 950 m.

Caecoserolis brinki (Kensley, 1978a). Natal, South Africa, 680 m.

Caecoserolis monodi (Cals, 1979). Kerguelen Is., 250 m.

Ceratoserolis trilobitoides (Eights, 1833). South Shetland Is. Kerguelen Is.; Patagonia; Cape Horn; Graham Region, off Victoria Land, Antarctic (Nordenstam 1933).

Septemserolis septemcarinata (Miers, 1875a). Crozet Is., 1-700 m. Marion & Prince Edward's Is.; Kerguelen Is. (Beddard 1884b). South Shetland Is.; South Georgia Is.; Antarctic Peninsula (Brandt 1988).

Serolis antarctica Beddard, 1884a. Crozet Is., 2516-2928 m. Brazil, 732 m (Beddard 1884b). Mozambique, 1611-1629 m (Schotte 1992).

Serolis aspera Sheppard, 1933. South Georgia Is. Kerguelen Is., 50 m (Kussakin & Vasina 1980a).

Serolis zoiphila Stechow, 1921. Kerguelen Is., 2516-2928 m.

Spinoserolis latifrons (White, 1847). Kerguelen Is., 1-378 m. Crozet Is. (Miers 1875b). New Zealand (Bastida & Torti 1969).

Family Sphaeromatidae

Afrocerceis kenyensis Müller, 1995. Kenya, intertidal.

Agostodina munta Bruce, 1994a. Western Australia, 52-82 m.

Agostodina shara Bruce, 1994a. Western Australia, 36-44 m.

Artopoles natalis Barnard, 1920. Natal, South Africa, 12 m.

Bathycopea typhlops Tattersall, 1905. Ireland. Natal, South Africa, 680 m (Kensley 1978a). New South Wales; Tasmania; Victoria, 135-1277 m (Bruce 1991a).

Cassidias africana Barnard, 1920. Natal, South Africa, 80 m. * Mozambique, 110-112 m. * Somalia, 59-90 m.

Cassidina extenda Joshi & Bal, 1992. India, shallow.

Cassidina pulchra Chilton, 1924. Chilka Lake, India.

* *Cassidinidea* sp. nov. Mahé Is., Seychelles, intertidal.

Cassidinidea quadricarinata Pillai, 1954. Travancore, India. Madagascar (Roman 1970).

Cassidinopsis emarginata (Guérin Méneville, 1843). Falkland Is. Kerguelen Is. (Studer 1884). Crozet Is. (Stephenson 1947). * Heard Is.

Cassidinopsis maculata (Studer, 1884). Kerguelen Is.

Cerceis bicarinata Barnard, 1936. Andaman Is.

Cerceis biforamina Javed & Yousuf, 1996. Pakistan, intertidal.

Cerceis granulata Pillai, 1954. Travancore, India.

Cerceis tuberculata Müller, 1991h. Sri Lanka, intertidal.

Cercosphaera dilkera Bruce, 1994b. Western Australia, 0.5-52 m.

Cercosphaera wirritin Bruce, 1994b. Western Australia, 1-23 m.

Cilicaea antennalis Miers, 1884. Western Australia, shallow.

Cilicaea beddardi Stebbing, 1905. Sri Lanka.

Cilicaea canaliculata (Thomson, 1879). New Zealand, 531 m. Bay of Bengal (Barnard 1936).

Cilicaea latreillei Leach, 1818. Type locality unknown. Red Sea; Indonesia (Studer 1884). Sri Lanka (Stebbing 1905). Kenya (Stebbing 1910a). East London to Natal, South Africa, 26-104 m; Australia; Philippines (Kensley 1978d). * Mozambique, 140 m. * Somalia.

Cilicaeopsis sculpta Baker, 1928. Western Australia.

Cilicaeopsis whiteleggei (Stebbing, 1905). Sri Lanka.

Clianella amblysinus (Pillai, 1954). Travancore, India.

* *Cymodoce* sp. nov. A. Persian Gulf, 1-4 m.

* *Cymodoce* sp. nov. B. Mahé Is., Seychelles, 0.5-1 m.

Cymodoce acanthiger Barnard, 1914b. East London, South Africa, 600 m.

Cymodoce africana Barnard, 1914b. Cape Peninsula & East London, South Africa, 262-580 m.

Cymodoce alis Barnard, 1955. False Bay to Port Elizabeth, South Africa, 15-36 m.

Cymodoce amplifrons (Stebbing, 1902). Natal, South Africa, 48 m. Port Elizabeth to Natal, South Africa, 27-48 m (Kensley 1978d).

Cymodoce bicarinata Stebbing, 1904a. Maldive Is. Sri Lanka, Stebbing 1905. Zanzibar (Stebbing 1910a).

Cymodoce comans Barnard, 1914b. False Bay to Agulhas, South Africa, 10-40 m. Natal, South Africa, 5 to 19 m (Kensley 1978d).

Cymodoce cryptodoma Barnard, 1920. Natal, South Africa, 80 m.

Cymodoce davieae Kensley & Buxton, 1984. Port Elizabeth, South Africa, 33 m.

Cymodoce erythraea Nobili, 1906b. Red Sea.

Cymodoce excavans Barnard, 1920. False Bay, South Africa, 25-40 m.

Cymodoce inornata Whitelegge, 1902. SE Australia, 99-101 m. Sri Lanka (Stebbing 1905). Philippines, 110 m (Richardson 1910).

Cymodoce lis Barnard, 1955. Algoa Bay, South Africa.

Cymodoce longistylis Miers, 1884. Torres Strait. Nicobar Is. (Barnard 1936). Kerala & Madras, India (Pillai 1961). * Phuket Is., Thailand.

Cymodoce madrasensis (Srinivasan, 1959). India.

Cymodoce mammifera Haswell, 1881. Queensland, Australia. Western Australia (Tattersall 1922). Travancore, India (Pillai 1954). Madagascar (Roman 1970).

Cymodoce natalensis Barnard, 1920. Natal, South Africa, 80 m. Algoa Bay to Natal,

South Africa, 80 m (Kensley 1978d).

Cymodoce pelsarti Tattersall, 1922. Western Australia.

Cymodoce picta Brocchi, 1875. St. Paul Is.

Cymodoce pilosa (H. Milne Edwards, 1840). Mediterranean. Gulf of Suez (Stebbing 1910b).

Cymodoce radiata Barnard, 1957a. East London, South Africa, 150 m.

Cymodoce richardsoniae Nobili, 1906b. Red Sea.

Cymodoce setulosa (Stebbing, 1902). Algoa Bay, South Africa, 20-30 m. Saldanha Bay to Natal, South Africa, 7-42 m (Kensley 1978d).

Cymodoce tetrathele Barnard, 1920. East London, South Africa, 85 m.

Cymodoce truncata Leach, 1818. England. Gulf of Suez (Stebbing 1910b). E Atlantic, Mediterranean (Monod 1933).

Cymodoce umbonata Barnard, 1914b. False Bay, South Africa, 100 m.

Cymodoce uncinata Stebbing, 1902. Algoa Bay, South Africa, 49 m.

Cymodoce unguiculata Barnard, 1914b. Saldanha Bay, Table Bay, South Africa, intertidal. South Australia (Hale 1929).

Cymodoce valida (Stebbing, 1902). Algoa Bay, South Africa, 18-29 m. Saldanha Bay to Natal, South Africa, 8-160 m (Kensley 1978d).

Cymodoce velutina Kensley, 1975a. Still Bay, South Africa, 120 m.

Cymodoce zanzibarensis Stebbing, 1910a. Zanzibar. Mozambique (Barnard 1955).

Cymodocella algoense (Stebbing, 1875). Algoa Bay, South Africa, intertidal. Nahoon River, South Africa, 2 m (Kensley & Bursey 1996).

Cymodocella cancellata Barnard, 1920. East London, South Africa, 44 m.

Cymodocella pustulata Barnard, 1914b. Table Bay & False Bay, South Africa, intertidal. Lüderitz, Namibia to East London, South Africa, intertidal-10 m (Kensley 1978d).

Cymodocella sublevis Barnard, 1914b. Table Bay, South Africa, intertidal. Lüderitz, Namibia to East London, South Africa, intertidal-18 m (Kensley 1978d). * Madagascar.

Diclidocella bullata Bruce, 1995. South Australia, Western Australia, 3-17 m.

* *Dynamenella* sp. nov. A. Off Sumatra.

* *Dynamenella* sp. nov. B. Nosy Be, Madagascar, intertidal.

Dynamenella bullejiensis Javed & Ahmed, 1988a. Pakistan, intertidal.

Dynamenella dioxus Barnard, 1914b. Table Bay & False Bay, South Africa, intertidal. Lüderitz, Namibia to False Bay, South Africa (Kensley 1978d).

Dynamenella granulata Javed & Ahmed, 1988a. Pakistan, intertidal.

Dynamenella kraussi Barnard, 1914b. Table Bay to Natal, South Africa, intertidal.

Dynamenella mossambica Ortiz, Berze Freire & Wasikete, 1990. Mozambique.

Dynamenella navicula Barnard, 1940. Algoa Bay & East London, South Africa.

Dynamenella quilonensis Pillai, 1954. Travancore, India, estuarine.

Dynamenella savignii (H. Milne Edwards, 1840). Egypt. Natal, South Africa (Krauss 1843).

Dynamenella scaptocephala Messana, 1990. Somalia.

Dynamenella taurus Barnard, 1940. Port Nolloth & East London, South Africa.

Dynoides indicus Müller, 1991h. Sri Lanka, intertidal.

Dynoides serratisinus Barnard, 1914b. Natal, South Africa, intertidal. Natal, South Africa to Mozambique, intertidal (Kensley 1978d).

Eterocerceis somala Messana, 1990. Somalia, 3 m.

Exosphaeroma antikraussi Barnard, 1940. Table Bay, Algoa Bay, South Africa, intertidal. Saldanha Bay to Algoa Bay, South Africa (Kensley 1978d).

Exosphaeroma estuarium Barnard, 1951. Natal, South Africa, estuarine.

Exosphaeroma gigas (Leach, 1818). Unknown. Kerguelen Is. (Vanhöffen 1914). Tasmania; New Zealand; South Africa (Nierstrasz 1931).

Exosphaeroma hylecoetes Barnard, 1940. Table Bay, Plettenberg Bay, East London, South Africa, estuarine.

Exosphaeroma kraussi Tattersall, 1913. St. Sebastian Bay, South Africa. Lüderitz, Namibia to East London, South Africa, intertidal, estuarine (Kensley 1978d).

Exosphaeroma laeviusculum (Heller, 1868). Java. N Namibia to Keurbooms River, South Africa, intertidal, shallow infratidal (Kensley 1978d).

Exosphaeroma parva Chilton, 1924. Chilka Lake, India.

Exosphaeroma planum Barnard, 1914b. Table Bay, South Africa, intertidal. Lüderitz, Namibia to Port Alfred, South Africa, 4-100 m (Kensley 1978d).

Exosphaeroma porrectum Barnard, 1914b. Table Bay, South Africa, intertidal. Lüderitz, Namibia to Port Elizabeth, South Africa, intertidal-5 m (Kensley 1978d).

Exosphaeroma serventii Baker, 1928. Western Australia.

Exosphaeroma varicolor Barnard, 1914b. Saldanha Bay, Table Bay, South Africa, intertidal. N Namibia to Hermanus, South Africa, intertidal to 17 m (Kensley 1978d). South Australia (Hale 1929).

* *Heterodina* sp. nov. Persian Gulf, 3-3.5 m.

Ischyromene australis (Richardson, 1906). Table Bay, South Africa. Lüderitz, Namibia to Hermanus, South Africa, intertidal (Kensley 1978d).

Ischyromene brunnea (Vanhöffen, 1914). St. Paul & Amsterdam Is., intertidal to 4 m.

Ischyromene eatoni (Miers, 1875b). Kerguelen Is. Marion Is., Crozet Is. (Kensley 1980b).

Ischyromene glauerti (Baker, 1928). Western Australia.

Ischyromene huttoni (Thomson, 1879). New Zealand. Lüderitz, Namibia to Natal, South Africa (Kensley 1978d).

Ischyromene intermedia (Baker, 1926). Western Australia.

Ischyromene macrocephala (Krauss, 1843). Table Bay, South Africa. Table Bay to East London, South Africa, intertidal to shallow infratidal (Kensley 1978d).

Ischyromene ovalis (Barnard, 1914b). False Bay, South Africa, intertidal. Lüderitz, Namibia to East London, South Africa, intertidal (Kensley 1978d).

Ischyromene sapmeri (Kensley, 1976). St. Paul & Amsterdam Is., intertidal to 60 m.

Ischyromene scabricula (Heller, 1868). 'Cape', South Africa. Lüderitz, Namibia to Keurbooms River, South Africa, intertidal (Kensley 1978d).

Isocladus excavatus (Baker, 1910). Western Australia.

Juletta mirandae Bruce, 1993. Western Australia, 25-27 m.

Kranosphaera haptomela Bruce, 1992. NW Shelf, Australia, 39-52 m.

Moruloidea darwini (Cunningham, 1871). Tierra del Fuego. Kerguelen Is. (Vanhöffen 1914). Crozet Is. (Kensley 1980b).

Moruloidea fraudatrix (Kussakin & Vasina, 1982a). Kerguelen Is., 152 m.

Moruloidea ornata (Kussakin & Vasina, 1982a). Kerguelen Is., 138 m.

Moruloidea tumida Harrison, 1984. Western Australia.

Oxinasphaera australis (Baker, 1928). Western Australia.

* *Oxinasphaera* sp. nov. A. Mozambique, 73-88 m.

* *Oxinasphaera* sp. nov. B. Somalia, 9-175 m.

* *Oxinasphaera* sp. nov. C. Zanzibar, intertidal-7 m.

* *Oxinasphaera* sp. nov. D. Zanzibar, intertidal-7 m.

Oxinasphaera frailea Bruce, 1997a. Western Australia; Northern Territory, 3-30 m.

Oxinasphaera islaya Bruce, 1997a. Western Australia, 6 m; South Australia, 12 m.

Oxinasphaera kensleyi Bruce, 1997a. South Africa, 90 m.

Oxinasphaera tuberculosa (Stebbing, 1873). Western Australia; South Australia; Victoria; Tasmania, 5-33 m (Bruce 1997a).

Paracassidina bakeri Bruce, 1994a. Western Australia, 6-83 m.

Paracassidina fuscina Bruce, 1994a. Western Australia, 36-54 m.

Paracassidina incompta Bruce, 1994a. Western Australia, 36-81 m.

Paracassidina kutyo Bruce, 1994a. Western Australia, 42 m.

Paracassidina pectinata Baker, 1911. Western Australia.

Paracassidinopsis perlata (Roman, 1974). Madagascar. * Mahé Is., Seychelles, 0.5-3 m.

Paracerceis caudata (non Say, 1818). Madagascar (Roman 1970).

* *Paracilicaea* sp. nov. Mahé Is., Seychelles, 2.5-5 m.

Paracilicaea clavus Barnard, 1955. Mozambique, intertidal.

Paracilicaea cordylina Kensley, 1984b. Natal, South Africa, 280-454 m.

Paracilicaea dakini (Tattersall, 1922). Western Australia.

Paracilicaea eupyga (Nobili, 1906b). Red Sea.

Paracilicaea flexilis Baker, 1928. Western Australia.

Paracilicaea hanseni Stebbing, 1910a. Zanzibar. Gulf of Kutch, India (Monod 1971).

Paracilicaea keijii Javed, 1990. Pakistan.

Paracilicaea mossambica Barnard, 1914b. Mozambique, intertidal. Madagascar (Roman 1970). * Aldabra Atoll, intertidal-10 m. * Mahé Is., Seychelles, 0.5-1 m. * Mauritius. * Red Sea.

Paracilicaea pubescens (H. Milne Edwards, 1840). South Australia. Zanzibar, 18 m, Stebbing 1910a. Indonesia (Nierstrasz 1931).

Paracilicaea setosa Müller, 1995. Kenya, intertidal.

Paracilicaea teretron Barnard, 1955. Mozambique, intertidal. Madagascar (Roman 1970).

Paracilicaea watamuae Müller, 1995. Kenya, intertidal.

Paradella dianae (Menzies, 1962b). Baja California. Western Australia; Queensland; Puerto Rico (Harrison & Holdich 1982). Hong Kong (Bruce 1986c). Arabian Sea (Javed & Ahmed 1987). Florida, USA (Kensley & Schotte 1989).

Paradella harrisoni Müller, 1995. Kenya, intertidal.

Paraimene tuberculata Javed & Ahmed, 1988b. Karachi coast, Pakistan, intertidal.

Paraleptosphaeroma indica Müller, 1990b. Réunion Is. * Madagascar, intertidal.

Parisocladus perforatus (H. Milne Edwards, 1840). St. Paul Is. N Namibia to East London, South Africa (Kensley 1978d) St. Paul & Amsterdam Is., intertidal to 3 m (Kensley 1976).

Parisocladus stimpsoni (Heller, 1868). 'Cape', South Africa. Lüderitz, Namibia to East London, South Africa, intertidal (Kensley 1978d).

Pistorius sasyamai Nunomura, 1990. Phuket Is., Thailand.

Pseudocerceis seleneides Messana, 1988. Somalia.

Pseudosphaeroma barnardi Monod, 1931b. Hout Bay, South Africa. Hout Bay to Knysna, South Africa, estuarine (Kensley 1978d).

Sphaeramene polytylotos Barnard, 1914b. Plettenberg Bay, False Bay, South Africa, intertidal. Lüderitz, Namibia to Plettenberg Bay, intertidal-22 m (Kensley 1978d).

Sphaeroma annandalei annandalei Stebbing, 1911. W Bengal. Natal, South Africa, intertidal (Kensley 1978d).

Sphaeroma annandalei travancorensis Pillai,1955. India.
Sphaeroma bigranulatum Budde Lund,1908. Zanzibar.
Sphaeroma conglobator (Pallas,1766). Gulf of Suez.
Sphaeroma globicauda Dana,1853. India. Laccadive Is. (Stebbing 1904a).
Sphaeroma granti Walker & Scott,1903. Gulf of Aden.
Sphaeroma hookeri Leach,1814. East Africa.
Sphaeroma serratum (Fabricius, 1787). Locality unknown. Gulf of Suez (Monod 1933). Natal, South Africa; Mozambique, intertidal; North Atlantic; Mediterranean Sea; Black Sea (Kensley 1978d). * Madagascar, intertidal. * Phuket Is., Thailand.
Sphaeroma sieboldii Dollfus, 1889. Japan. Madagascar (Roman 1970).
Sphaeroma terebrans Bate, 1866. India. Queensland, Australia (Baker 1926). Brazil; Florida, USA; Travancore, India; Sri Lanka; Zanzibar (Pillai 1955). Madagascar (Roman 1970). Knysna, South Africa to Mozambique, intertidal (Kensley 1978d). * Aldabra Atoll.
Sphaeroma triste Heller, 1861. Nicobar Is. Malay Peninsula (Pillai 1961).
Sphaeroma tuberculata Brocchi, 1875. St. Paul Is.
Sphaeroma tuberculata Purosotham & Rao, 1971. Tuticorin, India.
Sphaeroma walkeri Stebbing, 1905. Sri Lanka. India; Suez, Egypt; New South Wales, Australia (Pillai 1961). South Africa; Mozambique, estuarine-5 m (Kensley 1978d).
* *Sphaeromopsis* sp. nov. Mahé Is., Seychelles.
Sphaeromopsis amathitis Holdich & Jones, 1973. Kenya, intertidal. Somalia (Messana 1990). * Persian Gulf.
Sphaeromopsis minutus Javed & Yousuf, 1995. Pakistan, intertidal.
Sphaeromopsis reticulata (Stebbing, 1910b). Red Sea.
Syncassidina aestuaria Baker, 1928. Western Australia.
Tholozodium ocellatum Eleftheriou, Holdich & Harrison, 1980. India, intertidal.
Waiteolana gibbera Harrison, 1984. Western Australia.
Xynosphaera colemani Bruce, 1994b. Nosy Be, Madagascar, 1 m.
Zuzara furcifer Barnard, 1920. Algoa Bay. False Bay to Algoa Bay, South Africa (Kensley 1978d).

SUBORDER GNATHIIDEA
Family Gnathiidae
Bathygnathia porca (Kensley, 1980b). Crozet Is., 245-250 m.
* *Caecognathia* sp. nov. Mauritius, 43 m.
Caecognathia antarctica (Studer, 1884). Patagonia. Kerguelen Is.; coast of Antarctica (Monod 1926a). Prince Edward Is. & Crozet Is., 150-285 m (Kensley 1980b).
Caecognathia wagneri (Monod, 1925). Antarctic. Kerguelen Is. (Kussakin & Vasina 1980b).
* *Elaphognathia* sp. nov. A. Aldabra Atoll, intertidal-6 m.
* *Elaphognathia* sp. nov. B. Somalia, 27-90 m.
Elaphognathia insolita (Stebbing, 1905). Sri Lanka.
Elaphognathia wolffi (Müller, 1989b). Kenya.
Euneognathia gigas (Beddard, 1886). Kerguelen Is., 229 m. Crozet Is. (Kensley 1980b).
* *Gnathia* sp. nov. A. Inhambane, Mozambique 190 m.
* *Gnathia* sp. nov. B. Phuket Is., Thailand, 15 m.
* *Gnathia* sp. nov. C. Red Sea, 0.5-3 m.

* *Gnathia* sp. nov. D. Somalia, 31-39 m.
* *Gnathia* sp. nov. E. Aldabra Atoll, 1 m.
* *Gnathia* sp. nov. F. Persian Gulf, 1-2.5 m.
* *Gnathia* sp. nov. G. Nosy Be, Madagascar, 1.5 m.
* *Gnathia* sp. nov. H. Phuket Is., Thailand, intertidal-10 m.
* *Gnathia* sp. nov. I. Somalia, 27-31 m; Zanzibar, intertidal.
* *Gnathia* sp. nov. J. Aldabra Atoll, intertidal-20 m.
* *Gnathia* sp. nov. K. Zanzibar, 7-25 m.
Gnathia africana Barnard, 1914a. Table Bay, False Bay, South Africa, intertidal. Lambert's Bay to Algoa Bay, South Africa (Kensley 1978d).
Gnathia aldabrensis Schoenichen, 1908. Aldabra Atoll [praniza only].
Gnathia arabica Schotte, 1995. Arabian Sea, 47-79 m.
Gnathia disjuncta Barnard, 1920. Off Knysna, South Africa, 80 m.
Gnathia firingae Müller, 1991e. Réunion Is., intertidal.
Gnathia piscivora Paperna & Por, 1977. Gulf of Aqaba, Red Sea.
Gnathia rhinobatis (Kossmann, 1880). Red Sea. Gulf of Suez (Monod 1933).
Gnathia taprobanensis Monod, 1926a. Sri Lanka.
Gnathia tuberculosa (Beddard, 1886). Kerguelen Is. 55-228 m.

SUBORDER MICROCERBERIDEA
Family Microcerberidae
Coxicerberus andamanensis (Coineau & Rao, 1972). Andaman Is., interstitial.
Coxicerberus anfindicus (Messana, Argano & Baldari, 1978). Maldive Is., interstitial.
Coxicerberus enckelli (Messana, Argano & Baldari, 1978). Andaman Is., interstitial.
Coxicerberus interstitialis (Chappuis, Delaware Deboutteville & Paulian, 1956). Réunion Is., interstitial.
Coxicerberus pauliani (Chappuis & Delamare Deboutteville, 1956). Madagascar, interstitial.
Coxicerberus singhalensis Enckell, 1970. Sri Lanka, interstitial.
Coxicerberus tabai Wägele, 1982. Red Sea.

SUBORDER VALVIFERA
Family Amesopodidae
Amesopous richardsonae Stebbing, 1905. Sri Lanka. * Somalia, 150 m. * Bay of Bengal, 66 m. * Mozambique, 88 m. * Northern Territory, Australia, 14 m.

Family Arcturidae
Acantharcturus sp. nov. Southern Mozambique, 1610 m.
Agularcturus granulatus Kensley, 1984b. East London, South Africa, 90 m.
Antarcturus bicornis Kensley, 1984b. Off Natal, South Africa, 560-620 m.
Antarcturus furcatus furcatus (Studer, 1882). Kerguelen Is., 110 m. Heard Is., 135 m. S Indian Ocean, 3015 m (Beddard 1886). Crozet Is., 152-395 m (Kussakin & Vasina 1982b).
Antarcturus glacialis (Beddard, 1886). S Indian Ocean, 3015 m.
Antarcturus kladophoros Stebbing, 1908. Mossel Bay, South Africa.. Saldanha Bay to Natal, South Africa, 12-150 m (Kensley 1978d).
* *Arcturella* sp. nov. S of East London, South Africa, 80 m.
Arcturella brevipes Barnard, 1920. Agulhas Bank, False Bay, South Africa, 40 m. False

Bay to Algoa Bay, South Africa, 22-40 m (Kensley 1978d).
Arcturella lineata (Stebbing, 1873). Algoa Bay. Lüderitz, Namibia to East London, South Africa, 20-200 m (Kensley 1978d).
Arcturella lobulata Barnard, 1925b. Natal, South Africa, 80 m. False Bay to Natal, South Africa, 80-120 m (Kensley 1978d).
Arcturella longipes Barnard, 1920. Table Bay, 40 m, Cape St. Francis, 150 m, South Africa.
Arcturella pustulata Barnard, 1920. Natal, South Africa, 26-80 m.
Arcturina cylindralis Pillai, 1963a. Quilon, India, intertidal.
Arcturina hexagonalis Barnard, 1925b. Mossel Bay, South Africa, 80 m. False Bay to Mossel Bay, South Africa, 5 to 87 m (Kensley 1978d).
Arcturina triangularis Barnard, 1957b. Mossel Bay & Still Bay, South Africa, 9-54 m.
* *Arcturinoides* sp. nov. Persian Gulf, 1.3-2.3 m.
Arcturinoides gibbosus Müller, 1989c. Gulf of Aden. * Zanzibar, 16 m.
Arcturinoides sexpes Kensley, 1977. Natal, South Africa, 26-60 m. * Mozambique, 73-92 m.
Astacilla amblyura Stebbing, 1905. Gulf of Manaar, intertidal. India, intertidal (Pillai 1963a).
Astacilla corniger (Stebbing, 1873). Algoa Bay. Saldanha Bay to East London, South Africa, 13-160 m (Kensley 1984b).
Astacilla eminentia Kensley, 1984b. Zululand to Transkei, South Africa, 560-775 m.
Astacilla gibbossa Pillai, 1954. Travancore, India. Quilon, India (Pillai 1963a).
Astacilla longispina (Kensley, 1978a). East London, South Africa, 150-200 m. * Mozambique, 347 m.
Astacilla tranquilla (Kensley, 1975a). East London, South Africa, 80-90 m.
Austroarcturus foveolatus Kensley, 1975a. South Africa, 71-200 m.
Chaetarcturus aculeatus (Kussakin, 1967). Patagonia, 500 m. Crozet Is., 90-340 m (Kensley 1980b). Marion & Prince Edward Is. (Branch et al. 1991).
Chaetarcturus brunneus brunneus (Beddard, 1886). Crozet Is., 2928 m.
Chaetarcturus franklini (Hodgson, 1902). Antarctic. Falkland Is. (Nordenstam 1933). Crozet Is., 150- 280 m (Kussakin & Vasina 1982b).
Dolichiscus studeri (Beddard, 1886). Kerguelen Is., 40-229 m.
Dolichiscus subantarcticus Kussakin & Vasina, 1980b. Kerguelen Is., 460 m.
Fissarcturus hirticornis (Monod, 1926b). Antarctic, 120-500 m. Marion & Crozet Is. (Kensley 1980b).
Holidotea unicornis Barnard, 1920. Knysna, Agulhas Bank, 90-100 m. Agulhas Bank, South Africa, 60-240 m (Kensley 1978d).
Neastacilla sp. nov. Kuwait, Persian Gulf, 1-3 m.
Neastacilla bacillus (Barnard, 1920). Knysna, Natal, South Africa, 80-110 m. False Bay to Natal, South Africa, 5-400 m (Kensley 1984b). * Mozambique, 110 m.
Neastacilla kerguelensis (Vanhöffen, 1914). Kerguelen Is., 183 m. Crozet Is., 173 m (Kussakin & Vasina 1982b).
Neastacilla marionensis (Beddard, 1886). Marion Is., 200 m. Kerguelen Is.; Marion, Prince Edward & Crozet Is. (Kensley 1980b).
* *Neoarcturus* sp. nov. S Mozambique, 1610 m.
Neoarcturus barnardi (Kensley, 1984b). East London, South Africa, 90 m.
Neoarcturus biserialis (Kensley, 1978a). Natal, South Africa, 1300 m. * Mozambique, 2750 m.

Neoarcturus dayi (Kensley, 1977). Saldanha Bay to East London, South Africa, 75-240 m.

Neoarcturus halei (Kensley, 1984b). Zululand to Transkei, South Africa, 150-775 m.

Neoarcturus longispinus (Kensley, 1984b). East London, South Africa, 150-775 m.

Neoarcturus mawsoni (Hale, 1946). Kerguelen Is., 45-47 m. Weddell Sea (Brandt 1990).

Neoarcturus nordenstami (Kensley, 1984b). East London, South Africa, 90 m.

Neoarcturus oudops Barnard, 1914a. Off Cape Point, South Africa, 1380 m. Zululand to Transkei, South Africa, 400-850 m (Kensley 1984b).

Neoarcturus quadriconus (Kensley, 1975a). False Bay to East London, South Africa, 22-80 m.

Neoarcturus stebbingi (Beddard, 1886). Kerguelen Is., 60 m. South Georgia Is., 75-310 m (Nordenstam 1933). South Africa, 54 m (Nierstrasz 1941).

Neoarcturus youngi (Kensley, 1978a). Natal, South Africa, 550-680 m.

Oxyarcturus beliaevei (Kussakin, 1967). Kerguelen Is., 1580-1601 m. Off Cape Point, South Africa, 3000 m (Kensley, 1978d). Heard Is., Davis Sea (Brandt 1990).

Oxyarcturus spinosus (Beddard, 1886). Marion Is., 2475 m. Kerguelen Is.; South Georgia Is.; S Atlantic; Antarctic (Nordenstam 1933).

Pleuroprion chuni (zur Strassen, 1902). Agulhas Bank, South Africa, 200 m.

Rectarcturus laevis (Kensley, 1975a). False Bay to Agulhas Bank, South Africa, 48-81 m.

Rectarcturus ornatus (Kensley, 1975a). Agulhas Bank & Transkei, South Africa, 150-670 m.

Rectarcturus similis (Barnard, 1925b). Saldanha Bay to Cape Point, South Africa, 73-220 m. Lüderitz, Namibia, to East London, South Africa, 15-24 m (Kensley 1978d).

Spinarcturus natalensis Kensley, 1978a. Natal, South Africa, 550-680 m.

Family Austrarcturellidae

Austracturella aphelura Poore & Bardsley, 1992. Western Australia, 39-120 m.

Austrarcturella corona Poore & Bardsley, 1992. Western Australia, 134 m.

Austrarcturella pictila Poore & Bardsley, 1992. Western Australia, 43-142 m.

Family Chaetiliidae

Stegidotea latipoda Poore, 1990. Northwest Shelf, Western Australia, 78 m.

Stegidotea pinnata Poore, 1985. Bass Strait, Tasmania; NW Shelf, Western Australia, 42-140 m.

Family Holognathidae

Cleantioides albaniensis Poore & Lew Ton, 1990. South Australia; Western Australia, 0-25 m.

Cleantioides natalensis (Barnard, 1925b). Natal, South Africa. Bay of Bengal, India; Myanmar (Barnard 1936). Madagascar (Roman 1970).

Cleantis granulosa Heller, 1861. St. Paul Is., intertidal. Tierra del Fuego (Ohlin 1901). Argentina (Nordenstam 1933). St. Paul & Amsterdam Is, littoral (Kensley 1976).

Family Idoteidae

Crabyzos longicaudatus Bate, 1863. South Australia, 8 m. Falkland Is. (Miers 1881). Victoria (Hale 1929). Western Australia; Tasmania (Poore & Lew Ton 1993).

Engidotea cristata Poore & Lew Ton, 1993. South Australia, 15 m, Western Australia, subtidal.

Engidotea lobata (Miers, 1881). South Africa, intertidal.

Euidotea bakeri (Collinge, 1917). South Australia, intertidal. Tasmania, New South Wales, Victoria, South Australia, 0-17 m (Poore & Lew Ton 1993).

Euidotea halei Poore & Lew Ton, 1993. Western Australia, Tasmania, Victoria, South Australia, intertidal-20 m.

Euidotea peronii (H. Milne Edwards, 1840). Australia. South Australia, 4-5 m (Hale 1924c). New Zealand; Cape, South Africa (Nierstrasz 1941).

Glyptidotea lichensteini (Krauss, 1843). Table Bay, South Africa. Lüderitz, Namibia to Transkei, South Africa (Kensley 1978d).

Idotea balthica (Pallas, 1772). NE Atlantic. Red Sea (Miers 1881).

Idotea brevicorna H. Milne Edwards, 1840. Western Australia. Irian Jaya (Poore & Lew Ton 1993).

Idotea indica H. Milne Edwards, 1840. Malabar, India. N Namibia to Table Bay, South Africa, intertidal to shallow infratidal (Kensley 1978d).

Idotea metallica Bosc, 1801. Unknown. Seychelles Mauritius Plateau (Stebbing 1910a). Table Bay, South Africa; Chinde, Mozambique (Barnard 1914a). St. Paul & Amsterdam Is. (Kensley 1976).

Idotea rugosa H. Milne Edwards, 1840. India.

Paridotea aquarii Poore & Lew Ton, 1993. Western Australia, shallow.

Paridotea reticulata Barnard, 1914b. Table Bay, South Africa. St. Paul & Amsterdam Is., 0-5 m (Kensley 1976). Lüderitz, Namibia to False Bay, South Africa (Kensley 1978d).

Paridotea ungulata (Pallas, 1772). Antarctic region (Miers 1881). Falkland Is.; Chile; New Zealand; Brazil (Nierstrasz 1941). Walvis Bay, Namibia to East London, South Africa (Kensley 1978d).

Synidotea fluviatilis Pillai, 1954. Quilon & Cochin, India, estuarine.

Synidotea hirtipes (H. Milne Edwards, 1840). South Africa. Red Sea (Monod 1933). Lüderitz, Namibia to Algoa Bay (Kensley 1978d).

Synidotea setifer Barnard, 1914a. Algoa Bay, 60 m. Agulhas Bank to Algoa Bay, South Africa, 7-80 m (Kensley 1978d).

Synidotea variegata Collinge, 1917. India. Suez Canal; Indochina (Pillai 1963a). Madagascar (Roman 1970). Algoa Bay, South Africa to Mozambique, shallow infratidal; India; Sri Lanka (Kensley 1978d).

Synidotea watsonae Poore & Lew Ton, 1993. Western Australia, 7-35 m.

Synidotea worleiensis Joshi & Bal, 1959. India, intertidal.

Family Pseudidotheidae

Arcturides cornutus (Studer, 1882). Kerguelen Is., 210 m. Prince Edward Is. (Sheppard 1957). Kerguelen Is., 138-203 m (Kussakin & Vasina 1982b). Crozet Is., 143-178 m; Prince Edward & Marion Is., 97-650 m (Kensley 1980b).

ACKNOWLEDGEMENTS

I am grateful to all those individuals whose collecting efforts resulted in an accumulation of isopod material in the collections of the National Museum of Natural History, including participants in the International Indian Ocean Expedition. Many of these records are included in the accompanying species list. Dr Niel Bruce of Queensland, Australia, who made large collections from Zanzibar and Phuket, Thailand, available for study, also read an earlier draft of this paper, and made many useful comments and additions. Marilyn Schotte of the Department of Invertebrate Zoology, National Museum of Natural History, contributed substantially to the compilation of the species list and bibliography; I am very grateful for her help.

REFERENCES

Abele, L.G. 1982. Biogeography. In D.E. Bliss (ed.), *The Biology of the Crustacea, vol. 1, Systematics, the Fossil Records, and Biogeography*: pp. 241-304. New York: Academic Press.

Ahmed, M.M. 1970a. New Isopoda (Flabellifera) from Iraq and Arabian Gulf III. *Ichthyoxenus asymmetrica* sp. nov. *Bulletin of the Iraq Natural History Museum* 4(2): 33-36.

Ahmed, M.M. 1970b. New Isopoda (Flabellifera) from Iraq and Arabian Gulf IV. *Nerocila heterozota* sp. nov. *Bulletin of the Iraq Natural History Museum* 4(3): 558.

Ahmed, M.M. 1971. New Isopoda from Iraq and Arabian Gulf. *Mitteilungen aus dem Zoologische Museum in Berlin* 47: 77-83.

Alluaud, C. & R. Jeannel 1914. In R. Jeannel & E.-G. Racovitza (eds), *Biospeologica 23. Enumération des grottes visitées 1911-1913. Archives de Zoologie Expérimentale et Générale* 53:325-558.

Avdeev, V.V. 1977. Three new species of parasitic isopods (Flabellifera, Anilocridae) from the collection of the Laboratory of Parasitology of Sea Animals of TINRO. *Fisheries, Oceanography, Hydrobiology and Fish Parasitology of the Pacific Ocean* 1011: 139-144.

Avdeev, V.V. 1978. Parasitic isopods of the family Cymothoidae (Crustacea, Flabellifera) from the Red Sea. *Biologiya Morya, Vladivostok* 4: 30-35.

Avdeev, V.V. 1979. New species of the genus *Cymothoa* (Isopoda, Cymothoidae) from the coastal region of the northern and northwestern Australia. *Parazitologiya* 13: 50-55.

Baker, W.H. 1908. Notes on some species of the isopod family Sphaeromatidae from the South Australian coast. *Transactions of the Royal Society of South Australia* 32: 138-162.

Baker, W.H. 1910. Notes on some species of the isopod family Sphaeromidae, from the South Australian coast. Part II. *Transactions of the Royal Society of South Australia* 34: 75-88.

Baker, W.H. 1911. Notes on some species of the isopod family Sphaeromidae from southern Australian seas. Part III. *Transactions of the Royal Society of South Australia* 35: 89-93.

Baker, W.H. 1926. Species of the isopod family Sphaeromidae, from the eastern, southern, and western coasts of Australia. *Transactions and Proceedings of the Royal Society of South Australia* 50: 247-277.

Baker, W.H. 1928. Australian species of the isopod family Sphaeromidae. (Continued). *Transactions and Proceedings of the Royal Society of South Australia* 52: 49-61.

Bal, D.V & U.N. Joshi 1959. Some new isopod parasites on fishes. *Journal of the Bombay History Society* 56(3): 1-569.

Barnard, K.H. 1914a. Contributions to the crustacean fauna of South Africa. 1. Additions to the marine Isopoda. *Annals of the South African Museum* 10: 197-230.

Barnard, K.H. 1914b. Contributions to the crustacean fauna of South Africa. 3. Additions to the marine Isopoda, with notes on some previously incompletely known species. *Annals of the South African Museum* 10: 325-442.

Barnard, K.H. 1920. Contributions to the crustacean fauna of South Africa. 6. Further additions to the list of marine Isopoda. *Annals of the South African Museum* 17: 319-438.

Barnard, K.H. 1924. Description of a new genus and species of isopod crustacean belonging to the family Bathynomidae, procured in South African Marine Survey. *Fisheries and Marine Biological Survey* Report No. 4 (1923). Special Report 11:1-4.

Barnard, K.H. 1925a. A revision of the family Anthuridae (Crustacea Isopoda) with remarks on certain morphological peculiarities. *Journal of the Linnean Society (Zool.)* 36: 109-160.

Barnard, K.H. 1925b. Contributions to the crustacean fauna of South Africa. 9. Further additions to the list of Isopoda. *Annals of the South African Museum* 20: 381-412.

Barnard, K.H. 1926. Report on a collection of Crustacea from Portuguese East Africa. *Transactions of the Royal Society of South Africa* 13: 119-129.

Barnard, K.H. 1935. Report on some Amphipoda, Isopoda, and Tanaidacea in the collections of the Indian Museum. *Records of the Indian Museum* 37: 279-319.

Barnard, K.H. 1936. Isopods collected by the RIMS 'Investigator'. *Records of the Indian Museum* 38: 147-191.

Barnard, K.H. 1940. Contributions to the crustacean fauna of South Africa. 12. Further additions to the Tanaidacea, Isopoda and Amphipoda, together with keys for the identification of hitherto recorded marine and fresh water species. *Annals of the South African Museum* 32: 381-543.

Barnard, K.H. 1951. New records and descriptions of new species of isopods and amphipods from South Africa. *Annals and Magazine of Natural History* (12)4: 698-709.

Barnard, K.H. 1955. Additions to the fauna list of South African Crustacea and Pycnogonida. *Annals of the South African Museum* 43: 1-107.

Barnard, K.H. 1957a. Three additions to the fauna list of South African Crustacea. *Annals and Magazine of Natural History* (12)10· 814-816.

Barnard, K.H. 1957b. Additions to the fauna list of South African Crustacea. *Annals and Magazine of Natural History* (12)10: 1-12.

Barnard, K.H. 1958. Further additions to the crustacean fauna list of Portuguese East Africa. *Memorias do Museu Dr. Alvarez de Castro* 4: 1-23.

Barnard, K.H. 1960. Isopoda parasitic on Madagascar fish. *Mémoires de L'Institut Scientifique de Madagascar* (F)3: 93-95.

Bastida, R. & M.R. Torti 1969. Campagne de la Calypso au Large des Côtes Atlantiques de l'Amerique du Sud (1961-1962). 18. Crustaceos Isopodos: Serolidae. *Résultats Scientifiques des Campagnes de la 'Calypso'* 9:61-105.

Bate, C.S. 1863. On some new Australian species of Crustacea. *Proceedings of the Zoological Society of London* 1863: 498-505.

Bate, C.S. 1866. Carcinological gleanings. *Annals and Magazine of Natural History* (3)17: 24-31.

Becker, G. & B. Kampf 1958. Funde der holzzerstorenden Isopodengattung *Limnoria* an der Festlandkuste Indiens und Neubeschreibung von *Limnoria indica. Zeitschrift für angewandte Zoologie* 45:1-9.

Beddard, F.E. 1884a. Preliminary notice of the Isopoda collected during the voyage of HMS Challenger. Part 1. *Serolis. Proceedings of the Zoological Society of London* 23: 330-41.

Beddard, F.E. 1884b. Report on the Isopoda collected by the HMS Challenger during the years 1873-1876. Part 1. The Genus *Serolis. Report of the Voyage of HMS Challenger* (11)33: 1-85.

Beddard, F.E. 1885. Preliminary notice of the Isopoda collected during the voyage of HMS 'Challenger'. Part 2. Munnopsidae. *Proceedings of the Zoological Society of London* 1185: 916-925.

Beddard, F.E. 1886. Report on the Isopoda collected by HMS Challenger during the years 1873-76. Part 2. *Report of the Voyage of HMS Challenger* 17: 1-178.

Bianconi, G. 1869. Specimina zoologica Mosambicana. *Memorie dell'Accademia delle Scienze dell'Istituto di Bologna* 9: 199-222.

Bleeker, P. 1857. Recherches sur les Crustacés de l'Inde Archipelagique II. Sur les Isopodes cymothoadiens de 'Archipel Indien. Batavia. *Verhandelingen van het Natuurkund Vereeniging* 2(5): 20-40.

Bonnier, J. 1900. Contribution à l'etude des Epicarides Les Bopyridae. *Travaux de l'Institut Zoologique de Lille et du Laboratoire de Zoologie Maritime de Wimereux* 8: 478.

Boone, P.L. 1918. Descriptions of ten new isopods. *Proceedings of the United States National Museum* 54: 591-603.

Bosc, L. 1801. Histoire naturelle des Crustacés. In G.L Buffon, *Histoire Naturelle de Buffon, classée d'apres le système de Linne* Paris: R. R. Castel.

Bourdon, R. 1967. Sur deux nouveaux epicarides (Isopoda) parasites de crustaces decapodes. *Zoologische Mededelingen Leiden* 42:167-174.

Bourdon, R. 1972. Epicaridea de Java, Ile Maurice et Afrique du sud (Crustacea, Isopoda). *Steenstrupia* 2: 105-119.

Bourdon, R. 1976a. Les bopyres des porcellanes. *Bulletin du Muséum National d'Histoire Naturelle, Paris* (3) no. 359, Zoologie 252: 353-392.

Bourdon, R. 1976b. Epicarides de Madagascar. I. *Bulletin du Muséum National d'Histoire Naturelle, Paris* (3), no. 371, Zoologie 259: 353-392.

Bourdon, R. 1979. Epicarides de Madagascar. II. *Bulletin du Muséum National d'Histoire Naturelle, Paris* serie 4, no. 1, section A, no. 2: 471-506.

Bourdon, R. 1980a. *Aporobopyrus dollfusi* n. sp. (Crustacea, Epicaridea, Bopyridae) parasite de la Mer rouge. *Bulletin du Muséum National d'Histoire Naturelle, Paris* (4)2 Section A(1): 237-244.

Bourdon, R. 1980b. Les espèces du genre *Bopyrella* (J. Bonnier) (Crustacea, Isopoda, Bopyridae). Bulletin du Muséum National d'Histoire Naturelle, Paris (4)2 (Section A)(1): 185-236.

Bourdon, R. 1981. Résultats de campagnes MUSORSTOM. 1. Philippines (18-28 mars 1976). Crustacés isopodes. 1. Bopyridae parasites des Peneides. *Mémoirs ORSTOM* No. 91. 1981: 237-260.

Bourdon, R. 1982a. *Orbione digitata* spec. nov. (Isopoda: Bopyridae), parasite de *Solenocera choprai* Nataraj (Decapoda: Penaeidae). *Zoologische Mededelingen* 56(15): 193-196.

Bourdon, R. 1982b. Sur deux nouveaux Isopodes Epicarides du genre *Bopyrella* Bonnier. *International Revue der Gesamten Hydrobiologie* 67: 749-756.

Bourdon, R. 1983. Expédition Rumphius II (1975) crustacés parasites, commensaux, etc. (Th. Monod, ed.) VIII. Crustacés isopodes (3e partie: Epicarides Bopyridae). *Bulletin du Muséum d'Histoire Naturelle, Paris* 4 ser., 5(A)(3): 845-869.

Bourdon, R. & A.J. Bruce 1979. *Bopyrella saronae* sp. nov., a new bopyrid parasite of the shrimp *Saron marmoratus* (Olivier). *Crustaceana* 37: 191-197.

Bowman, T.E. & T.M. Iliffe 1991. *Annina fustis*, a new isopod from Phang Nga, Thailand (Crustacea Isopoda: Cirolanidae). *Proceedings of the Biological Society of Washington* 104: 247-252.

Bowman, T.E. & R.N. Mariscal 1968. *Renocila heterozota*, a new cymothoid isopod, with notes on its host, the anemone fish, *Amphiprion akalliopisos*, in the Seychelles. *Crustaceana* 14: 97-104.

Bowman, T.E. & I.U. Tareen 1983. Cymothoidae from fishes of Kuwait (Arabian Gulf) (Crustacea: Isopoda). *Smithsonian Contributions to Zoology* 382-30.

Branch, M.L., C.L. Griffiths, B. Kensley & J. Sieg 1991. The benthic Crustacea of subantarctic Marion and Prince Edward Islands: Illustrated keys to the species and results of the 1982-1989 University of Cape Town Surveys. *South African Journal of Antarctic Research* 21: 3-44.

Brandt, A. 1988. Antarctic Serolidae and Cirolanidae (Crustacea: Isopoda): new genera, new species and redescriptions. In R. Fricke (ed.), *These Zoologicae* 10: 7-143. Konigstein: Koeltz Scientific Books.

Brandt, A. 1990. *Antarctic Valviferans (Crustacea, Isopoda, Valvifera). New genera, new species, and redescriptions*. Leiden: E.J. Brill.

Briggs, J.C. 1974. *Marine Zoogeography*. New York: McGraw-Hill.

Brocchi, M. 1875. Liste des crustacés isopodes provenant de l'ile Saint Paul et deposes dans les galeries du Muséum d'Histoire naturelle, par MM. Velain et de Lisle. *Bulletin du Société Philomathique de Paris* (Ser. 6)12: 97-101.

Bruce, A.J. 1965. An interesting new phryxid (Epicaridea) parasitic on a pontoniinid shrimp. *Annals and Magazine of Natural History* (8)13: 385-390.

Bruce, A.J. 1968. A new species of *Hypophryxus* (Isopoda, Bopyridae) from an Indo-Pacific pontoniinid shrimp. *Crustaceana* 14: 1-20.

Bruce, A.J. 1972. *Orophryxus shiinoi* gen. nov., sp. nov., an unusual phryxid (Crustacea, Isopoda,

Bopyridae) parasitic upon a pontoniinid shrimp from Zanzibar. *Parasitology* 64: 445-450.

Bruce, A.J. 1973a. *Mesophryxus ventralis* gen. nov., sp. nov. a phryxid bopyrid parasitic upon the pontoniinid shrimp *Harpiliopsis beaupresi* (Audouin). *Parasitology* 66: 515-523.

Bruce, A.J. 1973b. *Prophryxus globicaudatus* gen. nov. sp. nov., a hemiarthrinid bopyrid parasite of pontoniinid shrimps of the genus *Coralliocaris*. *Parasitology* 67: 365-373.

Bruce, A.J. 1974. *Allophryxus malindiae* gen. nov., sp. nov., a hemiarthrinid bopyrid parasitic upon the pontoniid shrimp *Corallocaris superba* (Dana). *Parasitology* 68: 127-134.

Bruce, N.L. 1981. New records of Cirolanidae (Crustacea:Isopoda) from the Tanzanian Coast of East Africa. *Journal of Crustacean Biology* 1: 457-460.

Bruce, N.L. 1982. On the genus *Corallana* (Dana 1853) (Isopoda, Corallanidae) with description of a new species from Western Australia. *Crustaceana* 42: 241-249.

Bruce, N.L. 1983. Aegidae (Isopoda: Crustacea) from Australia with descriptions of three new species. *Journal of Natural History* 7: 757-788.

Bruce, N.L. 1986a. Cirolanidae (Crustacea: Isopoda) of Australia. *Records of the Australian Museum*, Supplement 6: 1-239.

Bruce, N.L. 1986b. Revision of the isopod crustacean *Mothocya* Costa, in Hope, 1851 (Cymothoidae), crustacean parasites of marine fishes. *Records of the Australian Museum* 39: 355-412.

Bruce, N.L. 1986c. New records of isopod crustaceans (Flabellifera) from Hong Kong. In B. Morton (ed.), *Proceedings of the Second International Marine Biological Workshop: The Marine Flora and Fauna of Hong Kong and southern China*: pp. 549-554. Hong Kong: Hong Kong University Press.

Bruce, N.L. 1987a. Australian *Pleopodias* (Richardson 1910) and *Anilocra* (Leach 1818) (Isopoda: Cymothoidae), crustacean parasites of marine fishes. *Records of the Australian Museum* 39: 85-130.

Bruce, N.L. 1987b. Australian Species of *Nerocila* (Leach 1818) and *Creniola* n. gen. (Isopoda: Cymothoidae), crustacean parasites of marine fishes. *Records of the Australian Museum* 39: 355-412.

Bruce, N.L. 1987c. Australian *Renocila* (Miers 1880) (Isopoda: Cymothoidae), crustacean parasites of marine fishes. *Records of the Australian Museum* 39: 169-182.

Bruce, N.L. 1990. The genera *Catoessa, Elthusa, Enispa, Ichthyoxenus, Idusa, Livoneca* and *Norileca* n. gen. Isopoda, Cymothoidae), crustacean parasites of marine fishes, with descriptions of eastern Australian species. *Records of the Australian Museum* 42: 247-300.

Bruce, N.L. 1991a. New records of marine isopod crustaceans (Sphaeromatidae, Cirolanidae) from southeastern Australia. *Memoirs of the Museum of Victoria* 52: 263-275.

Bruce, N.L. 1991b. Two new species of *Renocila* from off the tropical Australian coasts (Isopoda: Cymothoidae), crustacean parasites of marine fishes. *The Beagle, Records of the Northern Territory Museum of Arts and Sciences* 8: 159-168.

Bruce, N.L. 1992. A new genus of hemibranchiate sphaeromatid isopod crustacean from tropical Western Australia. *Journal of Natural History* 26: 1263-1272.

Bruce, N.L. 1993. Two new genera of marine isopod crustaceans (Flabellifera Sphaeromatidae) from southern Australia, with a reappraisal of the Sphaeromatidae. *Invertebrate Taxonomy* 7: 151-171.

Bruce, N.L. 1994a. The Cassidininae (Hansen 1905) (Crustacea: Isopoda: Sphaeromatidae) of Australia. *Journal of Natural History* 28: 1077-1173.

Bruce, N.L. 1994b. Four new genera of marine isopod crustaceans (Sphaeromatidae) from eastern and southern Australia. *Memoirs of the Museum of Victoria* 54:399-438.

Bruce, N.L. 1995. The taxonomy and phylogeny of tube-tailed sphaeromatid isopods, Crustacea, with descriptions of new species and a new genus from southern Australia. *Ophelia* 43: 127-180.

Bruce, N.L. 1997a. A new genus of marine sphaeromatid isopod crustacean from Australia and the Indo-Pacific region. *Memoirs of the Museum of Victoria* 56: 145-234.

Bruce, N.L. 1997b. Order Isopoda: Sea lice. In M.D. Richmond (ed.), *A guide to the seashores of eastern Africa and the western Indian Ocean islands*: pp. 198-201. Stockholm: SIDA/SAREC.

Bruce, N.L. & T.E. Bowman 1989. Species of the parasitic isopod genera *Ceratothoa* and *Glossobius* (Crustacea: Cymothoidae) from the mouths of flying fishes and halfbeaks (Beloniformes). *Smithsonian Contributions to Zoology* 489: 1-37.

Bruce, N.L. & W. Javed 1987. A new genus and two new species of cirolanid isopod Crustacea from the northern Indian Ocean. *Journal of Natural History* 21: 1451-1460.

Bruce, N.L. & D.A. Jones 1978. The systematics of some Red Sea Isopoda (Family Cirolanidae) with descriptions of two new species. *Journal of Zoology, London* 185: 395-413.

Brunnich, M. 1783. *Spicilegia Zoologica e Museis Naturae Curiosorum in itineribus apud exteros reportata. fol.* 1765-99.

Brusca, R.C. 1981. A monograph on the Isopoda Cymothoidae (Crustacea) of the eastern Pacific. *Zoological Journal of the Linnean Society* 73: 117-199.

Brusca, R.C. 1983. A monograph on the isopod family Aegidae in the tropical eastern Pacific I. The genus *Aega. Allan Hancock Monographs in Marine Biology* 12:1-39.

Budde-Lund, G. 1908. Isopoda von Madagaskar und Ostafrika. Mit Diagnosen verwandter Arten. In A. Voeltzkow, *Reise in Ostafrika in den Jahren 1903-1905. Wissenschaftliche Ergebnisse* 2: 265-308.

Calman, W.T. 1910. On two new species of wood-boring Crustacea from Christmas Island. *Annals and Magazine of Natural History* (8)5: 181-186.

Cals, P. 1979. Tectonique des plaques et speciation dans les grands fonds oceaniques. Endemisme et originalité d'un crustace holobenthique abyssal du bassin d'Enderby *Serolis monodi*, n. sp. *Comptes Rendus Hebdomadaires des Séances de l'Académie des Sciences, Paris* 288: 1031-1034.

Carvacho, A. 1977. Sur une importante collection d'isopodes des Iles Kerguelen. *Comité National Francais des Recherches Antarctiques, Paris* 42: 173-191.

Chappuis, P.-A. 1927. *Die Tierwelt der unterindischen Gewaesser*. Stuttgart.

Chappuis, P-A. 1959. Crustacés des eaux douces de l'ile Amsterdam: *Protocharon antarctica* n. sp. Mémoires de L'Institut Scientific de Madagascar 12 A: 13-24.

Chappuis, P-A. & C. Delamare Deboutteville 1952. Nouveaux isopodes (Crustacea) du sable des plages du Rousillon. *Comptes Rendus Hebdomadaires des séances de l'Académie des Sciences* 234:2014-2016.

Chappuis P-A. & C. Delamare Deboutteville 1956. Présence de la sous famille des Microcerberinae à Madagascar: *Microcerberus pauliani* n. sp. (Crustaces Isopodes). *Mémoires de L'Institut Scientifique de Madagascar* Ser. A 10: 81-88.

Chappuis P-A., C. Delamare Deboutteville & R. Paulian 1956. Recherches sur la faune interstitielle des sediments marins et d'eau douce à Madagascar. Crustacés des eaux souterraines littorales d'une résurgence d'eau douce à La Réunion. *Mémoires de L'Institut Scientifique de Madagascar* Ser. A(11): 51-78.

Chilton, C. 1884. Additions to the sessile eyed Crustacea of New Zealand. *Transactions and Proceedings of the New Zealand Institute* 16: 249-265.

Chilton, C. 1924. Fauna of the Chilke Lake. *Memoirs of the Indian Museum* 5: 875-895.

Chopra, B. 1923. Bopyrid isopods parasitic on Indian Macrura. *Records of the Indian Museum* 25: 411-550.

Chopra, B. 1930. Further notes on bopyrid isopods parasitic on Indian Decapoda Macrura. *Records of the Indian Museum* 32:113-147.

Cleret, J.J. 1971. Isopoda. Deux espèces nouvelles d'Isopodes Asellotes à l'Ile Marion. In E.M. van Zinderen Bakker, J.M. Winterbottom & R.A. Dyer (eds), *Marion and Prince Edwards Island. Report of the South African Biological and Geological Expedition 1965-1966*: pp. 379-383. Cape Town: A.A. Balkema.

Cleret, J.J. 1973. Isopodes et Tanaidaces. *Comité National Français des Recherches Antarctiques* 32: 31-33.

Codreanu, R., M. Codreanu & R.B. Pike 1960. Sur un nouveau type d'Entoniscien, *Diogenion vermifactus* n.g. n.sp., parasite du Pagure *Diogenes senex* Heller de la mer Rouge. *Comptes Rendus Hebdomadaires des séances de l'Académie des Sciences* 251: 439-441.

Codreanu, R., M. Codreanu & R.B. Pike 1965. Sur deux bopyriens parasites de pagures recueillis par M.A. Horridge dans la mer Rouge et sur leur asymetrie. *Crustaceana* 9: 225-244.

Cohen, B.F. & G.C.B. Poore 1994. Phylogeny and biogeography of the Gnathiidae (Crustacea, Isopoda) with descriptions of new genera and species, most from southeastern Australia. *Memoirs of the Museum of Victoria* 54: 271-397.

Cohen, D.M. 1973. Zoogeography of the fishes of the Indian Ocean. In B. Zeitzschel (ed.), *The Biology of the Indian Ocean. Ecological Studies: Analysis and Synthesis*, volume 3: pp. 451-463. Berlin: Springer-Verlag.

Coineau, N. 1971. Les isopodes interstitiels documents sur leur ecologie et leur biologie. *Mémoires du Muséum d'Histoire Naturelle* Serie A 64: 1-170.

Coineau, N. & G.C. Rao 1972. Isopodes et amphipodes des sables intertidaux des îles Andaman et Nicobar (Golfe du Bengale). *Vie et Milieu* Serie A 23 1A: 65-100.

Collinge, W.E. 1917. *Paridotea bakeri* sp. n. from South Australia. *Journal of Zoological Research* 2: 112.

Cooke, W.J. 1977. On the occurrence of the commensal asellote *Caecijaera horvathi* (Menzies 1951) in Hawaii. *Crustaceana* 33:105-106.

Cookson, L. 1991. Australian species of Limnoriidae (Crustacea: Isopoda). *Memoirs of the Museum of Victoria* 52(2): 137-262.

Cookson, L. & G.C.B. Poore 1994. New species of *Lynseia* and transfer of the genus to Limnoriidae (Crustacea: Isopoda). *Memoirs of the Museum of Victoria* 54:179-189.

Cunningham, R.O. 1871. Notes on the Reptiles, Amphibia, Fishes, Mollusca, and Crustacea obtained during the voyage of HMS Nassau in the years 1866-1869. *Transactions of the Linnean Society of London* 27: 465-502.

Dana, J.D. 1853. Crustacea. Part II. In C. Wilkes (ed.), *United States Exploring Expedition 1838-42 under the command of C. Wilkes*. Philadelphia.

Danforth, C.G. 1972. New bopyrids (Isopoda) from New Guinea. *Bulletin of the Southern California Academy of Sciences* 70: 99-102.

Delamare Deboutteville, C. 1960. *Biologie des eaux souterraines littorales et continentales*. Paris: Herman.

Delamare Deboutteville, C & P-A. Chappuis 1956. Recherches sur la faune interstitielle des sediments marins et d'eau douce à Madagascar. VIII. *Angliera phreaticola* Chappuis et Delamare, le premier Microparasellinae souterrain de Madagascar (Crustacés Isopode). *Mémoires de L'Institut Scientifique de Madagascar* Serie A 10: 89-94.

Delaney, P. 1989. Phylogeny and biogeography of the marine isopod family Corallanidae (Crustacea, Isopoda, Flabellifera). *Contributions in Science, Natural History Museum of Los Angeles County* 409: 1-75.

Devi, S.L. 1982. Bopyrid parasites of prawns at Kakinda. *Journal of the Marine Biological Association of India* 24: 23-32.

Dollfus, A. 1889. Sur quelques isopodes du Musée de Leyde. *Notes from the Leyden Museum*. 11: 91-94.

Dollfus, A. 1891. Crustaces isopodes. In *Mission Scientifique de Cape Horn 1882-1883. Tome VI, Zoologie*: pp. 55-76. Paris: Gauthier-Villars.

Duvernoy, G.L. 1841. Sur un nouveau genre de l'orde des Crustacés isopodes et un l'espèce type de ce genre, le Kepone type. *Annales des Sciences Naturelles* (2)15: 10.

Eigths, J. 1833. Description of a new crustaceous animal found on the shores of the South Shetland Islands, with remarks on their natural history. *Transactions of the Albany Institute* 2: 53-69.

Ekman, S. 1967. *Zoogeography of the Sea*. London: Sidgwick & Jackson.

Eleftheriou, A. & D.A. Jones 1976. The genus *Eurydice* on the west coast of India. *Journal of Zool-*

ogy London 178: 385-394.

Eleftheriou, A., D. Holdich & K. Harrison 1980. The systematics and ecology of a new genus of isopod (Sphaeromatidae) from the west coast sandy beaches of India. *Estuarine and Coastal Marine Science* 11: 251-262.

Enckell, P.H. 1970. Isopoda Asellota and Flabellifera from Ceylon. *Arkiv for Zoologi* 22: 557-569.

Fabricius, J.C. 1787. *Mantissa Insectorum, sistens eorum species nuper detectas; adjectis characteribus genericus, differentiis specificis, emendationibus, observationibus*. Hafniae.

Fryer, G. 1965. A new isopod of the genus *Lironeca* parasitic on a cichlid fish of Lake Tanganika. *Revue Zoologique et Botanique Africaine* 71: 376-384.

Ganapati, P.N. & M.V.L. Rao 1960. On some crustacean wood borers from Andamans. *Current Science* 29: 275-276.

Ganamuthu, C.P. 1954. Two new sand dwelling isopods from the Madras seashore. *Annals and Magazine of Natural History* (12)7: 257- 274.

George, P.C. 1946. *Megacepon choprai*, gen. et sp. nov., a bopyrid isopod from the gill chamber of *Sesarma tetragonum* (Fabr.). *Records of the Indian Museum* 44: 385- 390.

Giard, A. & J. Bonnier 1888. Sur deux nouveaux genres d'Epicarides. *Comptes Rendus hebdomadaires des Séances de l'Académie des Sciences* 2 juillet 1888: 44-47.

Guérin-Méneville, F.E. 1832. Crustacés. In *Expédition scientifique de Morée* (sous la direction de M. Bory de Saint-Vincent) 3. Paris:

Guérin-Méneville, F.E. 1836. Description du genre *Pterelas* et de l'espece *P. webbi*. *Magasin de Zoologie, Paris* 6: 7.

Guérin-Méneville, F.E. 1843. Crustacés. In Cuvier G. (ed.), *Iconographie du Regne animal de G. Cuvier, ou représentation d'apres nature de l'une des espèces les plus remarquable et so non encore figurées de chaque genre d'animaux*. 3: 1-48. Paris: Bailliere.

Hale, H.M. 1924c. Notes on Australian Crustacea. No. III. Transactions of the Royal Society of South Australia 48: 209-225.

Hale, H.M. 1925. Review of Australian isopods of the cymothoid group. Part 1. *Transactions of the Royal Society of South Australia* 49: 128-185.

Hale, H.M. 1926. Review of the Australian isopods of the cymothoid group. Part 2. *Transactions of the Royal Society of South Australia* 50: 201-234.

Hale, H.M. 1929. The Crustaceans of South Australia. Part II. In *Handbook of the Flora and Fauna of South Australia*: pp. 202-308. British Science Guild. Adelaide: Harrison Weir.

Hale, H.M. 1940. Report on the cymothoid Isopoda obtained by FIS 'Endeavor' on the coasts of Queensland, New South Wales, Victoria, Tasmania, and South Australia. *Transactions of the Royal Society of South Australia* 64: 288-304.

Hale, H.M. 1946. Isopoda Valvifera. *Reports of the B.A.N.Z. Antarctic Expedition 1929-1931* (B) 5(3): 161-212.

Hale, H.M. 1952. Isopoda. Families Cymothoidae and Serolidae. *Reports of the B.A.N.Z. Antarctic Research Expedition 1929-1931*. (B) 6: 21-36.

Haller, G. 1880. Uber einige neue Cymothoinen. *Archiv für Naturgeschichte* 46: 375-395.

Hansen, H.J. 1890. Cirolanidae et familiae nonullae propinquae Musei Hauniensis. Et bidrag til kundskaben om nogle familier af isopode krebsdyr. *Kongelige Danske Videnskabernes Selskab Skrifter, Copenhagen, 6 Raekke, Naturvidenskabelig og Mathematisk* 3: 239-426.

Hansen, H.J. 1897. Reports on the dredging operations off the west coast of Central America to the Galapagos Islands, to the west coast of Mexico, and in the Gulf of California, in charge of Alexander Agassiz, carried on by the US Fish Commission Steamer Albatross during 1891, Lieut. Commander Z. L. Tanner, US Navy commanding. XXII. The Isopods. *Bulletin of the Museum of Comparative Zoology, Harvard College* 31(5): 95-129.

Hansen, H.J. 1905. On the morphology and classification of the Asellota group of Crustacea, with descriptions of the genus *Stenetrium* (Haswell), and its species. *Proceedings of the Zoological*

Society of London (1904)2: 302-331.

Harger, O. 1883. Reports on the results of dredgings by the US Coast Survey Steamer Blake. XXIII. *Bulletin of the Museum of Comparative Zoology, Harvard* 11: 91-104.

Harrison, K. 1984. Some sphaeromatid isopods (Crustacea) from southern and south western Australia, with description of a new genus and two new species. *Records of the Western Australian Museum* 11: 259-286.

Harrison, K. & D. Holdich 1982. Revision of the genera *Dynamenella, Ischyromene, Dynamenopsis,* and *Cymodocella* (Crustacea: Isopoda, including a new genus and five new species of eubranchiate sphaeromatids from Queensland waters. *Journal of Crustacean Biology* 2: 84-119.

Haswell, W.A. 1881. On some new Australian Marine Isopoda. Part I. *Proceedings of the Linnean Society of New South Wales* 5: 470-481.

Haswell, W.A. 1884. On a new crustacean found inhabiting the tubes of *Vermilia* (Serpulidae). *Proceedings of the Linnean Society of New South Wales* 9: 676-680.

Hayward, P.J. 1983. Biogeography of *Adeonella* (Bryozoa, Cheilostomata): a preliminary account. *Bulletin of Marine Science* 33:582-596.

Heller, C. 1861. Vorlaufiger Bericht über die wahrend der Weltumseglung der k.k. Fregatte Novara gesammelten Crustaceen. *Verhandlungen zoologisch botanischen Gesellschaft in Wien* 11: 495-498.

Heller, C. 1868. Crustaceen. In *Reise des Oesterreischen Fregatte 'Novara' um die Erde.* Zoologischer Theil 2(8): 1-280.

Hilgendorf, F. 1878. Die von Hrn. W. Peters on Moçambique gesammelten Crustaceen. *Monatsberichte der Koniglich Preussischen Akademie der Wissenschaften zu Berlin* 1878: 782-851.

Hilgendorf, F. 1893. Bermerkungen über 2 Isopoden, die japanische Susswasser Assel und eine neue Munna art. *Sitzungs Bericht Gesellschaft Naturforschender Freunde zu Berlin* 1893: 1-3.

Hobbins, C. & D.A. Jones 1993. New species of deep sea isopods from the Red Sea and North West Indian Ocean: Families Cirolanidae and Corallanidae. *Senckenbergiana maritima* 23(4/6): 115-134.

Hocutt, C.H. 1987. Evolution of the Indian Ocean and the drift of India: A vicariant event. *Hydrobiologia* 150: 203-223.

Hodgson, T.V. 1902. Crustacea. In *Report on the collections of natural history made in Antarctic regions during the voyage of the 'Southern Cross'*. pp. 228-261. London.

Hodgson, T.V. 1910. *Crustacea. IX. Isopoda. National Antarctic Expedition 1901-1904. Natural History. 5. Zoology and Botany.* London: British Museum..

Holdich, D.M. & D.A. Jones 1973. The systematics and ecology of a new genus of sand beach isopod (Sphaeromatidae) from Kenya. *Journal of Zoology, London* 171: 385-395.

Holthuis, L. 1949. The Isopoda and Tanaidacea of the Netherlands, including the description of a new species of *Limnoria. Zoologische Mededelingen* 30(12): 163-190.

Horst, R. 1910. *Palaegyge buitendijki*, n. sp., parasitic on *Palaemon carcinus* Fabr. from Java. *Notes of the Leyden Museum* 32: 67-70.

Javed, W. 1990. A new species of hemibranchiate sphaeromatid isopod of the genus *Paracilicaea* from Pakistan with the study of developmental stages of the male. *Hydrobiologia* 199:21-28.

Javed, W. & R. Ahmed 1987. On the occurrence of *Paradella dianae* (Menzies 1962) a genus and species of Sphaeromatidae (Isopoda, Flabellifera) in the Arabian Sea. *Crustaceana* 53: 215-216.

Javed, W. & R. Ahmed 1988a. Two new species of the genus *Dynamenella* from the northern Arabian Sea (Isopoda). *Crustaceana* 55: 234-241.

Javed, W. & R. Ahmed 1988b. *Paraimene tuberculata*, a new genus and species of Isopoda (Sphaeromatidae) from Karachi, Pakistan. *Hydrobiologia* 169: 371-378.

Javed, W. & R. Yasmeen 1989. *Atarbolana setosa*, a new cirolanid isopod from the northern Arabian Sea. *Crustaceana* 56: 78-82.

Javed, W. & R. Yasmeen 1990. A new species of cirolanid isopod of the genus *Neocirolana* from Pakistan with a review of the genus. *Crustaceana* 58: 67-73.

Javed, W. & R. Yasmeen 1992. Two genera and species of Isopoda new to the fauna of the northern Arabian Sea. *Pakistan Journal of Marine Sciences* 1: 11-22.

Javed, W. & R. Yasmeen 1995. A new cirolanid isopod of the genus *Cirolana* from the Pakistan coast (northern Arabian Sea). In M.F. Thompson & N.M. Tirmizi (eds), *The Arabian Sea: Living Marine Resources and the Environment*: pp. 119-124. Vanguard Press: Lahore, Pakistan.

Javed, W. & F. Yousuf 1995. A new species of *Sphaeromopsis* (Holdich & Jones 1973) from Pakistan waters (Isopoda, Sphaeromatidae). *Pakistan Journal of Marine Sciences* 4: 51-58.

Javed, W. & F. Yousuf 1996. A new species of *Cerceis* (Isopoda, Sphaeromatidae) from the Arabian Sea. *Crustaceana* 69:155-161.

Jayadev Babu, S. & P.J. Sanjeeva Raj 1980. [no title] *Proceedings of the Symposium on Coastal Aquaculture* 3: 820-823.

Johnston, G. 1834. Illustrations in British Zoology. *Magazine of Natural History*. 7: 230-235.

Jones, D.A. 1971. The systematics and ecology of some sand beach isopods (Crustacea: Eurydicidae) from the coast of Kenya. *Journal of Zoology, London* 165: 201-227.

Jones, D.A. 1974. The systematics and ecology of some sand beach isopods (Family Cirolanidae) from the coasts of Saudi Arabia. *Crustaceana* 26: 201-211.

Jones, D.A. 1976. The systematics and ecology of some isopods of the genus *Cirolana* (Cirolanidae) from the Indian Ocean region. *Journal of Zoology, London* 178: 209-222.

Jones, D.A. 1979. The ecology of sandy beaches in Penang, Malaysia, with special reference to *Excirolana orientalis* (Dana). *Estuarine and Coastal Marine Science* (1979)9: 677-682.

Jones, D.A. 1982. New isopods of the genus *Lanocira* (Corallanidae) from the Indian Ocean region. *Crustaceana* 42: 65-75.

Jones, D.A. 1983. On the status of the cirolanid isopod *Annina* (Budde-Lund 1908) and *Excirolana* (Richardson 1912). *Crustaceana* 45: 309-312.

Jones, D.A. & J.D. Icely 1981. *Excirolana bowmani*, a new mangrove-boring isopod from Kenya (Isopoda, Cirolanidae). *Crustaceana* 40:266-271.

Jones, D.S. & G.J. Morgan 1993. An annotated checklist of Crustacea from Rottnest Island, Western Australia. In F.E. Wells, D.I. Walker, H. Kirkman & R. Lethbridge (eds), *Proceedings of the Fifth International Marine Biological Workshop: the marine flora and fauna of Rottnest Island, Western Australia* 1: 135-162.

Joshi, U.N. & D.V. Bal 1959. Some of the littoral species of Bombay isopods, with detailed description of two new species. *Journal of the University of Bombay New Series* 27B: 57-69.

Joshi, U.N. & D.V. Bal 1992. *Cassidina extenda*, a new species of isopod of Bombay. *Proceedings of the Indian Academy of Sciences* 56B(6): 372-376.

Keable, S.J. 1997. The Cirolanidae (Crustacea: Isopoda) of Darwin Harbour, Northern Territory, with additional records from northern Australia and Papua New Guinea. In J.R. Hanley, G. Caswell, D. Megirian & H.K. Larson (eds), *Proceedings of the Sixth International Marine Biological Workshop. The marine flora and fauna of Darwin Harbour, Northern Territory, Australia*: pp. 245-278.

Kensley, B. 1974. Bopyrid Isopoda from southern Africa. *Crustaceana* 26: 259-266.

Kensley, B. 1975a. Marine Isopoda from the continental shelf of South Africa. *Annals of the South African Museum* 67: 35-89.

Kensley, B. 1975b. Five species of *Jaeropsis* from the southern Indian Ocean. (Crustacea, Isopoda, Asellota). *Annals of the South African Museum* 69: 367-380.

Kensley, B. 1975c. A new species of anthurid isopod from the Cape. *Zoologica Africana* 10: 209-213.

Kensley, B. 1976. Isopodan and Tanaidacean Crustacea from the St. Paul and Amsterdam Islands, southern Indian Ocean. *Annals of the South African Museum* 69: 261-323.

Kensley, B. 1977. New records of marine Crustacea Isopoda from South Africa. *Annals of the South African Museum* 72: 239-265.

Kensley, B. 1978a. The South African Museum's Meiring Naude Cruises. Part 7, Marine Isopoda. *Annals of the South African Museum* 74: 125-157.

Kensley, B. 1978b. Redescription of *Parabathynomus natalensis* (Barnard) (Crustacea, Isopoda, Cirolanidae). *Annals of the South African Museum* 75: 25-31.

Kensley, B. 1978c. The South African Museum's Meiring Naude Cruises. Part 8. Isopoda Anthuridea. *Annals of the South African Museum* 77:1-25.

Kensley, B. 1978d. *Guide to the marine isopods of southern Africa.* Cape Town: Trustees of the South African Musem.

Kensley, B. 1978e. Two new species of the genus *Pseudanthura* (Richardson) (Crustacea: Isopoda: Anthuridae). *Proceedings of the Biological Society of Washington* 91: 222-233.

Kensley, B. 1978f. A new marine isopod family from the south-western Indian Ocean. *Annals of the South African Museum* 75:41-50.

Kensley, B. 1979. A second genus in the marine isopod family Bathynataliidae. *Annals of the South African Museum* 79: 35-41.

Kensley, B. 1980a. Anthuridean isopod crustaceans from the International Indian Ocean Expedition, 1960-1965, in the Smithsonian Collections. *Smithsonian Contributions to Zoology* 304: 1-37.

Kensley, B. 1980b. Marine isopods from Marion, Prince Edwards, and Crozet Islands (Crustacea, Isopoda). *Annals of the South African Museum* 82: 155-185.

Kensley, B. 1982. Revision of the Southern African Anthuridea. (Crustacea: Isopoda). *Annals of the South African Museum* 90: 95-200.

Kensley, B. 1984. The South African Museum's Meiring Naude Cruises. Part 15. Marine Isopoda of the 1977, 1978, 1979 Cruises. *Annals of the South African Museum* 93: 213-301.

Kensley, B. 1989. Marine isopod crustaceans from the St. Paul and Amsterdam Islands, southern Indian Ocean. *Bulletin du Muséum National d'Histoire Naturelle, Paris* 4 ser. 11(A)(1): 147-164.

Kensley, B. 1998. Estimates of species diversity of free-living marine isopod crustaceans on coral reefs. *Coral Reefs* 17:83-88.

Kensley, B. & M. Bursey 1996. Rediscovery of *Cymodocella algoense* From South Africa (Crustacea: Isopoda: Sphaeromatidae). *Proceedings of the Biological Society of Washington* 109: 91-96.

Kensley, B. & C. Buxton 1984. Inshore small-mesh trawling survey of the Cape south coast. Part 5. Crustacea, Stomatopoda, Isopod Decapoda. *South African Journal of Zoology* 19: 189-193.

Kensley, B. & G.C.B. Poore 1982. Anthurids from the Houtman Abrolhos Islands, Western Australia (Crustacea: Isopoda: Anthuridae). *Proceedings of the Biological Society of Washington* 95: 625-636.

Kensley, B. & R. Reid 1984. *Arabanthura enigmatica*, a new species of anthurid isopod from the Arabian Gulf. *Proceedings of the Biological Society of Washington* 97: 674-680.

Kensley, B. & M. Schotte 1987. New records of marine isopod Crustacea from the Caribbean, the Florida Keys, and the Bahamas. *Proceedings of the Biological Society of Washington* 100: 216-247.

Kensley, B. & M. Schotte 1989. *Guide to the Marine Isopod Crustaceans of the Caribbean.* Washington: Smithsonian Press.

Kensley, B. & M. Schotte 1994. A new genus and species of cirolanid isopod from the western Indian Ocean (Crustacea: Peracarida). *Proceedings of the Biological Society of Washington* 107: 283-290.

Kensley, B., M. Ortiz & M. Schotte 1997. New records of marine Isopoda from Cuba (Crustacea: Peracaridea). *Proceedings of the Biological Society of Washington* 110: 74-98.

Klausewitz, W. 1972. The zoogeographical and palaeo-geographical problem of the Indian Ocean and the Red Sea according to the ichthyofauna of the littoral. *Journal of the Marine Biological Association of India* 14:697-706.

Klausewitz, W. 1976. The zoogeography of the littoral fishes of the Indian Ocean based on the distribution of the Chaetodontidae. *Revue des Travaux de l'Institut des Pêches Maritimes* 40: 636-640.

Klausewitz, W. 1978. Zoogeography of the littoral fishes of the Indian Ocean, based on the distribution of the Chaetodontidae and Pomacanthidae. *Senckenbergiana biologia* 59:25-39.

Kossmann, R. 1880. Malacostraca (2. Theil, Anomura, Macrura, Schizopoda, Stomatopoda, Isopoda, Laemodipoda, Amphipoda). *Zoologische Ergebnisse einer im Auftrage der Koniglich Akademie der Wissenschaft zu Berlin ausgefuhrten Reise in die Kustengebiete des Rothen Meeres, zweite Halfte, erste Lieferung*: 67-140.

Kossmann, R. 1881. Studien uber Bopyriden. I. *Gigantione moebii* und Allegemeines über die Mundwerkzeuge der Bopyriden. II. *Bopyrina virbii*, Beitrage zur Kenntnis der Anatomie und Metamorphose der Bopyriden. *Zeitschrift für Wissenschaftliche Zoologie* 35: 652-680.

Krauss, F. 1843. *Die sudafrikanischen Crustaceen*. Stuttgart: E. Schweizerbart'sche Verlagsbuchhandlung.

Kurian, C.V. 1973. Cumacea of the Indian Ocean. In B. Zeitzschel (ed.), *The Biology of the Indian Ocean*: pp. 329-330. Berlin: Springer-Verlag.

Kussakin, O.G. 1967. Isopoda and Tanaidacea from the coastal zones of the Antarctic and Subantarctic. *Issledovaniya Fauny morei* 4: 220-380. [in Russian]

Kussakin, O.G. & G. Vasina 1980a. Additions to the Marine Isopoda and Gnathiida of Kerguelen Islands (southern Indian Ocean). *Tethys* 9: 355-369.

Kussakin, O.G. & G. Vasina 1980b. Two new isopod species of antarctic genus. *Biologiya Morya* 2: 31-37. [In Russian]

Kussakin, O.G. & G. Vasina 1982a. Addition to the fauna of benthic Isopoda and Gnathiidea (Crustacea) of subantarctic waters of the Indian Ocean. I. Isopoda (Flabellifera and Anthuridea). *Tethys* 10: 261-273.

Kussakin, O.G. & G. Vasina 1982b. Additions to the fauna of benthic Isopoda and Gnathiida (Crustacea) of subantarctic waters of the Indian Ocean. 2. Isopoda (Valvifera and Asellota). *Tethys* 10: 315-336.

Lanchester, W.F. 1902. On the Crustacea collected during the 'Skeat' Expedition to the Malay Peninsula. *Proceedings of the Zoological Society of London* 2: 377-378.

Leach, W.E. 1814. Crustaceology. In Brewster's Edinburgh Encyclopedia Volume 7: 383-439.

Leach, W.E. 1815. A tabular view of the external characters of four classes of animals, which Linne arranged under Insecta; with the description of the genera comprising three of these classes into order, etc., and descriptions of several new genera and species. *Transactions of the Linnean Society of London* 2: 306-400.

Leach, W.E. 1818. *Dictionaire des Sciences Naturelles*. Paris and Strasbourg.

Lockington, W.N. 1877. Description of seventeen new species of Crustacea. *Proceedings of the California Academy of Science* 7:44-46.

Markham, J.C. 1980. A new species of *Dicropleon* and new records of three other species of bopyrid isopods from the Indian Ocean. *Bulletin of Marine Science* 30: 623-633.

Markham, J.C. 1982. Bopyrid isopods parasitic on decapod crustaceans in Hong Kong and southern China. In B.S. Morton & C.K.Tseng (eds), *Proceedings of the First International Marine Biological Workshop: The Marine Flora and Fauna of Hong Kong and Southern China* 1: 325-391. Hong Kong: Hong Kong University Press.

Markham, J.C. 1985. Additions to the bopyrid fauna of Thailand. *Zoologische Verhandelingen* 224: 1-63.

Markham, J.C. 1991. Redescriptions and new records of Isopoda Bopyridae (Crustacea) from Thailand. *The Raffles Bulletin of Zoology* 39: 289-297.

Markham, J.C. 1994. Crustacea Isopoda: Bopyridae in the MUSORSTOM collections from the tropical Indo-Pacific I. Subfamilies Pseudioninae (in part), Argeiinae, Orbioninae, Athelginae and Entophilinae, In A Crosnier (ed.), *Résultats des Campagnes MUSORSTOM*, vol. 10. *Mémoires du Muséum National d'Histoire Naturelle, Paris* 161: 225-253.

Menzies, R.J. 1951a. A new genus and new species of asellote isopod, *Caecijaera horvathi*, from Los Angeles Long Beach Harbor. *American Museum Novitates* 1542: 1-7.

Menzies, R.J. 1951b. New marine isopods, chiefly from Northern California, with notes on related forms. *Proceedings of the United States National Museum* 101(3273): 105-156.

Menzies, R.J. 1957. The marine borer family Limnoriidae (Crustacea, Isopoda). *Bulletin of Marine*

Science of the Gulf and Caribbean 7: 101-200.

Menzies, R.J. 1962a. Reports of the Lund University Chile Expedition 1948-1949. 42. The Zoogeography, ecology, and systematics of the Chilean marine isopods. *Lunds Universitets Arsskrift N. F. Avd.* 2 57(11): 1-162.

Menzies, R.J. 1962b. The marine isopod fauna of Bahia de San Quintin, Baja California, Mexico. *Pacific Naturalist* 3: 337-348.

Menzies, R.J. & P.W. Glynn 1968. The common marine isopod Crustacea of Puerto Rico: A handbook for marine biologists. *Studies on the Fauna of Curaçao and other Caribbean Islands* 27(104): 1-133.

Menzies, R.J. & G.A. Schultz 1968. Antarctic isopod Crustacea. II. Families Haploniscidae, Acanthaspidiidae, and Jaeropsidae, with diagnoses of new genera and species. *Antarctic Research Series* 11: 141-184.

Messana, G. 1984. A new locality for *Annina lacustris* (Budde-Lund 1908). Some considerations taxonomy and reproduction of the genera *Annina* and *Excirolana* (Isopoda, Cirolanidae). *Revue de Zoologie Africain* 98: 766-770.

Messana, G. 1988. Researches on the coast of Somalia. *Pseudocerceis seleneides* n. sp., a new marine isopod (Sphaeromatidae, Dynameninae) from the Somalian coast. *Crustaceana* 55: 306-312.

Messana, G. 1990. Researches on the coast of Somalia. On some Dynameninae (Crustacea Isopoda Sphaeromatidae) from the coast south of Mogadishu. *Tropical Zoology* 2: 243-253.

Messana, G., R. Argano & F. Baldari 1978. *Microcerberus* (Crustacea Isopoda Microcerberidea) from the Indian Ocean. *Monitore zoologico italiano* ns supplemento 10(3): 69-79.

Mezhov, B.V. 1988. The first findings of Macrostylidae (Isopoda, Asellota) in the Indian Ocean. *Zoologichesky Zhurnal* 67: 983-994.

Mezhov, B.V. 1989. Additions to the fauna of macrostylids in the Indian Ocean. *Zoologichesky Zhurnal* 68: 60-69.

Miers, E.J. 1875a. Descriptions of three additional species of Crustacea (and Pycnogonida) from Kerguelen's Land and Crozet Island, with remarks upon the genus Paramoera. *Annals and Magazine of Natural History* (4)16: 115-118.

Miers, E.J. 1875b. Description of new species of Crustacea (and Pycnogonida) collected at Kerguelen's Island by the Rev. A.E Eaton. *Annals and Magazine of Natural History* (4)16: 73-76.

Miers, E.J. 1877. On a collection of Crustacea, Decapoda and Isopoda, chiefly from South America, with descriptions of new genera and species. *Proceedings of the Zoological Society of London* 1877: 653-679.

Miers, E.J. 1880. On a collection of crustacea from the Malaysian Region. Part I. Crustacea Oxyrhyncha and Cyclometopa except Telphusidea. *Annals and Magazine of Natural History* (5)5: 226-472.

Miers, E.J. 1881. Revision of the Idoteidae, a family of sessile-eyed Crustacea. *Journal of the Linnean Society of London* 16: 1-88.

Miers, E.J. 1884. Crustacea. In *Report on the zoological collections made in the Indo-Pacific during the voyage of H.M.S. Alert, 1881-1882*: pp. 178-322. London: British Museum (Natural History).

Miller, M.A. 1941. The isopod Crustacea of the Hawaiian Islands, II. Asellota. *Occasional Papers of the Bernice P. Bishop Museum* 16: 305-320.

Milne Edwards, H. 1840. *Histoire Naturelle des Crustaces* 3: 1-638. Paris: Roret.

Monod, T. 1924. On a few isopods from Ceylon. *Spolia zeylanica* 13: 97-101.

Monod, T. 1925. Isopodes et Amphipodes de l'Expédition antarctique belge (S.Y. Belgica). *Bulletin du Muséum National d'Histoire Naturelle, Paris* 31: 159-162.

Monod, T. 1926a. Les Gnathiidae. Essai monographique (morphologie, biologie, systematique). *Mémoires de la Société des Sciences Naturelles du Maroc* 12: 1- 667.

Monod, T. 1926b. Tanaidaces, Isopodes et Amphipodes. Résultats du Voyage de la Belgica. *Rapports Scientifiques, Zoologie* 1-67.

Monod, T. 1931a. Tanaidaces et Isopodes subantarctique de la collection Kohl Larsen du Senckenberg Museum. *Senckenbergiana* 13: 10-30.

Monod, T. 1931b. Tanaidaces et Isopodes aquatiques de l'Afrique Occidentale et septentrionale. 3 Partie. Sphaeromatidae. *Mémoires Société des Sciences Naturelles du Maroc* 29: 7-91.

Monod, T. 1933. Tanaidacea et Isopoda. *Mission Robert Ph. Dollfus en Egypte. Mémoires de l'Institute d'Egypte* 21: 161-264.

Monod, T. 1968. II. Isopoda. In Gordon, I & T. Monod (eds), Sur quelques Crustacés des eaux douces de Zanzibar. *Bulletin de l'Institut Fondamental d'Afrique Noire* Series A 30: 497-517.

Monod, T. 1971. Sur deux isopodes marins du Golfe de Kutch (Inde). *Bulletin du Muséum National d'Histoire Naturelle, Paris* (2)42(5): 944-956.

Monod, T. 1974. Nouvelles observations sur le genre *Bagatus* (Crustacea Isopoda). *Bulletin du Muséum National d'Histoire Naturelle, Paris* (242) Zoologie 166: 1121-1135.

Morris, B.F. & D.D. Mogelberg 1973. Identification manual to the pelagic Sargassum fauna. *Bermuda Biological Station, Special Publication* 11:1-63.

Müller, H-G. 1989a. *Acanthocope pentacornis* n. sp. from the deep sea of the Gulf of Aden (Crustacea: Isopoda: Munnopsidae) *Senckenbergiana Maritima* 20(3/4):125-129.

Müller, H-G. 1989b. *Gnathia wolffi* n. sp., a coral reef inhabiting isopod from Kenya, with a key on the *Gnathia ferox* complex (Cymothoidea: Gnathiidae). *Bonner zoologische Beiträge* 40: 63-67.

Müller, H-G. 1989c. *Arcturinoides gibbosus* n. sp. from the Gulf of Aden (Crustacea: Isopoda: Arcturidae) *Senkenbergiana Maritima* 20(5/6): 195-199.

Müller, H-G. 1990a. Anthuridea from coral reefs at Réunion Island, southern Indian Ocean (Crustacea: Isopoda). *Senckenbergiana Biologia* 70(4/6): 359-395.

Müller, H-G. 1990b. Sphaeromatidae from Réunion Island, southern Indian Ocean, with description of a new species of *Paraleptosphaeroma* (Buss & Iverson 1981) (Crustacea: Isopoda). *Revue Suisse de Zoologie* 97: 635-645.

Müller, H-G. 1991a. New species and records of *Amakusanthura, Cyathura*, and *Haliophasma* from Sri Lanka (Crustacea: Isopoda: Anthuridae). *Revue Suisse de Zoologie* 98: 589-612.

Müller, H-G. 1991b. Coral-reef inhabiting Joeropsidae from Réunion Island, southern Indian Ocean (Crustacea: Isopoda: Asellota). *Senckenbergiana Biologia* 71(1/3): 155-168.

Müller, H-G. 1991c. Four new species of shallow-water Asellota. from the Gulf of Aden (Crustacea: Isopoda). *Senckenbergiana Maritima* 21(5/6): 205-214.

Müller, H-G. 1991d. Two new species of *Joeropsis* Koehler from a sabellid reef at the south-west coast of Sri Lanka. *Zoologische Abhandlungen, Staatliches Museum für Tierkunde Dresden* 46: 121-130.

Müller, H-G. 1991e. Isopoda from coral reefs of Réunion Island: Cirolanidae and Gnathiidae (Crustacea). *Cahiers de Biologie Marine* 32: 371-386.

Müller, H-G. 1991f. New species and records of *Amakusanthura, Cyathura*, and *Haliophasma* from Sri Lanka (Crustacea: Isopoda: Anthuridae). *Revue Suisse de Zoologie* 98: 589-612.

Müller, H-G. 1991g. Stenetriidae from coral reefs at Réunion Island, southern Indian Ocean. Description of three new species. *Senckenbergiana Biologica* 71(4/6): 303-318.

Müller, H-G. 1991h. Two new species of *Cerceis* and *Dynoides* from a sabellid reef at Sri Lanka (Crustacea: Isopoda: Sphaeromatidae). *Zoologische Anzeiger* 226(5/6): 307-318.

Müller, H.-G. 1991i. Redescriptions of the corallanid isopods *Lanocira gardineri* (Stebbing 1904) and *Lanocira glabra* (Jones 1982) from the Watamu area, Kenya (Crustacea). *Zoologische Anzeiger* 227(1/2): 44-57.

Müller, H-G. 1992. The distribution of *Maresiella* (Fresi & Scipione 1980) in the world oceans, with descriptions of three new species. *Senckenbergina Biologia* 72(1/3): 205-217.

Müller, H-G. 1993. The genus *Mesanthura* (Barnard 1914) from coral reefs in Kenya and Malaysia, with descriptions of three new species (Crustacea: Isopoda: Anthuridae). *Mitteilungen Zoologische Museum Berlin* 69: 19-44.

Müller, H-G. 1995. Sphaeromatidae from the Watamu area, Kenya. Description of a new genus and four new species (Isopoda, Flabellifera). *Crustaceana* 68: 350-381.

Nair, V.R. & T.S.S. Rao 1973. Chaetognaths of the Arabian Sea. In B. Zeitzschel (ed.), *The Biology of the Indian Ocean*: pp. 293-317. Berlin: Springer-Verlag.

Negoescu, I. 1980. Littoral anthuridean isopods (Isopoda, Anthuridea) from the north western Indian Ocean. *Travaux Muséum d' Histoire Naturelle 'Grigore Antipa'* 22: 401-420.

Negoescu, I. 1981. New data about the fauna of Anthuridean isopods (Isopoda, Anthuridea) from the Libyan coast and the Suez Canal. *Travaux Muséum d'Histoire Naturelle 'Grigore Antipa'* 23: 109-122.

Neyman, A.A., M.N. Sokolova, N.G. Vinogradova & F.A. Pasternak 1973. Some patterns of the distribution of bottom fauna in the Indian Ocean. In B. Zeitzschel (ed.), *The biology of the Indian Ocean*. Berlin: Springer-Verlag. 549 pages.

Nicholls, G.E. 1929. Some new species of *Stenetrium* from Western Australia. *Proceedings of the Linnean Society of New South Wales* 54: 361-374.

Nicolet, H. 1849. Crustaceos. In Gay, C. (ed.), *Historia fiscia y politica de Chile (zool.)* 3: 1-547.

Nierstrasz, H.F. 1931. Die Isopoden der Siboga Expedition. III. Isopoda Genuina. II. Flabellifera. *Siboga Expeditie Monographie* 32e: 123-233.

Nierstrasz, H.F. 1941. Die Isopoden der Siboga Expedition. IV. Isopoda Genuina. III. Gnathiidea, Anthuridea, Valvifera, Asellota, Phreatocoidea. *Siboga Expedition Monographie* 32d: 235-305.

Nierstrasz, H. & G. Brender á Brandis 1923. Die Isopoden der Siboga Expedition. II. Isopoda Genuina. I. Epicaridea. *Siboga Expeditie Monographie* 32b: 57-121.

Nierstrasz, H. & G. Brender á Brandis 1929. Neue Epicaridea. *Zoologischer Anzeiger* 85: 295-302.

Nierstrasz, H. & G. Brender á Brandis 1931. Papers from Dr Th. Mortensen's Pacific Expedition 1914-16. 57. Epicaridea 2. *Videnskabelige Meddelelser fra Dansk Naturhistorisk Forening i Kjobenhavn* 91: 147-225.

Nierstrasz, H. & G. Brender á Brandis 1932. Alte und neue Epicaridea. *Zoologischer Anzeiger* 101: 90-100.

Nobili, G. 1906a. Diagnoses préliminaires de Crustaces, Decapodes et Isopodes nouveaux recueilles par M. le Dr G. Seurat aux îles Touamotou. *Bulletin du Muséum National d'Histoire Naturelle, Paris* 12: 256-270.

Nobili, G. 1906b. Tre nuovi Sferomidi Eritrei del Museo Zoologico dell' Universita di Napoli. *Annuario del Museo zoologico della R. Universita di Napoli* N. S. 2(16): 1-7.

Nobili, G. 1906c. Nuovi Bopiridi. *Atti della Real Accademia della Scienze di Torino* 41: 1-18.

Nordenstam, A. 1933. Marine Isopoda of the families Serolidae, Idotheidae, Pseudidotheidae, Arcturidae, Parasellidae, and Stenetriidae mainly from the South Atlantic. *Further Zoological Results of the Swedish Antarctic Expedition 1901-1903* 3(1): 1-284.

Nordenstam, A. 1946. Marine Isopoda from Professor Dr Sixten Bock's Pacific Expedition 1917-1918. *Arkiv for Zoologi* 37A(7):1-31.

Nunomura, N. 1990. A new species of *Pistorius* (Crustacea, Isopoda) from Phuket Island, Thailand. *Bulletin of the Toyama Science Museum* 13: 59-64.

Ohlin, A. 1901. Isopoda from Tierra del Fuego and Patagonia. i.Valvifera. *Svenska Expeditionen till Magellanslanderna* 2: 261-306.

Ortiz, M., K. Berze-Freire & A. Wasikete 1990. Una nueva especie de isopodo (Flabellifera, Sphaeromatidae), de la Republica Popular de Mozambique. *Revista de Investigaciones Marinas* 11: 207-211.

Pallas, P.S. 1766. *Miscellanea Zoologica, quibus novae imprimis atque obscurae animalium species describuntur et observationibus iconibusque illustrantur*. Hagae comitum. Apud Petrum van Cleef. 224 p.

Pallas, P.S. 1772. *Spicilegia Zoologica* 1767-80. Berolini.

Paperna, I. & F.D. Por 1977. Preliminary data on the Gnathiidae (Isopoda) of the northern Red Sea, the Bitter Lakes and the eastern Mediterranean and the biology of *Gnathia piscivora* n. sp. *Rapports et Proces Verbaux des Réunions, Commission Internationale pour l'Exploration Scientifique de la Mer Mediterranée, Monaco* 24: 195-197.

Paulian, R. & C. Delamare Deboutteville 1956. Un cirolanide cavernicole à Madagascar. *Mémoires de l'Institut Scientifique de Madagascar* (A)9: 85-88.

Pearse, A.S. 1930. Parasites of Japanese Crustacea. *Annotationes Zoologicae Japonenses* 13(1): 1-8.

Perez, C. 1920. Sur un type nouveau d'Epicarides, *Rhopalione uromyzon* n.g. n.sp. parasite sous abdominal d'un Pinnothere. *Comptes Rendus des Séances Hebdomadaires de l'Académie des Science, Paris* 170: 1615-1617.

Pfeffer, G. 1887. Die Krebse von Sud-Georgien nach der Ausbeute der Deutschen Station 1882-83. 1er. Teil. *Jahrbuch der Hamburgischen wissenschaftlichen Anstalten* 4: 1-110.

Pillai, N.K. 1954. A preliminary note on the Tanaidacea and Isopoda of Travancore. *Bulletin of the Central Research Institute, University of Travancore, Trivandrum, Natural Sciences* 3(1)C:1-21.

Pillai, N.K. 1955. Wood boring Crustacea of Travancore. I. Sphaeromidae. *Bulletin of the Central Research Institute, Trivandrum* 4(1)C: 127-139.

Pillai, N.K. 1957. A new species of *Limnoria* from Kerala. *Bulletin of the Central Research Institute, Trivandrum* V(2)C: 149-157.

Pillai, N.K. 1961. *Monograph: Wood boring Crustacea of India*. Simla: Government of India Press.

Pillai, N.K. 1963a. South Indian Marine Isopods. *Journal of the University of Bombay* 31(3&5): 100-112.

Pillai, N.K. 1963b. Two new genera of parasitic isopods from Kerala. *Journal of the Zoological Society of India* 15(1-2): 66-72.

Pillai, N.K. 1963c. Isopod parasites of Indian mysids. *Annals and Magazine of Natural History.* (13)6: 739-743.

Pillai, N.K. 1965. Isopods of the family Sphaeromidae from the waters of South India. *Crustaceana* 9: 75-89.

Pillai, N.K. 1966. Littoral and parasitic isopods from Kerala: Family Anthuridae. 1. *Journal of the Bombay Natural History Society* 63: 152-161.

Pillai, N.K. 1967. Littoral and parasitic isopods from Kerala. Families Eurydice, Corallanidae and Aegidae 2. *Journal of the Bombay Natural History Society* 64: 267-283.

Pillai, N.K. & K.C. Eapen 1966. On a new genus of anthurid isopod *Indanthura* from South India. *Bulletin of the Department of Marine Biology and Oceanography, University of Kerala* 2:13-18.

Pires, A.M.S. 1982. Taxonomic revision of *Bagatus* (Isopoda, Asellota) with a discussion of ontogenetic polymorphism in males. *Journal of Natural History* 16: 227-259.

Poore, G.C.B. 1975. Australian species of *Haliophasma* (Crustacea: Isopoda: Anthuridae). Records of the Australian Museum 29: 503-533.

Poore, G.C.B. 1980. A revision of the genera of the Paranthuridae (Crustacea: Isopoda: Anthuridae) with a catalog of species. *Zoological Journal of the Linnean Society* 68: 53-67.

Poore, G.C.B. 1984. Redefinition of *Munna* and *Uromunna* (Crustacea: Isopoda: Munnidae), with descriptions of five species from coastal Victoria. *Proceedings of the Royal Society of Victoria* 96: 61-81.

Poore, G.C.B. 1985. Australian chaetiliids (Crustacea: Isopoda: Valvifera): a new genus, new species and remarks on the family. *Memoirs of the Museum of Victoria* 46: 153-172.

Poore, G.C.B. 1987. Lynseiidae (Isopoda: Flabellifera), a new monotypic family from Australia. *Journal of Crustacean Biology* 7: 258-264.

Poore, G.C.B. 1990. Two new species of isopod crustaceans belonging to Australian endemic genera (Serolidae and Chaetiliidae). *Memoirs of the Museum of Victoria* 51: 109-119.

Poore, G.C.B. 1994. New species of *Mariococcus brucei*, an unusual new genus and species of Sphaeromatidae from southern Australia (Crustacea: Isopoda). *Memoirs of the Museum of Victoria* 54: 171-178.

Poore, G.C.B. & T.M. Bardsley 1990. *Chelanthura* (Crustacea: Isopoda: Anthuridae), a new genus from southern Australia. *Memoirs of the Museum of Victoria* 51: 109-119.

Poore, G.C.B. & T.M. Bardsley 1992. Austrarcturellidae (Crustacea: Isopoda: Valvifera), a new family from Australasia. *Invertebrate Taxonomy* 6: 843-908.

Poore, G.C.B. & B. Kensley 1981. *Coralanthura* and *Sauranthura*, two new genera of anthurideans from Northeastern Australia. *Proceedings of the Biological Society of Washington* 94:503-513.

Poore, G.C.B. & J.Just 1990. *Pseudojanira invrestigatoris*, a new species from southern Australia: second species in the Pseudojaniridae (Isopoda: Asellota) with new morphological information and interpretations. *Journal of Crustacean Biology* 10: 520-527.

Poore, G.C.B. & H.M. Lew Ton 1988a. Antheluridae, new family of Crustacea (Isopoda: Anthuridea) with new species from Australia. *Journal of Natural History* 22: 489-506.

Poore, G.C.B. & H.M. Lew Ton 1988b. A generic review of the Hyssuridae (Crustacea: Isopoda) with a new genus and new species from Australia. *Memoirs of the Museum of Victoria* 49: 169-193.

Poore, G.C.B. & H.M. Lew Ton 1988c. *Amakusanthura* and *Apanthura* (Crustacea: Isopoda: Anthuridae) with new species from tropical Australia. *Memoirs of the Museum of Victoria* 49: 107-147.

Poore, G.C.B. & H.M. Lew Ton 1988d. More Australian species of *Haliophasma* (Crustacea: Isopoda: Anthuridae). *Memoirs of the Museum of Victoria* 22: 489-506.

Poore, G.C.B. & H.M. Lew Ton 1990. The Holognathidae (Crustacea: Isopoda: Valvifera) expanded and redefined on the basis of body plan. *Invertebrate Taxonomy* 4: 55-80.

Poore, G.C.B. & H.M. Lew Ton 1993. Idoteidae of Australia and New Zealand (Crustacea: Isopoda: Valvifera). *Invertebrate Taxonomy* 7: 197-278.

Purosotham & Rao 1971.

Qasi, M.H. 1959. Some bopyrid isopods of West Pakistan. *The Scientist, Karachi* 3:55-62.

Ramakrishna, G. & P. Ramaniah 1978. A new cymothoid of the genus *Nerocila* from Madras. *Bulletin of the Zoological Survey of India* 1: 177-180.

Rao, M.V.L. & P.N. Ganapati 1969. A new species of *Limnoria* from the Andaman Islands (Isopoda Flabellifera). *Crustaceana* 17: 225-230.

Richardson, H. 1903. Isopods collected at the Hawaiian Islands by the US Fish Commission Steamer Albatross. *US Fish Commission Bulletin* 1903: 47-54.

Richardson, H. 1904. Contributions to the Natural History of the Isopoda. *Proceedings of the United States National Museum* 27:1-89.

Richardson, H. 1905. A monograph on the isopods of North America. *Bulletin of the United States National Museum* 54: 1-727.

Richardson, H. 1906. Descriptions of new Isopod crustaceans of the family Sphaeromidae. *Proceedings of the United States National Museum* 31: 1-22.

Richardson, H. 1910. *Marine isopods collected in the Philippines by the US Fisheries steamer Albatross in 1907-1908*. Washington D.C.: Department of Commerce Bureau of Fisheries Document No. 736.

Robinson, D.J. & R.J. Menzies 1961. *Munna spinifera*, a new species of isopod crustacean from bathyal regions in the Gulf of Aden. *Crustaceana* 3: 110-114.

Roman, M-L. 1970. Ecologie et repartition de certains groupes d'Isopodes dans les divers biotopes de la région de Tuléar (sud-ouest de Madagascar). *Receuil des Travaux de la Station Marine d' Endoume Fascicule Hors Serie*, Supplement 10: 163-208.

Roman, M-L. 1974. Diagnose general et description de *Neocassidina perlata* n.g., n. sp., de Sphaeromatidae (Crustaces Isopodes) des recifs coralliens de Tuléar (S. O. de Madagascar). *Tethys* 5: 351-360.

Roman, M-L. 1979. *Tanaidaces et isopodes benthiques recifaux et littoraux du sud-ouest de Madagascar Autecologie Synecologie Chorologie*. Université de Droit, d'Economie et des Sciences d'Aix, Marseille III. 413 p. [unpublished dissertation]

Rosen, B.R. 1971. The distribution of reef coral genera in the Indian Ocean. In D.R. Stoddart & M. Yonge (eds), *Regional variation in Indian Ocean coral reefs*: 263-299. London: Academic Press.

Sars, M. 1859. Oversigt over de i den norsk arctiske region forskommende Krebsdyr. *Forhandlinger i Videnskaps selskabet i Christiana* 1858 99:122-163.

Sars, G.O. 1898. *An account of the Crustacea of Norway*, Vol 2. Isopoda. Parts 11-12, pp. 145-232. Bergen.

Sars, G.O. 1905. Pacifisches Plankton. Crustaceen. *Zoologische Jahrbucher Abteilung fur Systematik, Geographie und Biologie der Tiere* 21: 371-414.

Say, T. 1818. An account of the Crustacea of the United States. *Journal of the Academy of Natural Sciences of Philadelphia* 1: 393-401, 423-433.

Schioedte, J.C. & F.W.S. Meinert 1879-1880. De Cirolanis Aegas simulantibus commentatio brevis. *Naturhistorisk Tidsskrift* 12: 279-302.

Schioedte, J.C. & F.W.S. Meinert 1881. Symbolae ad Monographium Cymotharum Isopodum Familiae 2. Anilocridae. *Naturhistorisk Tidsskrift* (3)13: 1-166.

Schioedte, J.C. & F.W.S. Meinert 1883. Symbolae ad Monographium Cymotharum Crustaceorum Isopodum Familiae 3. Saophridae. 4. Ceratothoinae. *Naturhistorisk Tidsskrift* (3)13: 281-378.

Schioedte, J.C. & F.W.S. Meinert 1884. Symbolae ad Monographium Cymothoarum Isopodum Familiae 4. Cymothoidae. Trib. II. Cymothoinae. Trib. III. Livonecinae. *Naturhistorisk Tidsskrift* (3)14: 221-454.

Schoenichen, W. 1908. *Gnathia aldabrensis* n. sp. ein neuer Isopode aus dem Indischen Ozean. In A. Voeltzkow (ed.), *Reise in Ostafrika in den Jahren 1903-1905 mit Mitteln der Hermann und Elise geb.* Heckmann Wentzel Stiftung ausgefuhrt, Wissenschaftliche Ergebnisse, Band II, Systematische Arbeiten Heft 3.

Schotte, M. 1992. *Serolis apheles*, a new species (Isopoda: Serolidae) from the southwest Indian Ocean, and a range extension for *Serolis antarctica* (Beddard 1884). *Proceedings of the Biological Society of Washington* 105: 90-96.

Schotte, M. 1994. *Annina mannai*, a new isopod from Ganges River, West Bengal (Crustacea: Isopoda: Cirolanidae). *Proceedings of the Biological Society of Washington* 107: 268-273.

Schotte, M. 1995. *Gnathia arabica*, n.s. first record of a gnathiid from the Arabian Sea, and new records of isopods (Crustacea; Isopoda) from Pakistan in collections of the Smithsonian Institution. In Thompson, M.F. & N.S. Tirmizi (eds), *The Arabian Sea, Living Resources and the Environment*: pp. 111-118. Lahore, Pakistan: Vanguard Press.

Schultz, G.A. 1982. Species of Protallocoxoidea and Stenetroidea (Isopoda, Asellota) from the Antarctic and southern seas. *Biology of the Antarctic Seas* 10, *Antarctic Research Series* 32: 17-62.

Sheppard, E.M. 1933. Isopod Crustacea. 1. The family Serolidae. *Discovery Reports* 7: 253-362.

Sheppard, E.M. 1957. Isopoda Crustacea. Pt. II: The suborder Valvifera. Families: Idotheidae, Pseudidotheidae and Xenarcturidae fam. n. with a supplement to Isopod Crustacea, Part I. The family Serolidae. *Discovery Reports* 29: 141-198.

Shiino, S.M. 1933. Bopyrids from Tanabe Bay. *Memoirs of the College of Science, Kyoto Imperial University* (B) 8 3, Article 8): 249-300.

Shiino, S.M. 1934. Bopyrids from Tanabe Bay II. *Memoirs of the College of Science, Kyoto University* (B) 9 (4, Article 7): 257-287.

Shiino, S.M. 1937. Bopyrids from Tanabe Bay, IV. *Memoirs of the College of Science, Kyoto Imperial University* (B) 12 (3, Article 18): 479-493.

Shiino, S.M. 1939. Bopyrids from Kyusyu and Ryukyu. *Records of Oceanographic Works in Japan* 10: 79-99.

Shiino, S.M. 1941. Further notes on bopyrids from Kyusyu and Ryukyu. *Annotationes Zoologicae Japonenses* 20: 154-158.

Shiino, S.M. 1949. On two new bopyrid parasites belonging to the genus *Orbione*. *Bulletin of the Biogeographical Society of Japan* 14(10): 51-55.

Shiino, S.M. 1958. Note on the bopyrid fauna of Japan. Report. *Faculty of Fisheries. Prefectural University of Mie* 3: 29-73.

Shyamasundari, K., C.J. Kumari, K.H. Rao & A. Mary 1991. A new species of the genus *Heteranthura* Kensley (Crustacea: Isopoda: Anthuridae) from Visakhaptam coast. *Journal of the Bombay Natural History Society* 88: 262-264.

Siewing, R. 1959. *Angliera xarifae*, ein neuer Isopode aus dem Kustengrundwasser der Insel Abd-Kuri (Golf von Aden). *Zoologischer Anzeiger* 163: 365-370.

Sivertsen, E. & L.B. Holthuis 1980. The marine isopod Crustacea of the Tristan da Cunha Archipelago. *Gunneria* 35: 10128.

Southwell, T. 1915. Notes from the Bengal Fisheries Laboratory, Indian Museum. No. 2. On some Indian parasites of Fish, with a note on carcinoma in trout. *Records of the Indian Museum* 11: 3110330.

Srinivasan, T.K. 1959. On a new species of wood boring isopod, *Exosphaeroma madrasensis* from Madras. *Journal of the Timber Dryers' and Preservers' Association of India* 5(3): 20-23.

Stebbing, T.R.R. 1873. A sphaeromid from Australia, and two Arcturidae from South Africa. *Annals and Magazine of Natural History* (4)12: 279-302.

Stebbing, T.R.R. 1875. On some new exotic sessile-eyed crustaceans. *Annals and Magazine of Natural History* (4)15:184-188.

Stebbing, T.R.R. 1900. On Crustacea brought by Dr. Willey from the South Seas. *Willey's Zoological Results* 5: 605-690.

Stebbing, T.R.R. 1902. *South African Crustacea. Part 2. Marine Investigations of South Africa.* 1-92.

Stebbing, T.R.R. 1904a. Marine crustacean. 12. Isopoda, with description of a new genus. In J.S. Gardiner (ed.), *The Fauna and Geography of the Maldive and Laccadive Archipelagoes, being the account of the work carried on and of the collections made by an expedition during the years 1899 and 1900* (2): 699-720. Cambridge: Cambridge University Press.

Stebbing, T.R.R. 1904b. Gregarious Crustacea from Ceylon. *Spolia Zeylanica* 2(5): 1-29.

Stebbing, T.R.R. 1905. Report on the Isopoda collected by Professor Herdman at Ceylon, in 1902. In *Report to the Government of Ceylon on the pearl oyster fisheries in the Gulf of Manaar*. London: Royal Society.

Stebbing, T.R.R. 1907. A freshwater isopod from Calcutta. *Journal of the Linnean Society (Zoology)* 30: 39-42.

Stebbing, T.R.R. 1908. South African Crustacea, Part IV. Annals of the South African Museum 6: 1-96.

Stebbing, T.R.R. 1910a. The Percy Sladen Trust Expedition to the Indian Ocean in 1905, under the leadership of Mr J. Stanley Gardiner. Isopoda from the Indian Ocean and British East Africa. *Transactions of the Linnean Society of London, Zoology* (2)14: 83-122.

Stebbing, T.R.R. 1910b. Reports on the Marine Biology of the Sudanese Red Sea. XIV. On the Crustacea Isopoda and Tanaidacea. *Journal of the Linnean Society, Zoology* 31: 215-230.

Stebbing, T.R.R. 1910c. General Catalogue of South African Crustacea. Part V. *Marine Investigations in South Africa* 6: 281-593.

Stebbing, T.R.R. 1911. Indian Isopods. *Records of the Indian Museum* 6: 179-191.

Stebbing, T.R.R. 1923. *Crustacea of Natal*. Fisheries and Marine Biological Survey, Report No. 3, for the year 1922. Cape Town.

Stechow, E. 1921. Symbiosen zwischen Isopoden und Hydroiden. *Zoologischer Anzeiger* 53: 221-223.

Stephensen, K. 1947. Tanaidacea, Isopoda, Amphipoda and Pycnogonida. *Scientific Results of the Norwegian Antarctic Expedition 1927:* 28, Oslo 2(27): 1-90.

Stimpson, W. 1855. Descriptions of some new marine Invertebrata. *Proceedings of the Academy of Natural Sciences, Philadelphia* 7: 385-395.

Studer, T. 1882. Uber eine neue art *Arcturus* und eine neue Gattung der Idotheiden. *Sitzungsberichte der Gesellschaft naturforschender Freunde zu Berlin* 1882: 56-58.

Studer, T. 1883. Verzeichniss der während der Reise S.M.S. Gazelle an der westkuste von Afrika, Ascension und dem Kap der Guten Hoffnung gesammelten Crustaceen. *Abhandlungen der Preussische Akademie der Wissenschaften zu Berlin* (1882): 1-32.

Studer, T. 1884. Isopoden gesammelt wahrend der reise S.M.S. Gazelle um die Erde 1874-76. *Anhang zu den Abhandlungen der Koniglich Akademie der Wissenschaften zu Berlin* 1883: 1-28.

Suhm, R von Willemoes 1876. Preliminary Report. Voyage of HMS Challenger. *Proceedings of the Royal Society of London* 26: 569-592.

Tattersall, W.M. 1905. The marine fauna of the coast of Ireland. Part V. Isopoda. *Fisheries, Ireland, Scientific Investigations*. 1904(2): 1-90.

Tattersall, W.M. 1913. The Schizopoda, Stomatopoda, and Non-Antarctic Isopoda of the Scottish National Antarctic Expedition. *Transactions of the Royal Society of Edinburgh* 49: 865-894.

Tattersall, W.M. 1922. The Percy Sladen Trust Expeditions to the Abrohols Islands (Indian Ocean) under the leadership of Prof. W.J. Dakin, FLS, FZS. Amphipoda and Isopoda. *Journal of the Linnean Society of London* 35: 1-20.

Thielemann, M. 1910. Beitrage zur Kenntnis der Isopodenfauna Ostasiens. *Abhandlungen Akademie der Wissenschaften Mathematischen Physikalische Klasse in Munchen* 2(3): 1-109.

Thomson, G.M. 1879. New Zealand Crustacea, with descriptions of new species. *Transactions of the New Zealand Institute* 11:230-248.

Thomson, J.M. 1946. New Crustacea from the Swan River Estuary. *Journal of the Proceedings of the Royal Society of Western Australia* 30: 35-53.

Thomson, J.M. 1951. The fauna of Rottnest Island. X. Anthuridae. *Journal of the Royal Society of Western Australia* 35: 1-8.

Tiwari, K.K. 1952. On a new species of the rare cymothoid genus *Agarna* (Schioedte & Meinert), parasitic on the clupeid fish *Nematalosa nasus* (Bl.) in the Bay of Bengal. *Records of the Indian Museum* 50: 295-300.

Vanhöffen, E. 1914. Die Isopoden der Deutschen Südpolar Expedition 1901-1903. *Deutsche Südpolar Expedition 1901-1903* 25 (Zoologie) 7: 447-598.

Wägele, J.W. 1979. Morphologische Studien an *Eisothistos* mit Beschreibung von drei neuen Arten (Crustacea, Isopoda, Anthuridea). *Mitteilungen aus dem Zoologischen Museum der Universität Kiel* 1(2): 21-27.

Wägele, J.W. 1981a. A study of the Hyssuridae (Crustacea: Isopoda: Anthuridea) from the Mediterranean and the Red Sea. Israel Journal of Zoology 30: 47-87.

Wägele, J.W. 1981b. Study of the Anthuridae (Crustacea: Isopoda: Anthuridea) from the Mediterranean and the Red Sea. *Israel Journal of Zoology* 30: 113-159.

Wägele, J.W. 1982. On a new *Microcerberus* from the Red Sea and the relationship of the Microcerberidea to the Anthuridea (Crustacea, Isopoda, Anthuridea). *Zoologica Scripta* 11: 281-286.

Walker, A.O. 1901. Contributions to the Malacostracan fauna of the Mediterranean. *Journal of the Linnean Society of London, Zoology* 28, 182: 290-307.

Walker, A.O. 1903. Report on the Isopoda and Amphipoda collected by Mr George Murray, FRS, during the cruiser of the 'Oceana' in November 1898. *Annals and Magazine of Natural History* (7)12: 223-233.

Walker, A.O. & A. Scott 1903. Decapod and sessile-eyed crustaceans from Abd-Kuri. Macrura and Eridiophthalma. In Forbes, H. (ed.), *The Natural History of Sokotra and Abd-Kuri*: pp. 212-213. Liverpool: Liverpool Museum.

Weber, M. 1892. Die Susswasser Crustaceen des Indischen Archipels, nebst Bemerkungen uber die Susswasser Fauna im Allgemeinen. In *Zoologische Ergebnisse*: Vol. II. Leiden: E.J Brill.

Wells, F.E. (ed.) 1993. *The marine flora and fauna of Rottnest Island, Western Australia: Proceedings of the Fifth International Marine Biology Workshop, held at Rottnest Island in January 1991*. Perth: Western Australian Museum.

White, A. 1847. *List of specimens of Crustacea in the collection of the British Museum*. London: British Museum.

Whitelegge, T. 1902. Scientific results of the trawling expedition of HMCS 'Thetis', off the coast of New South Wales. Crustacea, Pt. IV. Isopoda III. *Memoirs of the Australian Museum* 4: 247-282.

Wolff, T. 1962. The systematics and biology of bathyal and abyssal Isopoda Asellota. *Galathea Reports* 6: 1-320.

Wolff, T. 1989. The genera of Santiidae (Kussakin 1988) with the description of a new genus and species (Crustacea, Isopoda, Asellota). *Steenstrupia* 15: 177-191.

Wolff, T. 1990. The appendages of *Limnoria stephenseni* Menzies (Isopoda, Flabellifera). *Bijdragen tot de Dierkunde* 60: 311-318.

Wyrtki, K. 1971. *Oceanographic Atlas of the International Indian Ocean Expedition.* Washington D.C.: National Science Foundation.

zur Strassen, O. 1902. Uber die Gattung und die Arcturiden der Deutschen Tiefsee Expedition. *Zoologischer Anzeiger* 25: 682-689.

2 Ecology and evolution

Parental care behavior in the wood-boring isopod *Sphaeroma terebrans*

MARTIN THIEL
Smithsonian Marine Station, Fort Pierce, Fla.. USA (Present address: Facultad de Ciencias del Mar, Universidad Catolica del Norte, Campus Guayacan, Coquimbo, Chile)

ABSTRACT

Several wood-boring crustaceans engage in parental care, i.e., juveniles remain in the parent's dwelling after having hatched from the female's brood pouch. It has been hypothesized that small juveniles are not capable of burrowing into wood, and therefore rely on parental assistance during their early life. In this study, I examined parental care in the wood-boring isopod *Sphaeroma terebrans*, which inhabits burrows in aerial roots of the red mangrove *Rhizophora mangle*. Offspring remained in the maternal burrow for up to 40 days after hatching from the female's brood pouch. The female remained in her burrow throughout this time. Most juveniles started burrowing to the outside from within the female's burrow but some left the mother's burrow. The latter started to build their own burrows in the immediate vicinity of the natal burrow on the maternal root rather than emigrating from it. Small orphan juveniles that were experimentally separated from their mothers were able to build their own burrow on aerial roots. However, they required several days to establish a burrow, whereas small sub-adults that had already reached independence built a new burrow within 48 hours. These results suggest that parental care in wood-boring crustaceans provides protection for small offspring during their early life stages when they cannot efficiently bore themselves. Thus, woody substrate may have contributed to the evolution of parental care in two ways: 1) It constitutes a stable microhabitat allowing for a moderately long (~ 40 days) persistence of parent-offspring groups, and 2) These isopods, as well as other crustacean borers, become efficient borers late in life, therefore requiring parental protection during the early phase of their life..

1 INTRODUCTION

In the marine environment, wood or wood-like substrates constitute important habitats for boring crustaceans. Several isopod and amphipod species establish temporary or permanent burrows in these substrates on which they also may feed. Substrates such as wood, kelp-stipes, sand- or limestone permit the establishment of stable burrows that may persist for time periods of months to several years (e.g., Menzies 1954, Rotramel 1975, Chess 1993). It is therefore not surprising that several boring isopods or amphipods also reproduce within their burrows. Small peracarid offspring may remain in the parental burrow after emerging from the maternal brood pouch (Henderson 1924, Menzies 1957, Conlan &

Chess 1992, Messana et al. 1994), i.e., these species engage in extended parental care. During this time period, small juveniles possibly obtain nourishment from their parents or they may be protected against intra- and interspecific enemies within the parental burrow. Since little is known about parental care in boring peracarids it cannot be said whether all juveniles enjoy parental care, or whether some have to leave at an early developmental stage. In the soft-bottom dwelling amphipod *Leptocheirus pinguis*, some juveniles leave the maternal burrow shortly after emerging from the female brood pouch while others remain there for several weeks (Thiel 1999a). This intra-clutch variation in parental care duration was attributed to spatial constraints in the maternal burrow (Thiel 1999a), a factor that may also affect parental care duration in boring peracarids. Juveniles that remain in the parental burrow for long time periods may have better survival chances than their siblings that (have to?) leave their parents shortly after emerging from the brood pouch.

Juvenile peracarids may not only live in the parental burrow for some time, but they may also start their own burrows from within the parental burrow. Drawings provided by Henderson (1924) and Menzies (1957) indicate that this occurs in some *Limnoria* species (Fig. 1). Similar to some soft-bottom dwelling amphipods and tanaids, where juveniles start their own burrows from within the maternal burrow (Thamdrup 1935, Bückle-Ramirez 1965, Thiel et al. 1997), these wood-boring peracarids would never have to come

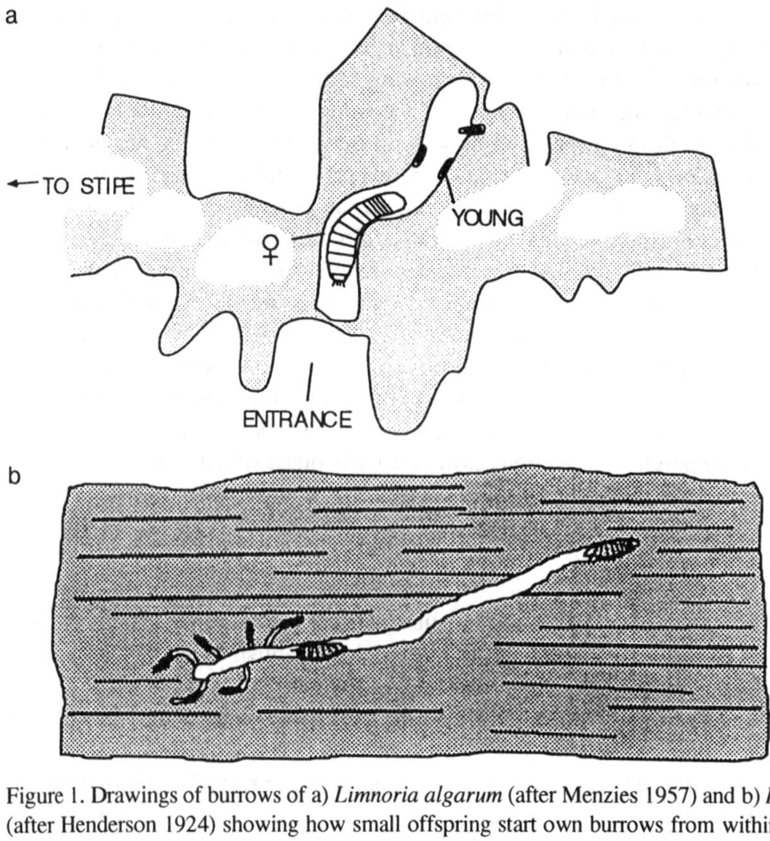

Figure 1. Drawings of burrows of a) *Limnoria algarum* (after Menzies 1957) and b) *Limnoria lignorum* (after Henderson 1924) showing how small offspring start own burrows from within the parental burrow.

to the substrate surface but could spend their entire life in the wood in which they were born. This would have important consequences for their population biology and possibly for intraspecific interactions, since many individuals within one piece of wood may be closely related to each other. It therefore is important to determine how long juveniles remain in their maternal burrows, how they survive without their parents, and where and how they start their own first burrows.

A common boring peracarid species in subtropical lagoonal environments is the isopod *Sphaeroma terebrans* that preferentially bores into aerial roots of the red mangrove *Rhizophora mangle*. Female *S. terebrans* have been found to host their offspring within their burrows (Messana et al. 1994, Thiel in review). Herein, I examined how long juveniles remain in the maternal burrow, and whether small juveniles, when taken from the maternal burrows, are able to establish their own burrows.

2 MATERIAL AND METHODS

Sphaeroma terebrans was collected in the Indian River Lagoon, which extends approximately 200 km along Florida's Atlantic coast. Aerial roots of the red mangrove *Rhizophora mangle* containing burrows of *Sphaeroma terebrans* were collected at the western shore of the lagoon. Upon return to the laboratory, the roots were carefully dissected and isopods were taken from their burrows. All isopods were kept for not more than 3 hours in seawater before being introduced to the respective experiments. Two experiments were conducted in running seawater, which was pumped from a small harbor basin that is in immediate vicinity (200 m distance) to the collecting site. One experiment (BORECAP) was designed to examine whether small juveniles taken from maternal burrows can bore as efficiently as small sub-adults. Three different treatments with ten replicates each were compared in this experiment: orphan juveniles that were placed in containers without their mother, orphan juveniles that were placed in containers together with their mother, and small subadults that had been collected from their own individual burrows. Five small juveniles (or subadults) were placed each in enclosed containers that contained a 15-cm long tip of a young aerial root *R. mangle* (Fig. 2A). Two sides of the containers had openings that were covered with a 0.5-mm mesh to allow water exchange but to prevent isopods from leaving the containers. The roots were examined every two days in order to count isopods that had not established a burrow and were still in the open. Some burrows were easily visible, but many were hidden under small algae or under cracks of the root bark, so that they could only be safely revealed at the end of the experiment. In order to minimize the disturbance of the isopods, it was not attempted to find isopod burrows on the experimental roots during the experiment. Thus, during the experiment, only isopods that were still visible in the containers or on the roots were counted. At the end of the experiment (day 8), the individual roots were carefully examined and all isopods that had established burrows were collected and preserved.

In a second experiment (PADUR), the duration of parental care in *S. terebrans* was examined. A pair of *S. terebrans* consisting of a male and a female with embryos was introduced to a small container with a 15-cm long tip of a young aerial root *R. mangle*. Only males and females that had been found together in one burrow in the field were used. In the females selected for this experiment, embryos could be seen through the ventral part of the female pereonites. All embryos were in stage C (see Holdich 1968) where limbs

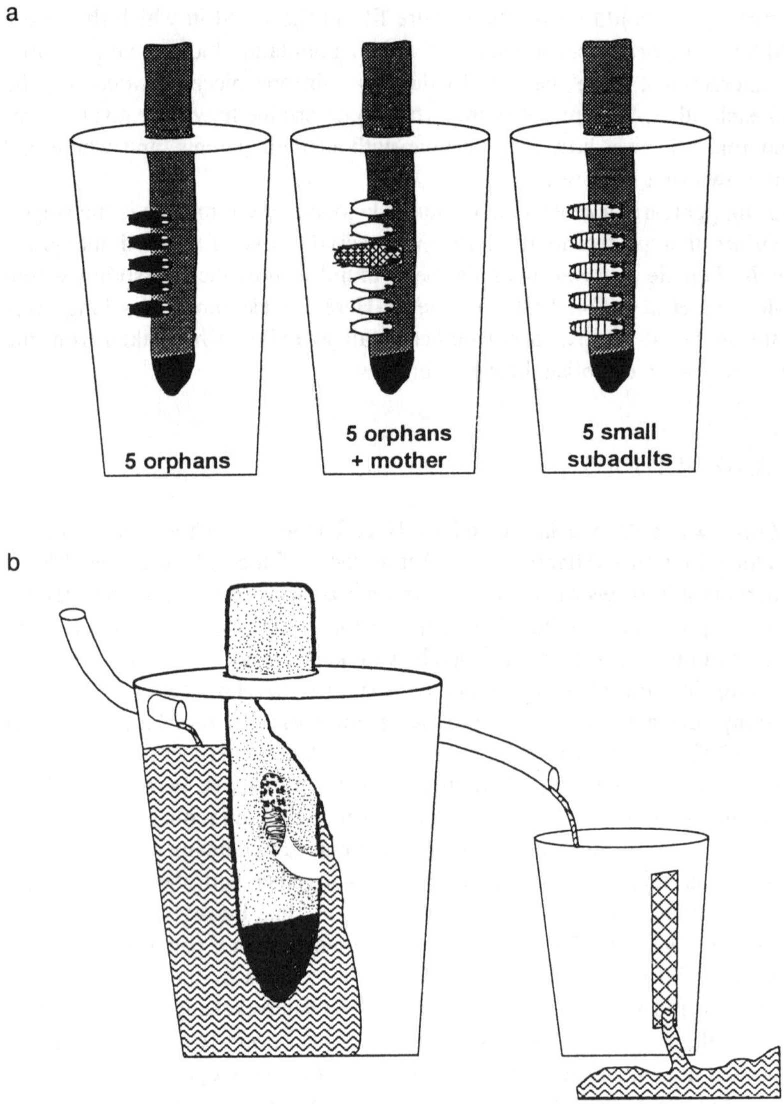

Figure 2. a) Sketch of experimental set-up for the BORECAP-experiment conducted to examine the boring capability of small *Sphaeroma terebrans*; 10 replicates per treatment. b) Sketch of experimental set-up for PADUR-experiment conducted to examine the duration of extended parental care in *Sphaeroma terebrans*; 7 replicate females gave birth and survived until the end of the experiment.

and eyespots are already clearly visible. At this stage, the embryos still contain considerable amounts of yolk, so that they appear yellowish/orange. Since it was not possible to peer into the burrows during the experiment, it cannot be said with certainty at what day the embryos actually emerged from the female's body. However, for other sphaeromatid isopods of similar sizes, it is known that embryos of stage C (stage III in Jensen 1955) require about 10-14 days of development before they emerge from the female's body (Jensen 1955). For the first two days of the PADUR-experiment, the isopods were housed in closed containers that allowed water exchange through a 0.5-mm mesh. After they had established their own burrows, the roots containing the females and males in their burrows were transferred to containers with a continuous flow of water. The water flowing out was filtered through a trap (0.5-mm mesh-size) to collect all juvenile isopods that had emigrated from the parental root (Fig. 2B). The traps were monitored every morning, and all isopods retained in the traps were counted and preserved. At the end of the experiment all roots were carefully dissected and all isopods were collected and preserved. Seven (out of eight) females survived until the end of the PADUR-experiment. The experiments were conducted between 4 and 12 December 1997 (BORECAP) and between 27 March and 19 May 1998 (PADUR).

3 RESULTS

3.1 *Boring capability of small* Sphaeroma terebrans *(BORECAP-experiment)*

Small juveniles remained visible on the mangrove roots for longer time periods than small sub-adults (Fig. 3A). Most sub-adults had established burrows two days after the start of the experiment, and were no longer visible at the surface of the roots. In contrast, juveniles that had been taken from the burrows of their mothers remained at the surface of the roots for several days. Only 8 days after the start of the experiment had most of them established their

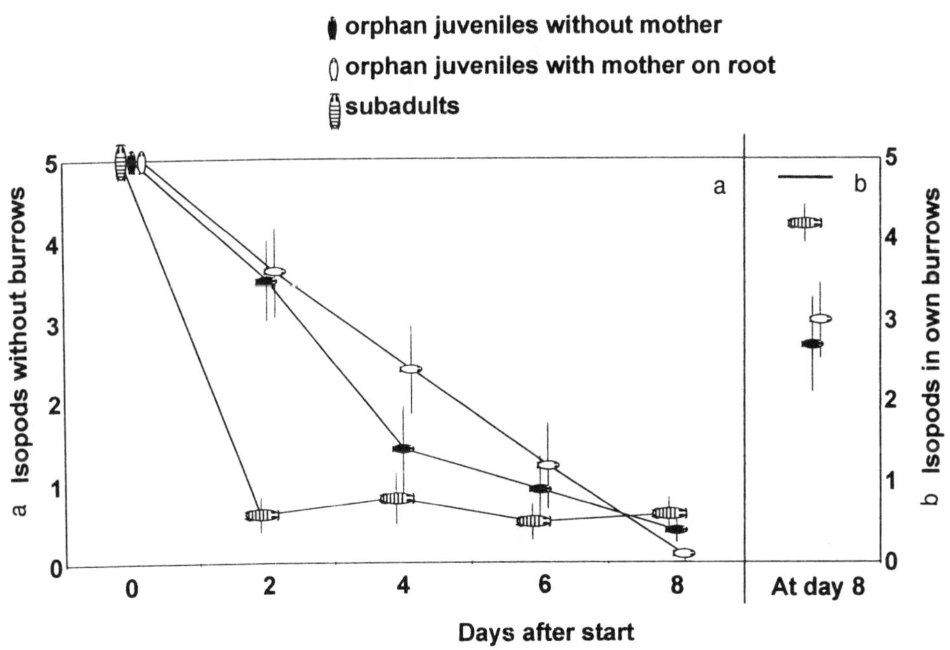

Figure 3. a) Average number (± SE) *Sphaeroma terebrans* counted without a burrow in each container during the BORECAP-experiment. b) Average number (± SE) *S. terebrans* found in burrows at the end of the BORECAP-experiment.

own burrows. Survival of sub-adult *S. terebrans* was higher than that of or-phan juveniles (Fig. 3B), but the differences were not significant (one-way ANOVA, $p < 0.05$). The presence of the mother had no effect on the behavior and survival of the orphan juveniles. Juveniles that were placed in the containers together with their mother did not reassociate with her during this experiment.

3.2 *Duration of parental care in* Sphaeroma terebrans *(PADUR-experiment)*

Seven female *Sphaeroma terebrans* hosted between 36 and 72 juveniles in their burrows during the experiment (Fig. 4). Most juveniles remained either in the maternal burrow or

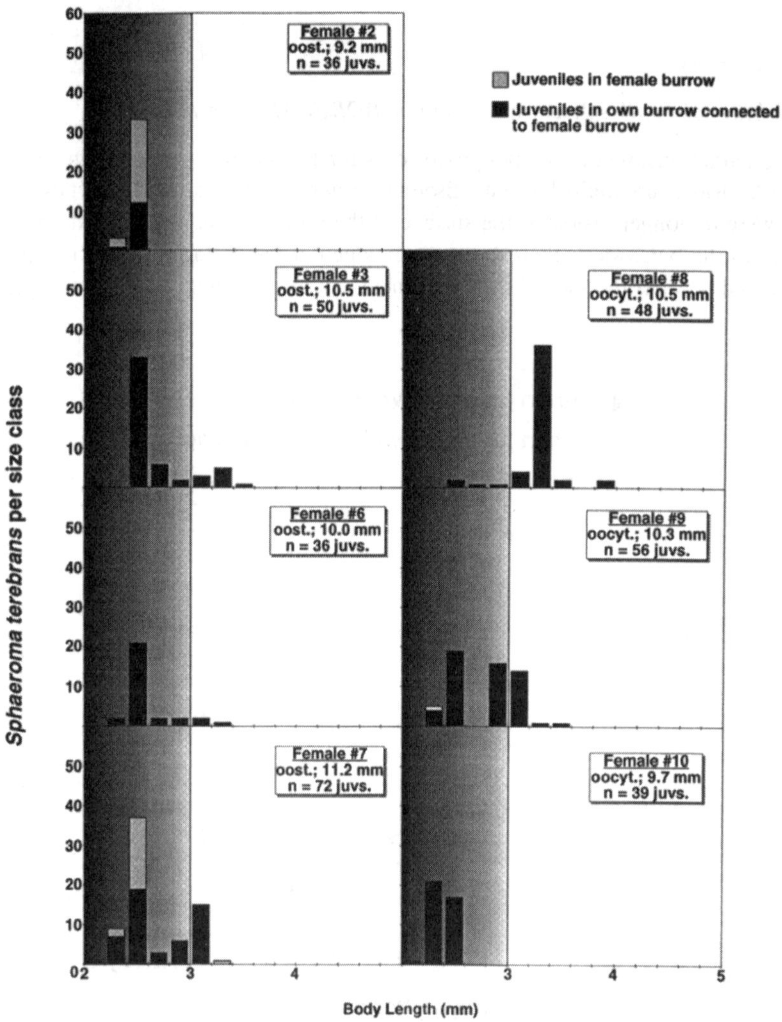

Figure 4. Juvenile *Sphaeroma terebrans* in the respective size classes at the end of the PADUR-experiment; all juveniles of 7 females are shown; shaded areas indicate sizes of juveniles frequently found in maternal burrows under natural conditions.

later in their own burrows on the maternal root, i.e., few juveniles emigrated from the containers. At the end of the experiment most juveniles were in their own burrows many of which had originated from the maternal burrow. In some instances, the females had elongated their own burrows, apparently to get away from the root region in which the juveniles had established their burrows. Juveniles reached sizes of 2-4 mm during the experiment (Fig. 4). Throughout the experimental period, most males and females remained in separate burrows. Only female number 3 was joined by the male in its burrow at the end of the experiment.

4 DISCUSSION

Small orphan juveniles were able to bore into mangrove roots, but they required considerably more time to build a burrow than sub-adult *Sphaeroma terebrans*. Females hosted their juveniles for relatively long time periods and allowed them to build their own burrows originating from the maternal burrow. Thus, females provided parental care for their offspring during a time period when these are not yet capable of rapidly boring into wood.

Wood, kelp-stipes, sand- and lime-stone rocks provide relatively stable habitats for boring crustaceans. While many other boring invertebrates use a combination of chemical and mechanical ways to bore into their substrates, crustaceans rely exclusively on mechanical devices, usually the mandibles. It is therefore not unlikely to expect a size-dependent boring efficiency in boring crustaceans. John (1970) experimentally demonstrated that adult *Sphaeroma terebrans* built burrows large enough to house them within 48 hours. A healthy adult was even found to establish a burrow large enough to accommodate the whole isopod within 3 hours (John 1968). In the present experiments, small sub-adults also had established new burrows after 48 hours, considerably faster than small orphan juveniles, most of which required > 4 d to build a burrow. Small isopods, just after being born are still relatively soft (Henderson 1924, own observations), and their boring devices and muscle attachment sites may not be rigid enough to permit efficient boring. Yet, while not being very efficient borers, small isopods managed to slowly establish burrows over the course of the experiment. While this slow boring proved sufficient for survival in the sheltered experimental containers, under natural conditions it may be less successful. Many boring isopods and amphipods are found in shallow coastal environments with a high abundance of fish predators. In addition, some boring crustaceans such as *Sphaeroma terebrans* preferentially bore in the intertidal zone where they are frequently exposed to air. Thus, for *S. terebrans* a burrow provides primarily a shelter against predation but also against desiccation (John 1970). Under natural conditions it may be necessary to build a new burrow as fast as possible. Small juveniles that are inefficient borers enjoy the privilege to reside in and even establish their own burrows from within the maternal burrow. During this period of extended parental care, they probably are well sheltered from threats such as predation and desiccation.

Although being in abundant supply, mangrove roots are relatively thin (root diameters range from ~10 to ~30 mm), and each root can only support a limited number of burrows and isopods. This is supported by observations made by Estevez & Simon (1975) who found an average of ~8 *Sphaeroma terebrans* per habitable root with a maximum of 54 isopods (34 of which were juveniles) per root. Perry & Brusca (1989) found for *S. peruvianum* densities of 3-4 holes root^{-1}, similar to what Villalobos et al. (1985) reported for *S.*

terebrans (4.1 individuals root^{-1}). Ribi (1982) mentioned an average value of 10.3 *S. terebrans* root^{-1}, indicating that an upper value may range somewhere between 10 and 20 isopods root^{-1}. Thus, it appears likely that not all offspring of a mature female (20-50 juveniles female^{-1} – see e.g., Venkata Krishnan & Nair 1973) can establish homes on the natal root.

Juveniles that start a burrow from the maternal burrow quickly reach the outside of an aerial root. At this point they may be faced with the decision to stay on their natal root or to leave and find another root suitable to build a burrow. In the PADUR-experiment very few juveniles emigrated from the natal root, even though they had reached relatively large sizes, and many of them had already started an independent life at the end of the experiment. Unfortunately, the experiment was terminated too early to decide at what size or density on the natal root juveniles might start to emigrate. It should be noted however, that females in the PADUR-experiment hosted much larger numbers of juveniles in their burrows than females in the field (Thiel 1999b). Similarly, in the lab-experiment, many juveniles reached considerable sizes on the maternal root which is rarely seen in the field (Thiel 1999b). Under natural conditions, single small subadults are sometimes found on the roots of mature females, but there is little indication that whole offspring cohorts remain on the natal root. The relatively long coexistence of females and their offspring on the experimental roots may have been due to the fact that both mothers and their offspring were kept in a relatively sheltered environment without disturbance by other isopods (Thiel 2000). In contrast to other boring peracarids (*Limnoria* spp. – Menzies 1954; *Peramphithoe stypotrupetes* – Chess 1993) where males and females cohabit for long time periods, in *S. terebrans* most males leave the female after copulation (Thiel 1999b). Therefore, in *S. terebrans* the female has to conduct all tasks connected with the burrow-life and extended parental care (ventilation, defense, increasing the burrow size) on its own. It is possible that under field conditions with regular exposure of burrows (at low tide) and disturbance by other isopods, maternal care activities are frequently disrupted, and juveniles leave their mother earlier than under sheltered laboratory conditions. In other boring isopods (*Limnoria* spp.), sub-adult or adult stages are regularly found to leave the natal piece of wood and migrate to new habitats (Henderson 1924, Eltringham & Hockley 1961). While Henderson (1924) speculated that migration could be a response to overcrowding, Eltringham & Hockley (1961) found little indication for this suspicion, but rather noted that large individuals regularly migrate.

As a consequence of limited space on aerial roots and due to the early break-up of mother-offspring units in the field, it appears unlikely that closed families of *Sphaeroma terebrans* would occur on one root. However, in boring isopods or amphipods that dwell in larger substrate units such as large pieces of wood, parents and offspring may live close together for long time periods, as has been suspected by several authors (e.g., Kühne & Becker 1964). Burrows are actively defended, particularly by reproductive females (Henderson 1924, Menzies 1954, Thomas 1979) but it is not known whether wood-dwelling peracarids are able to discriminate between kin and non-related individuals. Most interesting is an observation by Menzies (1954) who found that fecundity on densely populated pieces of wood is considerably lower than on pieces with few neighboring conspecifics. In view of these observations, it appears highly desirable to examine the genetic relationships and intraspecific behavior of wood-dwelling peracarids that engage in extended parental care.

ACKNOWLEDGEMENTS

I thank S. Reed for assistance during sampling of the isopods, H. Reichardt for superb maintenance of the flowing seawater system, and the other staff members of the Smithsonian Marine Station for providing ideal working conditions. Financial support for this study was received in form of a Smithsonian Marine Station Postdoctoral Fellowship. This is contribution # 474 of the Smithsonian Marine Station.

REFERENCES

Bückle-Ramirez, L.F. 1965, Untersuchungen über die Biologie von *Heterotanais oerstedi* Kröyer (Crustacea, Tanaidacea). *Zeitschr. Morph. Ökol. Tiere* 55: 714-782.

Chess, J.R. 1993. Effects of the stipe-boring amphipod *Peramphithoe stypotrupetes* (Corophioidea: Amphithoidae) and grazing gastropods on the kelp *Laminaria setchellii*. *J. Crust. Biol.* 13: 638-646.

Conlan, K.E. & Chess, J.R. 1992. Phylogeny and ecology of a kelp-boring amphipod, *Peramphithoe stypotrupetes*, new species (Corophioidea: Amphithoidae). *J. Crust. Biol.* 12: 410-422.

Eltringham, S.K. & Hockley, A.R. 1961. Migration and reproduction of the wood-boring isopod, *Limnoria*, in Southampton Water. *Limn. Oceanogr.* 6: 467-482.

Estevez, E.D. & Simon, J.L. 1975. Systematics and ecology of *Sphaeroma* (Crustacea: Isopoda) in the mangrove habitats of Florida. In Walsh, G.E., Snedaker, S.C. & Teas, H.J. (eds), *Proceedings of the International Symposium on biology and management of mangroves*: pp. 286-304. Gainesville: Institute of Food and Agricultural Sciences, University of Florida.

Henderson, J.T. 1924. The Gribble: a study of the distribution factors and life-history of *Limnoria lignorum* at St. Andrews, N.B. *Contr. Canadian Biol.* 2, part 1(14): 309-327.

Holdich, D.M. 1968. Reproduction, growth and bionomics of *Dynamene bidentata* (Crustacea: Isopoda). *J. Zool.* 156: 137-153.

Jensen, J.P. 1955. Biological observations on the isopod *Sphaeroma hookeri* Leach. *Vidensk. Meddel. Dansk. Naturh. Foren* 117: 305-339.

John, P.A. 1968. Habits, structure and development of *Sphaeroma terebrans* (a wood boring isopod). *Univ. Kerala Publ.* 1: 1-73.

John, P.A. 1970. Observations on the boring activity of *Sphaeroma terebrans* Spence Bate, a wood boring isopod. *Zool. Anz.* 185: 379-387.

Kühne, H. & Becker, G. 1964. Der Holz-Flohkrebs *Chelura terebrans* Philippi (Amphipoda, Cheluridae). *Beih. Zeitschr. angewandte Zool.* 1: 1-141.

Menzies, R.J. 1954. The comparative biology of reproduction in the wood-boring isopod crustacean *Limnoria*. *Bull. Mus. Comp. Zool. Harvard* 112: 363-388

Menzies, R.J. 1957. The marine borer family Limnoridae (Crustacea, Isopoda). Part I: Northern and Central America: systematics, distribution and ecology. *Bull. Mar. Sci. Gulf. and Carib.* 7: 101-200.

Messana, G., Bartolucci, V., Mwaluma, J. & Osore, M. 1994. Preliminary observations on parental care in *Sphaeroma terebrans* (Bate 1866) (Isopoda, Sphaeromatidae), a mangrove wood borer from Kenya. *Ethol. Ecol. Evol.*, special issue 3: 125-129.

Perry, D.M. & Brusca, R.C. 1989. Effects of root-boring *Sphaeroma peruvianum* on red mangrove forests. Mar. Ecol. Prog. Ser. 57: 287-292.

Ribi, G. 1982. Differential colonization of roots of *Rhizophora mangle* by the wood boring isopod *Sphaeroma terebrans* as a mechanism to increase root density. *Mar. Ecol.* 3: 13-19.

Rotramel, G. 1975. Filter-feeding by the marine boring isopod, *Sphaeroma quoyanum* (H. Milne Edwards 1840) (Isopoda, Sphaeromatidae). *Crustaceana* 28: 7-10.

Thamdrup, H.M. 1935. Beiträge zur Ökologie der Wattenfauna auf experimenteller Grundlage. *Meddel. Kommiss. Danm. Fisk. – Havunders. (Serie: Fisk.)* 10: 1-125.

Thiel, M. 1999d. Duration of extended parental care in marine amphipods. *J. Crust. Biol.* 19: 60-71.

Thiel, M. 1999b. Reproductive biology of a wood-boring isopod (*Sphaeroma terebrans*) with extended parental care. *Mar. Biol.* 135: 321-333.

Thiel, M. 2000. Juvenile *Sphaeroma quadridentatum* invading female-offspring groups of *Sphaeroma terebrans*. *J. Nat. Hist.* 34: 737-746.

Thiel, M., Sampson, S. & Watling, L. 1997. Extended parental care in two endobenthic amphipods. *J. Nat. Hist.* 31: 713-725.

Thomas, J.D. 1979. A redescription of the wood-rasping amphipod *Tropichelura gomezi* (Ortiz 1976) (Cheluridae) from the Florida Keys, with notes on its distribution and ecology. *Proc. Biol. Soc. Wash.* 92: 863-872.

Venkata Krishnan, R. & Nair, N.B. 1973. Observations on the brood size and brood composition of timberboring sphaeromatids of Cochin Waters. *Zool. Anz.* 191: 119-122.

Villalobos, C.R., Cruz, G.A. & Cruz, R.A. 1985. Notas sobre la biologia de *Sphaeroma terebrans* (Bate 1866) (Sphaeromatidae: Isopoda) en el manglar de Pochote, Provincia de Puntarenas, Costa Rica. *Brenesia* 24: 287-296.

Depth-size related patterns of marine isopods in the Nordic Seas

JÖRUNDUR SVAVARSSON & BJARNI KR. KRISTJÁNSSON
Institute of Biology, University of Iceland, Reykjavík, Iceland and Sandgerði Marine Centre, Garðvegur, Sandgerði, Iceland.

TORLEIV BRATTEGARD
Department of Fisheries and Marine Biology, University of Bergen, Bergen, Norway

ABSTRACT

Patterns of body size with depth were evaluated for marine isopods crustaceans in the Nordic Seas (the Iceland Sea, the Norwegian Sea), within the depth range of 18 to 3900 m. In all, 55 isopod species were considered, of which most species (85%) belong to the suborder Asellota. The pattern of body size with depth differed considerably depending upon groups of isopods considered. When all isopod species were included in the analysis no pattern in average maximum body size (calculation based on all species at each locality) was observed with depth. There was, however, a significant increase in the average maximum size of asellote isopods with increasing depth and also when munnopsid asellotes were considered only. There was further a significant correlation between the average maximum body size of both asellote isopods and munnopsid asellotes and the near-bottom temperature, while no pattern was observed when all isopods were considered. No correlation was found between the average maximum body size and the species richness of the isopods. There was, however, a highly significant correlation between the variation of the maximum size of the asellote isopods and the species richness. It is concluded that size-depth related patterns are more likely to occur within a homogenous group than a heterogeneous one as functional aspects may be more size determining than physiological factors, such as the temperature. We suggest that the temperature, in connection with food availability, constraints size-depth related patterns of asellote isopods in the Nordic Seas. The pattern of diversity of the deep Arctic Ocean and Nordic Seas may be reflected in large size distribution of asellotes at high diversity sites but narrower size range in low diversity areas.

1 INTRODUCTION

Patterns of body size of animals in time and space have widely been investigated (Atkinson 1994). These give insight into evolutionary patterns of different scales (Jablonski 1997), and body size may even be important as a indicator of diversity (Siemann et al. 1996), although that has recently been questioned (Finley et al. 1996).

Patterns of size with depth in the oceans have for long time been investigated (Rex & Etter 1998). These have been evaluated by studies focusing on patterns within communities or within taxa (species, genus, family, order) for many areas (see synopsis by Rex &

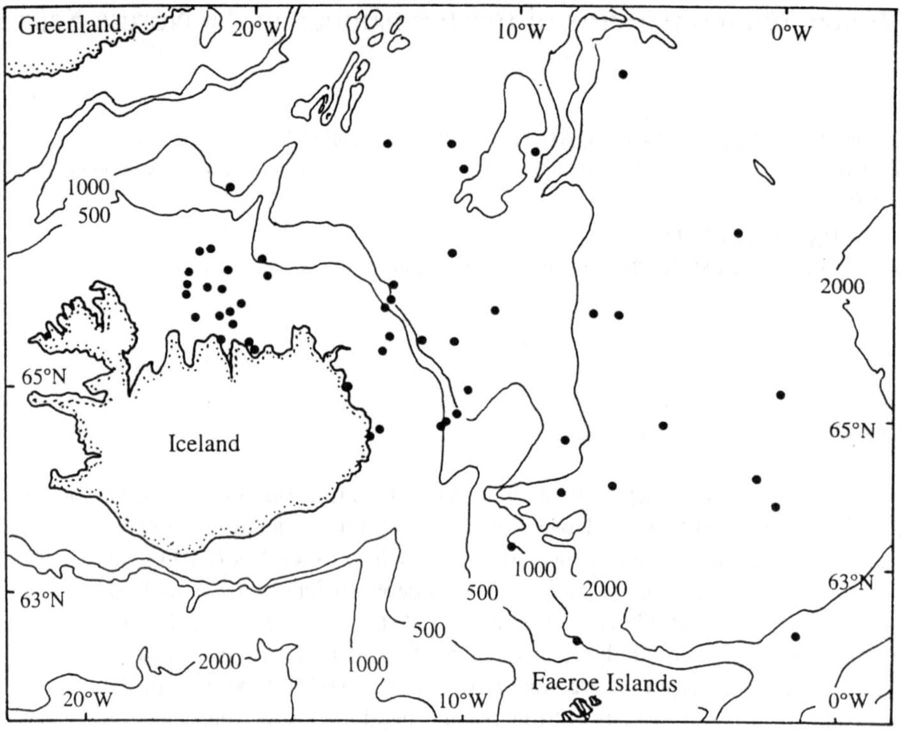

Figure 1. Localities studied in the Nordic Seas.

Etter 1998). Although much of the literature deals with similar or even the same taxa (soft bodied meiofauna, see Thiel 1975, Soltwedel et al. 1996) results have been contradicted. Rex & Etter (1998) concluded that 'it remains unclear whether consistent geographic trends in size actually exist in the deep-sea, except for the observation that deep-sea species seem on the average smaller that their shallow-water counterparts'. Part of the conflicting results observed might be due to a lack of standardization in the methods used (Rex & Etter 1998).

For the isopod crustaceans studies have focused on size pattern within families, genera, or species. Wolff (1962) showed that within some isopod genera there is a good relationship between size and depth. Large species of the respective genera occurred in deep waters of the world oceans, while small ones occurred in shallow waters. Wilson (1983) reported an increase in size within the asellote isopod *Eurycope iphthima* at the depth range of 2500 to 4800 m and Poulin (1995) showed on basis of phylogenetic approach that deep water groups tend to evolve larger body size. Studies on community basis are, however, lacking for the isopods.

The deep waters of the Arctic Ocean and the Nordic Seas are unique among the world ocean basins in having a low diversity fauna. This holds true for all groups studied (Bouchet & Warén 1979, Mannvik 1984, Svavarsson et al. 1990, Svavarsson 1997, see also Fig. 2b). This has earlier been explained on the basis of the young age of the fauna and due to the effects of the Greenland-Iceland-Faeroe Ridge, which isolates the deep Arctic and the deep Nordic Seas from the North Atlantic Ocean (Gurjanova 1938, Svavarsson et al. 1990, Svavarsson 1997). Furthermore, this environment has extremely

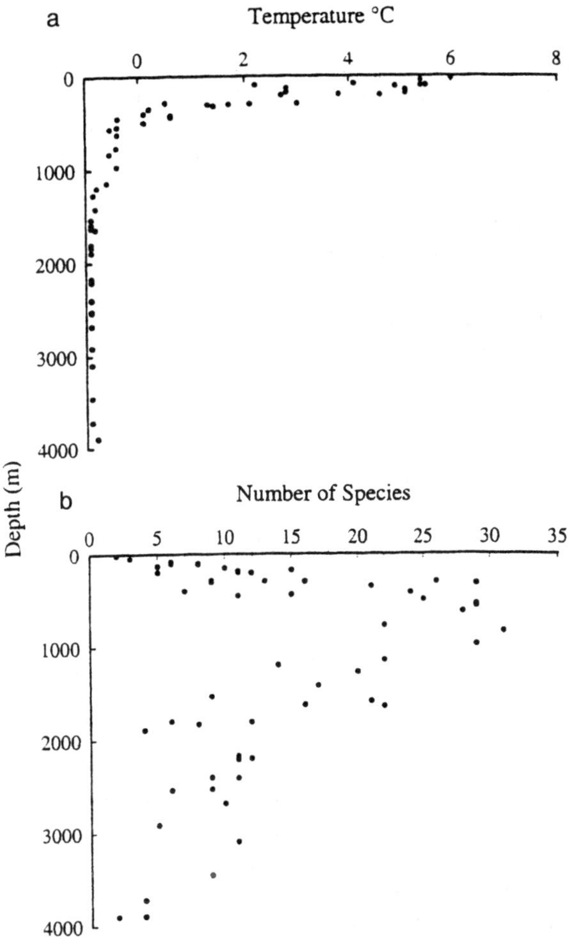

Figure 2. Temperature at different depths in the Nordic Seas (summer temperatures) (a) and number of species at different depths in the Nordic Seas (from Kluwer Academic Publishers Biodiversity and Conservation, vol. 6, 1987, page 1573, Diversity of isopods (Crustacea): new data from the Arctic and Atlantic Oceans, J. Svavarsson, fig. 1a: with a kind permission from Kluwer Academic Publishers) (b).

low temperatures (< 0° C below about 400-500 m, see also Fig. 2a), which may cause physiological strain to the animals. Body size may be affected by temperature (Atkinson 1994). Many exothermic animals produce larger offspring at low temperatures (Yampolsky & Scheiner 1996). Other factors, such as food availability, predation and competition may also effect the size of marine organisms.

Only a little is known of the patterns of body size on community basis in the shallow and deep Arctic Ocean and the Nordic Seas, apart from the study of Kendall et al. (1997) in Arctic shallow waters. Here we focus on isopod crustaceans and address the following question: are there differences in body sizes of isopods along depth, temperature and diversity gradients? We address this question by using maximum body size (body length) of all isopod species present in samples from the Nordic Seas and compare these at different depths, where different biodiversity and temperatures occur.

2 METHODS

The material from the Nordic Seas was collected during cruises of the Norwegian R/V *Håkon Mosby* during the years 1982 to 1994 and the Faroese R/V *Magnus Heinason* in 1993. Cruises from 1991 to 1994 were part of the ongoing BIOICE (Benthic Invertebrates of Icelandic waters; 1991-) project.

Samples were taken with a modified Rothlisberg-Pearcy epibenthic sled (RP sled; Rothlisberg & Pearcy 1976, Brattegard & Fosså 1991), which has 0.5 mm mesh net in the cod-end. The sled is towed along the bottom and collects animals lifted in turbulence in front of the sled. A closing mechanism ensures that the sled samples only along the bottom. The samples were elutriated and decanted over a 0.5 mm sieve. The sediment with remaining animals was later sieved over a series of sieves (8, 4, 2, 1 and 0.5 mm) and then preserved in 10 % formalin. Isopods were sorted out at the Sandgerði Marine Center and later identified to species.

Data from 56 localities in the Nordic Seas (Iceland Sea, southwestern Norwegian Sea) are presented, from a depth range of 18 to 3900 m (Fig. 1). There is a rapid decline in near-bottom temperature with depth in these waters (Fig. 2a). Near-shore surface temperature ranges from 2° in winter to 8° in summer (Ingólfsson 1996), whereas the near-bottom temperature below 400-500 m is < 0°C.

There were 55 isopod species sampled in the area. Prior to our analysis we estimated the pattern of size with depth *within* nine species at 14 to 30 localities (Table 1). These species were selected because of their ample depth range, or because of their high abundance. We measured the size (body length, from anterior end of the head to the posterior end of the pleotelson) of the largest individual at each station, but found no significant changes in size with increasing depth within each of those nine species (Table 1). As no pattern was observed within individuals of these species, we feel justified to use reported maximum body size for all occurrences of each of the species. Hence only a single value was used for each species. As an example the maximum size of the asellote, *Munnopsurus giganteus*, is 33 mm, and that number was used at all stations were *M. giganteus* was reported.

We used the maximum size recorded in the literature for each species or our measurement of the respective species, when the latter was larger than the recorded maximum size. All isopods collected in the area were included. In cases where the specimens were only identified to genus because of taxonomic problems or when specimens were too mu-

Table 1. List of species measured at 14 to 30 stations to evaluate pattern of maximum size at different stations.

Species	Spearman rank corr. (r)	*P*	Number of stations
Haplonciscus bicuspis	0.396	N.S.	14
Munna acanthifera	0.138	N.S	24
Heteromesus frigidus	0.055	N.S	16
Desmosoma strombergi	–0.255	N.S.	17
Echinozone arctica	0.166	N.S	20
Ilyarachna hirticeps	0.152	N.S.	30
Eurycope inermis	0.049	N.S.	19
Baeonectes muticus	–0.301	N.S.	16
Calathura brachiata	0.226	N.S.	30

tilated to be identified properly, we measured the largest specimen available at each station and used that value as maximum number. By using the maximum size of each species we standardized the data. By doing so we avoid potential seasonal changes in body sizes, which may occur as the samples partly occur from highly seasonal shallow waters. Furthermore, we avoid biases resulting from different sample sizes. A large sample is more likely to hold extremely large individuals than a small sample.

While analyzing trends we used average maximum sizes. Average maximum size was calculated for each station and was based on the maximum size of all species present at that station.

3 RESULTS

The maximum sizes of the species recorded were scattered, ranging from 1 to 66 mm (Fig. 3a). The scatter was somewhat less (about 1 to 33 mm) when only asellotes were considered (Fig. 3b).

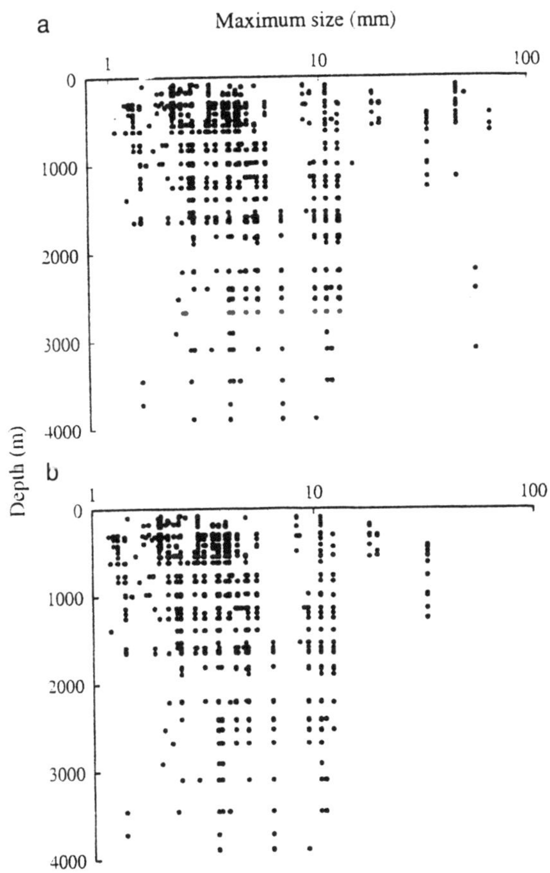

Figure 3. Scatter plot of maximum size of all isopods (a) and asellote isopods only (b) at all studied localities in the Nordic Seas.

When all isopod species were combined no pattern of average maximum size/station (each number is the average of maximum sizes of all species at that station) was observed with depth (Pearson correlation; r = –0.182, N.S., N = 56) (Fig. 4a). When asellote isopods were considered only, the average maximum size/station increased significantly with increased depth (Pearson correlation, r = 0.379, $p < 0.01$, N = 56) (Fig. 4b). When the munnopsid species (Munnopsidae sensu (Wilson 1989), the species richest family present within the suborder Asellota) were considered only, there was a significant increase in size with increased depth (Pearson correlation; r = 0.337, $p = 0.01$, N = 56) (Fig. 5).

There was no significant correlation (Pearson correlation, r = 0.211, N.S., N = 56) between the average maximum size/station and the temperature at each locality when all isopods were considered. When asellote isopods were considered only there was a significant pattern (Pearson correlation, r = –0.387, $p < 0.01$, N = 56). There was no significant correlation between the average maximum size/station and the species richness (data in Fig. 2b) at each station when all isopods were considered (r = 0.048, N.S., N = 56). Similar results were seen when asellotes were considered only.

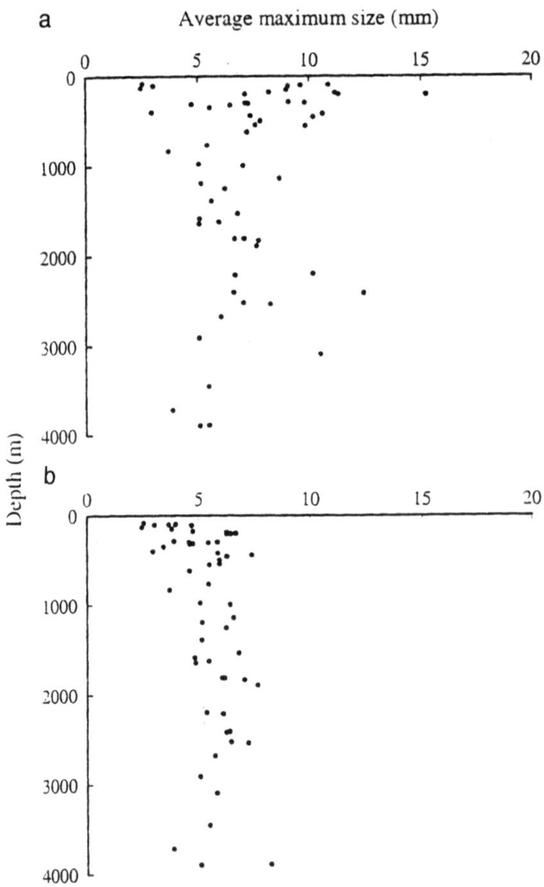

Figure 4. Average of the maximum size of all species (a) and asellote isopod species only (b) at each station in the Nordic Seas.

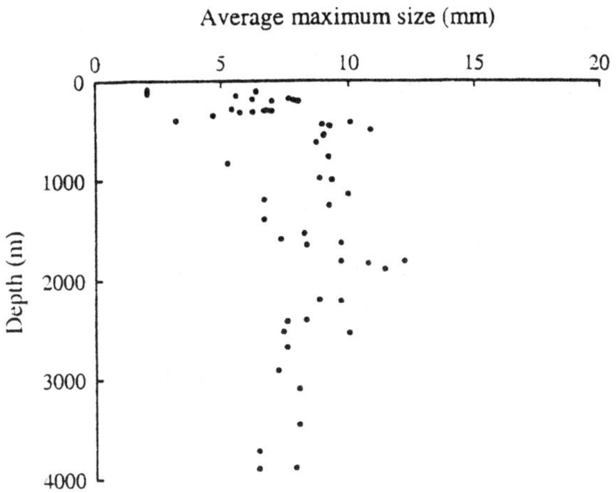

Figure 5. Average of the maximum size of munnopsid asellote isopod species at each locality in the Nordic Seas.

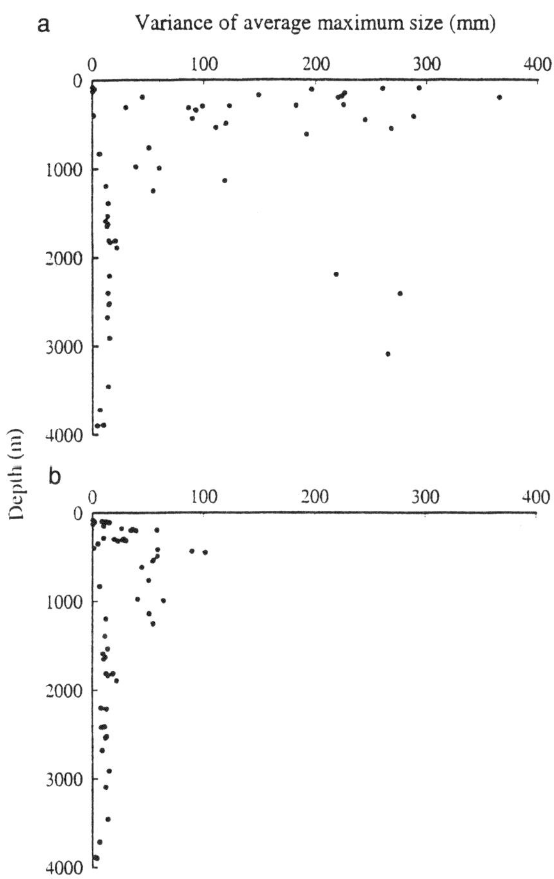

Figure 6. Variance of the maximum size of all isopod species (a) and asellote isopod species only (b) at each locality in the Nordic Seas.

The variance of the average maximum size/station decreased significantly with increasing depth when all isopods were considered (r = –0.397, $p < 0.01$, N = 56) (Fig. 6a) and further when asellotes were considered only (r = –0.359, $p < 0.01$, N = 56) (Fig. 6b). The pattern of variation of the average maximum size/station for the asellotes showed a peak at about 400 to 1200 m and was highly correlated to the species richness (r = 0.425, p = 0.001, N = 56). There was, however, no correlation between the variance of the average maximum size/station of all species and the species richness (r = 0.159, N.S., N = 56).

We also considered the average maximum size within different depth intervals, using the depth intervals 0-500 m, 500-1000 m, 1000-2000, 2000-3000 and >3000 m. Here the calculation of the average was based on average sizes at each station within the respective depth range. When all species were considered, there was no obvious pattern with depth (Fig. 7a). The average maximum size of the asellote isopods increased with increased depth (Fig. 7b). The average maximum size of the munnopsid asellotes was highest at the depth range of 1000 to 2000 m and declined towards shallower and deeper waters (Fig. 8). The average was though lowest at the depth interval of 0-500 m.

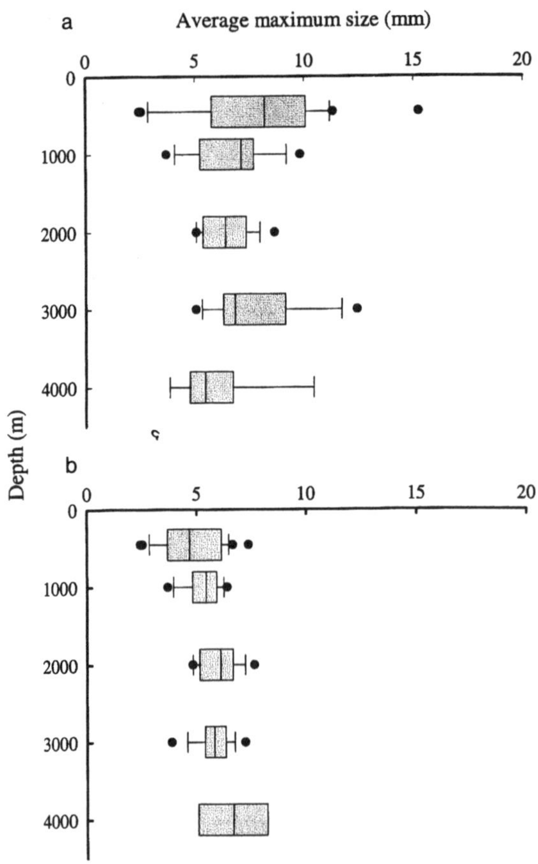

Figure 7. Average of the average maximum sizes at each station within depth intervals (0-500 m; 500-1000 m, 1000-2000 m, 2000-3000 m and 3000-4000 m) for all isopods (a) and for asellote isopods only (b). Outliers outside the 10th and 90th percentiles are shown.

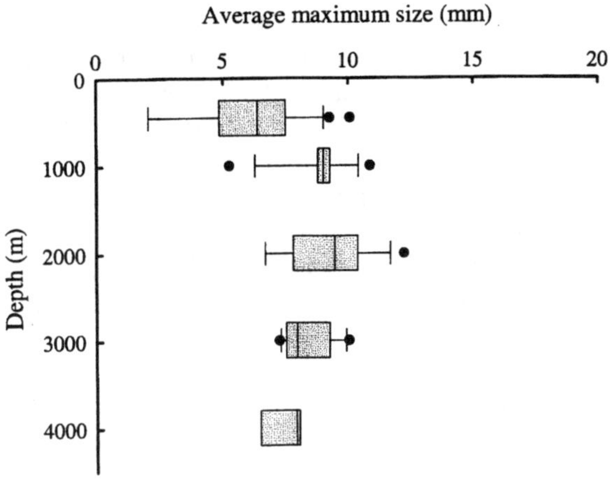

Figure 8. Average of the average maximum sizes at each station within depth intervals (0-500 m; 500-1000 m, 1000-2000 m, 2000-3000 m and 3000-4000 m) for munnopsid asellotes only. Outliers outside the 10th and 90th percentiles are shown.

4 DISCUSSION

The asellote isopods become larger with increasing depth, but no such pattern was observed when all isopods were considered. The pattern for the asellotes can mainly be seen in a lower proportion of very small species and higher proportion of fairly large species in deeper waters (> 2000 m), and further in a smaller size range of species in deeper waters. A number of studies have shown an increase in body size of individual taxa with increasing depth (Thurston 1979, Rex & Etter 1998), while studies on community basis have most frequently shown decease in the size of the fauna (Thiel 1975, Rex & Etter 1998).

The pattern of size with depth may be shaped by various factors. These may be constraints of some kind forcing animals to be larger under certain conditions and temperature may be such factor, acting upon the metabolism (Yampolsky & Scheiner 1996). Other contributing factors may be food resource availability, predation and competition. These factors may be inter-correlated and are therefore not easily isolated.

The temperature presumably causes different strain on the animals at different depths. In shallow waters (< 400 to 500 m) the species live with fluctuations in the temperatures (in shallowest waters 2-8°C) occurring above 0°C, while in deep waters the species live constantly at temperature < 0°C. Little is known of how isopod crustaceans adapt to temperatures < 0°C, and whether temperature between –0.6 and –0.9°C (at depths of about 1000 to 4000 m) have more strain on the species than temperature between 0 and –0.6°C (at depths about 500 to 1000 m). Animals living at temperatures below zero have been shown to develop anti-freezing substances, such as glycogenated proteins (Schreiber et al. 1992).

A low proportion of very small species in deep waters of the Nordic Seas may partly be explained due to food availability. There is a highly seasonal pulse of biogenic carbon to the deep waters of the Nordic Seas (Bauerfeind et al. 1994). This seasonal availability of organic matter is utilized by the Foraminifera (Protozoa), which react quickly to the seasonal pulse of food (Linke 1992). At other times much of the available energy in the deep waters is presumably located in the living tissues of the fauna and a large proportion of the deep-sea biomass is located in the foraminiferan protoplasm (Altenbach & Sarnthein 1989). Asellote isopods have previously been inferred as deposit feeders (Wolff 1962), but recent studies have shown the importance of foraminifera in their diet. Large and intermediately sized arctic deep-sea munnopsid isopods feed on both calcareous and agglutinating Foraminifera (Svavarsson et al. 1993). They may, due to their size, be able to swallow and/or crunch foraminiferans which are often of considerable size and utilize available food resource in a relatively food poor environment. Here the size of the asellote isopods may be an adaptive response.

Despite a general trend towards increased size some large species do occur only in shallow and intermediate waters and do not enter the deep waters. This includes the valviferan isopod *Arcturus baffini* (Sabine) (maximum size 66 mm), the anthuridean *Calathura brachiata* (maximum size 45.5 mm) and the asellote munnopsid *Munnopsurus giganteus* (maximum size 33 mm). It is not possible to evaluate whether it is a physiological constraint concerning their size that limits their distribution to shallow water, or whether their feeding habits may play an important role. The latter may hold for *A. baffini*, which is a filter feeder, and *Calathura brachiata*, which may be predator on peracarid crustaceans, if feeding habits resemble that of other paranthurideans (Wägele 1985, 1989). Physiological strain may limit the distribution of *Munnopsurus giganteus*. This species feeds on Foraminifera (Brandt & Negoescu 1997), which are abundant in deep arctic waters.

The diversity pattern in the Nordic Seas follows a right-skewed bell-shaped curve, increasing from 18 m towards a maximum at 310 to 1100 m, but then the diversity decreases and at 2000 to 4000 m a low diversity occurs (Fig. 2b). No correlation was observed between the average maximum size and the species richness. We found, however, a highly significant correlation between the variance of the average maximum sizes and the diversity of asellote isopods, while no significant pattern was found when all isopods were considered. This may indicate that a morphologically and ecologically homogenous group diversifies towards different sizes when the diversity is high, but heterogeneous groups (all isopods) may not evolve in the same way. Highest variance of the average maximum size and the highest diversity occurred at bathyal depths (300 to 1100 m) in the Nordic Seas. Rex & Etter (1998) found the bathyal environment to be an important site of diversification and the present findings may support that notion.

We found that the pattern of low diversity (below 1000 m) occurred simultaneously with significant reduction in the frequency of very small species and a relative increase of fairly large species. A large size of species may play an important role in limiting diversity in a food limited and a low density environment, as the finite amount of energy may support fewer larger than many smaller individuals (Rex & Etter 1998). In the deep Nordic Seas and the Arctic Ocean this may act as a barrier towards an increased diversity.

A low diversity fauna characterizes the deep Nordic Seas and the Arctic Ocean and the low diversity has been explained by a short evolutionary time of the fauna as a cold water fauna (Svavarsson et al. 1993). The present fauna is considered a fairly recent one, as partly seen in a low endemism at the generic level (Svavarsson et al. 1993). Whether the larger size of the deep water species has been an adaptation during colonization of the deep Nordic Seas and the Arctic Ocean or a prerequisite prior to a colonization of the deep cold waters remains to be evaluated. Limited information exists on the phylogeny of the deep-water asellote isopods at present for such an evaluation.

It is concluded that the differences in environmental factors (temperature possibly acting upon metabolism and respiration) and in food availability may be reflected in differences in size. These differences may not be evident when heterogeneous functional groups are considered as the functional aspects may be more size determining than other factors, such as the temperature. It is not yet known whether an increase in asellote isopod size with depth is restricted to the Arctic environment. Generally warmer waters in other parts of the world oceans may result in different strain and response to the size of the animals.

ACKNOWLEDGMENTS

We thank the Fiskirannsóknastofa of the Faeroe Islands for providing ship-time on R/V *Magnus Heinason* and the crew of R/Vs *Magnus Heinason* and *Håkon Mosby* for their hospitality. We thank the staff of the Sandgerði Marine Centre, in particular Sigrún Haraldsdóttir, for sorting isopods, Dr Hólmgeir Björnsson for statistical advice, Prof. Agnar Ingólfsson for his advice and comments and an anonymous reviewer for constructive and fruitful comments.

REFERENCES

Altenbach, A.V. & M. Sarnthein 1989. Productivity record in benthic Foraminifera. In W.H. Berger, V.S. Smetacek & G. Wefer (eds), *Productivity of the oceans: present and past:* 255-279. New York: Wiley.

Atkinson, D. 1994. Temperature and organism size – a biological law for ectotherms? In M. Begon & A.H. Fitter (eds), *Advances in ecological research*, 25:1-58. London: Academic Press.

Bauerfeind, E., B. v. Bodungen, K. Arndt & W. Koeve 1994. Particle flux, and composition of sedimenting matter, in the Greenland Sea. *J. Mar. Systems* 5: 411-423.

Bouchet, P. & A. Warén 1979. The abyssal molluscan fauna of the Norwegian Sea and its relation to other faunas. *Sarsia* 64: 211-243.

Brandt, A. & I. Negoescu 1997. Redescription of the arctic-boreal species *Calathura brachiata* (Stimpson 1853) (Isopoda, Anthuridea, Parathuridae). *Crustaceana* 70: 418-453.

Brattegard, T. & J.H. Fosså 1991. Replicability of an epibenthic sampler. *J. Mar. Biol. Ass. UK* 71: 153-166.

Finley, B.J., G.F. Esteban & T. Fenchel 1996. Global diversity and body size. *Nature* 383: 132-133.

Gurjanova, E. 1938. On the question of the composition and origin of the abyssal fauna of the Polar Basin. *Comptes Rendus (Doklady) de 'Academie des Sciences de URSS'* 20: 333-336.

Ingólfsson, A. 1996. The distribution of intertidal macrofauna on the coasts of Iceland in relation to temperature. *Sarsia* 81: 29-44.

Jablonski, D. 1997. Body-size evolution in Cretaceous molluscs and the status of Cope's Rule. *Nature* 385: 250-252.

Kendall, M.A., R.M. Warwick & P.J. Somerfield 1997. Species size distributions in Arctic benthic communities. *Polar Biol.* 17: 389-392.

Linke, P. 1992. Metabolic adaptations of deep-sea benthic Foraminifera to seasonally varying food input. *Mar. Ecol. Prog. Ser.* 81: 51-63.

Mannvik, H.-P. 1984. Slangestjerner (Echinodermata: Ophiuroidea) i Nord-Norge; deres bathymetriske fordeling og zoogeografi. Thesis, University of Tromsö, 86 pp (in Norwegian).

Poulin, R. 1995. Evolutionary influences on body size in free-living and parasitic isopods. *Biol. J. Linn. Soc.* 54: 231-244.

Rex, M.A. & R.J. Etter 1998. Bathymetric patterns of body size: implications for deep-sea biodiversity. *Deep-Sea Res.* II 45: 103-128.

Rothlisberg, P.C. & W.G. Pearcy 1976. An epibenthic sampler used to study the ontogeny of vertical migration of *Pandalus jordani* (Decapoda, Caridea). *Fisheries Bull.* 74: 994-997.

Schreiber, A., J. Svavarsson & V. Storch 1992. Blood proteins in bipolar Priapulida. *Polar Biol.* 12: 667-672.

Siemann, E., D. Tilman & J. Haarstad 1996. Insect species diversity, abundance and body size relationships. *Nature* 380: 704-706.

Soltwedel, T., O. Pfannkuche & H. Thiel 1996. The size structure of deep-sea meiobenthos in the north-eastern Atlantic: nematode size spectra in relation to environmental variables. *J. Mar. Biol. Ass. UK.* 76: 327-344.

Svavarsson, J. 1997. Diversity of isopods (Crustacea): new data from the Arctic and Atlantic Oceans. *Biodiv. Conser.* 6: 1571-1579.

Svavarsson, J., T. Brattegard & J.-O. Strömberg 1990. Distribution and diversity patterns of asellote isopods (Crustacea, Isopoda) in the deep Norwegian and Greenland Seas. *Prog. Oceanogr.* 24: 297-310.

Svavarsson, J., J.-O. Strömberg & T. Brattegard 1993. The deep-sea asellote (Isopoda, Crustacea) fauna of the Northern Seas: species composition, distributional patterns and origin. *J. Biogeogr.* 20: 537-555.

Thiel, H. 1975. The size structure of the deep-sea benthos. *Int. Revue Ges. Hydrobiol.* 60: 575-606.

Thurston, M.H. 1979. Scavenging abyssal amphipods from the north-east Atlantic. *Oceanogr. Mar. Biol.* 51: 55-68.

Wägele, J.W. 1985. Observations on nutrition and ultrastructure of digestive tract and fat body of the giant paranthurid *Accalathura gigantissima* Kussakin. *Polar Biol.* 4: 33-43.

Wägele, J.W. 1989. Evolution und phylogenetisches System der Isopoda. *Zoologica* 140: 1-262.

Wilson, G.D. 1983. Variation in the deep-sea isopod *Eurycope iphthima* (Asellota, Eurycopidae): depth related clines in rostral morphology and in population structure. *J. Crust. Biol.* 3: 127-140.

Wilson, G.D. 1989. A systematic revision of the deep-sea subfamily Lipomerinae of the isopod crustacean family Munnopsidae. *Bull. Scripps Inst. Oceanogr.* 27: 1-138.

Wolff, T. 1962. The systematics and biology of bathyal and abyssal Isopoda Asellota. *Galathea Rep.* 6: 1-320.

Yampolski, L.Y. & S.M. Scheiner 1996. Why larger offspring at lower temperatures? Demographic approach. *Amer. Nat.* 147: 86-100.

Isopod assemblages on the continental shelf and upper slope from the southwestern Atlantic

ANA MARIA SETUBAL PIRES-VANIN
Instituto Oceanográfico, University of São Paulo, Brazil

ABSTRACT

Composition and distribution of isopod assemblages on the shelf and upper slope from the São Paulo northern coast, southeastern Brazil, were studied from sixty-nine dredge hauls and fifty four van Veen grab samples taken between 10 and 600 m depth. Data from both samplers were binary transformed for community analysis. Multivariate methods for classification and ordination, as Cluster Analysis and Canonical Correspondence Analysis, were employed in such studies. Forty-eight species were discovered in the area, 21 of them new to science. From those new species 71% were found as deep as 50 m, and the Anthuridea was the most diverse group. Clustering analysis disclosed three distinct groups. Group 1 comprises the inner shelf assemblage, which is influenced by warm and light water – the Coastal Water (CW). Species from this group are *Politolana eximia, Politolana* sp, *Cristaserolis laevis, Paranthura* sp. 1 and *Colanthura* sp. Group 2 comprises the outer shelf plus shelfbreak assemblages, and is situated within a transitional area under both the CW and the South Atlantic Central Water (SACW) influence. *Accalathura crenulata* and *Joeeropsis* sp. dominate it. Group 3 characterizes the slope, is always within the SACW domain, and *Gnathia ricardoi, Bathygnathia magnifica, Eisothistos* sp., and *Metacirolana* sp. are the most frequent species. The isopod assemblage structure in the shelf and upper slope appears to be determined primarily by hydrodynamics and factors linked to depth and less controlled by sediment composition. It is suggested that the slope is probably an upper limit for deep-living isopods in the region.

1 INTRODUCTION

Benthic assemblages have been studied by several authors in different latitudes and oceans (Mills 1975, Ambrogi et al. 1990, Pires 1992, Sahade et al. 1998). The objective of such studies is generally the search for patterns in the species distribution and knowledge of the main factors controlling these distributions. The factors that define these patterns are physical disturbances or biological interactions such as competition, predation and behavior (Brandt & Pipenburg 1994, Franke & Janke 1998). More recently community studies have been considered as an important tool for evaluating ecosystem health (Thouzeau et al. 1991), since benthic organisms have an immediate response to contaminants deposited on the sea bed.

Marine isopod communities have been studied in a narrow range of habitats, mostly in slope and deep-sea waters of temperate and polar seas (Svavarsson et al. 1990, Zimmermann & Brandt 1992, Brandt & Piepenburg 1994, Poore et al. 1994). Studies dealing with isopod assemblages are scanty in the southwestern Atlantic. The only report deals with isopod zonation in rocky shores submitted to several degrees of wave exposure (Pires 1981b).

The southern Brazilian continental shelf is characterized by a complex hydrographic regime caused by the presence of three water masses. The warm and light Coastal Water – CW (T > 20°C, S < 36) moves along the coast nearshore. The warm and saline Tropical Water –TW (T > 20°C, S > 36) fills the shelf offshore. The cold and dense South Atlantic Central Water –SACW (T < 18°C, S < 36) flows from the slope towards the coast near the seabed (Castro Filho et al. 1987). The SACW entrance is seasonal, during the austral summer months, and varies in intensity from one year to another (Matsuura 1996). The regional hydrography and the primary production are highly affected by this meso-scale oceanographic process, with consequences not only in the distribution of the benthic fauna (Pires 1992, Pires-Vanin 1993), but also in the trophic dynamics of the whole ecosystem (Pires-Vanin et al. 1993). There is also evidence that the hydrology along with the seabed topography contribute to differential sedimentation rates and organic matter content in the area (Mahiques et al. 1999).

In the Brazilian Bight the continental shelf varies in width between 50 to 230 km, with about 120 km in front of Ubatuba. It can be divided into an inner portion as deep as 80 m with terrigenous sediment and an outer part, between 80-100 m, where carbonate content is high (Kowsmann & Costa 1979). The shelf break region is narrow, with nearly 30 km in width, presenting patches of mud among blocks and fragments of calcareous algae, coral and bryozoan exoskeletons (Zembruski 1979). Through analyses of sediments, the upper slope, down to 240 m depth, is mainly silt-clay with spots of sand and planktonic foraminiferan tests mixed (Kowsmann & Costa 1979). The main source of the fine grained sediments in the slope is the local reworking of ancient material from the shelf (Kowsmann et al. 1977).

The present work is part of a broad interdisciplinary oceanographic investigation to study the structure and dynamics of the continental shelf ecosystem offshore of Ubatuba, southeastern Brazilian Bight. It provides information on the composition, spatial distribution and community structure of isopod species present in the area. As mostly reducer organisms isopods constitute an important link in the marine food webs and contribute decisively to the organic matter recycling process. Furthermore, they present high diversity and clear habitat preference that makes them good material for community studies. The present paper attempts to fill some of the gaps in the knowledge of shelf isopod diversity and distribution in the southwestern Atlantic. It is the first work dealing with distribution of the benthic marine species in such a large depth range in the area.

2 MATERIAL AND METHODS

Benthic samples were obtained at 33 sites located between 10 and 600 m depth over the continental shelf and upper slope off Ubatuba in the southeastern Brazilian Bight (23°25'S, 44°52.2'W and 24°53'S, 44°34.7'W) (Fig. 1). Sampling was carried out seasonally on the inner shelf (10-50 m depth, sites 1 to 9) with the research trawler 'Veliger II'

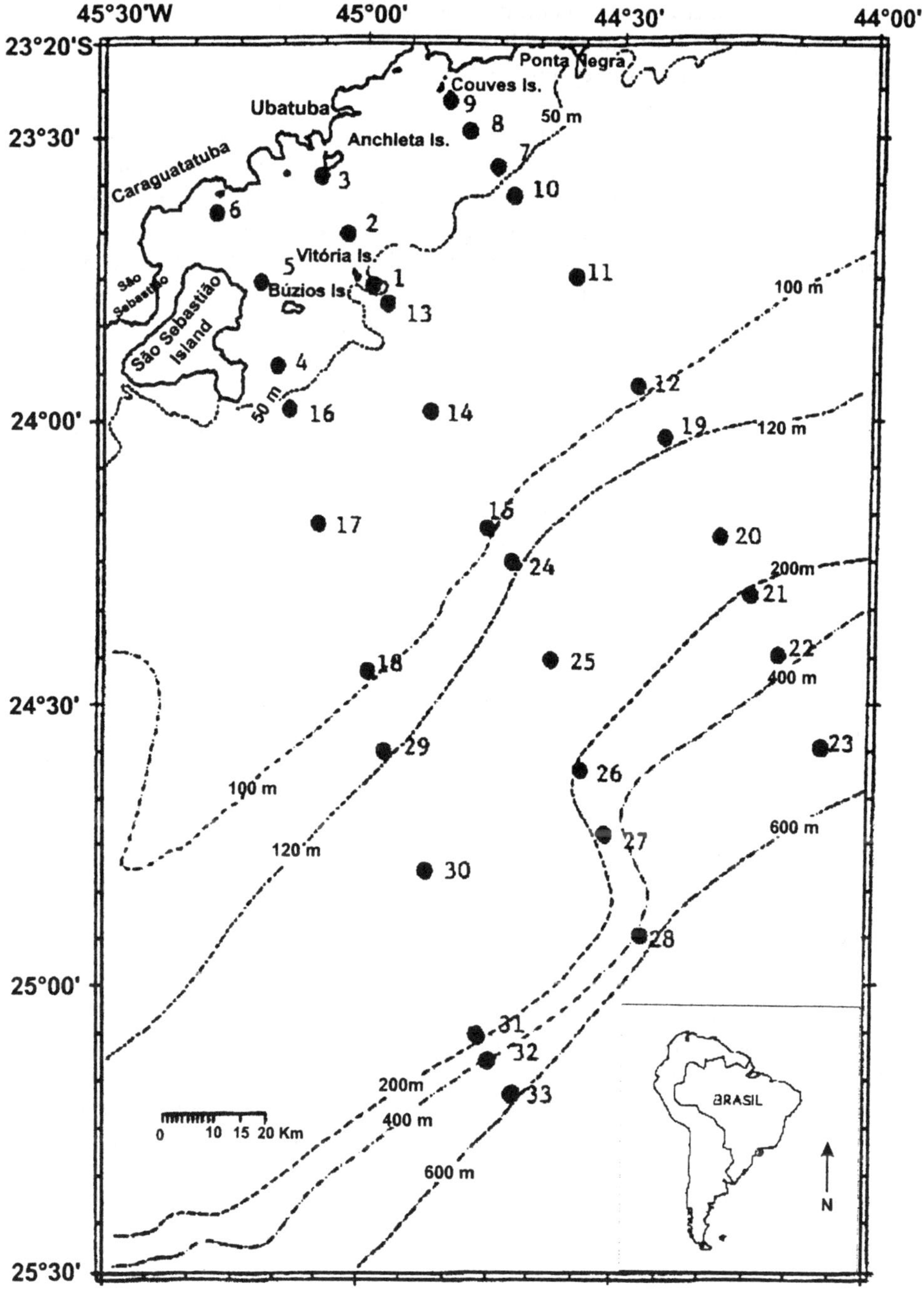

Figure 1. Map of Ubatuba region, southeastern Brazilian Bight, showing sampling stations and bathymetry.

and biannually (summer and winter) on the outer shelf (50-100 m depth, sites 10 to 18) and slope (120-600 m, sites 19 to 33) with the RV 'Prof. W. Besnard', of the University of São Paulo. A total of eight survey cruises were performed from October 1985 to July 1987, obtaining 123 samples.

The benthic fauna was collected with a 0.1 m^2 van Veen grab, a rectangular dredge and a beam trawl. A total of 54 van Veen grab samples and 69 dredge hauls were obtained. The nets were towed at 6 km/hr over the bottom, perpendicular to the coast, at 10, 30, 40, 50, 75, 100, 130, 180, 240, 320, 500 m depth, during 10 minutes. The samples were washed on board through 0.5 mm mesh and the sediment preserved in 70% alcohol. The isopods were sorted and identified under stereomicroscope. Data from each sampler were put altogether and binary transformed for community analysis.

Bottom temperature, salinity, and water oxygen concentration were measured at each sample unit. Sediment was collected with a van Veen grab only at the shelf locations for analysis of composition and granulometry. Sediments from the slope were classified through visual analysis. The oxygen was calculated according to Riley (1971).

A reduced matrix with the 23 most frequent species was used in multivariate analysis. Cut level employed considered frequency equal or higher than 10%. Cluster analysis was performed to describe similarity among samples (Q-mode) and species associations (R-mode) utilizing Kulczynski's similarity coefficient and the unweighted pair-group method average (UPGMA). Canonical Correspondence Analysis was employed to highlight the influence of the environmental factors upon the species distribution (ter Braak 1986). The model achieved was tested with the Monte Carlo permutation test. Species richness (S) was calculated for each species group obtained through cluster analysis in order to estimate means and standard deviations. The values of S for different groups were compared using one way analysis of variance (ANOVA), and the mean values were tested with T-test.

3 RESULTS

3.1 *Study area*

In the sampled period the South Atlantic Coastal Water laid from upper slope to about 70 m depth during the winter reaching shallower areas near the coast in summer (20-30 m depth). The Tropical Water always stayed offshore from the coast from surface to mid depths. Sandy bottoms dominate the shelf, graded by size with latitude and depth. Fine grained sediments were deposited both in the south and around 100 m depth along side the coast, whereas coarser grains were found in the northwest. The shelf break depths around 130 m presented a conspicuous hard bottom formed mainly by calcareous algae (*Halimeda* spp.) and coral exoskeletons. Mud mixed with sandy spots characterized the sea bed up to 240 m depth, and below 320 m we found only mud.

3.2 *Species composition*

A total of 48 species of Isopoda occurred in the data sets, of which 19 were nominal and 29 morphospecies (Table 1). Of these, 19 are new species (nearly 40%) and 11 occurred down to 100 m depth (60%). The fauna also presented high number of species that appeared only once (50%).

Table 1. Species of the isopods found in the Ubatuba shelf and upper slope, southeastern Brazilian Bight, and their respective depth range. The code refers to the species utilized in multivariate analysis.

Species	Depth range	Code
ANTHURIDEA		
Accalathura crenulata	70–240	ACCR
Amakusanthura sp	15–117	AMKU1
Apanthura sp1	130–136	
Apanthura sp2	130–136	
Apanthura sp3	15– 48	
Apanthura sp4	180–188	
Belura sp	35–117	BELU1
Chalixanthura sp	45– 48	
Colanthura sp	32– 44	COLA1
Eisothistos sp	134–188	EISO1
Malacanthura caribbica	15–117	MACA
Paranthura sp1	15– 48	
Paranthura sp2	15– 48	PARA1
Quantanthura brasiliensis		32–184
Xenanthura sp	30– 35	QUBR
ASELLOTA		
Joeropsis sp.	70- 79	JAAER1
Munidion sp.	136-240	
Muna sp.	70-117	
Munnogoniumv sp.	70-117	
Munnopsis sp.	100-117	
FLABELLIFERA		
Cirolana sp.	240	
Cristaserolis laevis	35- 48	CRLA
Cristaserolis similis	73-117	CRSI
Cymodoce ornata	76- 79	
Cymodoce sp.	48-188	CYMO1
Eurydice elongata	43- 48	
Excorallana costata	180	
Metacirolana sp.	134-248	META1
Natatolana sp.1	19-102	NATA1
Natatolana sp.2	35-248	NATA2
Natatolana sp.3	116-134	
Natatolana sp.4	134-320	
Natatolana sp.5	240	
Neoserolis usperta	43-100	NEUA
Politolana eximia	15-116	POEX
Politolana sp.	32- 50	POLI1
Rocinela oculata	510	
Rocinela signata	70- 79	
Serolis insignis	510-600	SEIN
Thysanoserolis completa	15- 50	THCO
GNATHIIDEA		
Bathygnathia magnifica	240-248	

Table 1. (Continued).

Species	Depth range	Code
Gnathia andrei	130	GNAN
Gnathia ricardoi	134-530	GNRI
Gnathia ubatuba	19-240	GNUB
VALVIFERA		
Erichsonella sp.	73- 75	
Macrochiridotea sp.	76- 79	
Synidotea sp.	32- 36	

Table 2. Number of genera (G) and species (S) found offshore of Ubatuba. IS = Inner Shelf; OS = Outer Shelf; SL = Slope; N = New species.

Suborder	IS			OS			SL		
	G	S	N	G	S	N	G	S	N
Anthuridea	8	9	7	5	5	2	4	6	4
Asellota	0	0	0	4	4	4	1	1	1
Flabellifera	7	9	4	6	9	4	6	9	6
Gnathiidea	1	1	0	1	1	0	2	4	0
Valvifera	1	1	0	2	2	2	0	0	0
Total	17	20	11	18	21	12	13	20	11

The number of genera and species in each sub-order are shown in Table 2. Flabellifera, mostly Cirolanidae, presented the greater number of genera and species, followed by Anthuridea. Asellota became more diversified at mid depths. Within the Flabellifera, Cirolanidae was the group least studied, with eight new species. Anthuridea is represented by 15 species, totaling 454 specimens, of which 12 are new species. The most frequent isopods were *Gnathia ubatuba, Malacanthura caribbica* and *Amakusanthura* sp. ranging from 73 to 40%.

3.3 *Communities*

The code of the 23 species employed in the multivariate analyses and the depth range of all species are presented in Table 1. The cluster analysis performed on the samples showed three main groups with two sub-groups within each group (Fig. 2). The first division isolates samples from the slope, down to 320 m depth, from samples belonging to the shelf. The second division separates samples from inner shelf from those related to outer shelf and shelf break.

Two sub-groups were delineated within the inner shelf. Group 1 encompassing sites between 30 and 50 m depth and group 2 having the shallowest places (between 15 and 24 m depth). The outer shelf locations form group 3 (50 to 100 m depth) and the shelf break sites, group 4 (130 to 240 m). Groups 5 and 6 are located in the slope area deeper than shelf break. Group 5 aggregates stations between 300 and 350 m depth and group 6 the deepest places, down to 500 m.

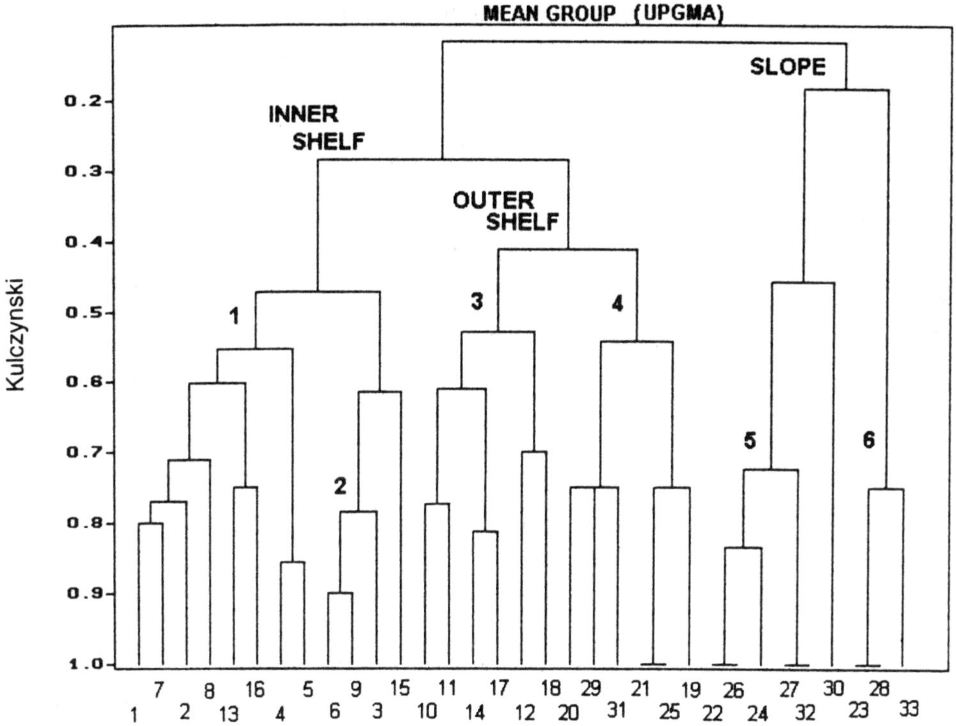

Figure 2. Q- mode dendrogram (UPGMA – Kulczynski), grouping the stations. Stations 1 to 9 are from inner shelf; stations 10 to 18 are from outer shelf; stations 19 to 33 are from the upper slope.

Cluster analysis performed on species disclosed three main groups, A, B and C, the first with two sub-groups (Fig. 3). The analysis shows a straight relation between species assemblages and bathymetric zones. Group C clusters species from the slope and shelf break, places that are under the influence of the South Atlantic Central Water (SACW). It is dominated by high frequency of *Gnathia ricardoi* (47%), species found between 134 and 530 m depth. Within this same group *Gnathia andrei* present a patchy distribution, occurring at 130 m depth in galleries within the large calcareous blocks of the algae *Halimeda* spp.

Group B assembles species found in both outer shelf and shelf break (from 70 to 240 m), under the SACW domain. It is characterized by *Accalathura crenulata*, a species common in more than 50% of the samples from the outer shelf. Group A gathers species from all the shelf. It is divided into sub-group A1, with species exclusive of inner shelf and under the domain of Coastal Water, and sub-group A2 having species that occur in the whole area (15 to 100 m). In the first sub-group *Politolana* sp. (67%) and *Paranthura* sp. (56%) are the most frequent species. In the latter association the anthurids *Malacathura caribbica* and *Amakusanthura* sp. are highly dominant, occurring in 66 and 83% of the samples, respectively.

The mean species richness (S') showed significant decrease with depth (ANOVA; F-ratio = 8.5227, $P < 0.01$). Mean values decrease from the shelf down to the slope. Higher values were found at the shelf (from 30 to 48 and 100 m depth), and extreme low values

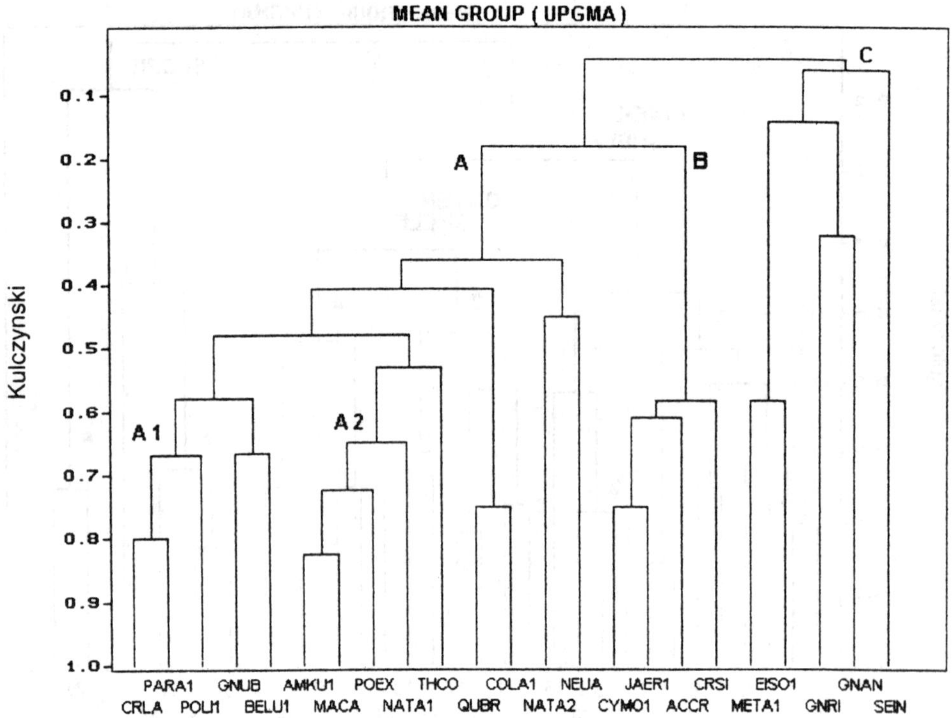

Figure 3. R- mode dendrogram (UPGMA – Kulczynski), grouping the species. Species names and respective codes are in Table 1.

were obtained at the deepest places (down to 500 m). There is no significant differences between inner shelf and outer shelf species richness, but values were significantly different between shelf and shelf break (T = 4.029; P = 0.005) and between shelf break and slope (T = 2.676; P = 0.010).

A general overview of the CCA ordination is presented in Figure 4. The environmental variables employed were depth, temperature, oxygen concentration, and sediment type. The explanatory variables selected were depth and temperature. The model explained 69% of the variance of species-environment relation, and first axis was responsible for 48.8% of this variation. The Monte Carlo permutation test showed significance for the first axis (P = 0.01). The species-environment correlation in axis I was 0.88. Depth was positively correlated to axis I (80%) and temperature was negatively correlated with this axis (71%) (Table 3).

The ordination diagram shows the same groups disclosed in the classification of species and stations (Fig. 4). A gradient of species replacement was observed along axis I, from cold and deep places to warm and shallow ones. Consequently, axis I can be interpreted as the depth gradient associated to water masses dynamics. It discriminates locations of the slope under the influence of the SACW from those of inner shelf under the CW domain. Outer shelf stations and some species were placed in the middle of the axis, indicating the presence of a water mixing zone.

Table 3. Weighted matrix of species-environment correlations produced by CCA. Only Axis 1 was significant with Monte Carlos' test. Significant values at P< 0.01% are in bold. SP = Species; Env = Environment; AX = Axis.

	SP AX1	SP AX2	ENV AX1	ENV AX2
SP AX1	1			
SP AX2	0.027	1		
ENV AX1	**0.881**	0	1	
ENV AX2	0	0.749	0	1
PROF	**0.803**	0.309	–0.194	0.832
TEMP	**–0.713**	0.440	0.907	–0.303

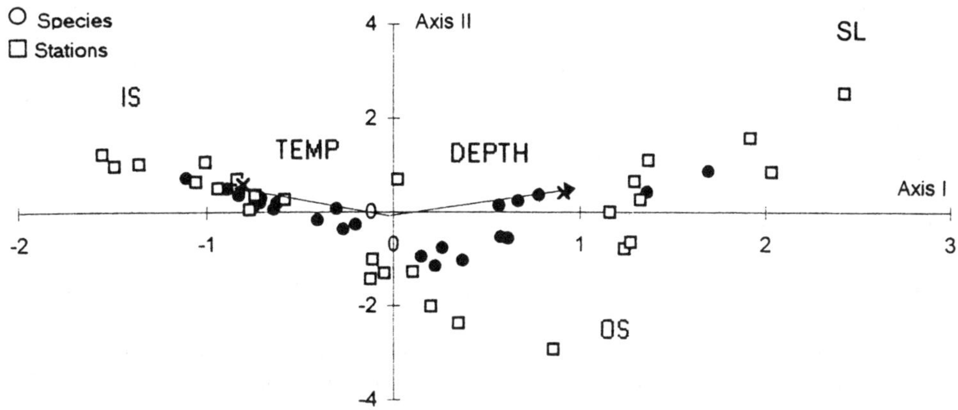

Figure 4. Ordination diagram obtained from the canonical correspondence analysis. IS = Inner shelf; OS = Outer shelf; SL = slope.

4 DISCUSSION

The Ubatuba region is situated in the western Atlantic within the Paulista Province (Palacio 1982). This zoogeographic area is characterized as a transition between the tropical north-northeast Caribbean Province and the temperate southern Magellanic Province. It is a region with fewer species when compared with the Caribbean Province but with high endemism.

Despite the fauna of Isopoda from continental shelf and slope is poorly known in the South Atlantic, the major part of the previously studied species is reported for the Paulista Province (Pires-Vanin 1998). Considering that about 100 species of free living marine isopods are recorded for Brazilian waters as stated by the last author, the important number of new species found in the present survey at Ubatuba shelf and upper slope (nearly 40%) highlight the necessity of intensive sampling effort in this area. Our results show that Flabellifera account for the greater number of new species found, regardless the special attention it had received in the past decades (Moreira 1971, 1974). Other groups with many new species were Anthuridea and Asellota since they were only studied in intertidal and shallow waters (Pires 1981a, 1985). As the Cirolanidae had been neglected, many new species in this family is also expected.

Another characteristic of the fauna is the high number of rare species found mainly at the shelf break and slope (50%). In such a case, we suggest that small population size or spotted distribution may account for the low frequency observed, in addition to the lack of sampling effort in these places. In fact, the existence of nearly 60% of new species below 100 m depth corroborates the last assumption. However, knowledge of the isopod fauna from deep shelves or slopes in subtropical waters is lacking not only in the southwestern Atlantic. An even more dramatic result was obtained in the Australian southeast slope where 90% of the species were unknown (Poore et al. 1994).

There is a clear division between shelf and slope environment in relation to location groups and species assemblages, and a shelf break area could be recognized. The species are distributed along a depth gradient, which is gradual at the shelf and shelf break but steep at the slope. The community composition presented a drastic change in the slope with the presence of bathyal species as *Serolis insignis* and *Bathygnathia magnifica*. The individuality of the slope was noticed for several species of many benthic groups also (Fariña et al. 1997, Sumida & Pires-Vanin 1997) and it seems that the area is an unique environment and an effective barrier for species distribution.

We have found a conspicuous group of species at the inner shelf and another one at the outer shelf plus shelf break. The first group is under the alternate influence of the CW and the SACW and consequently, subjected to bottom water temperatures that can vary from 25 to 14°C in a few days. The species found on the outer shelf and shelf break, however, are constantly affected only by cold water. In this respect these communities are more similarly structured and closely related as corroborated by the results of ordination.

The shelf break can be considered a transitional faunal area, due to modifications in the substrate composition and the fast increase of depth. The bottom is characterized by blocks of calcareous algae, serving as a shelter for many of the organisms sampled such as *Accalathura crenulata* and *Gnathia andrei*. Both species were more frequent or exclusive in that place, and although having a mixed fauna the shelf break shares more species with outer shelf (5) than with the slope (2). The shelf break has been reported as a transitional unit also by many authors working with several groups of invertebrates. Assemblages of crustacean decapods presented a zonation with depth in the shelf and upper slope from Galicia, northwestern Spain, and characterized the shelf break as a place with typical and neighboring species (Fariña et al. 1997). Sumida & Pires-Vanin (1997) working with 56 species of several taxa of macrobenthos from 130 to 600 m depth in southeastern Brazil found also shelf break as a unit distinguished by a dissimilar assemblage in relation to upper slope.

The main factors that delimitate macrobenthos distribution besides depth and changes in oceanography, are sediment and food. Although sediment type is an important factor for species distribution (Ambrogi et al. 1990, Brandt & Piepenburg 1994), in the present study it contributed to define the broad areas, as shelf, shelf break and slope. Species at inner shelf sub-areas were not clustered in spite of the high variability in sediment composition found there. Respecting food, quantity and quality seem to be adequate to maintain the macrofaunal productivity in the shelf area, as reported by Matsuura & Wada (1994). These authors working with stable isotopes of carbon and nitrogen showed that pelagic organic matter is the main food supply to macrobenthos. In addition, the pelagic material increases seasonally with the SACW intrusion in the shelf (Aidar et al. 1993). There is no available information regarding the food supply below 100 m depth for the study area.

Through the exposed above, it can be concluded that spatial zonation of isopods on the Ubatuba shelf and upper slope is controlled mainly by depth and watermass dynamics, and depths around 300 m seem to be an effective upper barrier for the deep-living species.

ACKNOWLEDGEMENTS

The author is thankful for financial support from the Brazilian National Council for Research (CNPq) through the research grant 30.5600/78-OC. Thanks are due also to the crew of the R/V 'Prof. W. Besnard' and 'Veliger II' for logistic assistance.

REFERENCES

Aidar, E., S.A. Gaeta, S.M.F. Gianesella-Galvão, M.B.B. Kutner & C. Teixeira 1993. Ecossistema costeiro tropical: nutrientes dissolvidos, fitoplancton e clorofila-a, e suas relações com as condições oceanográficas na região de Ubatuba. *Publ. esp. Inst. oceanogr., S Paulo* 10: 9-43.

Ambrogi, R., D. Bedulli & G. Zurlini 1990. Spatial and temporal patten in structure of macrobenthic assemblages. A three-year study in the northern Adriatic Sea in front of the Po River delta. *P.S.Z.N.I.: mar. Ecol.* 11: 25-41.

Brandt, A. & D. Piepenburg 1994. Peracarid crustacean assemblages of the Kolbeinsey Ridge, north of Iceland. *Polar Biol.* 14: 97-105.

Castro Filho, B.M., L.B. Miranda & S.Y. Miyao 1987. Condições hidrográficas na plataforma continental ao largo de Ubatuba; variações sazonais e em média escala. *Bolm Inst. oceanogr., S Paulo* 35: 135-151.

Fariña, A.C., J. Freire & E. González-Gurriarán 1997. Megabenthic decapod crustacean assemblages on the Galician continental shelf and upper slope (north-west Spain). *Mar. Biol.* 127: 419-434.

Franke, H-D. & M. Janke 1998. Mechanisms and consequences of intra-and interspecific interference competition in *Idotea baltica* (Pallas) and *Idotea emarginata* (Fabricius) (Crustacea: Isopoda): a laboratory study of possible proximate causes of habitat segregation. *J. Exp. Mar. Biol. Ecol.* 227: 1-21.

Kowsmann, R.O. & M.P.A. Costa 1979. Sedimentação quaternária da margem continental brasileira e das áreas oceânicas adjacentes. In *Projeto REMAC 8*. Rio de Janeiro: PETROBRÁS – CENPES – DINTEP.

Kowsmann, R.O., M.P.A. Costa, M.G.N. Coutinho & L.A.P. Gamboa 1977. Modelo da sedimentação holocênica da plataforma continental e do talude sul do Brasil. In *Projeto REMAC 2*. Rio de Janeiro: PETROBRÁS – CENPES – DINTEP.

Mahiques, M.M., Y. Mishima & M. Rodrigues 1999. Characteristics of the sedimentary organic matter on the inner and middle continental shelf between Guanabara Bay and São Francisco do Sul, southeastern Brazilian margin. *Cont. Shelf Res.* 19: 775-798.

Matsuura, Y. 1996. A probable cause of recruitment failure of the Brazilian sardine *Sardinella aurita* population during the 1974/1975 spawning season. *S. Afr. J. Mar. Sci.* 17: 29-35.

Matsuura, Y. & E. Wada 1994. Carbon and nitrogen stable isotope ratios in marine organic matters of the coastal ecosystem in Ubatuba, southern Brazil. *Cienc. Cultura, São Paulo* 46: 141-146.

Mills, E.L. 1975. Benthic organisms and the structure of marine ecosystems. *J. Fish. Res. Bd. Can.* 32: 1657-1663.

Moreira, P.S. 1971. Species of *Serolis* (Isopoda, Flabellifera) from southern Brazil. *Bolm Inst. oceanogr., S Paulo* 19: 85-144.

Moreira, P.S. 1974. A new record of *Serolis* (Crustacea, Isopoda, Flabellifera) from off southern Brazil. *Bolm Inst. oceanogr., S Paulo* 23: 89-101.

Palacio, F.J. 1982. Revisión zoogeográfica marina del sur del Brasil. *Bolm inst. oceanogr., S Paulo* 31: 69-92.

Pires, A.M.S. 1981a. Anthurids from intertidal and shallow infralittoral waters from southeastern Brazil. *J. Crust. Biol.* 1: 211-226.

Pires, A.M.S. 1981b. Ecological study on littoral and infralittoral Isopods from Ubatuba, Brazil. *Bolm Inst. oceanogr., S. Paulo* 30: 27-40.

Pires, A.M.S. 1985. The occurrence of *Munna* (Isopoda, Asellota) on the southern Brazilian coast, with description of two new species. *Crustaceana* 48: 64-73.

Pires, A.M.S. 1992. Structure and dynamics of the benthic megafauna from the continental shelf offshore of Ubatuba, southeastern Brazil. *Mar. Ecol. Progr. Ser.* 86: 63-76.

Pires-Vanin, A.M.S. 1993. A macrofauna bêntica da plataforma continental ao largo de Ubatuba, São Paulo, Brasil. *Publ. esp. Inst. oceanogr., S Paulo* 10: 137-158.

Pires-Vanin, A.M.S. 1998. Malacostraca – Peracarida. Marine Isopoda. Anthuridea, Asellota (pars), Flabellifera (pars) and Valvifera. In P.S. Young (ed.), *Catalogue of Crustacea of Brazil.* Rio de Janeiro: Museu Nacional.

Pires-Vanin, A.M.S., C.L.D.B. Rossi-Wongtschowski, E. Aidar, H.S.L. Mesquita, L.S.H. Soares, M. Katsuragawa & Y. Matsuura 1993. Estrutura e função do ecossistema de plataforma continental da região de Ubatuba, São Paulo: síntese dos resultados. *Publ. esp. Inst. oceanogr., S Paulo* 10: 217-231.

Poore, G.C.B., J. Just & B.C. Cohen. 1994. Composition and diversity of Crustacea Isopoda of the southeastern Australian continental slope. *Deep-Sea Res.* 41: 677-693.

Riley, J.P. 1971. The dissolved gases in the sea water. I. Gases other than Carbon Dioxide. In J. P.Riley & R. Chester (eds), *Introduction to marine chemistry.* London: Academic Press.

Sumida, P.Y.G. & A.M.S. Pires-Vanin 1997. Benthic associations of the shelf break and upper slope off Ubatuba-SP, south-eastern Brazil. *Est. Costl. Shelf Sci.*: 779-784.

Sahada, R., M. Tatían, J. Kowalke, S. Kühne & G.B. Esnal 1998. Benthic faunal associations on soft substrates at Potter Cove, King George Island, Antarctica. *Polar Biol.* 19: 85-91.

Svavarsson, J., T. Brattegard & J-O. Strömberg 1990. Distribution and diversity patterns of asellote isopods (Crustacea) in the deep Norwegian and Greenland Seas. *Prog. Oceanogr.* 24: 297-310.

ter Braak, C.J.F. 1986. Canonical, correspondence analysis: a new eigen vector technique for multivariate direct gradient analysis. *Ecology* 67: 1167-1179.

Thouzeau, G., G. Robert & R. Ugarte 1991. Faunal assemblages of benthic megainvertebrates inhabiting sea scallop grounds from eastern Georges Bank in relation to environmental factors. *Mar. Ecol. Prog. Ser.* 74: 61-82.

Zembruscki, S.G. 1979. Geomorfologia da margem continental sul-brasileira e das bacias oceânicas adjacentes. *Projeto REMAC 7.* Rio de Janeiro: PETROBRAS – CENPES – DINTEP.

Zimmermann A. & A. Brandt 1992. New results of the 'EPOS' leg 3 cruise to Antarctica: horizontal and vertical distribution of isopods (Crustacea) in the Weddell Sea. *Polar Biol.* 12: 121-128.

Origin, age, and evolutionary rate of the microparasellid isopod *Microcharon* from the ground water of the south-Riffian region (Morocco)

N. COINEAU
Université Pierre et Marie Curie, Paris 6, Observatoire Océanologique de Banyuls, UMR CNRS 76 28 Laboratoire Arago, Banyuls/Mer, France

C. BOUTIN
Université Paul Sabatier, Laboratoire de Zoologie, UMR CNRS 55 52, Toulouse, France

A. FAKHER EL ABIARI
Faculté des Sciences Semlalia, Laboratoire d'Hydrobiologie, Marrakech, Maroc

ABSTRACT

Issues of taxonomy, origins and evolution of the widely distributed isopod genus *Microcharon* (Microparasellidae) are reviewed. The purpose of this paper is to provide an understanding of the origin and of the age of *Microcharon* in the northeastern region of Morocco within an historical biogeographic context. We also discuss the morphological evolutionary rate within the genus *Microcharon* compared to species of other Moroccan regions and to the amphipod genus *Pseudoniphargus* from the same region resulting from similar vicariance processes. Matters of deep history are taken up in connection with biogeography for the taxon.

1 INTRODUCTION

The tiny threadlike isopods of the genus *Microcharon* (Microparasellidae) live in the interstitial water of coarse marine and coastal sands as well as of subterranean river sediments and continental groundwater. Although having a very feeble ability to disperse, since it is holobenthic and has no swimming larvae, the genus is widely distributed in the world from Polynesia, New Caledonia and Indian ocean islands, to Europe, Mediterranean, Caribbean and Galapagos (Delamare Deboutteville 1960, Coineau 1982, 1986, 1994). In the Mediterranean basin, more than 70 species are known, mainly in the northern and the southwestern regions (Pesce & Galassi 1988a, b, 1990, Pesce & Tetè 1978, Galassi 1991, Galassi et al. 1995). The genus is also diversified in Morocco: one species lives in the marine Mediterranean beaches, another one near Casablanca; different closely related species of the *messoulii* group are endemic in subterranean waters of the High Atlas rivers and in the Jbilet region. Other representatives of the genus have been collected in springs from the aquifer of the Haouz flood plain. In contrast, only one species, *Microcharon alamiae*, is widely distributed in the northeastern part of Morocco: the south-Rifian region (Fig. 1) (Messouli 1984, Boulanouar et al. 1994, 1995, 1997, Coineau 1994, Yacoubi-Khebiza et al. 1997).

Furthermore, at least seven to eight species of another interstitial stygobiont, the amphipod *Pseudoniphargus* (Crustacea), share the same distribution area in the northeastern

Moroccan region (Coineau & Boutin 1996). Therefore such a similar distribution pattern of the two groups in the latter area suggests differences either in the origin or in the morphological evolutionary rate of the two genera.

The purpose of this paper is to provide an understanding of the origin and of the age of *Microcharon* in the northeastern region of Morocco within an historical biogeographic context. We aim also at a discussion of the morphological evolutionary rate within the genus *Microcharon* compared to species of other Moroccan regions and to the amphipod genus *Pseudoniphargus* from the same region resulting from similar vicariance processes.

2 METHODS

According to Blondel (1986, 1995), the distribution pattern of extant species is an instantaneous state of the present geological period. It is to be considered as the result of the joint history of both the group and its environment 'at the scale of time, space, and change.'

Within historical biogeography or biogeographic reconstruction (Myers & Giller 1990), methods include: 1) The thorough description of the species in order to provide morphological characters for phylogenetic analyses; 2) The establishment of phylogenetic relationships and monophyly of the studied group; 3) The elaboration of the distribution pattern; 4) The search for geological and palaeogeographic events that could have acted as barriers responsible for evolutionary divergences; 5) The construction of a biogeographic historical scenario, and 6) Finally both cladistics and palaeogeography provide the possibility of a hypothesis on the origin, the age and the evolutionary history of the group, supported by all available data.

3 CLADISTICS

The Mediterranean freshwater species of *Microcharon* form a monophyletic group. Within known species from Morocco, representatives of the *messoulii* group lineage live in the interstitial groundwater in the High Atlas and in the Jbilet, while *Microcharon alamiae* was recently described from the Fes region. This latter species belongs to the so-called Eastern Mediterranean group as the sister taxon of the *motasi* group of species located in Romania (Coineau 1994, Boulanouar et al. 1995, 1997, Yacoubi-Khebiza et al. 1997, Coineau et al. 1999, in prep.).

4 DISTRIBUTION

The taxon *M. alamiae* occurs in the interstitial groundwater of:

- The Sebou River, northeastern Fes,
- The El Abiob River, in Sidi-Abdallah/Rhiata,
- The Moulouya River, in Guercif,
- The Kert River, near Telat-Azlaf,
- The Za River, near Guefalt (Fig. 1). In this latter region, the species was found in springs.

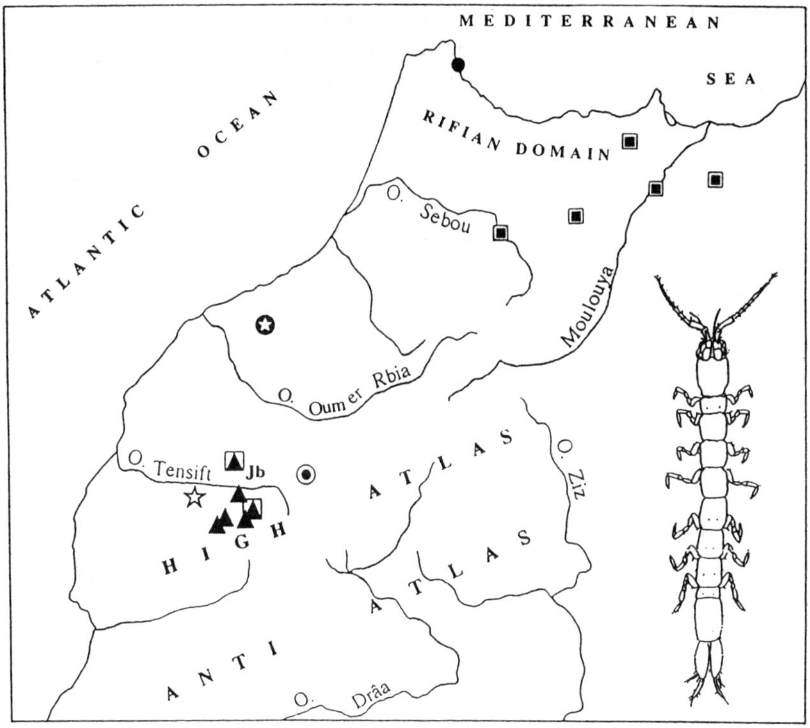

Figure 1. Distribution of the species of the genus *Microcharon* in Morocco. The species of the *messoulii* group are located in the Atlas region and in the Jbilet. *M. alamiae* is restricted to the north-eastern area.

5 ORIGIN AND AGE OF *MICROCHARON ALAMIAE*

As with other members of the family Microparasellidae and many inland crustacean stygobites (Notenboom 1988, 1991, Boutin et al. 1992, Holsinger 1986, 1994, Botosaneanu & Holsinger 1987, Stock 1977a, b, 1980, Wägele 1990), interstitial freshwater species of *Michrocharon* are derived from marine ancestors (Delamare Deboutteville 1960, Stock 1977a, Coineau 1971, 1986, 1992, 1994, Dole & Coineau 1987, Boulanouar et al. 1995, Yacoubi et al. 1997). Some extant members of the genus still occur in shallow marine sands. Many species live in brackish water of coastal beaches.

One of the scenarios providing an understanding of the colonization of interstitial inland subterranean water by marine surface ancestors and of their evolution, is the 'two step model of colonization and evolution' (Boutin & Coineau 1990, Notenboom 1991, Coineau & Boutin 1992, Holsinger 1994). The last author proposed a similar hypothesis described as the three-step model.

The first step is an active dispersion: the surface marine ancestor invaded the interstitial biotopes in littoral marine sands of the Tethys. The selection pressure in the new interstitial environment generated a faster physiological, morphological, and ethological evolutionary rate that resulted in the adaptation to interstitial life.

The second stage corresponds to the passive settlement of coastal interstitial organisms in fresh water during the Tethys regressions. Vicariance processes occurred at the end of these recessions when the gene flow was interrupted between the new limnostygobiont and the population remaining in the coastal sands. Stock (1980) called this second stage the 'Regression Model of Evolution'.

The different stages of this colonization of inland ground water by surface marine ancestors may occur at different geological epochs, until the Present. A series of eustatic cycles in the same region may have resulted in several waves of thalassoid immigrants that belong to the same phylogenetic group. After settlement in interstitial subterranean water, the new ancestral species evolves mainly through vicariance due to new geologic events acting as barriers within the distribution area (Platnik & Nelson 1978, Holsinger 1991, Coineau & Boutin 1992). As for *Microcharon*, orogenesis that results in new valleys due to erosion, and plate tectonics are viewed as vicariant events resulting in speciation. Each vicariant process generates closely related species, i.e. sister species.

In Morocco, all species of *Microcharon* occur in areas once covered by the Tethys both in the Atlas and in the northern regions. In the latter area, the Rifian domain did not exist as an emerged land in its present location, before the late Miocene period. The Tethys permanently occupied all the northern part of the present country. The Idrissid landmass was the most northerly emerged part of the country (Choubert & Faure-Muret 1962, Michard 1976, Rögl & Steininger 1984, Steiniger et al. 1985).

The Betico-Rifian domain results from the break up of the little Alboran-Kabilian-Calabrian plate. One of the microplates drifted southward and westward and reached its present location during the Tortorian period and formed the new Rifian domain (Dercourt et al. 1985, 1993, Rögl & Steininger 1984). This latter region was surrounded by the marine South-Rifian corridor (Fig. 2), which represents the southern part of the Betico-Rifian basin and is well documented (Alvinerie et al. 1992, cartes des Editions du Service géologique du Maroc 1992, Bernini et al. 1992, Dercourt et al. 1985, 1993).

All up-to-date known locations of *Microcharon alamiae* occur in the area covered by this Tortonian channel (Fig. 2). In Romania, the species of the sister group are located in the Panonian basin *(M. motasi, M. orghidani)* and the Dacic basin *(M. oltenicus)* (Serban 1964) in the area covered by the Central Paratethys during the Sarmatian and the Pontian (= Tortonian) times (Coineau 1994). Therefore, it is supposed that the common ancestor of the Moroccan and the Romanian species lived in the coastal interstitial during the Tortonian period and settled in subterranean fresh water during the following regressions at the end of this geologic period. This happened slightly earlier in Romania, about 9-8 MYBP, than in Morocco where the regression lasted until the Messinian epoch. In Ràmania, the ancestral new stygobiont continued to evolve in fresh water and generated a group of closely related species due to vicariant processes (Coineau 1994).

6 EVOLUTIONARY RATE IN *MICROCHARON*

As mentioned above, in Morocco, we know only one species of *Microcharon*, *M. alamiae*, in the northeastern region formerly covered by the Tortonian Sea where it is widely distributed (Figs 1 & 2). In contrast, we know at least seven to eight species (five are described) of the subterranean amphipod *Pseudoniphargus* that are also a product of marine ancestors left in fresh groundwater by the Tortonian regressions (Coineau & Boutin 1996, Fakher el Abiari et al. 1999). Their distribution is shown in Figure 3. Some of them are

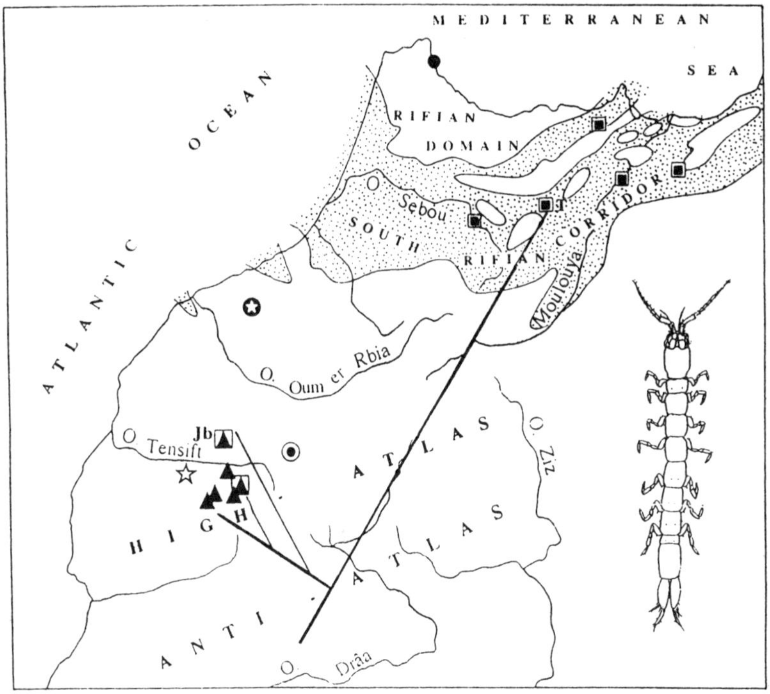

Figure 2. Distribution of the species of *Microcharon* and extension of the Tethyan south-Rifian corridor during the Tortonian (dotted area) after Choubert & Faure-Muret (1962), the maps of the Editions du Service Géologique du Maroc, Rabat (1992) and Alvinerie et al. (1992). In Morocco, the whole white southern part was emerged in the Tortonian period. Main lineages of the cladistic relationships.

located in exactly the same sites as *Microcharon.* A similar palaeogeographic scenario is envisaged for the establishment of the Tortonian ancestors of the Tethian shores in inland ground water.

This interpretation is complemented in *Pseudoniphargus* and helps to explain the further speciation through vicariance. During the general and progressive regression of the Tethys at the end of the Tortonian period, the South-Rifian channel became narrower in the Taza region due to the southward drift of the Rifian plate. Finally, the emerged lands of the Rifian and the African plates were connected. This connection resulted in two separated gulfs, the eastern one and the western embayment. Therefore, the littoral population of the *Pseudoniphargus* ancestor was divided in two isolated parts that evolved independently and might have resulted in the two main lineages of the cladogram (Fig. 3).

After this closure of the central region of the south-Rifian corridor, various paleogeographic events might have been biologically important for the interstitial coastal organisms in the eastern gulf. The gradual regression of the sea, deposits of detritic material due to the Rifian orogenesis, the Rifian-African compression resulted in many islands and archipelagos and in the isolation of small closed basins functioning like 'little dead-seas' (Boutin 1993), and finally gave rise to an anoxic environment. Such 'intricate palaeogeography favored also the diversification of facies', these basins 'were periodically sup-

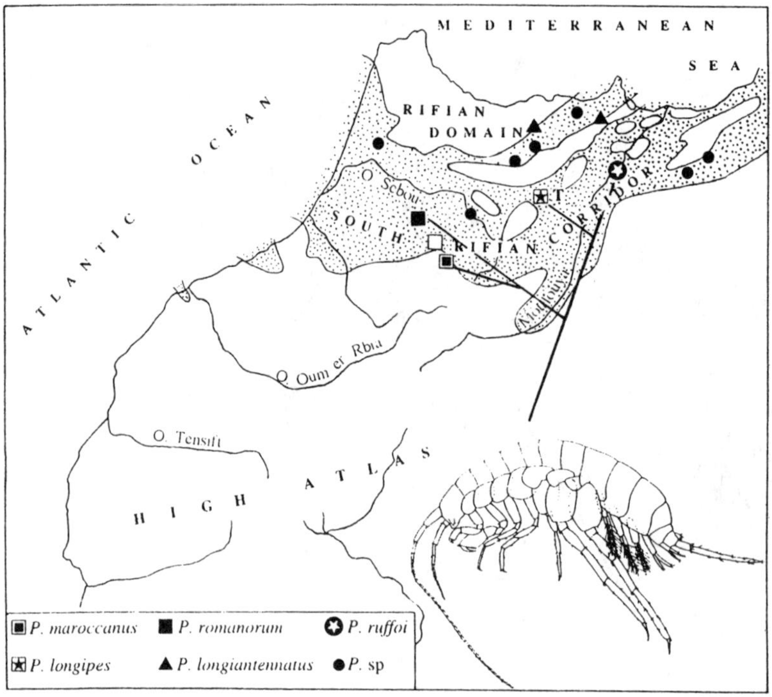

Figure 3. Distribution of the species of the amphipod *Pseudoniphargus* in the northern region of Morocco, extension of the Tethyan south-Rifian channel in the Tortonian Period (dotted area), and phylogenetic relationships of the known species. Small solid circles: several species to be described. Palaeogeographic outline: see figure 2.

plied with marine waters', and during the salinity crisis, in the Messinian, 'the intercalation of marine episodes could represent a marine reinvasion of the eastern part' (Alvinerie et al. 1992). Several different species could have derived from the ancestral populations in the small isolated basins. Later on, after the last marine recession, local orogenic events, river erosion, as well as a more recent water shed that occurred between S. Abdallah and Guercif resulted in a number of separated hydrographic systems. Consequently, isolated populations evolved separately in these isolated systems and resulted in closely related species such as *P. longipes* and *P. ruffoi*. Conversely, the western gulf was deeper and the ancestor of the western species might have settled in subterranean freshwater during the regression at the end of the Tortonian period for the less derived species (*P. maroccanus* and *P. romanorum*), and at the beginning of the Messinian time for others, even during the Pliocene in the most western region of the gulf.

Thus, the initial continuous area of the Tortorian coastal ancestor ofspecies of *Pseudoniphagus* that inhabited the interstitial water of the south-Rifian corridor was successively fragmented by vicariant geologic and palaeogeographic events, resulting in the different extant thalassoid species.

In contrast, in *Microcharon,* which occurs in some of the same locations and on each side of the above-mentioned Taza closure of the Tortonian South-Rifian corridor (Fig. 2), the same palaeogeographic events during the same geologic time did not result in several species. The similarly fragmented populations of *Microcharon* did not change their phenotype and did not give rise to different recognizable species. Therefore, the morphological evolutionary rate seems to be higher in the amphipods and lower in the isopod.

One hypothesis may explain these morphological evolutionary rate differences that could appear as being only an impression due to the important size difference between the two crustacean stygobites. *Microcharon* is a strict interstitial organism: it is very small (about 1 mm) as compared to *Pseudoniphargus* (4.5-8.5 mm) or other subterranean amphipods or isopods that live in more coarse grained sediment. Mechanical constraints due to strong perturbations after tectonic movements and isolation of hydrographic systems are probably less important for tiny organisms as *Microcharon* than for larger ones like *Pseudoniphargus*. Orogenesis generally results in uplifting and erosion of upper layers including the aquifer, and the aquifer level goes down, crossing new geological layers; there are changes in the biotope grain size for the organisms.

Moreover, in contrast to tiny organisms, longer and bigger phreatobionts exhibit a high number of characters easy to detect, such as chetotaxic coating, larger segments of body appendages, other morphological details. Therefore, true vicariant species of *Microcharon* might be hard, even impossible to separate based on their morphology. This hypothesis cannot be categorically excluded: each of the morpho-species of crustaceans always exhibits some intra-specific variability and can include sibling species that are difficult to distinguish. Only laboratory experiments could confirm or refute interfecundity of the different forms and ensure that they are not morphologically closely similar or identical species, but really separate species. A comparison of the different populations of *Microcharon* at the molecular level might provide a verification of their effective genetic isolation. The presence of some alleles in a population and the absence in another one might be enough to establish the complete separation and the beginning of a molecular evolution which distinguishes each population change through vicariance to new species. However, these molecular differences generally do not demonstrate intersterility of isolated populations because they can be observed within intraspecific polymorphism; they are not sufficient when establishing the speciation phenomenon.

However, the hypothesis based on small size is highly unlikely. Indeed, in spite of their small size, species of *Michrocharon* are numerous and morphologically well recognizable in the central region of Morocco.

Another alternative hypothesis deals with the requisite geological time for speciation. Perhaps a duration of 8-5,5 millions years since the settlement of the Tortonian-Messinian coastal ancestor in subterranean fresh water is too short for vicariant processes in *Microcharon*. In the High Atlas, the ancestors of the different species of the *messoulii* group settled earlier in fresh ground water than in the South-Rifian corridor. They were left in subterranean water by the Tethyan regressions at the end of the Senonian (80 MYBP) and of the Eocene (40 MYBP) periods. The Atlas orogenesis, from the late Eocene and later on, induced vicariant processes in newly separated valleys due to erosion. This resulted in several endemic species during a long geological time (about 40 MY) allowing speciation in the relatively stable freshwater environment (compared to the coastal milieu) for about 80 MA. Perhaps isolated basins also existed before the Atlas orogenesis from the Senonian regression (80 MY). In this case, it is possible to envisage a lower evolutionary rate within *Microcharon*, compared to the amphipod *Pseudoniphargus*.

We have already proposed that, based on the apomorphy rate exhibited by the same genus or family in *Pseudoniphargus*, *Metacrangonyx* (amphipods), Microparasellidae, and

Cirolanidae (isopods), the evolutionary rate of interstitial freshwater species is lower in relatively stable biotopes than in the littoral changing environment (Boutin & Coineau 1991, 1999). Moreover, we have also suggested that the evolutionary rate seems to be low in *Microcharon*, when describing a species that occur both downstream in the Atlas aquifers and in the Jbilet region (Figs 1 & 2). After the deposit of the ancestral population in the subterranean fresh water by the Eocene regression, we can hardly find very slight differences in the populations of the two regions. These differences may only represent geographic variations, or perhaps they may be considered as the first stages of a vicariant speciation in progress (Boulanouar et al. 1995).

In other parts of the Mediterranean basin, only the establishment in interstitial fresh groundwater of the first inland ancestor, and therefore of the main lineages, have been dated based on the Tethyan regressions. Further speciations due to vicariant events have not been considered (Dole & Coineau 19987, Coineau 1992, 1994), therefore, the evolutionary rate has not been documented. The case of speciation in *Microcharon* may be considered as an example showing that speciation originating in the present biodiversity, is the result of both the past geological and palaeogeographic evolution of the region and of the intrinsic ability of groups to change and diversify more or less rapidly.

ACKNOWLEDGEMENTS

We are grateful to Dr R. Brusca, the Columbia University, Oracle, Arizona, and Dr B. Kensley, Smithsonian Institution, Washington, who organized the Second International Isopod Conference in Amsterdam. The investigations were supported by the Action Intégrée Interuniversitaire No. 93/643, Contrat ACC-SV No. 7 'Diversité Fonctionnelle et Evolution en Milieu souterrain', and the Faculté des Sciences Semlalia of the University of Marrakech. Thanks are due to our colleagues Z. Oulbaz, M. Messouli, M. Boulanouar, and M. Yacoubi-Khebiza.

REFERENCES

Alvinerie, J., M.T. Antunes, B. Cahuzac, A. Lauriat-Rage, C. Montenat & C. Pujol 1992. Synthetic data on the paleogeographic history of northeastern Atlantic and Betic-Rifian basin, during the neogene (from Brittany, France, to Morocco). *Palaeogeogr., Paleoclim., Palaeoecol.* 95: 263-286.

Bernini, M., M. Bocaletti, J. El Mokhtarh, R. Gelat, S. Iaccarino, G. Moratti & G. Papani 1992. Données stratigraphiques nouvelles sur le Miocène supérieur du Bassin de Taza-Guercif (Maroc nord-oriental). *Bull. Soc. Géol. France* 163(1): 73-76.

Blondel, J. 1986. *Biogéographie évolutive*. Collection d'Ecologie, 20. Paris: Masson.

Blondel, J. 1995. *Biogéographie: approche écologique et évolutive*. Collection d'Ecologie, 27. Paris: Masson.

Botosaneanu, L. & R. Holsinger 1991. Some aspects concerning the colonization of the subterranean realm – especially of subterranean waters : a response to Rouch & Danielopol, 1987. *Stygologia* 6: 11-39.

Boulanouar, M., M. Yacoubi, M. Messouli & N. Coineau 1995. Un nouveau *Microcharon* (Isopoda, Janiroidea) du Maroc. Origine et biogéographie historique. *Contr. Zool.* 65: 53-64.

Boulanouar, M., M. Yacoubi-Khebiza, C. Boutin & N. Coineau 1997. Un nouvel Isopode Microparasellide (Crustacea Janiroidea) des eaux souterraines de la région péri-rifaine (Maroc). *Bull. Soc. Hist. Nat. Toulouse* 133: 21-27.

Boutin, C. 1993. Biogéographie historique des Crustacés Malacostracés stygobies du Maroc. Thèse Doct., Univ. Lyon I 262 p.

Boutin, C. 1994. Biogeography of the Metacrangonyctid amphipods in North Africa. *Hydrobiologia*. 287: 49-64.

Boutin, C. & N. Coineau 1990. 'Regression Model', 'Modèle biphase' d'évolution et origine des micro-organismes stygobies interstitiels continentaux. *Revue Micropaléont*. 33: 303-322.

Boutin, C. & N. Coineau 1991. Instabilité des conditions environnementales et vitesse de l'évolution. L'exemple des microcrustacés souterrains d'origine marine dans les pays méditerranéens. *Bull. Inst. Géol. Bassin d'Aquitaine* 50: 63-69.

Boutin C. & M. Messouli 1988b. *Metacrangonyx gineti* n. sp. d'une source du Haut-Atlas marocain et la famille des Metacrangonyctidae n. fam. (Crustacés Amphipodes stygobies). *Vie Milieu* 38: 67-84.

Boutin, C., M. Messouli & N. Coineau 1992. Phylogénie et biogéographie évolutive d'un groupe de Metacrangonyctidae, Crustacés Amphipodes stygobies du Maroc. II. Cladistique et paléobiogéographie. Avec l'examen comparatif de plusieurs logiciels de parcimonie. *Stygologia* 7: 159-177.

Boutin, C. & N. Coineau 1999. Evolutionary rates and phylogenetic age in some subterranean species. In Wilkens, Culver & Humphreys (eds), *Ecosystems of the world – subterranea Biota, VI. Evolution.* Elsevier Sci. (in press).

Cartes des Editions du Service Géologique du Maroc, Rabat 1992. Cartes des mouvements récents du Rif. *Notes Mém.* No. 365.

Choubert, G. & A. Faure-Muret 1962. Evolution du domaine atlantique marocain depuis les temps paléozoïques. In *Livre à la mémoire du Professeur Paul Fallot*. Mém. hors série, Soc. Géol. France, Paris. I: 447-527.

Coineau, N. 1982. *Microcharon* Kar. et *Angeliera* Chap. et Delam., Crustacés Isopodes Asellotes interstitiels Microparasellides. *Documents pour un atlas zoogéographique du Languedoc-Roussillon* 24: 1-4.

Coineau, N. 1986. Isopoda: Asellota: Janiroidea. In L. Botosaneanu (ed.), *Stygofauna mundi*. Leidein: E.J. Brill/W. Backhuys.

Coineau, N. 1992. Biogéographie évolutive du Crustacés interstitiel *Microcharon* (Isopoda, Janiroidea) dans l'ouest du Bassin Méditerranéen. *Bull. Inst. Océanogr. Monaco*, 9: 101-114.

Coineau, N. 1994. Evolutionary biogeography of the microparasellid isopod Microcharon (Crustacea) in the Mediterranean Basin. In: Biogeography of subterranean crustaceans: the effects of different scales. Summer Meeting of thc Crustacean Society, Charleston, June 1992. *Hydrobiologia* 287: 79-96.

Coineau, N. & C. Boutin 1992. Biological processes in space and time. Colonization, evolution and speciation in interstitial stygobionts. In A.I. Camacho (ed.), *The natural history of biospeleology*. Mus. Nac. Cienc. nat., CSIC, Madrid, *Monografias*. 7: 423-451.

Coineau, N. & C. Boutin 1996. Age and origin of the stygobiontic amphipod *Pseudoniphargus* (Crustacea) in Morocco, with the description of three new species. *Boll. Mus. civ. nat. Verona* 20: 503-520.

Coineau, N., M. Boulanouar, M. Yacoubi & B. Boutin in prep. Evolution and historical biogeography of the interstitial isopod *Microcharon* (Crustacea) in Morocco. *J. Crust. Biol.*

Dercourt, J. et al. 1985. Présentation de 9 cartes paléogéographiques au 1.20 000 000e s'étendant de l'Atlantique au Pamir, pour la période du Lias à l'Actuel. *Bull. Soc. Géol. France*. 8: 637-652, avec un atlas de 9 cartes couleurs.

Dercourt, J., E. Ricou & B. Vrielynck (eds) 1993. *Atlas Tethys Palaeoenvironmental maps*. Paris: Gauthier-Villars. 307 p., 14 maps, 1 pl.

Delamare Deboutteville, C. 1960. Biologie des eaux souterraines littorales et continentales. *Vie Milieu*, suppl. 9: 1-740.

Fakher el Abiari, A., Z. Oulbaz, M. Messouli & N. Coineau 1999. A new species of *Pseudoniphargus* (Crustacea Amphipoda) from subterranean water of north-eastern Morocco. *Contr. Zool.* .

Galassi, D.P. 1991. A new *Microcharon* Karaman from hyporheic waters of Greece: *Microcharon antonellae* n. sp. (Crustacea Isopoda: Microparasellidae). *Stygologia* 6: 201-207.

Galassi, D.P., G.P. Pesce & P. De Laurentiis 1995. The genus *Microcharon* Karaman (Isopoda Janiroidea) in the Balkan Peninsula: Biogeography of the *major* and the *profundalis* groups. *Mém. Biospéol.* 22: 25-34.

Holsinger, J.R. 1994. *Pattern and process in the biogeography of subterranean amphipods.* In Biogeography of subterranean crustaceans: the effects of different scales. *Summer Meeting of the Crustacean Society, Charleston, June 1992. Hydrobiologia* 287: 131-145.

Messouli, M. 1984. Recherches dela faune stygobie des sources de la région de Marrakech. Mém. DEA, Univ. Marrakech. 1-35.

Messouli, M. 1994. Evolution, phylogénie et biogéographie historique des Metacrangonyctidae, Crustacés Amphipodes stygobies du Nord de l'Afrique et des régions voisines. Thèse Doct. Etat, Univ. Marrakech. 1-309.

Michard, A. 1976. Eléments de géologie du Maroc. *Notes Mém. Serv. géol. Maroc*, 252: 1-408.

Myers, A & P.S. Giller 1990. *Analytical Biogeography. An integrated approach to the study of animal and plant distributions.* London: Chapman and Hall.

Notenboom, J. 1991. Marine regressions and the evolution of ground-water dwelling amphipods (Crustacea). *J. Biogeography* 18: 437-454.

Pesce, G.L. & D.P. Galassi 1988a. Microparasellids of Spain (Crustacea Isopoda: Janiroidea). *Stygologia* 4: 307-331.

Pesce, G.L. & D.P. Galassi 1988b. The genus *Microcharon* Karaman in Italy: an update and description of three new species (Crustacea Isopoda: Microparasellidae). *Stygologia* 4: 249-261.

Pesce, G.L. & D.P. Galassi 1990. First records in the family Microparasellidae in Turkey, and description of two new species of the genus *Microcharon* Karaman (Crustacea Isopoda: Janiroidea). *Stygologia* 5: 173-181.

Pesce, G.L. & P. Tetè 1987. Recherches en Afrique de l'Institut de Zoologie de l'Aquila (Italie). I. Microparasellides d'Algérie (Crustacea: Isopoda). *Rev. Zool. afr.* 92: 992-1002.

Platnik, H.J. & G. Nelson 1978. A method of analysis for historical biogeography. *Syst. Zool.* 27: 1-16.

Rögl, F. & F.F. Steininger 1984. Neogene Parathethys, Mediterranean and Indo-Pacific seaways. Implications for the paleobiogeography of marine and terrestrial biotas. In P. Brenchley (ed.), *Fossils and climate.* New York: John Wiley.

Serban, E. 1964. Nouvelles données sur les Microparasellides (Isopodes) de Roumanie: *Microcharon motasi* nov. sp., *Microcharon orghidani* nov.sp. et *Microcharon oltenicus* nov. sp. *Vie Milieu.* 15: 341-364.

Steininger, F.R. & F. Rögl 1985a. Paleogeography and palinspastic reconstruction of the Neogene of the Mediterranean and Paratethys. In J.E. Dixon & A.H.F. Robertson (eds), *The Geological Evolution of the Eastern Mediterranean. Special publ. Geolog. Soc.*, 17, Blackwell Sci. Publ.: 659-668.

Steininger, F.F., J. Senes, K. Kleeman & F. Rögl 1985. Neogene of the Mediterranean Tethys and Paratethys. Stratigraphic correlation tables and sediment distribution maps. Vol 1: 189 p., 10 cartes couleurs. In PICG UNESCO no 25 *Stratigraphic correlation Tethys-Paratethys Neogene*, 1974-1983, Institut of Paleontology, Univ. Wien.

Stock, J.H. 1977a. Microparasellidae (Isopoda Asellota) from Bonaire. *Stud. Fauna Curaçao* 51(168): 69-91.

Stock, J.H. 1977b. The taxonomy and zoogeography of the hadziid Amphipoda, with emphasis on the West Indian taxa. *Stud. Fauna Curaçao* 55: 1-130.

Stock, J.H. 1980. Regression model evolution as exemplified by the genus *Pseudoniphargus* (Amphipoda). *Bijdr. Dierk.* 50: 105-144.

Yacoubi Khebiza, M. 1996. Biocénoses aquatiques interstitielles des vallées du Haut-Atlas de Marrakech: Microdistribution, Dynamique de répartition, Ecologie et Biogéographie. Thèse Doct. d'Etat, Univ. Marrakech. 374 p.

Yacoubi-Khebiza, M., M. Boulanouar & N. Coineau 1997. Un nouveau Microparasellidae (Crustacé Isopode) du Haut-Atlas; aspects physogénétiques et biogéographiques. *Contr. Zool.* 66: 227-234.

The influence of genetic and extrachromosomal factors on population sex ratio in the marine isopod, *Paracerceis sculpta*

STEPHEN M. SHUSTER, GUENIVIER ZINSER & PAUL KEIM
Department of Biological Sciences, Northern Arizona University, Flagstaff, USA.

J. WILLIAM O. BALLARD
Field Museum, Roosevelt Road at Lake Shore Drive, Chicago, Illinois, USA

CLAY SASSAMAN
Department of Biology, University of California, Riverside, USA

ABSTRACT

In the marine isopod, *Paracerceis sculpta*, three male morphs are distinct at a single genetic locus, *Ams* (= *Alternative mating strategy*). In the laboratory, interactions among *Ams* alleles, alleles at another autosomal locus (*Tfr* = *Transformer*), primary sex factors (ZW = females; ZZ = males), and a possible extrachromosomal factor (ECF), appear to influence the frequencies of the male morphs and cause deviations in family sex ratio. To determine whether ECF is a feminizing *Wolbachia* endosymbiont, we probed isopod tissues using *Wolbachia*-specific primers and PCR. To identify ECF's effect on population sex ratio, we generated expected values using a stepwise model that included the effects of each known sex ratio-influencing factor. We then compared observed and expected sex ratios over a two-year sampling period. We found no evidence of DNA sequences characteristic of *Wolbachia* endosymbionts in the tissues of *P. sculpta*. However, simulations showed significant deviations of observed from expected population sex ratios when the presumed effects of ECF were excluded. When the effects of ECF were included in simulations, significant deviations in the expected population sex ratio occurred when P_{ECF} exceeded 0.50. Observed and expected population sex ratios fit best at low to intermediate P_{ECF}. Under these conditions, observed fluctuations in population sex ratio were explained by observed population frequencies of primary sex factors and *Ams* alleles, with oscillating frequencies of Tfr^2, an allele known to cause sex conversion of genetic males. At high frequencies of Tfr^2, the effects of ECF could be excluded from the model without affecting the sex ratio. Our results show: 1) *Wolbachia* endosymbionts are not responsible for sex ratio devations in *P. sculpta*; 2) ECF has a recognizable effect on population sex ratio, but unlike most extrachromosomal sex factors, ECF appears to persist at intermediate population frequency. Moreover, when fluctuations in Tfr^2 are allowed, the effects of ECF are no longer necessary to explain the population sex ratio. Thus, 3) Frequency-dependent selection on primary sex factor-*Ams*-*Tfr* interactions alone may be responsible for female-biased as well as variable sex ratios in the northern Gulf of California *P. sculpta* population.

1 INTRODUCTION

Three discrete male morphotypes coexist in *Paracerceis sculpta*, a Gulf of California isopod crustacean (Shuster 1987, 1992a). All male morphs possess mature external genitalia and sperm producing organs, and when isolated with females, the male morphs do not differ in their ability to sire young (Shuster 1989a, b). Although males' relative allocation of energy toward somatic and gonadal structures differ significantly (Shuster 1987, 1989a), average reproductive success among male types is equal over time (2.5 years; Shuster & Wade 1991a). Life history differences that may influence the relative contributions of α-, β- and γ- males to the population seem to cancel since male reproductive tenure varies inversely with maturation rate (Shuster & Wade 1991a). Moreover, relative fertilization success among males varies with the density of receptive females, as well as with the frequency of other male morphs in spongocoels (Shuster 1989b, Shuster & Wade 1991a). Thus, male polymorphism in this species appears to be maintained by frequency-dependent selection (Shuster & Wade 1991b). Such conditions are considered sufficient for genetic polymorphisms to persist in nature (Slatkin 1978, 1979).

Among the progeny of field-collected parents, alloalleles of the gene, *Pgm*, which encodes the enzyme *phosphoglucomutase*, segregate with male phenotype (Shuster & Sassaman 1997). These results indicate that a major gene (*Ams* = *Alternative mating strategy*), linked to *Pgm*, causes phenotypic differences among males. Male differences were not caused by allelic differences at *Pgm,* because linkage disequilibrium between *Pgm* alleles and the male morphs did not exist in the natural population, and because recombinant individuals were observed among lab-reared progeny. Shuster & Sassaman (1997) demonstrated that inheritance of *Ams* is Mendelian and that alleles at this locus exhibit directional dominance ($Ams^{\beta} > Ams^{\gamma} > Ams^{\alpha}$), consistent with an earlier population genetic model (Shuster & Wade 1991a). Shuster & Sassaman (1997) also showed that *Ams* expression is modified by another autosomal locus, *Tfr* (= *Transformer*), which interacts with *Ams*, primary sex determination loci and with a possible extrachromosomal factor (= ECF), to bias family sex ratio. Shuster & Levy (1999) confirmed that sex determination in *P. sculpta* involves female heterogamety (ZW = females; ZZ = males) as assumed in previous genetic models (Shuster & Wade 1991a, Shuster & Sassaman 1997).

The genetic model that explains observed variation in *P. sculpta* family sex ratios (Fig. 1; Shuster & Sassaman 1997) assumes *Tfr* to be a diallelic, autosomal locus whose alleles (Tfr^1, Tfr^2) interact with, but assort independently of alleles at *Ams* and at primary sex determination loci. According to this hypothesis, Tfr^1Tfr^1 has no effect on males of any *Ams* genotype, and no effect on $Ams^{\alpha}Ams^{\alpha}$ females. However, females bearing Tfr^1Tfr^1, as well as β- or γ- alleles at *Ams*, are assumed to mature as males, with phenotypes determined by their *Ams* allelic state. Tfr^2Tfr^2 is assumed to have no effect on females of any *Ams* genotype, and no effect on β- or γ- males. However, $Ams^{\alpha}Ams^{\alpha}$ males bearing Tfr^2Tfr^2 are assumed to mature as females. Tfr^1Tfr^2 heterozygotes are assumed to affect only individuals bearing two *Ams* genotypes: Tfr^1Tfr^2 females bearing β-alleles are assumed to mature as β-males, and Tfr^1Tfr^2, $Ams^{\alpha}Ams^{\alpha}$ males are assumed to mature as females, but only if they are descended from β-males. This latter effect assumes that the Tfr^2 allele interacts with an extrachromosomal factor, ECF, which appeared to be transmitted to individuals of both sexes descended from β-sires (Shuster & Sassaman 1997).

The pattern of sex ratio deviations reported in Shuster & Sassaman (1997) suggests five hypotheses regarding ECF and its possible effect on population sex ratio: First, ECF could

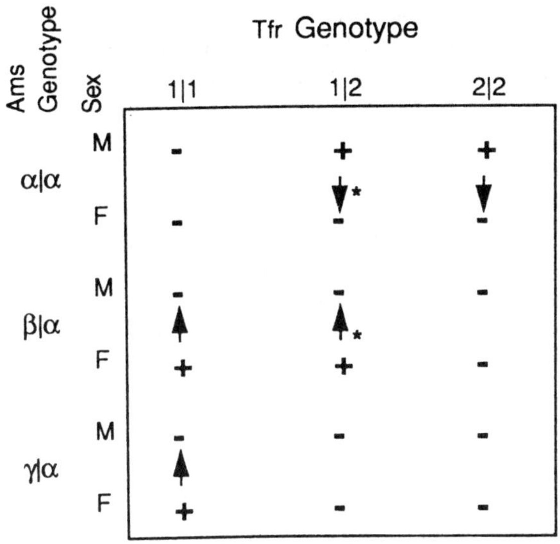

Figure 1: Effects of *Tfr* and ECF on *Ams* and primary sex determination loci: α, β,γ alleles at *Ams;* M = male; F = female; *Ti* = alleles at *Tfr*; (-) = no effect of *Tfr* on *Ams*-sex combination; (+) = sex change, with arrow indicating direction of change; (*) = effect produced by interaction of ECF with *Tfr*, *Ams*, and primary sex determination genotypes. The apparent effect of ECF is to enhance or suppress the expression of the *Tfr*2 allele depending on allelic states an individual's *Ams* and primary sex determination loci (redrawn from Shuster & Sassaman 1997).

be a *Wolbachia* endosymbiotic alpha-Proteobacteria (reviews in O'Neill et al. 1992, Rousset et al. 1992a, b). Such endosymbionts have been previously reported from 22 isopod species (Bouchon et al., 1998). If this hypothesis is correct, ECF should be detectable using *Wolbachia*-specific PCR primers (Rousset et al. 1992a, O'Neill et al. 1992). Second, if ECF is a *Wolbachia* species and causes feminization of certain *P. sculpta* males (as it does in other species), it should appear in *P. sculpta* population samples that were collected when females and/or β-males were in excess (β-males look and act like sexually receptive females, but in fact, are functional males; Shuster 1992). Third, since extrachromosomal factors tend to be maternally inherited as well as horizontally transferred, the population frequency of ECF is expected to be high if not fixed (Werren & Jaenike 1995, Juchault & Rigaud 1995). Fourth, if ECF is *not* involved in sex ratio deviations, known *Ams-Tfr*-primary sex factor interactions alone are expected to explain observed sex ratio variation in *P. sculpta*. Fifth, peaks in female abundance (i.e. intervals when R is large, where $R = N_{females}/N_{males}$) should coincide with increased frequencies of Tfr^2, as well as with peaks in ECF incidence, since both of these factors appear to cause α-males to mature as females. If this last hypothesis is true, it suggests that frequency-dependent selection on primary sex factor-*Ams-Tfr*- ECF interactions may be responsible for observed deviations in sex ratio, as proposed in Shuster & Sassaman (1997).

In this paper, we test hypotheses 1 and 2 by examining the tissues of adult isopods for the presence of *Wolbachia* endosymbionts, using primers specific to *Wolbachia* strains isolated from crustacean and insect hosts (Rousset et al. 1992a, b, O'Neill et al. 1992,

Zhou et al. 1998). Isopod tissue samples used in these experiments were collected and frozen during periods in which the population sex ratio was female-biased, and when the frequency of β-males was unusually high (November 1992, Shuster, unpubl. data). We test hypotheses 3 through 5 using a population genetic model in which all factors known to influence sex determination in *P. sculpta* are included in a stepwise manner. This procedure allowed us to assess each factor's relative influence on individual sexual phenotype as well as on population sex ratio.

We report six conclusions: 1) *Wolbachia* endosymbionts were *not* detectable in the tissues of *P. sculpta*, even when the frequency of such symbionts was expected to be high; 2) Observed population sex ratios in *P. sculpta* were not adequately explained by known interactions and estimated frequencies among *Ams*, *Tfr*, and primary sex factors, suggesting that an extrachromosomal sex-biasing factor (ECF) may indeed exist in *P. sculpta* populations; 3) ECF does not appear to exist at high frequencies in *P. sculpta* populations, thus ECF, if it exists, does not behave like most known maternally- and horizontally transmitted endosymbionts; 4) At all allowable frequencies of ECF ($P_{ECF} = 0\text{-}.5$), fluctuations in population sex ratio were explained by the observed population frequencies of primary sex factors and *Ams* alleles, combined with oscillating frequencies of Tfr^2, the allele that appears to cause sex conversion of genetic males; 5) When P_{Tfr2} was allowed to fluctuate, the amplitude of the expected oscillations was unaffected by P_{ECF}, even when $P_{ECF} = 0$, thus, the effects of ECF may not be necessary to explain observed deviations in population sex ratio; alternatively, the effects of ECF and Tfr^2 may be additive, allowing Tfr^2 to persist at lower population frequencies in the presence of ECF; 6) A significant correlation between the observed population sex ratio and estimates of P_{Tfr2} over a two year period suggests that fluctuations in population sex ratio may result from frequency-dependent selection on primary sex factor-*Ams*-*Tfr* interactions.

2 MATERIALS AND METHODS

2.1 *PCR Analyses*

If ECF is a *Wolbachia* endosymbiont in *P. sculpta* tissues, it should be widespread within isopod population samples containing an excess of females and β-males (Rigaud & Juchault 1995, Werren & Jaenike 1995, Shuster & Sassaman 1997). To simultaneously test these hypotheses, we retrieved seven adult isopods (2 α-, 1 β-, 1 γ-male, 3 females) from samples frozen at -80C in November 1992, when females comprised 61% of the total sample and β-males comprised 3.3% of the collected males ($N = 236$; usually, β-males make up 0-4% of samples; Shuster 1992a). We extracted total genomic DNA from each specimen using methods described in Kimberling et al. (1996). We visualized extracts on a 0.7% agarose gel, and used three primers specific for 16S and 23S rRNA gene sequences of *Wolbachia* bacteria to PCR amplify endosymbiont sequences from the total DNA extractions (Rousset et al. 1992a, O'Neill et al. 1992).

As controls, we performed the same extraction and amplification procedures with samples of: 1) PCR water, and 2) Ovarian tissue samples of *D. simulans* females collected from Riverside, California (DSR) that were known to be infected with *Wolbachia*; 3) Ovarian tissue samples of DSR females treated with tetracycline to remove *Wolbachia* (DSRT), and 4) A sample of non-endosymbiotic bacteria from laboratory stocks (*Bacillus cereus*). Reaction mixtures and cycling protocols were determined via optimization reactions, and PCR products were visualized on 1.5% agarose gels.

Additional *Wolbachia* primers became available after we had begun the experiments described above (Zhou et al. 1998), thus, we conducted a second independent analysis on seven additional isopods (2 α-, 1 β-, 1 γ-male, 3 females) retrieved from the same 1992 *P. sculpta* population sample. DNA from individual isopods was extracted using the PureGene Kit (Gentra) following the DNA Isolation from Fixed Tissue protocol. To confirm that each DNA extraction contained amplifiable DNA Domain III of the mitochondrial 12S rRNA was amplified with the primers SR-J-14233 (alias 12Sbi) and SR-N-14588 (alias 12Sai) designed by T. Kocher (Kocher et al. 1989) and listed in Simon et al. (1994). Two primer sets and three DNA concentrations were tested for the presence of *Wolbachia*. The V1 and V6 primers and conditions of O'Neill et al. (1992) were employed for the conserved 16S rDNA locus while the 81F and 691R primers and conditions of Zhou et al. (1998) were employed for the more variable *Wolbachia* surface protein gene (Wsp). A DNA extraction without an arthropod added provided the negative DNA control. The *D. simulans* DSR and DSRT lines described above provided positive and negative controls for the presence of amplifiable *Wolbachia* DNA.

2.2 *Simulation analyses*

2.2.1 *The model*

We designed a population genetic model to examine the known effects of primary sex factors, *Ams-Tfr* interactions and ECF on population sex ratio in *P. sculpta*. We assumed that the population frequency of each sex, P_i, must equal the product of the k frequencies of all heritable factors that influence adult morphology and sex, summed across all j genotypes that produce each sexual phenotype. Thus, the population frequency of each sex equals:

$$P_i = \sum_{j=1}^{n} \left\{ [P_{(\mathrm{SD})k}][P_{(Ams)k}][P_{(Tfr)k}][P_{(\mathrm{ECF})k}] \right\}_j \quad (1)$$

where $P_{(\mathrm{SD})k}$ is the kth population frequency of the jth primary sex determination genotype, $P_{(Ams)k}$ is the kth population frequency of the jth *Ams* genotype, $P_{(Tfr)k}$ is the kth population frequency of the jth *Tfr* genotype, and $P_{(\mathrm{ECF})k}$ is the kth population frequency of the jth ECF state. Note that only certain genotypes seem to be affected by ECF (Fig. 1), thus P_{ECF} for non-affected genotypes equals 1.0. For example, the frequency of genotypic females existing in the population as phenotypic γ-males is expected to equal: (1) the genotypic frequency of ZW individuals (ZW = females; ZZ = males, Shuster & Levy 1999), times (2) the genotypic frequency of $Ams^{\gamma}Ams^{\alpha}$ individuals (Shuster & Wade 1991a), times (3) the genotypic frequency of $Tfr^{1}Tfr^{1}$ individuals, times (4) 1, since ECF has no effect on this genotype (Shuster & Sassaman 1997; Table 1). The frequency of these phenotypic γ-males is then added to the frequencies of all other genotypes producing phenotypic males (Table 1), to obtain the population frequency of all male phenotypes, P_{males}. This fraction is equal to $1 - P_{\mathrm{females}}$.

2.2.2 *Calculation of expected frequencies*

Since evidence of *Tfr* and ECF was discovered in 1987-1988 samples (Shuster & Sas-

Table 1. Estimated genotype frequencies for genetic and extrachromosomal factors influencing sex ratio in *Paracerceis sculpta*; $P_\alpha = .917$; $P_\beta = .016$; $P_\gamma = .066$; $P_{Tfr1} = .6$; $P_{Tfr2} = .4$; $P_{ECF} = .5$.

Sex Class	1° SD	P_{SD}	*Ams* Genotype	P_{Ams}	*Tfr* Genotype	P_{Tfr}	ECF State	P_{ECF}	Pop. Freq.	Sex Ratio
Males	ZZ	0.5	$\alpha \mid \alpha$	0.84	$1 \mid 1$	0.36	+/–	1.0	0.15	0.36
		0.5		0.84	$1 \mid 2$	0.48	–	0.5	0.10	
		0.5	$\beta \mid \alpha$	0.03	$1 \mid 1$	0.36	+/–	1.0	0.01	
		0.5		0.03	$1 \mid 2$	0.48	+/–	1.0	0.01	
		0.5		0.03	$2 \mid 2$	0.16	+/–	1.0	0.00	
	ZW	0.5	$\beta \mid \alpha$	0.03	$1 \mid 1$	0.36	+/–	1.0	0.01	
		0.5		0.03	$1 \mid 2$	0.48	+	0.5	0.00	
	ZZ	0.5	$\gamma \mid \alpha$	0.12	$1 \mid 1$	0.36	+/–	1.0	0.02	
		0.5		0.12	$1 \mid 2$	0.48	+/–	1.0	0.03	
		0.5		0.12	$2 \mid 2$	0.16	+/–	1.0	0.01	
	ZW	0.5	$\gamma \mid \alpha$	0.12	$1 \mid 1$	0.36	+/–	1.0	0.02	
Female	ZZ	0.5	$\alpha \mid \alpha$	0.84	$1 \mid 2$	0.48	+	0.5	0.10	0.64
		0.5		0.84	$2 \mid 2$	0.16	+/–	1.0	0.07	
	ZW	0.5	$\alpha \mid \alpha$	0.84	$1 \mid 1$	0.36	+/–	1.0	0.15	
		0.5		0.84	$1 \mid 2$	0.48	+/–	1.0	0.20	
		0.5		0.84	$2 \mid 2$	0.16	+/–	1.0	0.07	
	ZW	0.5	$\beta \mid \alpha$	0.03	$1 \mid 2$	0.48	–	0.5	0.01	
		0.5		0.03	$2 \mid 2$	0.16	+/–	1.0	0.00	
	ZW	0.5	$\gamma \mid \alpha$	0.12	$1 \mid 2$	0.48	+/–	1.0	0.03	
		0.5		0.12	$2 \mid 2$	0.16	+/–	1.0	0.01	

saman 1997), we compared observed and expected population sex ratios in these collections to test the population model. We used information available from previous studies to calculate expected population frequencies of each known sex factor and then included their effects in the model stepwise, first in the aggregate 1987-1988 sample, and then in each constituent monthly sample.

In simulations of *aggregate* 1987-1988 population samples, we made three assumptions. First, we assumed equal frequencies of each primary sex determination genotype; $P_{ZW} = P_{ZZ} = .5$. Second, we assumed that all individuals bear one of three *Ams* genotypes: $Ams^\alpha Ams^\alpha$, $Ams^\beta Ams^\alpha$, and $Ams^\gamma Ams^\alpha$, at population frequencies equaling .84, .04, and .12, respectively ($N = 771$). These frequencies were estimated using the observed frequencies of α-, β- and γ-males in the aggregate 1987-1988 sample, and the methods described in Shuster & Wade (1991a) to calculate the frequencies of *Ams* alleles; this approach demonstrates that > 99% of all individuals in *P. sculpta* populations possess one of these three genotypes. Third, we assumed *Tfr* genotype frequencies to exist in the Hardy-Weinberg frequencies estimated by Shuster & Sassaman (1997): Tfr^1Tfr^1 (.36), Tfr^1Tfr^2 (.48), Tfr^2Tfr^2 (.16; N = 36); these frequencies were determined from laboratory crosses involving individuals collected in March 1988 (Shuster & Sassaman 1997). In the simulations described below, we varied the frequency of ECF to examine its effect on the aggregate population sex ratio. In each simulation, we compared observed and expected frequencies of males and females in aggregate population samples using a $2 \times 2\ X^2$ test.

In the simulations of *monthly* 1987-1988 samples, we assumed the same frequencies for primary sex factors and for *Tfr* alleles as in the aggregate population samples above. However, since we could identify the numbers of males of each morphotype in each monthly collection, we estimated the actual frequencies of *Ams* alleles in each of these samples using the method of Shuster & Wade (1991a). We then incorporated them into the equations for each P_j phenotypic class, as well as into each P_i sex class. Our expected frequencies for each sex class arose from the combined effects of several heritable factors, whose expected frequencies arose from specific genetic hypotheses. Thus, we compared observed and expected frequencies of males and females across all 1987-1988 samples using a X^2 test with v-1 degrees of freedom, where v equaled the total number of comparisons.

2.2.3 *Testing hypotheses 3-5*

If ECF is an extrachromosomal sex factor and thus can be maternally inherited as well as horizontally transferred, it should exist at *high frequency* in natural *P. sculpta* populations (Werren & Jaenike 1995, Juchault & Rigaud 1995). To examine the effect of high and low ECF frequencies on the aggregate 1987-1988 population sex ratio, we included all known sex factors in the model, and varied the frequency of ECF from 0 to .90 in .1 frequency increments.

To identify which *Tfr* allele frequencies would be necessary to approximate the observed sex ratios in the monthly samples at P_{ECF} = 0, we assumed the following conditions: (a) an equal primary sex ratio; (b) frequencies of *Ams* alleles equal to those calculated from 1987-1988 samples; (c) P_{ECF} equal to 0; (d) frequencies for Tfr^1 and Tfr^2 in each P_i equation that were (1) sufficient to bring observed and expected sex ratios in each 1987-1988 collection as close as possible, without (2) causing significant deviations between the observed and expected frequencies of α-, β- and γ-males. That is, *Tfr* allele frequencies were adjusted so that the expected frequencies of each sex generated using Equation (1) approached the observed frequencies of each sex in the 1987-1988 samples. Since this process also caused changes in the expected frequencies of α-, β- and γ-males, these expected frequencies were compared with the observed male morph frequencies using a X^2 test after each *Tfr* allele adjustment. Adjustments continued until observed and expected sex ratios in each 1987-1988 collection were as close as possible, without causing significant deviations between the observed and expected frequencies of α-, β- and γ-males.

Next, to identify which *Tfr* allele frequencies were necessary to approximate the observed sex ratios in the monthly samples at higher ECF frequency, we repeated the above conditions (a+b+d) nine additional times with P_{ECF} varied from .1 to .9 in .1 frequency increments. As described above, we compared observed and expected frequencies of the male morphs after each simulation using X^2 tests with $df = v$-1, where v equaled the number of comparisons.

3 RESULTS

3.1 *PCR Analyses*

Of the 11 specimens examined using the first set of DNA primers, PCR water and the γ-male specimen showed no PCR products, whereas PCR products were isolated from all other samples with both primer combinations. The *B. cereus* sample also amplified 16S and 23S sequences characteristic of this bacteria (Genbank 1997) using these same primers. This result suggests that these primers were not *Wolbachia*-specific. Indeed, both the *Wolbachia*-positive and the *Wolbachia*-negative *Drosophila* samples amplified sequences. Increasing primer annealing temperature did not increase specificity in our hands. Furthermore, analysis of isopod amplicons by DNA sequencing revealed no similarity to known *Wolbachia* sequences. Sequence tracings appeared to represent a mixture of DNA templates. Thus, we found no evidence of *Wolbachia* endosymbionts in the *P. sculpta* population using this approach.

The second independent analysis supported the above result that the isopods are not infected with *Wolbachia*. All isopod and *Drosophila* DNA extractions contained amplifiable DNA as determined by the ability of the 12Sai and 12Sbi primers to amplify a region of mtDNA. The control containing a water extract was negative. The *D. simulans* line from Riverside that was known to be infected with *Wolbachia* showed distinct 16S rDNA and Wsp amplicons indicating the presence of *Wolbachia* endosymbionts. However, the isopod samples and the *Wolbachia* negative controls showed no trace of amplification with either primer set. These results suggest that ECF, if present in *P. sculpta,* is not related to *Wolbachia* or any of its close prokaryote relatives. ECF may have been present in our samples in very low concentrations. However, this alternative is unlikely as we tested for the presence of *Wolbachia* at three distinct DNA concentrations.

3.2 *Simulation analyses*

The observed frequency of females in the aggregate 1987-1988 *P. sculpta* sample was .64 ($R = 1.77$). This value was significantly greater than the expected population sex ratio (.5) if sex in this species was determined by chromosomal sex factors alone (Shuster & Levy 1999; $X^2 = 60.0$, $P<.001$, $N = 771$). Not surprisingly, the expected frequency of females assuming only chromosomal sex determination also underestimated the observed frequency of females in each monthly 1987-1988 sample ($X^2 = 82.6$, $df = 11$, $P<.001$, $N = 771$)

When the effects of *Ams-Tfr* interactions (Shuster & Sassaman 1997) were included in the population model (Eq. 1), the expected frequency of females in the aggregate 1987-1988 sample was .54. This frequency was higher than in the previous simulation, but still differed significantly from the observed frequency of females (.64) in the pooled sample ($X^2 = 30.7$, $P<.001$; $N = 771$). The model also significantly underestimated the observed frequency of females when monthly samples were compared ($X^2 = 49.5$, $df = 11$, $P<.001$, $N = 771$), although this fit was closer than that observed above.

When the effects of ECF were included in the population model, we found significant deviations between observed and expected aggregate population sex ratios with ECF frequencies less than .4 and greater than .6 (Table 2). ECF frequencies below and above these boundaries generated significant excesses of males, and significant excesses of females, respectively. This result is *not* consistent with the usual tendency of maternally and horizontally transmitted endosymbionts to rapidly sweep through and become fixed within natural host populations (Werren & Jaenike 1995). However, it is consistent with other findings showing that some maternally and horizontally transmitted sex factors fail to become fixed in their host populations (Juchault & Rigaud 1995).

We generated four specific results when we assumed (a) an equal primary sex ratio, (b) frequencies of *Ams* alleles equal to those calculated from 1987-1988 samples, (c) P_{ECF} =

Table 2. Expected male and female frequencies generated by equation (1) using 1987-1988 aggregate population frequencies for 1°SD, *Ams*, and *Tfr*, with variable ECF frequencies ($N = 771$).

P_{ECF}	P_{males}	$P_{females}$	X^2	P
Observed	0.36	0.64	0.0	n.s.
0.0	0.46	0.54	30.7	<< 0.001
0.1	0.44	0.56	19.7	< 0.001
0.2	0.42	0.58	11.2	< 0.01
0.3	0.40	0.59	5.0	< 0.05
0.4	0.38	0.62	1.2	n.s.
0.5	0.36	0.64	0.0	n.s.
0.6	0.34	0.62	1.5	n.s.
0.7	0.32	0.68	5.8	< 0.05
0.8	0.30	0.70	13.5	< 0.01
0.9	0.28	0.72	24.8	< 0.001
1.0	0.26	0.74	40.5	<< 0.001

Table 3. Results of X^2 tests ($df = 17$) of observed and expected frequencies of α-, β-, and γ-males from 1987-1988 monthly population samples with $P_{ZW} = P_{ZZ} = 0.5$, P_{Ams} are determined for each month using the methods of Shuster & Wade (1991a), P_{Tfr2} is adjusted to produce a population sex ratio equal to $R_{observed}$, and P_{ECF} is varied from 0-0.9.

P_{ECF}	X^2	P
0.0	15.03	> 0.50
0.1	16.86	> 0.40
0.2	18.94	> 0.30
0.3	21.14	> 0.20
0.4	23.92	> 0.10
0.5	27.06	> 0.05
0.6	30.64	< 0.05
0.7	32.49	< 0.02
0.8	35.27	< 0.01
0.9	39.65	< 0.01

0-.9 and (d) frequencies of Tfr^2 (P_{Tfr2}) sufficient to make the observed and expected sex ratios in each 1987-1988 collection as close as possible, without causing significant deviations between observed and expected frequencies of the three male morphs. First, we found significant deviations in the expected frequencies of α-, β- and γ-males when P_{ECF} exceeded .5, a result corroborating our aggregate sample results above indicating that ECF exists at intermediate frequency in the *P. sculpta* population (Table 3). Second, we found no significant deviation between the observed and expected frequencies of the three male morphs at any P_{ECF} below .5, indicating that increased frequencies of the Tfr^2 allele eliminated the need to include ECF in the population model. Third, at all frequencies of ECF below .5, fluctuations of Tfr^2 appeared similar in amplitude (Fig. 2), indicating that while the effects of ECF and Tfr^2 may be additive, they do not interact. Fourth, a significant correlation existed between R_{obs} and P_{Tfr2} in all simulations (Spearman's $r_S = .9$, $P = .03$, $N = 6$ for each simulation; Fig. 3), indicating that oscillations in sex ratio could coincide with oscillations in the frequency of Tfr^2, an allele known to cause genetic males to ma-

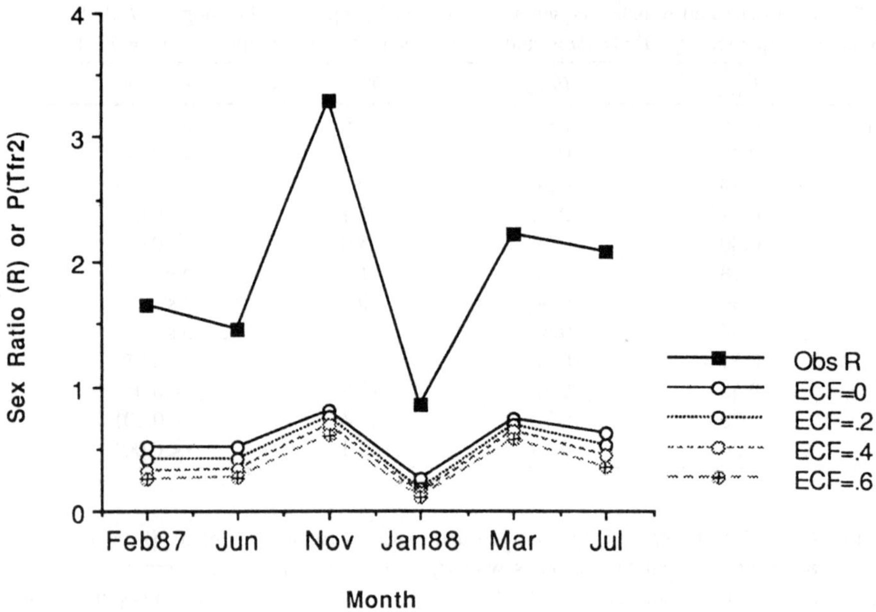

Figure 2. Observed sex ratio (*R*; solid line with filled squares) and expected frequencies for *Tfr*[2] (open circles with dotted lines) under conditions described in Table 3 and with $P_{ECF} = 0 - 0.6$.

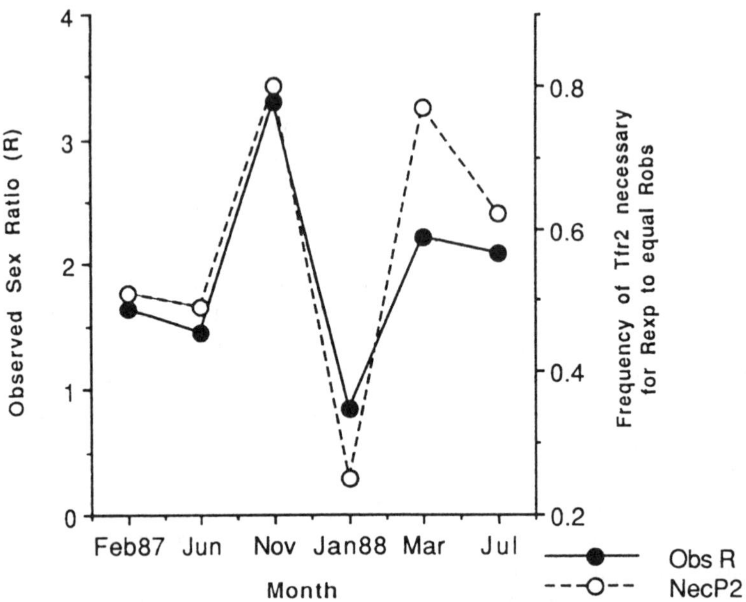

Figure 3. Simulation results to determine the frequency of *Tfr*[2] (P_{Tfr2}) necessary to match observe population sex ratio and *Ams* allele frequencies when the effects of ECF are excluded ($P_{ECF} = 0$). The correlation between R_{obs} and P_{Tfr2} is significant (Spearman's $r_S = .9$, $P = .03$, $N = 6$).

ture as phenotypic females (Shuster & Sassaman 1997). Fifth, the fact that the expected frequencies of *Tfr*^{2} must oscillate to approximate the observed variation in population sex ratio is consistent with the hypothesis that frequency-dependent selection may act on alleles at the *Tfr* locus (via primary sex factor- *Ams-Tfr* interactions).

4 DISCUSSION

Our results unambiguously show that ECF is not related to known *Wolbachia* endosymbionts or their close prokaryote relatives. Our first set of 16S and 23S primers (O'Neill et al. 1992, Rousset et al. 1992) amplified DNA of likely prokaryotic origin in all but one isopod. However, this DNA consisted of a mixture of non-*Wolbachia* sequences, and since these primers also amplified 16S and 23S sequences from *Bacillus cereus*, they appear to be somewhat nonspecific. Newer *Wolbachia*-specific primers (Zhou et al. 1998) showed no evidence of amplified DNA in isopod samples likely to contain endosymbiotic sex factors. Thus, while some endosymbiont may yet be found to cause sex ratio biases in *P. sculpta*, ECF is clearly not a *Wolbachia* species.

However, some of our simulation results suggest that an effect consistent with that attributed to ECF does exist and can have a measurable influence on population sex ratio. Simulations of aggregate as well as monthly population samples lacking the effects of ECF showed significant deficits of females. Moreover, differences between observed and expected population sex ratios disappeared when the presumed effects of ECF were included. In aggregate samples with variable ECF incidence, observed and expected population sex ratio deviated least at intermediate ECF frequency (.4-.6). This result is consistent with the observed frequencies of extrachromosomal factors affecting sex ratios in other isopod populations (Juchault & Rigaud 1995), but is at odds with theoretical predictions of endosymbiont frequencies in host populations (Werren & Jaenike 1995).

In simulations with the observed frequencies of *Ams* alleles and the sex ratio of each monthly sample included, variable frequencies of *Tfr*^{2}, combined with frequencies of ECF above .5, showed significant excesses of females, corroborating evidence that ECF cannot exist at high population frequency. However, we found no significant deviation in observed and expected sex ratios, or in *Ams* allele frequencies, when *Tfr*^{2} frequencies were allowed to vary and ECF ranged from 0 to .5. In fact, the deviation of observed and expected frequencies of *Ams* alleles decreased as the frequency of ECF decreased, and was smallest when P_{ECF} equaled 0 (Table 3). We found the same significant correlation between population sex ratio (R) and P_{Tfr2} in each of these simulations, and the amplitude of oscillations in P_{Tfr2} was consistent in each case; as P_{ECF} decreased, P_{Tfr2} necessarily increased. While high, oscillating frequencies of *Tfr*^{2} were sufficient to explain fluctuating and female-biased population sex ratios in *P. sculpta* without ECF, this last result was not consistent with the results of aggregate population sex ratio simulations, or the apparent influence of ECF in laboratory experiments (Shuster & Sassaman 1997).

In the laboratory, families sired by β-males were often female-biased or deficient in α-males (Shuster & Sassaman 1997). These deviations evidently occurred because ZZ, $Ams^{\alpha}Ams^{\alpha}$, Tfr^1Tfr^2 individuals matured as females instead of α-males. However, the tendency to produce female-biased families appeared to have been inherited by all β-male descendants. This effect, while evidently not caused by *Wolbachia* endosymbionts, was undeniable in a significant number of β-lineages, and was the source of the ECF-as-

endosymbiont hypothesis (Shuster & Sassaman 1997). This effect now has two possible explanations: (1) A still unknown, non-*Wolbachia* endosymbiont, which is not transferred within and among lineages like most endosymbionts (Werren & Jaenike 1995), interacts with *Ams-Tfr* and primary sex determination genotypes described as in Shuster & Sassaman (1997). Perhaps because β-males exist at low population frequencies, and perhaps because ECF is favored only when it co-occurs with Ams^{β}, it too persists only at low population frequency. Alternatively, (2) ECF may not exist at all, but rather some part of the genetic syndrome that produces β-males, or some unknown maternal effect, caused the observed family sex ratio deviations within β-lineages. Clearly, further study of the inheritance of family sex ratio in *P. sculpta* is warranted.

Many studies of population sex ratio in other species document female-biased sex ratios and sex ratio fluctuations in nature (Komdeur et al. 1997, Whiteman 1997, Atlan et al. 1997, Hurst & Pomiankowski 1998, Wilkinson et al. 1998). These deviations are usually attributed to greater pre-adult mortality among males than females, and are therefore assumed to represent a form of Fisherian sex ratio adjustment in which males are produced in quantities necessary to compensate for sexual differences in viability (Fisher 1930). The model presented here shows that a primary sex determination mechanism capable of generating a 50:50 sex ratio can be modified by interactions with autosomal sex ratio distorters, and that these factors can generate female-biased population sex ratios and sex ratio fluctuations without requiring differential male mortality. The model also shows that interactions between *Ams-Tfr* and primary sex determination factors can routinely produce fluctuating sex ratios in the natural *P. sculpta* population, and that such fluctuations can result from differential mating success among the three male morphotypes in this species. The *Ams-Tfr* model indicates that sexual selection and sex ratio distortion are genetically interrelated phenomena in this species. This result provides a compelling context for the co-occurrence of sexual selection and sex ratio distortion in a diversity of animal taxa (Komdeur et al. 1997, Whiteman 1997, Atlan et al. 1997, Hurst & Pomiankowski 1998, Wilkinson et al. 1998).

ACKNOWLEDGEMENTS

We are grateful to the editors of this volume for the opportunity to present our work at the Second International Isopod Conference in Amsterdam. We are also grateful to J. Schupp for his technical advice on the effects of primer annealing temperature on PCR products, as well as to T. Rigaud, M.J. Wade, B. Sinervo, K. Johnson and H.C. Wildey for their comments on earlier drafts of this manuscript. Logistical support for the field component of this research was provided by el Centro Intercultural de Estudios de Desiertos y Océanos (CEDO) in Puerto Peñasco, Sonora, México. Permission to collect and study *P. sculpta* in México (Oficio No. A00702-06296 and DAN 02384) was authorized by el Instituto Nacional de Ecología, Dirección General de Approvechamiento Ecologico de los Recursos Naturales, México, D.F. M. Turrelli provided the *D. simulans* lines from Riverside, CA. Financial assistance was provided by NSF grants OCE 84-01067, BSR 91-0644 and DEB 97-26504 to SMS, BSR 87-00112 to CS and SMS, and DEB-9702824 to JWOB, as well as by Organized Research grants, the Department of Biological Sciences and the Cowden Foundation at Northern Arizona University.

REFERENCES

Atlan, A., H. Mercot, C. Landre & C. Monchamp-Moreau 1997. The *sex ratio* trait in *Drosophila simulans*: Geographical distribution of distortion and resistance. *Evolution* 51: 1886-1895.

Bouchon, D., T. Rigaud & P. Juchault 1998. Evidence for widespread *Wolbachia* infection in isopod crustaceans: molecular identification and host feminization. Proc. R. Soc. Lond. B 265: 1081-1090.

Conover, D.O. & D.A. vanVoorhees 1990. Evolution of a balanced sex ratio by frequency-dependent selection in a fish. *Science* 250: 1556-1557.

Fisher, R.A. 1930. *The Genetical Theory of Natural Selection*, Clarendon Press, Oxford.

Frank, S.A. 1990. Sex allocation theory for birds and mammals. *Ann. Rev. Ecol. Syst.* 21: 13-56.

Godfray, H.C.J. & J.H. Werren 1996. Recent developments in sex ratio studies. *Trend. Ecol. Evol.* 11: 59-63

Hamilton, W.D. 1979. Wingless and fighting males in fig wasps and other insects. In M.S. Blum & N.A. Blum (eds), *Sexual Selection and Reproductive Competition in Insects*: pp. 167-222. New York: Academic Press.

Heath, D.J. & J.R. Ratford 1990. The inheritance of sex ratio in the isopod, *Sphaeroma rugicauda. Heredity* 64: 419-25.

Hurst, L.D. 1993. The incidences, mechanisms and evolution of cytoplasmic sex ratio distorters in animals. *Biol. Rev.* 68: 121-193.

Hurst, L.D. & A. Pomiankowski 1998. Sexual selection: the eyes have it. *Nature* 391: 223.

Juchault, P, T. Rigaud & J.-P. Mocquard 1992. Evolution of sex-determining mechanisms in a wild population of *Armadillidium vulgare* Latr. (Crustacea: Isopoda): competition between two feminizing parasitic sex factors. *Heredity* 69: 382-390 .

Juchault, P. & T. Rigaud 1995. Evidence for female heterogamety in two terrestrial crustaceans and the problem of sex chromosome evolution in isopods. *Heredity* 75: 466-471.

Kimberling, D., A. Ferreria, S.M. Shuster & P. Keim 1996. RAPD marker estimation of genetic structure among isolated northern leopard frog populations in the south-western USA. *Molec. Ecology.* 5: 521-529.

Kocher, T.D., W.K. Thomas, A. Meyer, S.V. Edwards, S. Pääbo, F.X. Villablanca & A.C. Wilson 1989. Dynamics of mitochondrial DNA evolution in animals: amplification and sequencing with conserved primers. *Proc. Natl. Acad. Sci.* USA 86: 6196-6200.

Komdeur, J., S. Daan, J. Tinbergen & C. Mateman 1997. Extreme adaptive modification in sex ratio of the Seychelles warbler's eggs. *Nature* 385: 522-525.

O'Neill, S.L., R. Giordano, A.M.E. Colbert, T.L. Karr & H.M. Robertson 1992. 16S rRNA phylogenetic analysis of the bacteria endosymbionts associated with cytoplasmic incompatilibity in insects. *Proc. Natl. Acad. Sci. USA* 89: 2699-2702.

Rigaud, T. & P. Juchault 1993. Conflict between feminizing sex ratio distorters and an autosomal masculinizing gene in the terrestrial isopod, *Armadillidium vulgare* Latr. *Genetics*: 133: 247-252

Rousset, F., D. Bouchon, B. Pintureau, P. Juchault & M. Solignac 1992a. *Wolbachia* endosymbionts responsible for various alterations of sexuality in Arthropods. *Proc. R. Soc. Lond.* B: 250: 91-98.

Rousset et al. 1992b. Molecular identification of *Wolbachia*, the agent of cytoplasmic incompatibility in *Drosophila simulans*, and variability in relation with host mitochondrial types. *Proc. R. Soc. Lond.* B. 247: 163-168.

Shuster, S.M. 1987. Alternative reproductive behaviors: three male morphs in *Paracerceis sculpta*, an intertidal isopod from the northern Gulf of California. *J. Crust. Biol.* 7: 18-327.

Shuster, S.M. 1989a. Female sexual receptivity associated with molting and differences in copulatory behavior among the three male morphs in *Paracerceis sculpta*, (Crustacea: Isopoda). *Biol. Bull.* 177: 331-337.

Shuster, S.M. 1989b. Male alternative reproductive behaviors in a marine isopod crustacean

(*Paracerceis sculpta*): The use of genetic markers to measure differences in fertilization success among α-, β- and γ-males. *Evolution* 34: 1683-1698.

Shuster, S.M. 1990. Courtship and female mate selection in a semelparous isopod crustacean (*Paracerceis sculpta*). *Anim. Behav.* 40: 390-399.

Shuster, S.M. 1991a. The ecology of breeding females and the evolution of polygyny in *Paracerceis sculpta,* a marine isopod crustacean. In: R. Bauer and J. Martin (eds), *Crustacean Sexual Biology*: pp. 91-110. New York: Columbia University Press.

Shuster, S.M. 1991b. Changes in female anatomy associated with the reproductive molt in *Paracerceis sculpta* (Holmes), a semelparous isopod crustacean. *J. Zool. (Lond.).* 225: 365-379.

Shuster, S.M. 1992. The reproductive behaviour of α-, β- and γ-males *in Paracerceis sculpta*, a marine isopod crustacean. *Behaviour* 121: 231-258.

Shuster, S.M. 1992b. The use of artificial cavities as breeding habitat by *Paracerceis sculpta* (Crustacea: Isopoda). *J. Exp. Mar. Biol. Ecol.* 165: 75-89.

Shuster, S.M. & L. Levy 1999. Sex-linked inheritance of a cuticular pigmentation marker in a marine isopod, *Paracerceis sculpta* Holmes (Crustacea: Isopoda: Sphaeromatidae). *J. Heredity* 90: 304-307.

Shuster, S.M. & C. Sassaman 1997. Genetic interaction between male mating strategy and sex ratio in a marine isopod. *Nature* 388: 373-376.

Shuster, S.M. & M.J. Wade 1991a. Equal mating success among male reproductive strategies in a marine isopod. *Nature* 350: 606-610.

Shuster, S.M. & M.J. Wade 1991b. Female copying and sexual selection in a marine isopod crustacean. *Anim. Behav.* 42: 1071-1078

Simon, C., F. Frati, A. Beckenbach, B. Crespi, H. Liu & P. Flook 1994. Evolution, weighting and phylogenetic utility of mitochondrial gene sequences and a compilation of conserved polymerase chain reaction primers. *Ann. Entomol. Soc. Amer.* 87: 651-701.

Slatkin, M. 1978. On the equilibration of fitnesses by natural selection. *Amer. Natur.* 112: 845-859.

Slatkin, M. 1979. The evolutionary response to frequency and density dependent interactions. *Amer. Natur.* 14: 384-398.

Werren, J.H., U. Nur & C-I. Wu 1988. Selfish genetic elements. *Trends. Ecol. Evol.* 3: 297-302.

Werren, J.H. & J. Jaenike 1995. *Wolbachia* and cytoplasmic incompatibility in mycophagous *Drosophila* and their relatives. *Heredity* 75: 320-326.

Whiteman, H.H. 1997. Maintenance of polymorphism promoted by sex-specific fitness payoffs. *Evolution* 51: 2039-2044.

Wilkinson, G.S., D.C. Presgraves & L. Crymes 1998. Male eyespan in stalk-eyed flies indicates genetic quality by meiotic drive suppression. *Nature* 391: 276-279.

Zhou, W., F. Rousset & S. O'Neill. 1998. Phylogeny and PCR-based classification of *Wolbachia* strains using Wsp gene sequences. *Proc. R. Soc. Lond.* B 265: 509-515.

3 Morphology

Morphology of respiratory organs in South American Oniscidea ('Philosciidae')

ANDREAS LEISTIKOW
*Universität Bielefeld, Abteilung für Morphologie und Systematik der Tiere, Germany * Ruhr-Universität Bochum, Lehrstuhl für Spezielle Zoologie, Gebäude NDEF, Germany*

PAULA BEATRIZ DE ARAUJO
Departamento de Zoologia, IB, Universidade Federal do Rio Grande do Sul, CEP, Porto Alegre, RS, Brazil

ABSTRACT

Respiratory organs are widespread among the Oniscidea. Especially within the Crinocheta, different developmental stages can be found in several phyletic lineages as an adaptation to their terrestrial habitat. From the New World, there are few genera known to possess pleopodal lungs. The genera of Balloniscidae, *Balloniscus* and *Plataoniscus* are known to have pleopodal lungs; nonetheless, their structure is virtually undescribed. Among the philosciids, *Benthana*, *Benthanoscia longicaudata*, *Atlantoscia floridana* are said to have respiratory areas on their pleopod exopods. All these taxa are found in temperate and eastern South America, where recently a new species of *Atlantoscia* was discovered. In this work, the lungs of *Balloniscus sellowii*, *Benthana picta*, *Atlantoscia floridana*, and a new species of *Atlantoscia* are examined by means of a Scanning Electron Microscope and semi-thin sections. The morphology of the respiratory structures of the South American species is described and discussed.

1 INTRODUCTION

The occurrence of respiratory organs is one of several adaptations of Oniscidea to their terrestrial life histories. Most of the primitive representatives lack specialized respiratory organs. Mainly the thin ventral cuticle of the pleopods manages the oxygen exchange. Among the more derived taxa, the small species, which live in moist places or caves, have no particular respiratory organs. Moisture is one of the prominent factors for respiration in a gill-like fashion. The shore-dwelling genus *Ligia* is the genus comprising some of the largest species of Oniscidea, all lack respiratory organs save the above described that represent the plesiomorphic character state. Hoese (1982) in his voluminous work on respiratory organs of the Oniscidea gave detailed descriptions of several, mainly Palearctic genera, another important work on respiratory organs is Ferrara et al. (1991), which deals with the pleopodal lungs of Eubelidae. Pleopodal lungs thus can be found in the Oniscidae, Porcellionidae, Trachelipodidae, Armadillidae, Armadillidiidae and Eubelidae. As can be evidenced from these works, among the Crinochaeta, pleopodal lungs have evolved several times in different phyletic lineages. The most basal species with pleopodal lungs or at least particular respiratory areas on their pleopods are the philosciid genera *Philoscia* and *Tiroloscia*, *Aphiloscia*, and *Congophiloscia*, which occur in the western part of the

Old World. Their phylogenetic affinities to other lung-bearing taxa are not established. Schmalfuss (1990) postulated a close relationship of *Philoscia* with *Oniscus*, which bear the same type of uncovered lungs.

Particular pleopodal lungs can be found within the family Balloniscidae, which is exclusively neotropical. Their structure has hot been described in detail until now. Respiratory areas are postulated for at least three South American genera of the paraphyletic assembly 'Philosciidae', i.e. *Benthana*, *Benthanoscia*, and *Atlantoscia*. Since for all the neotropical genera structural investigations are lacking, this work will deal with the above mentioned genera.

2 MATERIAL AND METHODS

Ethanol-preserved material of *Atlantoscia* sp., *A. floridana*, *Benthana picta* and *Balloniscus sellowiii* was used for the investigations. The specimens for SEM examination were dried in an ascending ethanol concentration with stays of 2 × 1h in 80, 90 and 96% ethanol, then transferred in 1,1,1,3,3,3-hexamethyl-disilazane. The dried appendages were coated with 30 nm gold with a Hummer sputter. The preparations were then examined with a Hitachi Scanning Electron Microscope. The specimens for semi-thin sections stayed in 70% ethanol and were mounted in unicryl resin, the unicryl blocks were cut with a Leica Ultracut in slides of 250µm thickness.

3 RESULTS

3.1 *Atlantoscia floridana*

The lateral area of pleopod 1 exopod has a structure supposed to be a respiratory area (Pl. 1A. ra). It is set apart from the rest of the ramus by a sinuous ridge. This area is less thick than the main part of the exopod. However, no distinct primary folding of the cuticle is discernible, nor are pectinate scales or microsculpture on the edge of the respiratory area (Pl. 1B, rla).

In cross section, the dorsal cuticle is much thicker than the ventral one, accompanied by a layer of connective tissue (Fig. 1). The thickness of the cuticle is characteristic: the ventral cuticle is rather thin because it has a respiratory function (as in all terrestrial isopods and their aquatic relatives cf. Hoese 1982). At the lateral edge, the dorsal cuticle is bent ventrally, forming a ridge, the subsequent region of about 20 µm breadth bears a much thinner cuticle, almost as thin as the intervening ventral cuticle. The haemolymphatic sinus is rather small. In some areas, the ventral cuticle comes in contact with the connective tissue of the dorsal cuticle.

3.2 *Benthana picta*

Possible respiratory areas were found solely in pleopod 1 (Pl. 1C, ra). A lateral area of semilunar shape on the caudal surface of the exopod is set apart from the main body by a sinuous ridge (Pl. 1D, rla). This area is half as thick as the exopod, the cuticle is very slightly folded. The ridge of the respiratory area lacks any microsculpture.

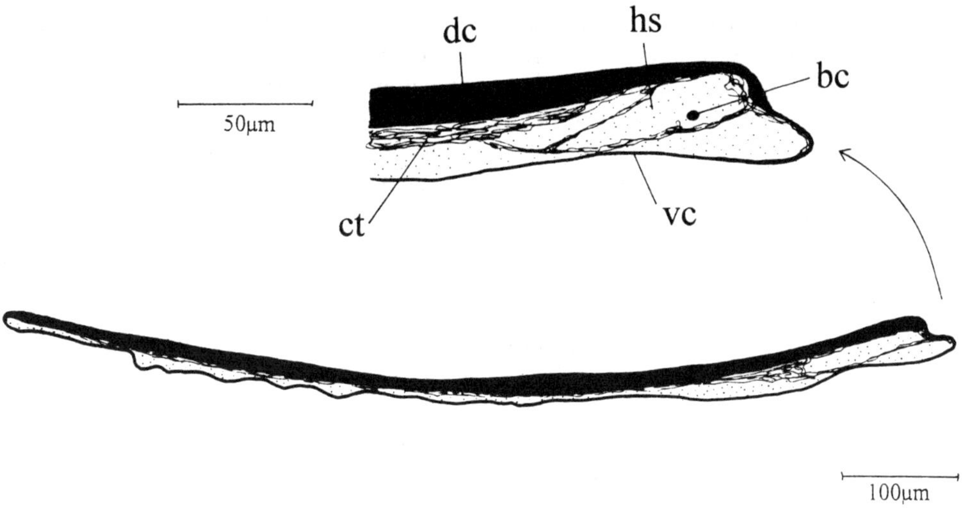

Figure 1. *Atlantoscia floridana* cross section through pleopod 1 exopod, lateral margin with respiratory area on the right, bc = blood cell, dc = dorsal cuticle, ct = connective tissue, hs = haemolymphatic sinus, rc = respiratory cuticle, vc = ventral cuticle.

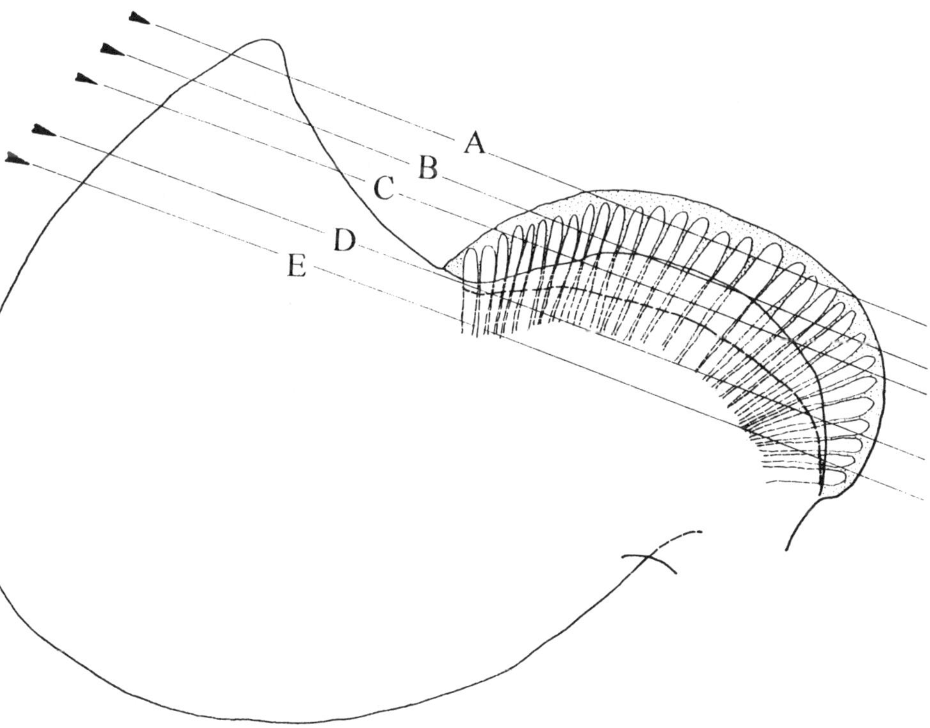

Figure 2. *Atlantoscia* sp. overview of cutting planes shown in Figure 3.

In cross section, the cuticle of this area is half as thick as the dorsal cuticle, but at least as thick as the ventral cuticle (Fig. 4C). Since the ventral cuticle has a respiratory function in all isopods as stated above, such a function has to be postulated for this dorsal portion, although it may be rather ineffective to fit the term of a pleopodal lung. In this species, the volume of haemolymphatic sinus of the exopod is much bigger than in the preceding one, bordered by a thick dorsal cuticle, the medial part is packed with connective tissue.

3.3 *Atlantoscia* sp.

Partially covered lungs are present in all five pairs of pleopods. They are situated at the latero-proximal edge of the dorsal surface of the exopod (Pl. 2A). The shape of the uncovered part is semicircular, with the cuticle radially folded in the respiratory area (primary folds, Pl. 2 B/C, rf), and separated by narrow grooves (Pl. 2C, rg). Medial to this uncovered area, the dorsal cuticle is raised and forms a semicircular fold (Pl. 2B-D, rla), the *Trachealfeldleiste* of Verhoeff (1917/1921) or 'ridge of lung atrium', a more appropriate than the term 'dorsal ridge' of Unwin (1932). The ridge of lung atrium and the area bordering it do not bear any pectinate scales nor cuticular microsculpture.

In cross section, the relation between the dorsal and ventral cuticle is the same as in the preceding species. The cuticle of respiratory function is especially thin and formed by a layer of respiratory tissue. It is strongly folded, forming rather deep grooves between the folds.

A series of semi-thin sections (Fig. 3) gives hints of the three-dimensional structure of the lung: the distal part consists of foldings of the respiratory cuticle that take a radial course. More proximally, the thick dorsal cuticle spreads over the folds, which are still open to the pleoventral cavity (Fig. 3C). In the following state, the outer folds of the lung are still free while the central ones have contact with the dorsal cuticle and form tubules which are surrounded by a haemolymphatic sinus (Fig. 3D). The tubules, which still run radially, do not anastomose.

3.4 *Balloniscus sellowii*

In morphological aspects, the lungs are of a partially uncovered type (Pl. 3A. ul). They have a rather broad area of the medial margin of the exopods in all five pleopods has the cuticle folded, and these folds are connected to each other forming a network of cuticular clasps (Pl. 3 B/C, cc). The most lateral area of the lung (Pl. 3B, lml) has a cuticle which bears small cuticular protrusions looking like small knobs. The dorsal cuticle forms a sinuous fold covering the proximal part of the lung (Pl. 3D, rla), with the surface covered by small pectinate setae, which make the area around the ridge of lung atrium water repellent and also may be of anti-cohesive function (Pl. 3D, ps).

In cross section the pleopod exopod is characterized by the thick dorsal cuticle that is rather concave and therefore, forming the ridge of lung atrium, the uncovered part of the lung is formed by primary folds that are interspersed with cuticular tubules. These tubules can also be found more proximally, covered by the dorsal cuticle. They are imbedded in the haematolymphatic sinus.

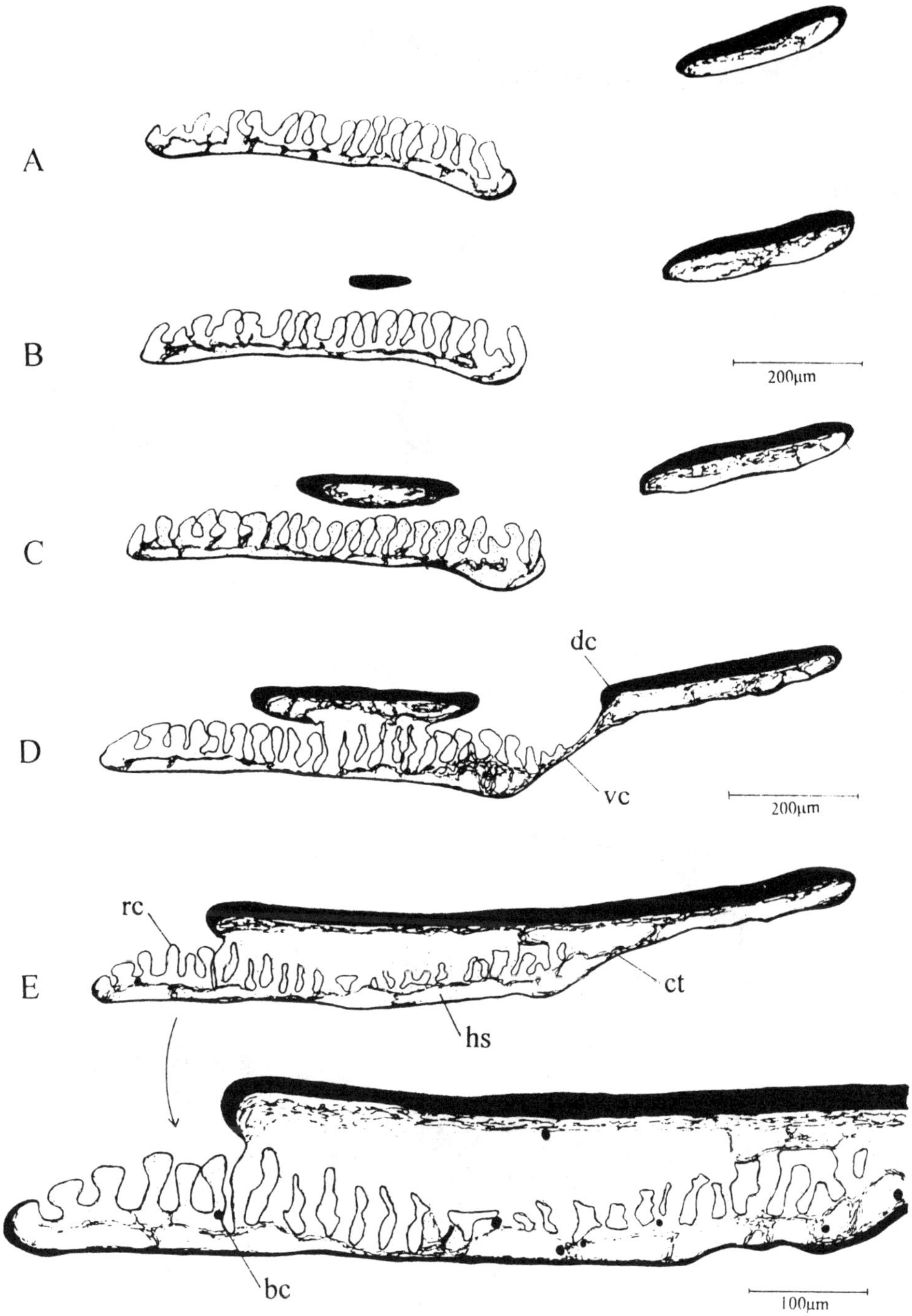

Figure 3. *Atlantoscia* sp. cross section through pleopod 1 exopod, lateral margin with semi-covered lung on the left, A distally to E proximally. Abbreviations, see Figure 1.

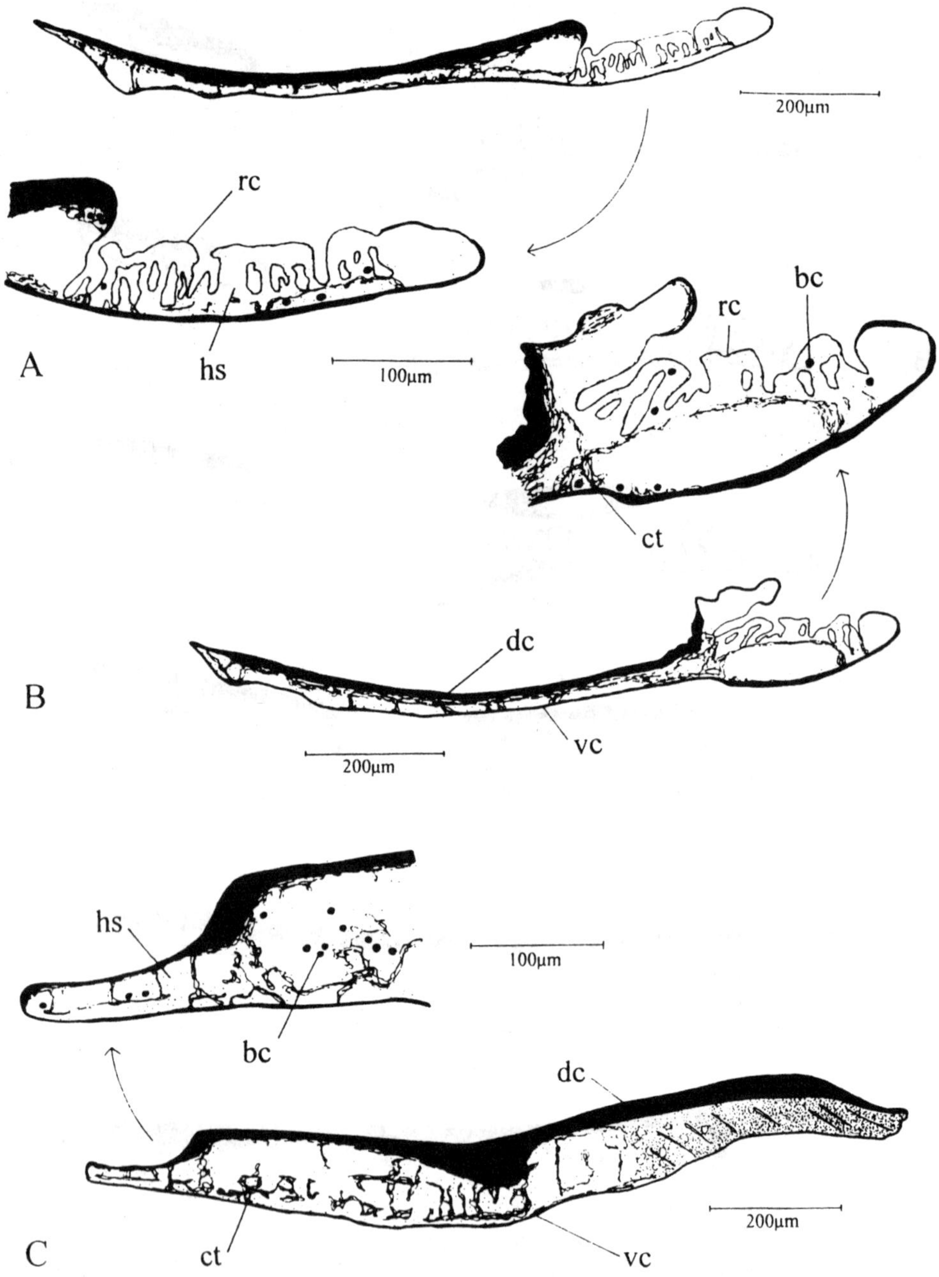

Figure 4. (A) *Balloniscus sellowii* cross section through pleopod 2 exopod, lateral margin with semi-covered lung on the right; (B) *Balloniscus sellowii* cross section through pleopod 5 exopod, lateral margin with semi-covered lung on the right; (C) *Benthana picta* cross section through pleopod 1 exopod, lateral margin with respiratory area on the left. Abbreviations, see Figure 1.

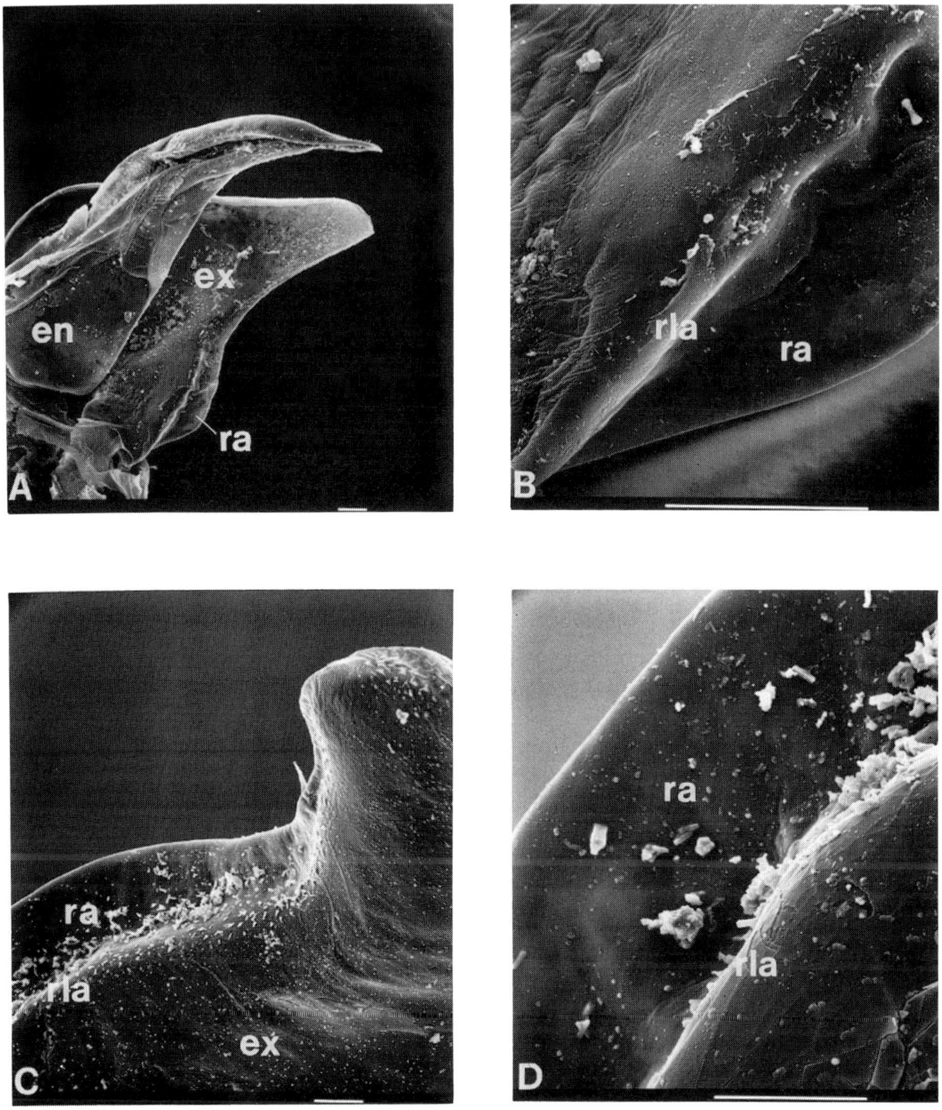

Plate 1. (A) *Atlantoscia floridana*, male pleopod 1 in dorsal view, scale bar: 50µm; (B) *Atlantoscia floridana*, detail of lateral part of exopod, scale bar: 50µm; (C) *Benthana picta*, female pleopod 1 exopod in dorsal view, scale bar: 50 µm; (D) *Benthana picta*, detail of lateral part of exopod, scale bar: 50 µm. Abbreviations: at air tubule, cc = cuticular clasps, ecl = entrance to covered part of lung, en = endopod, ex = exopod, lml = lateral margin of the lung, ps = pectinate setae, ra = respiratory area, rf = radial folding, rg = radial groove, rla = ridge of the lung atrium, ul = uncovered part of the lung.

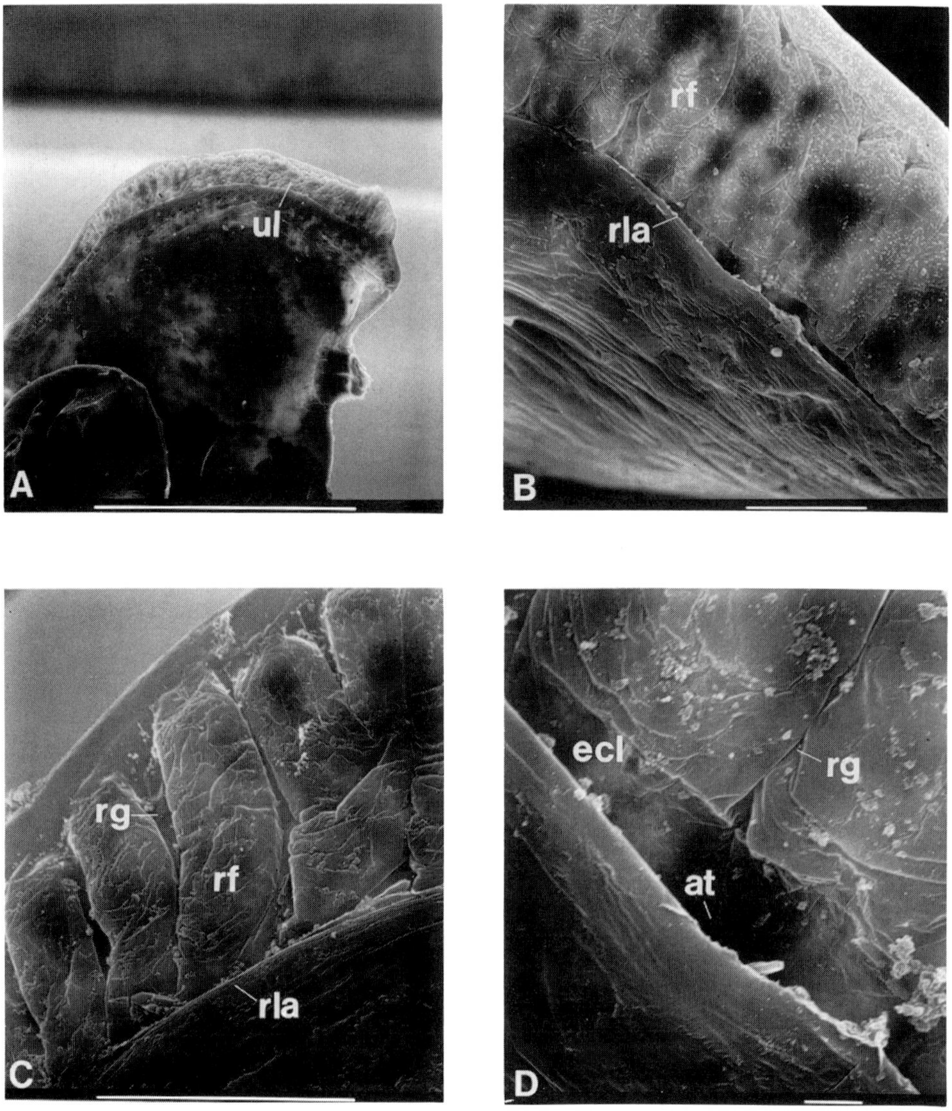

Plate 2. *Atlantoscia* sp. (A) female pleopod 2 exopod, scale bar: 500µm; (B-C) detail of primary folding of the lung of pleopod 2, scale bar: 50µm; (D) cuticular fold of ridge of the lung atrium, forming entrance to the covered part of the lung; scale bar: 5µm. Abbreviations, see Plate 1.

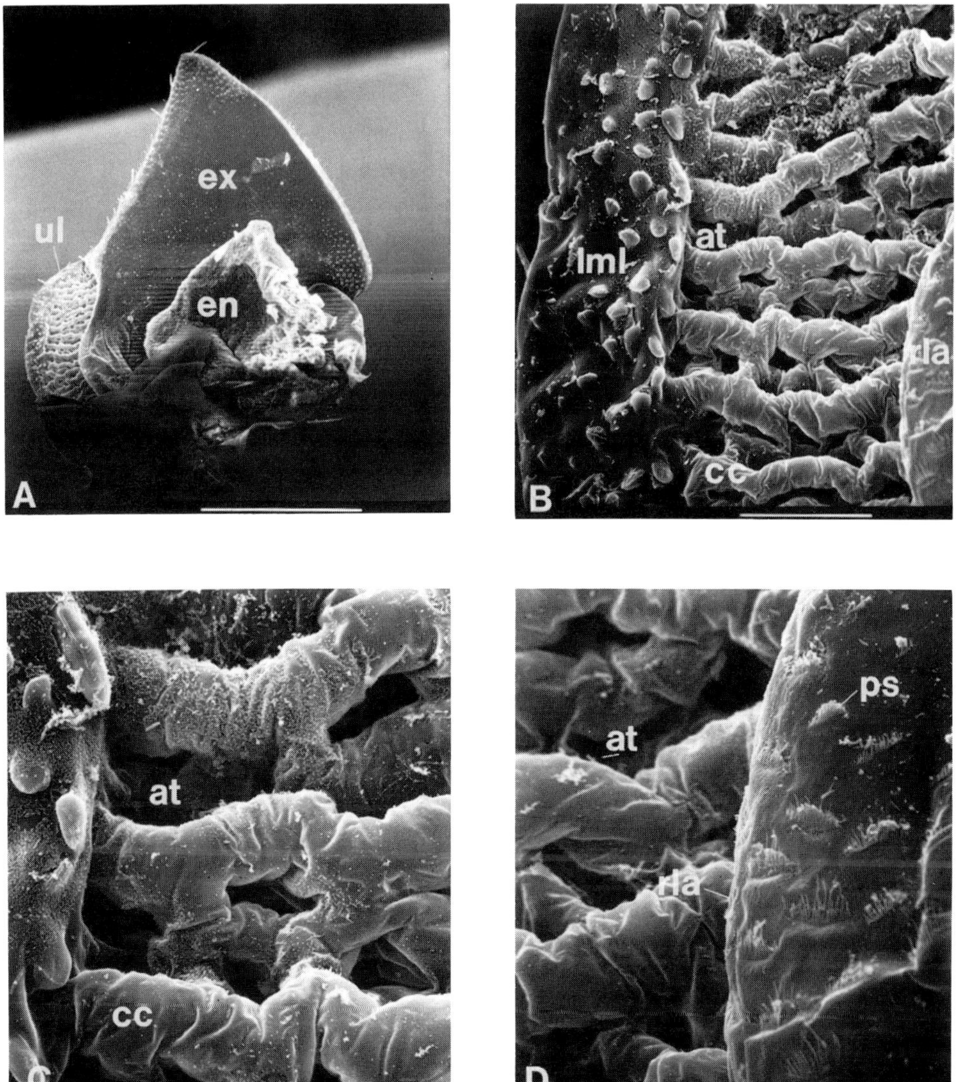

Plate 3. *Balloniscus sellowii* (A) female pleopod 2 exopod, scale bar: 500μm; (B) uncovered part of the lung, scale bar: 50μm; (C) detail of air tubules and cuticular clasps of the uncovered lung, scale bar: 5 μm; (D) ridge of the lung atrium with pectinate scales, scale bar: 5 μm. Abbreviations, see Plate 1.

4 DISCUSSION

The development of respiratory areas is very different in the four species examined herein, they reflect a morphological transition from simple unfolded, uncovered lungs to semi-covered with anastomosing lung tubules. This morphological transition does not represent a phylogenetic transition, as can be seen when the morphological data of several authors are compared (e.g. Lemos de Castro 1958a, Vandel 1963, Ferrara & Taiti 1981, Araujo et al. 1996, most recently Araujo & Leistikow 1999).

Atlantoscia floridana and *Benthana picta* represent the most primitive character state with the cuticle distally of the ridge of lung atrium thinned but lacking any prominent primary folding. Hoese (1982) postulated that for an effective gas exchange a maximum thickness of the cuticle would be 0.2μm. In *A. floridana*, the cuticle measures about 2.3 μm to 3.7μm and therefore may not fulfill an effective respiratory function. However, since the ventral cuticle is of the same thickness, it might have at least additional respiratory function. In *B. picta* the thickness is similar. In this species, the respiratory area is much better demarcated by a more prominent ridge of lung atrium. The species differ in the structure of the exopods, with *A. floridana* having rather thin pleopods with the ventral cuticle close to the dorsal connective tissue. In *B. picta*, the haematolymphatic sinus is much more voluminous and the dorsal cuticle forms a prominent ridge dividing it into a lateral and medial part.

The next evolutionary step is represented by *Atlantoscia* sp. with the uncovered part of its lung with radial primary folds, and a covered part consisting of radial tubules continuing the folds, formed by covering of the latter by the dorsal cuticle. The ridge of lung atrium lacks microsculpture, thus representing the plesiomorphic character state even found in *A. floridana* and *B. picta*. These relations are much complicated in *Balloniscus sellowii*. In this species, the radial folds anastomose and form a complex network of tubules both in the covered and uncovered part of the lung. The water repellent function of the ridge of lung atrium is improved by pectinate scales. This stage of pleopodal lungs is the most derived of the above discussed species. The lungs of *Atlantoscia* sp. are comparable to those uncovered lungs of *Trachelipus ratzeburgi* (Hoese 1982). The extension of the tubules within the haemolymphatic sinus of the exopod is intermediate to the eubelid species *Eubelum instrenuum* and *Gerutha pila* (Ferarra et al. 1891), in the lack of anastomoses more similar to that seen in *E. instrenuum*. The lungs of *B. sellowii* are closer to the *Gerutha*-type with frequently anastomosing tubules and microsculpture on the ridge of lung atrium.

The lung structures are presumably an adaptation to the life in more xeric environments than other genera of similar size. *Balloniscus* especially lives in the pastures of La Pampa in Argentina, where rain is seasonal and drier months occur. Other genera living in the moist environments of tropical rain forests, like *Ischioscia* do not have specialized lung structures. The moisture of their habitat is sufficient to warrant respiration through the ventral cuticle of the pleopod exopods. There might be a selection pressure leading to the development of respiratory devices in the habitats of the species dealt with, since the genera like *Ischioscia* are totally absent in southeastern South America.

ACKNOWLEDGEMENTS

The authors wish to thank Mrs R. Feist, Universität Bielefeld and Dr W. Kobusch, Ruhr-Universität Bochum, for technical support for semi-thin sections and SEM preparations, Mr Ch. Schmidt, Ruhr-Universität Bochum, for the provision with the semi-thin sections of *Balloniscus sellowii*. For the material supply the authors are indebted to Dr E. da Conceio Guerreiro Couto, Universidade Federal de Sergipe, and Dr M. von Tschirnhaus, Universität Bielefeld.

REFERENCES

Araujo, P.B., Buckup, L. & Bond-Buckup, G. 1996. Isópodos terrestres (Crustacea, Oniscidea) de Santa Catarina e Rio Grande do Sul, Brazil. *Iheringia*. Sér. Zool. 81: 111-138.

Araujo, P.B. & Leistikow, A. 1999. A new species of philosciid Oniscidea bearing pleopodal lungs and further Oniscidea from Brazil (Crustacea : Isopoda). *Contrib. Zool.* 68: 109-111.

Ferrara, F., Paoli, P. & Taiti, S. 1991. Morphology of the pleopodal lungs in the Eubelidae (Crustacea, Oniscidea). In Juchault & Monod (eds), *Biol. Terrestr. Isopods* (Poitiers) 3: 9-16.

Ferrara, F., Paoli, P., & Taiti, S. 1994. Philosciids with pleopodal lungs? The case of the genus *Aphiloscia* Budde-Lund, 1908 (Crustacea: Isopoda: Oniscidea) with a description of six new species. *J. Nat. Hist.* 28: 1231-1264.

Hoese, B. 1982. Morphologie und Evolution der Lungen bei den terrestrischen Isopoden (Crustacea, Isopoda, Onisoidea). *Zool. Jb. Anat.* 109: 487-501.

Unwin, E.E. 1932. On the structure of the respiratory organs of the terrestrial Isopoda. *Pap. Proc. Roy. Soc. Tasmania* 1931: 37-104.

Verhoeff, K.W. 1917. Zur Kenntnis der Entwicklung der Trachealsysteme und der Untergattungen von *Porcellio* und *Tracheoniscus*. *Sitz.-ber. Ges. naturforsch. Freunde Berlin* 1917: 195-223.

Verhoeff, K.W. 1921. Über die Atmung der Landasseln, zugleich ein Beitrag zur Kenntnis der Entstehung der Landtiere. *Z. wiss. Zool.* 118: 365-447.

Some unusual morphological features of Amazonian fish parasites (Cymothoidae)

VERNON E. THATCHER
Instituto Nacional de Pesquisas da Amazônia, Manaus, Amazonas, Brazil

ABSTRACT

Amazonian freshwater cymothoids have been found to possess certain morphological characters that are unreported for marine forms. In the genus *Anphira*, for example, the first coxal plates are free and moveable. In all other known Cymothoidae, the first coxal plates are fused to the pereion and the following six are free. In the genus *Braga*, as redefined by Thatcher, 1996, the pleopods are multilaminate. Pleopod 1 consists of two lamellae while pleopod 2 has four and pleopods 3 to 5 have three lamellae each. The mouth parts of South American freshwater cymothoids also differ from those of their marine cousins. The mandibles of the freshwater forms lack cutting shearing blades and incisors. The maxillules of freshwater species usually have five recurved spines (3-4 terminal and 1-2 subterminal) whereas the marine typically have only four, all terminal. The mouth parts of Amazonian fish parasitic isopods differ from one species to another but they do show a closer relationship with each other than any do to the marine species. These morphological features are illustrated and explained.

1 INTRODUCTION

For many years, the only information available on South American cymothoids was that found in the monograph of Schiödte & Meinert (1884). This consisted of descriptions in Latin and small line drawings. Then, around the middle of the century, a few papers began to appear, notably those of Lemos de Castro & Machado Filho (1946), Lemos de Castro (1959), and Leigh-Sharpe (1937). More recently, new studies by Trilles (1973), Bowman (1986), Bastos & Thatcher (1997), Thatcher (1988, 1991, 1993a, 1993b, 1995a, 1995b, 1996, 1997), Thun & Brusca (1980), and Thatcher & Carvalho (1988) have cast new light on the morphology and relationships of these fish parasites. The present report reviews our knowledge of Amazonian cymothoids and points out some of their unusual morphological features.

2 MATERIAL AND METHODS

Freshwater cymothoids were removed from their fish hosts and fixed in 70% ethanol.

They were later studied by means of a dissecting microscope and gross measurements taken with measuring calipers. Some mouth parts, pleopods and pereiopods were removed for more detailed study. Such small parts were dehydrated and cleared in pure phenol, passed through methyl salicylate and mounted in Canada balsam. Drawings of small parts were made with the aid of a compound microscope equipped with a drawing tube. Sizes were taken with a measuring ocular and are given in millimeters (mm) unless indicated as micrometers (µm).

3 RESULTS

3.1 *Bosses and tubercles on the cephalon*

One of the striking features and main distinguishing characteristics of the genus *Asotana* are the projections on the cephalon. These were pointed out by Leigh-Sharpe (1937) and by Thatcher (1988). In *Asotana magnifica,* there are flattened supraocular tubercles and pointed pre-ocular tubercles (Plate 2, B,C,D). Also, the anterior margin of the cephalon is splayed and forms three roughened bosses (Plate 2, C). The female of this species is always found over the tongue of the piranha with the cephalon near the esophagus (Plate 2, E). Thatcher (1988) suggested that the tubercles and bosses could serve to irritate the esophagus and cause the fish to regurgitate. The isopod may be feeding on 'vomitus.'

3.2 *Coxal plates on all seven pereionites*

Thatcher (1993) proposed the new genus and species, *Anphira branchialis,* from the branchial chambers of piranhas (Plate 1, E). The main distinguishing characteristic of the genus was the free first coxal plates. In all other known genera of Cymothoidae, the first coxal plate is fused to the first pereionite. In a subsequent paper, Thatcher (1995), described *Anphira xinguensis* from a different serrasalmid fish species and again affirmed the presence of a free first coxal plate. Our studies indicate that at least three additional species of *Anphira* exist in Amazonia but these have not yet been described because of a lack of sufficient specimens.

3.3 *Pleonites fused to the pleotelson or not (in penetrating cymothoids)*

Szidat (1948) described *Riggia paranensis,* a body cavity parasite of curimatid fishes, from Parana State, Southern Brazil. He distinguished his new genus from *Artystone trysibia* (Schiödte 1866), which also penetrates, on the fact that the pleonites are fused to the pleotelson in the former and not in the latter. This feature has been confirmed as can be seen by comparing *Riggia paranensis* (Plate 1, F) to *Artystone minima* Thatcher & Carvalho, 1988 (Plate 1, G).

3.4 *Foot-shaped mandibles and five spined maxillumes*

Thatcher (1997) pointed out that the mandibles of marine cymothoids normally are equipped with cutting plates, pointed incisors or both. The South American freshwater forms lack these structures and the mandible is frequently foot or slipper-shaped (Fig. 1,

Plate 1. Entire female specimens. A. *Braga nasuta* Schiödte & Meinert, 1881; dorsal; scale = 10 mm. B. *Braga cichlae* Schiödte & Meinert, 1881; dorsal; scale = 10 mm. C. *Braga amapaensis* Thatcher, 1996; dorsal; scale = 5 mm. D. *Braga amapaensis* Thatcher, 1996; lateral; scale = 5 mm. E. *Anphira branchialis* Thatcher, 1994; lateral; scale = 5 mm. F. *Riggia paranensis* Szidat, 1948; dorsal; scale = 5 mm. G. *Artystone minima* Thatcher & Carvalho, 1988; dorsal, of cleared specimen; scale = 2 mm. H. *Vanamea symmetrica* (Van Name, 1925, Thatcher, 1994); dorsal; scale = 5 mm.

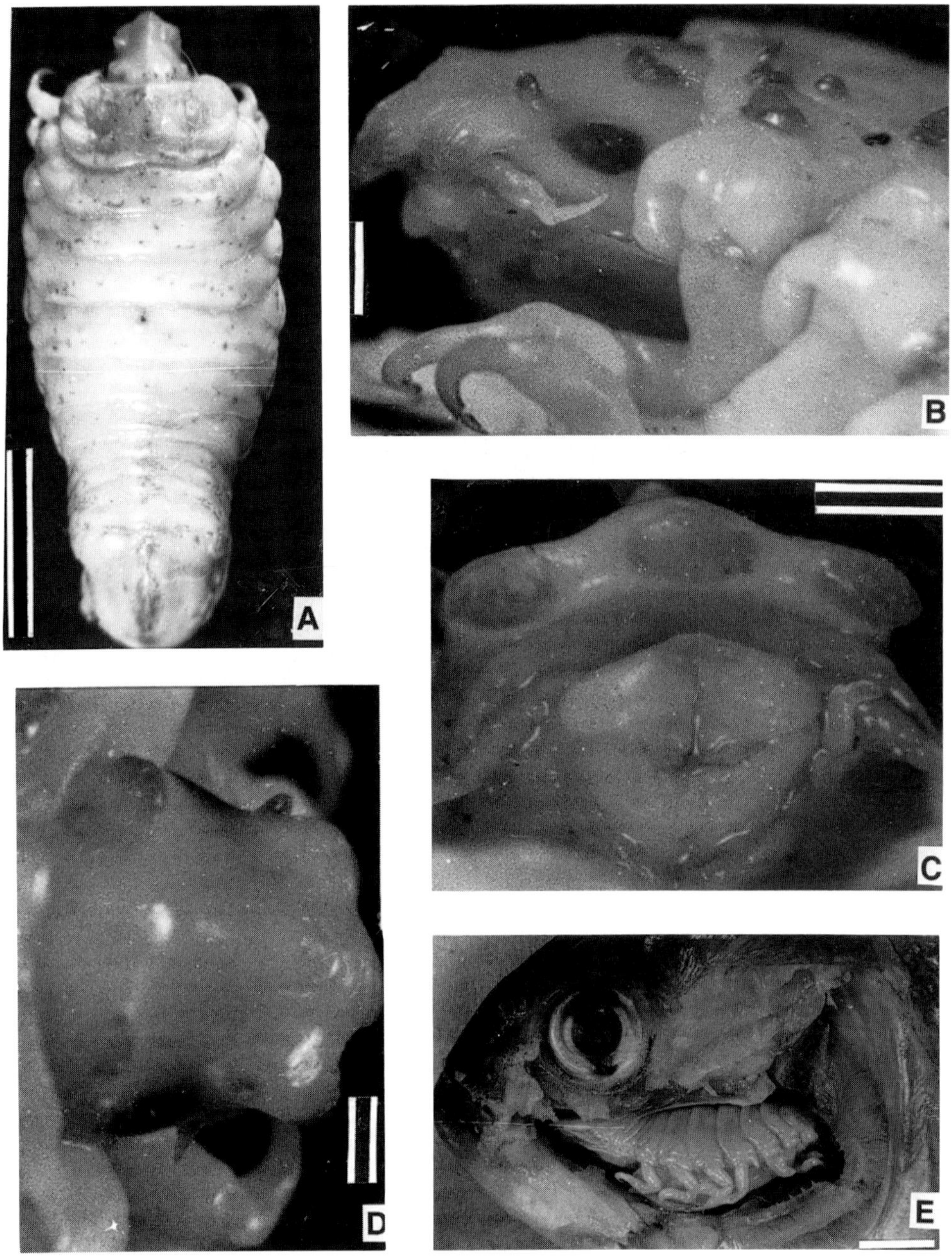

Plate 2. *Asotana magnifica* Thatcher, 1988. A. Entire specimen; dorsal; scale = 10 mm. B. Cephalon; lateral; scale = 1 mm. C. Cephalon; frontal; scale = 1 mm. D. Cephalon; dorsal; scale = 1 mm. E. Entire specimen on tongue of a piranha (operculum removed). Scale = 10 mm.

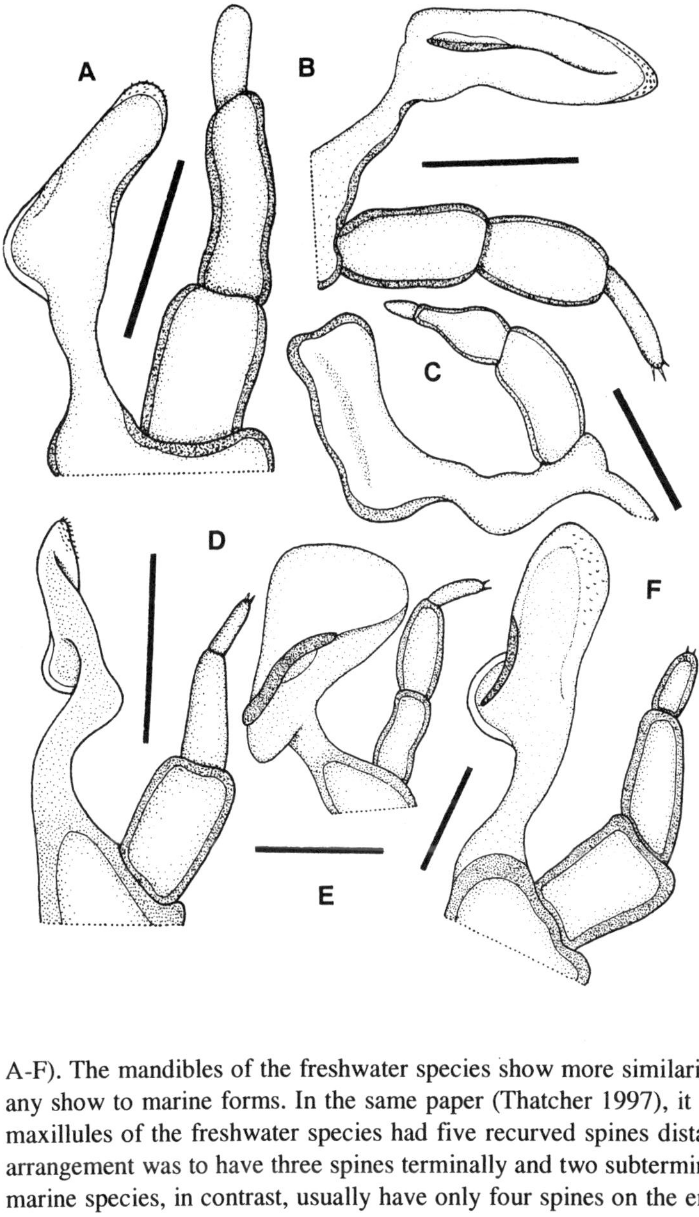

Figure 1. Mandibles and mandibular palps of Amazonian cymothoids. A. *Anphira xinguensis* Thatcher, 1995. B. *Vanamea symmetrica*. C. *Braga amapaensis*. D. *Riggia brasiliensis* Szidat & Schubart, 1959. E. *Asotana magnifica*. F. *Braga nasuta*. Scale for 18 = 1 mm; all others 0.5 mm.

A-F). The mandibles of the freshwater species show more similarity with each other than any show to marine forms. In the same paper (Thatcher 1997), it was mentioned that the maxillules of the freshwater species had five recurved spines distally. The most frequent arrangement was to have three spines terminally and two subterminally (Fig. 2, A-F). The marine species, in contrast, usually have only four spines on the end of the maxillule and all are terminal. On the basis of these two characteristics, it was suggested that these freshwater genera and species should constitute a separate subfamily to be called Artystonenae.

3.5 *Multi-laminate pleopods*

Thatcher (1995, 1996) described the multi-laminate pleopods found only in the genus *Braga* (Schiödte & Meinert 1881). In this genus as redefined by Thatcher (1996), pleopod 1 has two lamellae, 2 has four and 3-5 have three lamellae each. The lamellae of pleopod

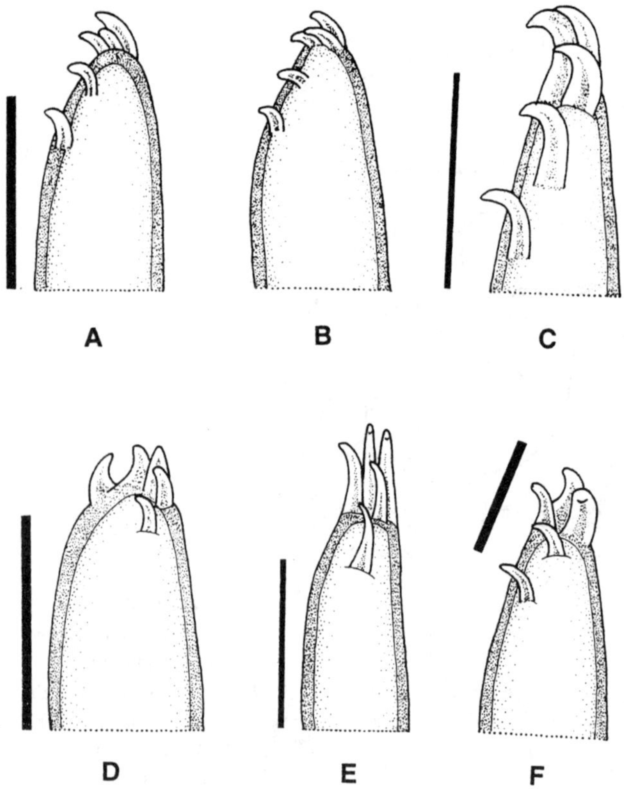

Figure 2. Maxillules of Amazonian cymothoids. A. *Braga cichlae.* B. *Braga nasuta.* C. *Vanamea symmetrica.* D. *Asotana magnifica.* E. *Artystone minima.* F. *Riggia brasiliensis.* Scale for 24 = 50 µm, all others 100 µm.

2 of several species are shown in Figure 3, A-F. The extra lamella (accessory gill) on the endopod is believed to have evolved from the appendix masculinum. At least two undescribed species of *Braga* occur in Amazonia (Fig. 3, A,B) and these will be described elsewhere when enough specimens become available.

4 CONCLUSIONS

Until recently, it had been difficult to define or distinguish the genera of South American freshwater Cymothoidae. Based on recent observations, we can now define at least some of these genera. For example: *Asotana* has head tubercles; *Anphira* is provided with free first coxal plates; *Riggia* has the pleonites fused to the pleotelson; and *Braga* has multilaminate pleopods. It is to be hoped that future studies including those with a scanning electron microscope will provide additional characters to define the genera and species of this group.

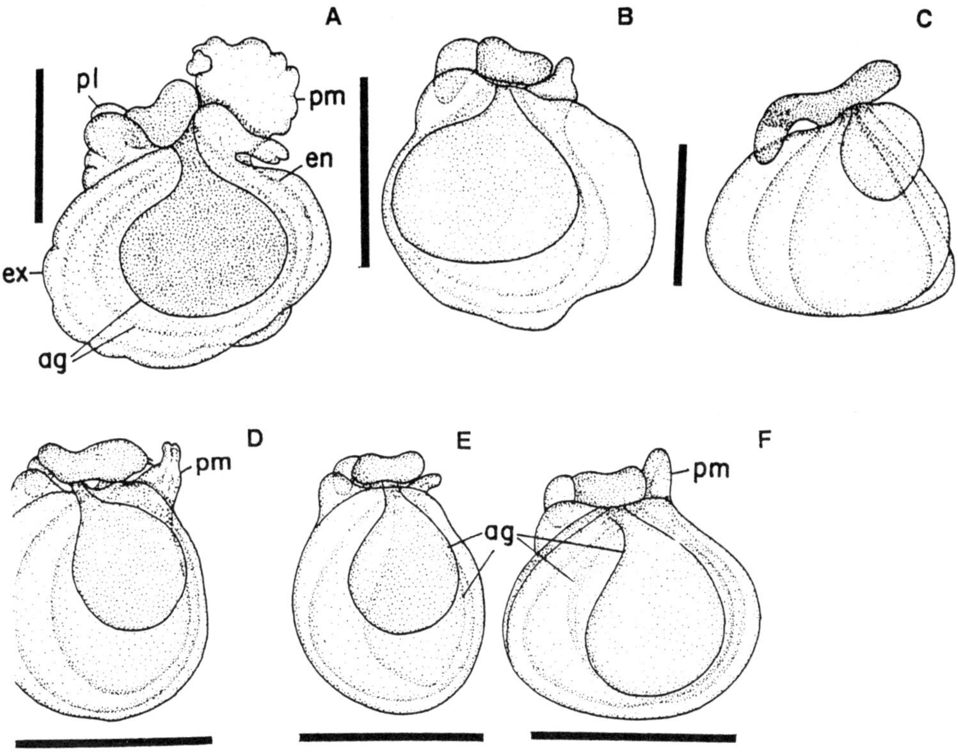

Figure 3. Second pleopods of Amazonian cymothoids of the genus *Braga*. A.. *Braga* sp. (undescribed). B. *Braga* sp. (undescribed). C. *Braga amapaensis*. D. *Braga* nasuta. E. *Braga patagonica* Schiödte & Meinert, 1884. F. *Braga cichlae*. Scale for 28 = 2 mm, all others = 5 mm. Abreviations used: ag = accessory gill lamella; en = endopod; ex = exopod; pl = lateral peduncular lobe; pm = proximomedial lobe of endopod.

REFERENCES

Bastos, P.B. & V.E. Thatcher 1997. A redescription of *Riggia paranensis* Szidat, 1948, (Isopoda, Cymothoidae) based on 32 specimens from curimatid fish of Rio de Janeiro, Brazil, with a redefinition of the genus. *Mem. Inst. Oswaldo Cruz* 92: 755-760.

Bowman, T.E. 1986. *Paracymothoa tholoceps*, a new freshwater parasitic isopod from Southern Venezuela (Flabellifera: Cymothoidae). *Proc. Biol. Soc. Wash*99: 753-756.

Leigh-Sharpe, W.H. 1937. *Badroulboudour splendida* n. g. et sp., a new parasitic isopod from Ecuador. *Parasitol.* 29: 391-394.

Lemos de Castro, A. 1959. Sobre as espécies sul-americanas do género *Braga* Schiödte & Meinert, 1881, (Isopoda, Cymothoidae). *Arq. Museu Nac., Rio de Janeiro, Brazil* 49: 69-92.

Lemos de Castro, A. & J. Pereira Machado Filho 1946. *Artystone trysibia* Schiödte, um crustáceo parasita de peixe d'agua doce do Brasil, com descrição do alótipo macho (Isopoda, Cymothoidae). *Rev. Brasil. Biol.* 6: 407-413.

Thatcher, V.E. 1988. *Asotana magnifica* n. sp. (Isopoda, Cymothoidae) an unusual parasite (commensal?) of the buccal cavities of piranhas (*Serrasalmus* spp.) from Roraima, Brazil. *Amazoniana* 10: 239-248.

Thatcher, V.E. 1991. Amazon Fish Parasites. *Amazoniana* 11: 263-571.

Thatcher, V.E. 1993a. *Vanamea* gen. nov. for *Livoneca symmetrica* Van Name, 1925, (Crustacea, Isopoda, Cymothoidae) and a redescription of the species based on specimens from Brazilian piranhas. *Acta Amazonica* 23: 287-296.

Thatcher, V.E. 1993b. *Anphira branchialis* gen. et sp. nov. (Crustacea, Isopoda, Cymothoidae) a gill cavity parasite of piranhas (*Serrasalmus* spp.) in the Brazilian Amazon. *Acta Amazonica* 23: 297-307

Thatcher, V.E. 1995a. *Anphira xinguensis* sp. nov. (Isopoda, Cymothoidae) a gill chamber parasite of an Amazonian serrasalmid fish, *Ossubtus xinguense* Jégu, 1992. *Amazoniana* 13: 293-303.

Thatcher, V.E. 1995b. Comparative pleopod morphology of eleven species of parasitic isopods from Brazilian fish. *Amazoniana* 13: 305-314.

Thatcher, V.E. 1996. *Braga amapaensis* n. sp. (Isopoda: Cymothoidae) a mouth cavity parasite of the Amazonian fish, *Acestrorhynchus guyanensis* Menzies with a redefinition of the genus *Braga*. *Amazoniana* 14: 121-129.

Thatcher, V.E. 1997. Mouthpart morphology of six freshwater species of Cymothoidae (Isopoda) from Amazonian fish compared to that of three marine forms, with the proposal of Artystonenae subfam. nov. *Amazoniana* 14: 311-322.

Thatcher, V.E. & M.L. Carvalho 1988. *Artystone minima* n. sp. (Isopoda, Cymothoidae) a body cavity parasite of the pencil fish (*Nannostomus beckfordi* Guenther) from the Brazilian Amazon. *Amazoniana* 10: 255-265.

Thun, M. & R.C. Brusca 1980. On the status of the eastern Pacific cymothoid fish parasite *Braga occidentalis* (Boone) and its synonymy with *B. patagonica* Schioedte & Meinert (Crustacea: Isopoda: Cymothoidae). *Bull. So. Calif. Acad. Sci.* 79: 130-132.

Trilles, J.P. 1973. Notes documentaires sur les Isopodes Cymothoadiens parasites de poissons d'eau douce de l'Amérique du Sud. *Bull. Mus. Natl. Hist. Nat.* 88: 239-272.

Microscopic anatomy of the integument and digestive system during the molt cycle in *Ligia italica* (Oniscidea)

JASNA ŠTRUS
Department of Biology, University of Ljubljana, Ljubljana, Slovenia

ANDREJ BLEJEC
National Institute of Biology, Ljubljana, Slovenia

ABSTRACT

The molt cycle of *Ligia italica* was divided into four stages with altogether eight substages. The average duration of the molt cycle was determined in laboratory animals. Males and non-breeding females molt every 15 days, breeding females molt every 34 days. Molt-related changes in the ultrastructure of the anterior sternal integument and digestive system were investigated with scanning and transmission electron microscopy. Calcium storage and translocation in the integument and the midgut glands during the molt cycle were examined with cytochemical methods. The total calcium concentrations in the midgut glands and haemolymph were determined. Calcium is resorbed from the old cuticle prior to ecdysis and stored as calcium granules in the ecdysal space. In the intramolt animal the resorbed calcium is transported, via the haemolymph, to the posterior part of the integument for the mineralization of the new cuticle. The epithelial cells of the digestive tract with high glycogen content and variously shaped apical parts direct the differentiation of the cuticular structures in the foregut and the hindgut during the premolt stage. The ultrastructure of the midgut and midgut glands also changed during the molt cycle. Changes in the lipid content and distribution were observed in the large cells of the midgut glands throughout the molt cycle. Numerous calcium granules in the small cells of the hepatopancreas of premolt animals may be the source of calcium employed for the mineralization of the cuticle of the digestive tract.

1 INTRODUCTION

Isopods are one of the few crustacean groups which have successfully invaded terrestrial habitats. They have even conquered the most extreme biotopes such as caves and deserts. They are well adapted to a terrestrial lifestyle due to several anatomical and physiological features related to respiration, excretion, sensory perception, reproduction and growth. Growth is intimately related to molting in most crustaceans. Some of them molt only several times during their life span while others molt frequently. The advantage of frequent molting in isopods was discussed by several authors and has been mainly ascribed to the regeneration of sensory structures (Price & Holdich 1980a) and to reproductive strategies (Steel 1980).

The purpose of this paper is to review the isopod molt cycle and to describe the functional morphology of the integument and digestive system during its occurrence in the amphibious isopod *Ligia italica*.

1.1 *Previous investigations on molt cycles in isopods*

The first detailed description of the decapod molt cycle, based on the hardness of cuticle (Drach & Tchernigovtzeff 1967) is still used as the basis for dividing the crustacean cycle into four stages. The biphasic molt cycle of isopods was first described by Stevenson (1961). The presence of white deposits on the first four anterior sternites in isopods during premolt was described by several authors (Tait 1917, Verhoeff 1940, Wieser 1964). The first detailed description of the molt cycle in *Porcellio scaber* and *Oniscus asellus* was based on the morphological and histological characteristics of the integument (Messner 1965). The molt cycle of *Oniscus asellus* was divided into fifteen stages based on the histology of the integument of the first four anterior sternites (Steel 1982). Recently our laboratory proposed a simple and rapid procedure for molt stage determination in *Porcellio scaber* for use in routine experiments (Zidar et al. 1998).

1.2 *Previous investigations on the integument during the molt cycle in isopods*

Isopods permanently synthesize the new cuticle of the integument and digestive tract due to the frequent molting. The functional morphology of the crustacean cuticle during the molt cycle and the hormonal control of molting were extensively studied in decapods (Roer & Dillaman 1993, Compère 1995, Chang 1995). The functional morphology of the isopod cuticle during molting was described by Price & Holdich (1980b). A detailed ultrastructural study of the sternal integument of *Porcellio scaber* during the molt cycle was performed by Ziegler (1997). Literature on the cuticle structure of the digestive tract and the ultrastructure of the digestive glands during the isopod molt cycle is very scarce (Storch & Štrus 1989).

2 MATERIALS AND METHODS

2.1 *Experimental animals*

Specimens of *L. italica* were collected from a single site on the seashore of Piran Bay in the Gulf of Trieste. Individuals of both sexes with a body length of 7 to 10 mm were kept singly in Petri-dishes with moistened filter paper. They were maintained at ambient room temperature (20°C) at a natural photoperiod from April to August and fed standardized fish food. They were inspected under the steromicroscope three times per week and the stage of the molt cycle was determined for each individual. Exuvia and food remnants were removed by changing the filter paper once weekly.

2.2 *Ultrastructural observations*

Tissues for ultrastructural analysis were dissected from narcotized and decapitated animals. They were fixed in 3.5% glutaraldehyde in 0.1M cacodylate buffer for two hours at 4°C. After rinsing in buffer the tissues were postfixed in osmium ferrocianide (Karnowsky

1971). After dehydration in a series of graded ethanols the tissues were either critical point dried or contrasted en bloc in 1% uranyl acetate, and embedded in Spurr (Spurr 1969). Dried samples were mounted on stubs, sputter-coated with gold and examined with a Jeol 840A scanning electron microscope. Ultrathin sections were cut, stained with 2.66% lead citrate, and viewed with a Zeiss EM 9S-2 and a Jeol 1200EX transmission electron microscopes.

2.3 *Cytochemical location of calcium*

Cytochemical detection of calcium with lead acetate was performed according to the method of Carasso & Favard (1966). The distribution of calcium ions in the cuticle was demonstrated by the K-pyroantimonate method (Wick & Hepler 1982) and confirmed with X-ray microanalysis and elemental mapping. The calcium content of haemolymph and the midgut glands was determined spectrophotometrically.

3 RESULTS

3.1 *Duration of the molt cycle in* L. italica

The molt cycle of *L. italica* was divided into four stages: premolt, molt, postmolt, and intermolt. The stages were further subdivided into eight substages based on the morphology of the anterior sternal integument (Table 1, Pl. 1A). The first clearly visible sign of the premolt is the appearance of calcium deposits on the first four sternites (Pl. 1B). Three groups of thirty animals of both sexes each were observed during the five months. The duration of the molt cycle and the number of exuviations were determined for each experimental animal and the average duration of the molt cycle was calculated for males, non-breeding females and breeding females (Table 2). Males and non-breeding females molt every 15 days while breeding females molt every 34 days.

Table 1. The description and duration of substages of the molt cycle in males and non-breeding females of *L. italica* kept under constant laboratory conditions.

Stage	Substage	Morphological character	Duration (days)
Premolt, pro ecdysis (PE)	1	Accumulation of fat bodies in the center of the first four sternites	Up to 2 days
	2	Accumulation of calcium deposits which cover the fat bodies	2 to 3 days
	3	Calcium deposits cover the whole sternal part, animals stop eating	Up to 1 day
Molt, ecdysis (E)	4	Exuviation of the posterior exoskeleton	Several hours
	5	Resorption of calcium deposits (intramolt)	Up to 1 day
	6	Exuviation of the anterior exoskeleton	Several hours
Postmolt, met ecdysis (ME)	7	The new exoskeleton is soft and transparent, animals start to eat	1 to 2 days
Intermolt, di ecdysis (DE)	8	The exoskeleton is hard and pigmented	Up to 8 days

Table 2. The duration of the molt cycle in *L. italica*.

Experimental group	Month	Mean (days)	Standard error (days)	Number of measurements
Males	April	15.75	0.91	16
	May	15.47	0.80	17
	June	16.20	1.05	10
Females	April	14.00	0.33	11
	May	14.86	0.83	21
	June	15.80	0.79	15
Breeding females	April	31.83	1.01	6
	May	37.78	0.92	9
	June	33.00	0.82	8

3.2 *The ultrastructure of the anterior sternal integument during the molt cycle*

The sternal integument of intermolt isopods consists of flat to cuboidal epithelium covered by a three-layered cuticle with more than twenty sublayers. The thin superficial epicuticle is not mineralized, the exocuticle with helically-arranged chitin microfibrils is impregnated with calcium salts, and the lamellated endocuticle is calcified only in its distal part (Pl. 1C, D). In premolt animals, the old cuticle separates from the underlying epithelium (apolysis); the epithelial cells contain large amounts of glycogen and secrete a new cuticle (Pl. 1E, F, G). Calcium is resorbed prior to exuviation and accumulates in the ecdysal space of the first four anterior sternites as calcium granules (Pl. 2A). Round crystalline structures with a diameter of 10 μm, surrounded by numerous small spherules with a diameter of less than 1 μm, were observed in the ecdysal space of sternites viewed with the scanning electron microscope (Pl. 2B, C). Ultrastructural analysis and the cytochemical reactions with lead acetate and uranyl acetate demonstrated the presence of calcium ions in the granules (Pl. 2D, E). In intramolt animals, the large crystalline structures were not observed and numerous small spherules were associated with the inner lamellae of the old cuticle (Pl. 2F).

3.3 *The ultrastructure of the digestive tract during the molt cycle*

The digestive tract of most terrestrial isopods is entirely covered with the cuticle, with the exception of a short endodermal midgut which is present in amphibious species *Ligia italica* and *Ligidium hypnorum* (Štrus et al. 1995). The cuticle of the foregut is cast with the anterior exoskeleton while the cuticle of the hindgut is partly digested and shed with the posterior exuvium during the molt cycle of *L. italica*. The epithelial cells of the filtering region in the stomach of the premolt *L. italica* are Y-shaped and form the template for the secretion of cuticular bars of the primary filter (Pl. 3A, B). The epithelial cells of the foregut atrium secrete the cuticle and are filled with glycogen or have an electron dense cytoplasm with numerous vacuoles (Pl. 3C). The cuticle of the hindgut is very complex in the anteriormost region at the junction with the foregut (Pl. 3D). The cuticular surface of the circular fold which prevents the regurgitation of food is arranged in shingle-like plates (Pl. 3E). The cuticle of the anterior hindgut is thin in postmolt animals and covers the surface of epithelial cells that have deep and dense apical invaginations and large vacuoles

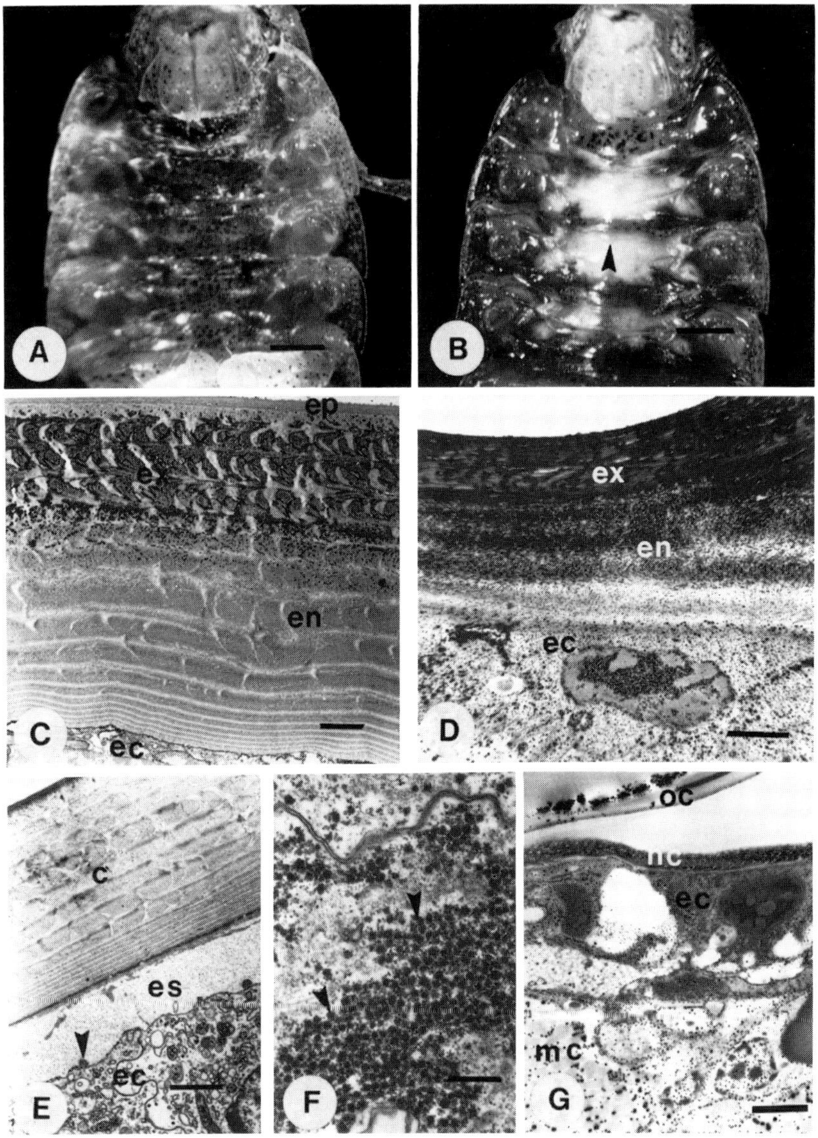

Plate 1. A) *L. italica*, the anterior sternal part of the integument in an intermolt animal. Scale bar: 1mm. B) *L. italica*, the anterior sternal part of the integument with calcium deposits (arrowhead) in early premolt animal. Scale bar: 1mm. C) Ultrastructure of the cuticle in intermolt animal, ep-epicuticle, ex-exocuticle, en-endocuticle, ec-epithelial cell; lead acetate incubation. Scale bar: 2μm. D) Ultrastructure of the integument in intermolt animal with antimony precipitates at the sites of calcium distribution, ex-exocuticle, en-endocuticle, ec-epithelial cell; K-pyroantimonate-containing fixative. Scale bar: 2μm. E) Ultrastructure of the integument in premolt animal, uranyl acetate en bloc staining, c-cuticle, es-ecdysal space, ec-epithelial cell secretes material for a new cuticle (arrowhead). Scale bar: 2μm. F) Glycogen deposits (arrowheads) in the epithelial cells of premolt animals. Scale bar: 0.5μm. G) The formation of a new cuticle in premolt animal, oc-old cuticle, nc new cuticle, ec-epithelial cell, mc-muscle cell; K-pyroantimonate-containing fixative. Scale bar: 5μm.

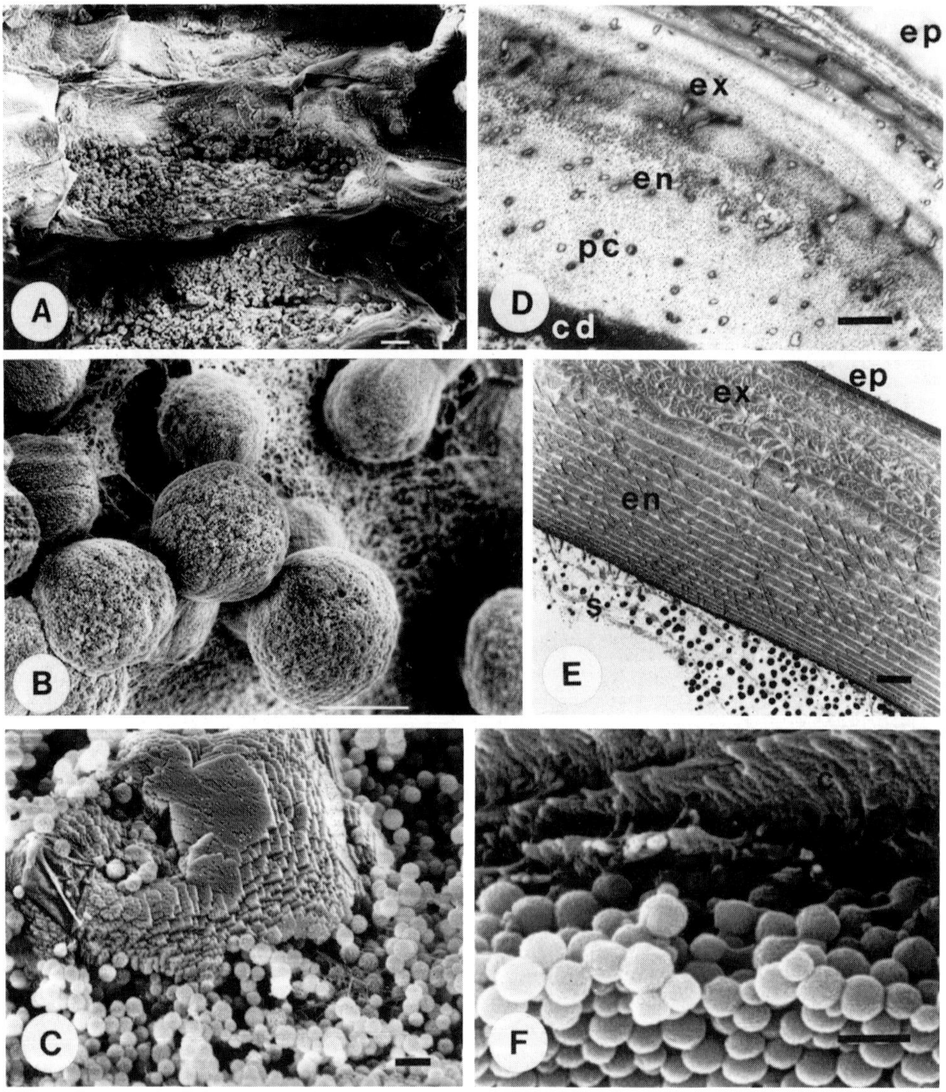

Plate 2. A) Scanning electron micrograph of the inner surface of the sternal cuticle in a premolt animal. Scale bar: 100µm. B) Crystalline structures on the cuticle surface. Scale bar: 10µm. C) A crystalline structure surrounded by small spherules. Scale bar: 1µm. D) Ultrastructure of the cuticle in intramolt animal with calcium deposits at the base, ep-epicuticle, ex-exocuticle, en-endocuticle, cd-calcium deposits, pc-pore canals; lead acetate incubation. Scale bar: 1µm. E) Ultrastructure of the old cuticle in intramolt animal with calcium granules in the ecdysal space, ep-epicuticle, ex-exocuticle, en-endocuticle, s-spherules; lead acetate incubation. Scale bar: 1µm. F) Scanning electron micrograph of the inner surface of sternal cuticle with spherules in intramolt animal, c-cuticle. Scale bar: 1µm.

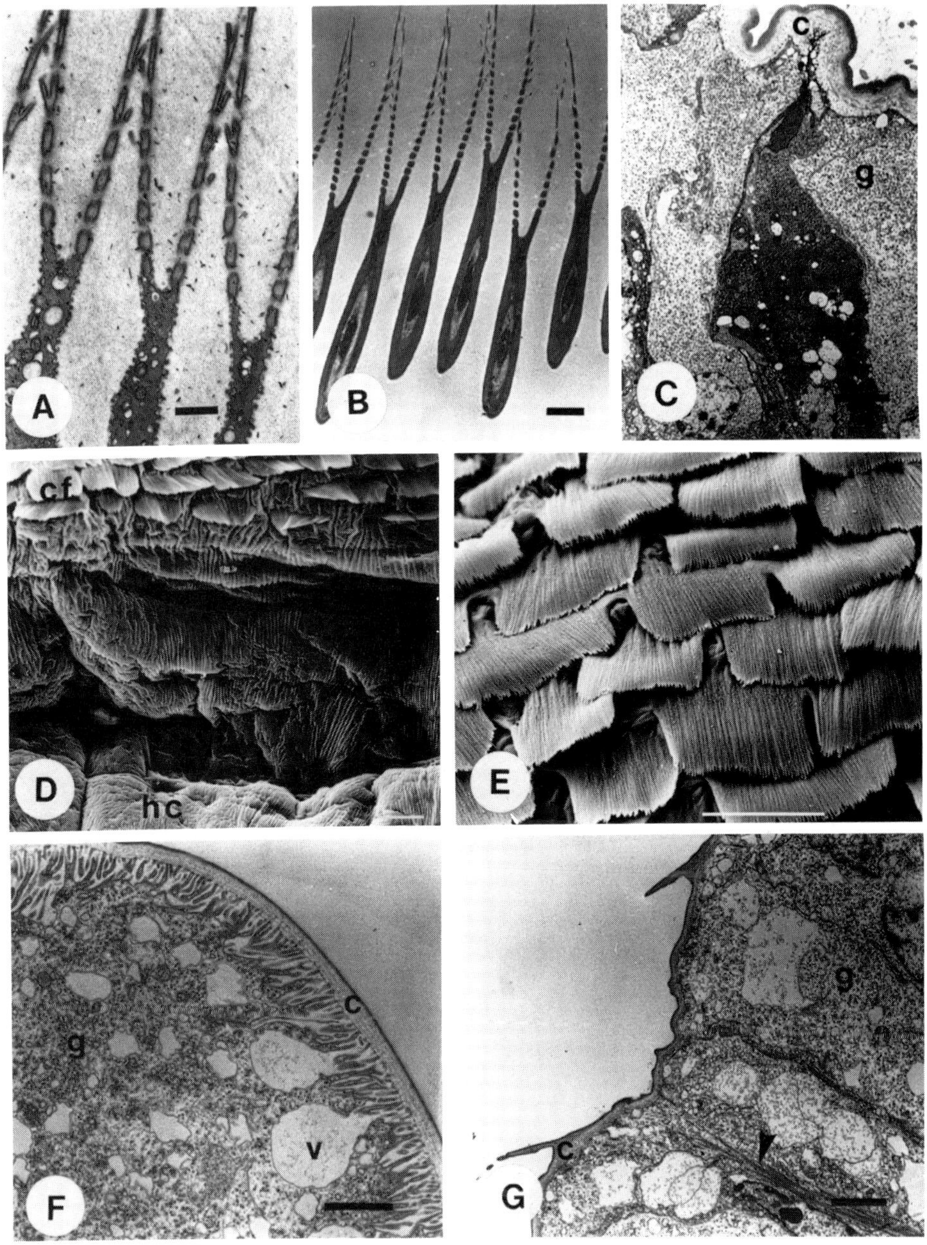

Plate 3. A) Y-shaped apical parts of the epithelial cells of the primary filter in the foregut of premolt animals. Scale bar: 1μm. B) Y-shaped cuticular bars of the primary filter in the foregut of intermolt animals. Scale bar: 2μm. C) Epithelial cells of the atrium in the foregut of postmolt animals, c-cuticle, g-glycogen. Scale bar: 5μm. D) The cuticle of the anterior part of the hindgut in intermolt animals, cf-circular fold, hc-hindgut cuticle. Scale bar: 10μm. E) The shingle-like cuticular plates of the circular fold in the anterior hindgut. Scale bar: 10μm. F) The epithelial cell of the anterior hindgut in a postmolt animal, c-cuticle, v-vacuole, g-glycogen. Scale bar: 2μm. G) The epithelial cells of the circular fold secrete the new cuticle in postmolt animals, c-cuticle, g-glycogen, filaments (arrowhead). Scale bar: 5μm.

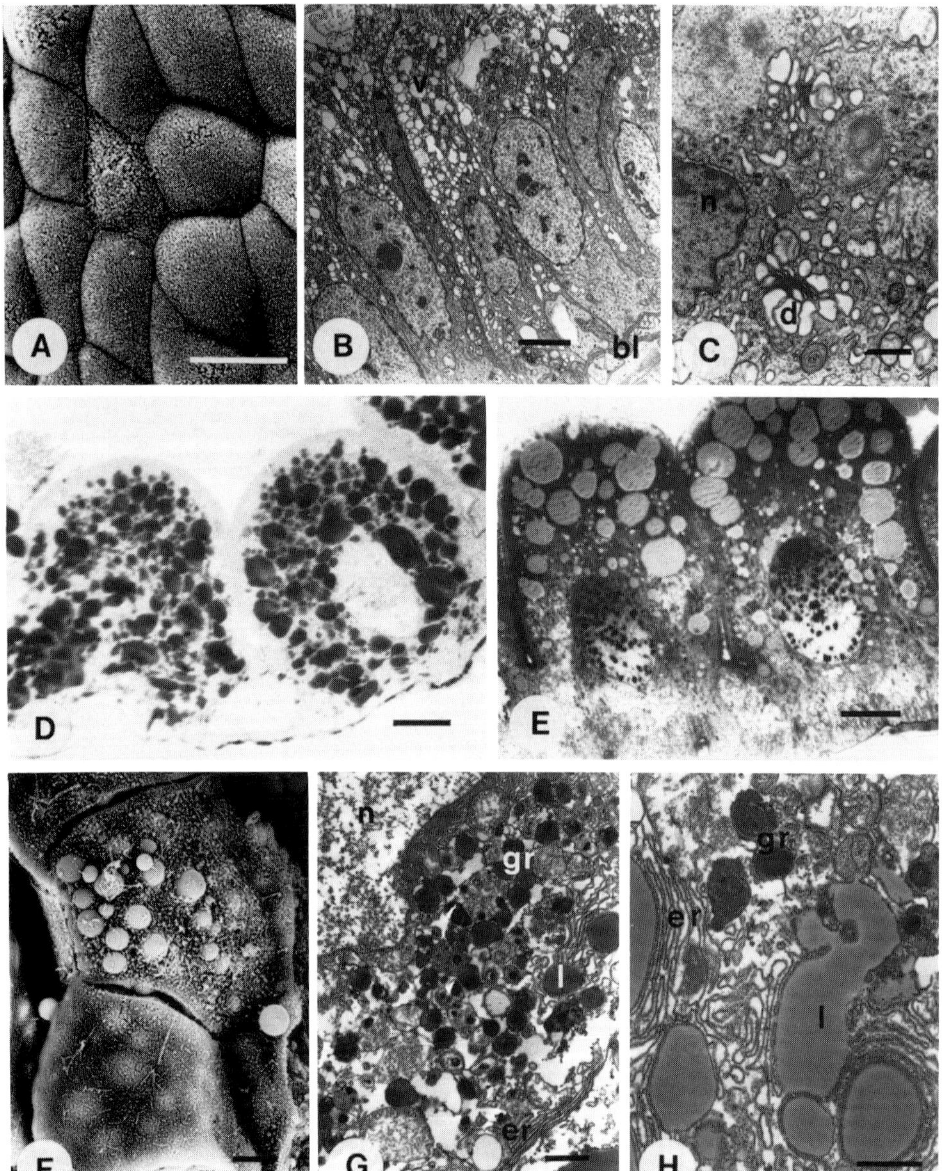

Plate 4. A) The cells of the midgut with a prominent microvillar surface. Scale bar: 10μm. B) The epithelial cells of the midgut in postmolt animals, bl-basal lamina, v-vacuoles. Scale bar: 5μm. C) The ultrastructure of the midgut cell in postmolt animal, d-dyctiosome, n-nucleus. Scale bar: 1μm. D) Large cells of the midgut glands are filled with lipid droplets in premolt animals. Scale bar: 5μm. E) Large cells of the midgut glands in postmolt animals. Scale bar: 5μm. F) Lipid droplets at the luminal surface of the midgut glands in postmolt animals. Scale bar: 10μm. G, H) The ultrastructure of the small midgut gland cell in the premolt animal, gr-granules, l-lipids, er-endoplasmic reticulum. Scale bars: 1μm.

(Pl. 3F). The epithelial cells of the circular fold contain large vacuoles and bundles of filaments in postmolt animals (Pl. 3G).

3.4 *The ultrastructure of the midgut and midgut glands during the molt cycle*

The midgut in *L. italica* is a short endodermal section of the digestive tract between the foregut and hindgut. The cells with a prominent microvillar border contain lipids, glycogen, and mitochondria and are similar to the large cells of the midgut glands in intermolt and premolt animals (Pl. 4A). In postmolt animals, the cells are filled with vacuoles, and contain numerous dictyosomes and secretory granules (Pl. 4B, C). The large and the small cells of the midgut glands differ in size and ultrastructure during the molt cycle. In premolt animals, the large cells with dark cytoplasm contain small lipid droplets that are distributed homogeneously throughout the cell (Pl. 4D). In the large cells of postmolt animals, the lipid droplets fuse into large vacuoles which accumulate apically and are released into the lumen of the gland (Pl. 4E, F). The small cells of premolt animals contain lipid droplets surrounded by parallel cisternae of the smooth endoplasmic reticulum and numerous electron dense granules that contain calcium (Pl. 4G, H). The amount of calcium granules in the small cells of postmolt and intermolt animals is much lower.

3.5 *Calcium content of the haemolymph and midgut glands during the molt cycle*

Haemolymph and midgut gland calcium concentrations were determined in each of ten premolt, intramolt, and intermolt animals (Fig. 1). The mean concentration (standard deviation) of haemolymph calcium was 13.6 ± 2.11 mmol/l in intermolt animals, 16.14 ± 2.78 mmol/l in premolt animals and 21.24 ± 5.71 mmol/l in intramolt animals. The mean calcium concentration in midgut glands was very similar in intermolt and premolt animals, 49.60 ± 31.78 mmol/kg and 42.53 ± 27.67 mmol/kg respectively, and much lower in intramolt animals, 23.38 ± 6.45 mmol/kg.

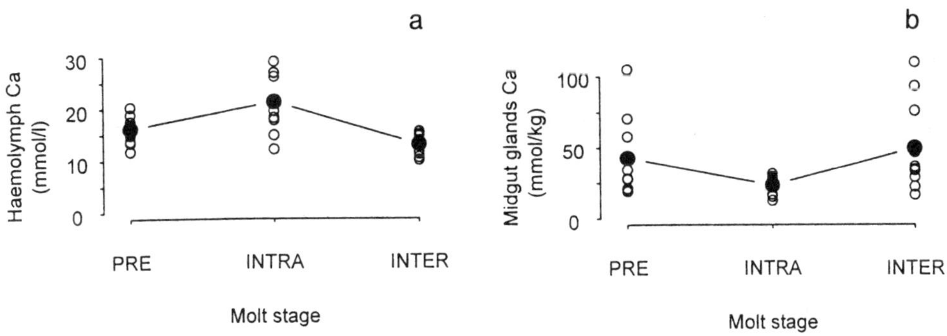

Figure 1. Haemolymph (a) and midgut gland (b) calcium concentrations during premolt (PRE, N = 10), intramolt (INTRA, N = 10) and intermolt (INTER, N = 10) in *L. italica* (o individual data, • mean value).

4 DISCUSSION

Compared to the other terrestrial isopods the molt cycle of the non-breeding adults of the amphibious species *L. italica* is much shorter under constant laboratory conditions. The literature on the molt cycle of the terrestrial species *Porcellio scaber* and *Oniscus asellus* show that the non-breeding adults of these two species molt once monthly (Steel 1982, Zidar et al. 1998). The duration of the molt cycle of the marine isopod *Idothea baltica* is thirty days (Maissiat & Graf 1973) and of the amphibious species *Sphaeroma serratum* forty days (Tchernigovtzeff & Ragage-Willigens 1968). Since the molt cycle influences the structure and function of the organism, its behavior and its communication with the environment, it is important to recognize the effects of frequent molting on isopod anatomy and physiology.

The sternal calcium deposits are the most prominent sign of the premolt stage in all isopods. Their size and shape vary even between individuals of the same species. Storage and translocation of calcium in the integument of *Oniscus asellus* during the molt cycle was described by Steel (1993). During the bi-phasic molt, calcium resorbed from the posterior integument is stored in the anterior sternal deposits during the premolt and used for the calcification of the new posterior cuticle after posterior ecdysis. During the intramolt the resorbed calcium is transported to the haemolymph and employed for the calcification of the anterior cuticle. Some of the calcium is resorbed from the cast exuvia that are consumed by the animals. This was also confirmed in *Porcellio scaber* where calcium carbonate deposits appear before the secretion of the new cuticle in the anterior sternal integument and the deposition of the anterior exocuticle begins before the posterior molt. During the intramolt, deposition of the anterior cuticle is retarded (Ziegler 1997).

Endogenous calcium sources play an important role in the mineralization of the new cuticle in terrestrial isopods (Graf 1978). The ultrastructural features of the integument in *L. italica* in different stages of the molt cycle are very similar to those described for the terrestrial species. The ultrastructure of the anterior sternal integument during this molt cycle was described by Štrus & Compère (1996). In *L. italica,* calcium is resorbed from the anterior and posterior cuticles during the premolt and stored in the ecdysal space as large calcium carbonate granules. Smaller spherules associated with the granules resemble in size those described for *L. italica* and *L. oceanica* by Ziegler & Miller (1997). The large granules are resorbed during the intramolt and calcium is transported through the epithelium to the haemolymph and after posterior ecdysis to the sites of mineralization of the new cuticle. This is also confirmed by an increased haemolymph calcium content in intramolt specimens of *L. italica*. The rise in the haemolymph calcium concentration was also observed in premolt decapods (Greenaway 1985). The postmolt specimens of *L. italica* do not consume cast exuvia, so calcium that was not resorbed from the cuticle is lost for the animal. The shortage of calcium in the environment is not a limiting factor for this amphibious species, which may use exogenous calcium sources for the mineralization of the new cuticle. During the molt cycle the cuticle of the digestive tract is cast with the exoskeleton.

The vacuolated epithelial cells of the digestive tract contain large amounts of glycogen in premolt animals, a source of glucose for the synthesis of chitin microfibrils. The variously shaped apical parts of the cells form the template for the secretion of the complex cuticle of the filters in the stomach and of the circular fold in the anterior hindgut. In the postmolt animals the new cuticle is soft and thin. The cells with a dark cytoplasm contain calcium ions that may be used for the mineralization of the cuticle of the digestive tract. In decapods, the calcium is stored in the stomach in the form of gastroliths (Stevenson 1985). No gastrolith-like structures were observed in the foregut of the premolt specimens of *L. italica*. It is very probable that the calcium granules stored in the midgut glands serve as an endogenous calcium source for mineralization of the new cuticle. The isopod digestive tract plays an important role in the mechanical degradation of food, sorting of food particles, and their transport. Frequent molting enables the regeneration of cuticular structures in the foregut and the renewal of the hindgut cuticle, which is colonized with different microorganisms (Drobne et al. 1996). The role of isopod gut microflora in degradation of leaf litter and cellulose digestion was discussed by several authors (Drobne 1995, Zimmer & Topp 1998).

The functional morphology of the digestive system in the intermolt *L. italica* was described by Štrus et al. (1995). The main role of the endodermal midgut and hepatopancreas in the intermolt animals is the secretion of the digestive enzymes and food absorption. The ultrastructure of the midgut cells indicates their role in the synthesis of the components of the cuticle in the premolt and in the synthesis and secretion of digestive enzymes in the postmolt and intermolt. The accumulation of lipids is a prominent feature of the large cells of the midgut glands. Lipids are an important source of energy in crustaceans, and several authors (Giese 1966, Donadey 1983) have described their role in the oxidative metabolism of these animals. The highest amount of lipid droplets was observed in the midgut glands of *L. italica* during the premolt, their amount was reduced in the postmolt and increased again during the intermolt. The size and distribution of the lipid droplets changed during the molt cycle. Lautier & Lagarrigue (1986) and Chang & O'Connor (1983) demonstrate that in decapod crustacean phospholipids and unsaturated fatty acids accumulate in the midgut glands of the premolt animals. Neutral lipids obtained from food are converted to phospholipids in the midgut glands of the animals. Lipids are an important energy source for the isopods during the premolt when they stop feeding. This is also supported by the fact that only small amounts of glycogen were detected in the midgut glands of the premolt animals.

Small cells of the midgut glands are mainly involved in the excretion and storage of waste products. They accumulate metals and store them as granules of different size and composition (Prosi et al. 1983). Cytochemical analysis demonstrated the presence of calcium granules in the small cells of the midgut glands in the premolt specimens of *L. italica*. Their amount was reduced in the intramolt and postmolt animals. This was supported by the results of spectro-photometrical analysis which demonstrated that calcium concentrations decreased in the midgut glands of the intramolt animals. Accumulation of calcium granules in the midgut glands of premolt animals and their depletion in postmolt animals suggested that they are the storage sites for the endogenous calcium employed in the mineralization of the cuticle of the digestive tract. The activity of the enzyme alkaline phosphatase that was observed at the base of the small gland cells and cells of the hindgut in premolt animals demonstrated that there was a transepithelial flow of ions between the two compartments. High haemolymph calcium content in the intramolt animals compared to the premolt and intermolt animals showed that calcium was transported via the haemolymph to the sites where it was used for cuticle mineralization. In isopods, calcium is not stored in the haemolymph, as was reported for amphibious or freshwater decapods (Greenaway & Farrelly 1991).

CONCLUSIONS

The molt cycle affects the anatomy and physiology of isopods. Frequent molting in the amphibious isopod *L. italica* enables the regeneration of the cuticular structures of the integument and digestive tract. In the breeding females, it is synchronized with the reproductive cycle. The significance of frequent molting for the excretion of waste products and its possible role in detoxification remains to be investigated.

The main ultrastructural and biochemical changes of the integument, digestive system and haemolymph have been described in different molt stages of *L. italica*. Changes in glycogen and lipid content of the integument and digestive system as well as calcium storage and translocation were investigated during the molt cycle. The most prominent changes were observed in the premolt when the animal prepares for ecdysis. Based on the ultrastructural observations only, it is impossible to define the onset of the premolt stage. In future, it would be useful to consider the hormonal control of molting in isopods and to determine the concentration of ecdysteroids in the haemolymph of molting isopods.

Consideration of molt-related changes in the structure and biochemistry of different organs is important in any study of isopod physiology, particularly in studies where the feeding activity of the animal is measured, as in toxicity testing and ecotoxicology.

ACKNOWLEDGEMENTS

The authors would like to thank Dr Kazimir Draölar from Department of Biology, University of Ljubljana for processing and viewing the samples with scanning electron microscope and Ms R. Mummert from Zoologisches Institut, University of Heidelberg for technical assistance with transmission electron microscope. The work was financially supported by the Slovenian Ministry of Science and Technology, project No. J1-7392-0487-96.

5 REFERENCES

Carasso, N. & P. Favard 1966. Mise en évidence du calcium dans les myonemes pédonuclaires de ciliés péritriches. *J. Microscopy* 5: 759-770.

Chang, E.S. & J.D.O'Connor 1983. Metabolism and transport of carbohydrates and lipids. In: D.E.Bliss (ed.), *The Biology of Crustacea* 5: 263-287. New York: Academic Press.

Chang, E.S. 1995. Physiological and biochemical changes during the molt cycle in decapod crustaceans: an overview. *J. Exp. Mar. Biol. Ecol.* 193, 1-14.

Compère, P. 1995. Fine structure and morphogenesis of the sclerite epicuticle in the Atlantic shore crab *Carcinus maenas*. *Tissue Cell*: 27:525-538.

Donadey, C. 1973. Contribution a l'étude ultrastructurale et histophysiologique des caecums digestif des crustacés isopodes. These Doc. d'Etat Sci. Nat. C.N.R.S. 7662: 1-157.

Drach, P. & C.Tchernigovtzeff 1967. Sur la méthode de determination des stades d'intermue et son application générale aux crustacés. *Vie Milieu (A, Biol. mar.)* 18: 595-607.

Drobne, D. 1995. Bacteria adherent to the hindgut of terrestrial isopods. *Acta Microbiol. Immunol. Hungarica* 42: 45-52.

Drobne, D., P. Zidar, J. Štrus, M. Rupnik & M. Janc 1996. Scanning Electron Microscopy of Microorganisms in Isopod Gut. *J. Comp. Assist. Microscopy* 8: 271-272.

Giese, A.C. 1966. Lipids in the economy of marine invertebrates. *Physiol. Rev.* 46: 244-298.

Graf, F. 1978. Les sources des calcium pour les Crustacés venant de muer. *Arch Zool. exp. Gen.* 119: 113-161.

Greenaway, P. 1985. Calcium balance and moulting in the Crustacea. *Biol. Rev.* 60: 425-454.

Greenaway, P. & C. Farelly 1991. Trans-epidermal Transport and Storage of calcium in *Holthuisana transversa* (Brachyura: Sundathelphusidae) during premoult. *Acta Zool.* 72: 29-40.

Karnovsky, M.J. 1971. Use of ferrocyanide reduced osmium tetroxide in electron microscopy. *J. Cell Biol.* 6: 201-208.

Lautier, L. & J.G. Lagarrigue 1986. Lipid metabolism of the male crab *Pachygrapsus marmoratus* during the moulting cycle. *Biochem. Syst. Ecol.* 14: 675-680.

Maissiat, J. & F.J. Graf 1973. Action de l'ecdysterone sur l'apolysis et l'ecdysis de divers Crustacés Isopodes. *J. Insect Physiol.* 19: 1265-1276.

Messner, B. 1965. Ein morphologisch-histologischer Beitrag zur Häutung von *Porcellio scaber.* Latr. und *Oniscus asellus* L. (Isopoda, Crustacea). *Crustaceana* 9: 285-301.

Price, J.B. & D.M. Holdich 1980a. The Formation of the Epicuticle and Associated Structures in *Oniscus asellus* (Crustacea, Isopoda). *Zoomorph.* 94: 321-332.

Price, J.B. & D.M. Holdich 1980b. An ultrastructural study of the integument during the moult cycle of the woodlouse, *Oniscus asellus* (Crustacea, Isopoda). *Zoomorph.* 95: 250-263.

Prosi, F., V.Storch & H.H. Janssen. 1983. Small cells in the midgut glands of terrestrial Isopoda: Sites of heavy metal accumulation. *Zoomorph.* 102: 53-64.

Roer, R. & R. Dillaman 1993. In Horst, M.N., & Freeman, J.A. (eds), *The Crustacean Integument. Morphology and Biochemistry*, pp.79-111, CRC Press, Boca Raton, FL.

Spurr, A.R. 1969. A low viscosity embedding medium for electron microscopy. *J. Ultrastruct. Res.* 26: 31-43.

Steel, C.G. H. 1980. Mechanisms of coordination between moulting and reproduction in terrestrial isopod Crustacea. *Biol.Bull.* 159: 206-218.

Steel, C.G.H. 1982. Stages of the intermoult cycle in the terrestrial isopod *Oniscus asellus* and their relation to biphasic cuticle secretion. *Can. J. Zool.* 60: 429-437.

Steel, C.G.H. 1993. Storage and translocation of integumentary calcium during the moult cycle of the terrestrial isopod *Oniscus asellus* (L.). *Can. J. Zool.* 71: 4-10.

Stevenson, J.R. 1961. Polyphenol oxidase in the integumental glands in relation to the moulting cycle of the isopod crustacean *Armadillidium vulgare. Biol. Bull.* 121: 554-560.

Stevenson, J.R. 1985. Dynamics of the integument. In D.E.Bliss (ed.), *Biology of Crustacea* 9: 2-41, New York: Academic Press.

Storch, V. & J. Štrus 1989. Microscopic anatomy and ultrastructure of the alimentary canal in terrestrial isopods. *Monitore zool. ital.* (N.S.) Monogr. 4:105-126.

Štrus, J., D. Drobne & P. Licar 1995. Comparative anatomy and functional aspects of the digestive system in amphibious and terrestrial isopods (Isopoda: Oniscidea). *Crust. Issues* 9: 15-23.

Štrus, J.& P. Compère 1996. Ultrastructural analysis of the integument during the moult cycle in *Ligia italica* (Crustacea, Isopoda). *Pflugers Arch - Eur. J. Physiol 431(Suppl)*: R251-252.

Tait, J. 1917. Experiments and observations on Crustacea. Part II. Moulting of Isopods. *Proc. Roy. Soc. Edinb.* 37: 59-68.

Tchernigovtzeff, C. & J. Ragage-Willigens 1968. Determination des stades d'intermue chez *Sphaeroma serratum* (Isopode Flabellifere). *Arch. Zool. exp. Gen.* 109/2: 305-318.

Verhoeff, K.W. 1940. Uber die Doppelhäutung der Land-Isopoden. *Z. Morph. Okol. Tiere* 37: 126-143.

Wieser, W. 1964. Uber die Häutung von *Porcellio scaber* Latr. *Verha. Dtsch. Zool. Ges., Kiel,* 178-195.

Wick, S.M. & P.D. Hepler 1982. Selective localization of intracellular Ca with potassium pyroantimonate. *J. Histochem. Cytochem.* 30: 1190-1204.

Zidar, P., D. Drobne & J. Štrus 1998. Determination of moult stages of *Porcellio scaber* (Isopoda) for routine use. *Crustaceana* 72: 1-9.

Ziegler, A. 1997. Ultrastructural changes of the anterior and posterior sternal integument of the terrestrial isopod *Porcellio scaber* Latr. (Crustacea) during the moult cycle. *Tissue & Cell* 29: 63-76.

Ziegler, A. & B. Miller 1997. Ultrastructure of $CaCO_3$ deposits of terrestrial isopods (Crustacea, Oniscidea) *Zoomorph.* 117: 181-187.

Zimmer, M. & W. Topp 1998. Microorganisms and cellulose digestion in the gut of the woodlouse *Porcellio scaber*. *J. Chem. Ecol.* 24:1397-1408.

Contributors

Araujo, Paula Beatriz de: Departamento de Zoologia, IB, Universidade Federal do Rio Grande do Sul, Av. Paulo Gama, pr. 12106. CEP 90040-060, Porto Alegre, RS, Brazil.

Ballard, O.: Field Museum, Roosevelt Road at Lake Shore Drive, Chicago, IL 60605 USA.

Blejec, Andrej: National Institute of Biology, Vecna pot 111, 1111 Ljubljana, Slovenia.

Boutin, C.: Université Paul Sabatier, Laboratoire de Zoologie, UMR CNRS 55 52, 118 route de Narbonne, F-61262 Toulouse cedex 4, France.

Brattegard, Torleiv: Department of Fisheries and Marine Biology, University of Bergen, P.O. Box 7800, N-5020 Bergen, Norway.

Brusca, Richard C.: Columbia University Biosphere 2 Center, P.O. Box 689, Oracle, AZ 85623; & Department Ecology & Evolutionary Biology, University of Arizona, Tucson, AZ 85721, USA.

Coineau, N.: Université Pierre et Marie Curie, F-70006 Paris; & Observatoire Océanologique de Banyuls, UMR CNRS 76 28 Laboratoire Arago, F-66650 Banyuls/Mer, France.

Fakher el Abiari, A.: Faculté des Sciences Semlalia, Laboratoire d'Hydrobiologie, BP S. 15, Marrakech, Morocco.

Ferrari, Frank D.: Department of Invertebrate Zoology, National Museum of Natural History, Smithsonian Institution, Washington, D.C. 20560, USA.

Keable, Stephen J.: Center for Evolutionary Research, Australian Museum, 6-8 College St., Sydney South, New South Wales, Australia 2000.

Keim, Paulo: Department of Biological Sciences, Northern Arizona University, Flagstaff, AZ 86011-5640, USA.

Kensley, Brian: Department of Invertebrate Zoology, National Museum of Natural History, Smithsonian Institution, Washington, DC 20560, USA.

Kristjánsson, Bjarni K.: Institute of Biology, University of Iceland, Grensásvegur 12, 108 Reykjavík, Iceland, & Sandgerði Marine Centre, Garðvegur 1, 245 Sandgerði, Iceland.

Leistikow, Andreas: Universität Bielefeld, Abteilung für Morphologie und Systematik der Tiere, Morgenbreede 45, D-33615 Bielefeld; & Ruhr-Univerität Bochum, Lehrstuhl für spezielle Zoologie, Universitätsstraße 150, D-44780 Bochum, Germany.

Markham, John C.: Arch Cape Marine Laboratory, Arch Cape, OR 97102-0105, USA.

Pires-Vanin, Ana Maria Setubal: Universidade de São Paulo, Instituto Oceanográfico, P.O. Box 66149, São Paulo, SP 05315-970, Brazil.

Poore, Gary C. B.: Museum Victoria, GPO Box 666E, Melbourne, Vic. 3001, Australia.

Riseman, Sarah F.: University of Charleston, Grice Marine Biological Laboratory, 205 Fort Johnson, Charleston, SC 29412, USA.

Sassaman, Clay: Department of Biology, University of California, Riverside, CA 92521 USA.

Shuster, Stephen M.: Department of Biological Sciences, Northern Arizona University, Flagstaff, AZ 86011-5640, USA.

Svavarsson, Jörundur: Institute of Biology, University of Iceland, Grensásvegur 12, 108 Reykjavík, Iceland; & Sandgerði Marine Centre, Garðvegur 1, 245 Sandgerði, Iceland.

Strus, Jasna: Department of Biology, University of Ljubljana, Vecna pot 111, 1111 Ljubljana, Slovenia.

Thatcher, Vernon E.: Depto. de Zoologia, Universidade Federal do Paraná, Caixa Postal 19020, 81531-990 Curitiba, PR, Brasil.

Thiel, Martin: Facultad de Ciencias del Mar, Universidad Católica del Norte, Larrondo 1281, Campus Guayacan, Coquimbo, Chile.

William, J.: Field Museum, Roosevelt Road at Lake Shore Drive, Chicago, IL 60605 USA

Wilson, George D.F. Wilson: Center for Evolutionary Research, Australian Museum, 6-8 College St., Sydney South, New South Wales, Australia 2000.

Zinser, Guenivier: Department of Biological Sciences, Northern Arizona University, Flagstaff, AZ 86011-5640, USA.

Index